Lecture Notes in Compute

Commenced Publication in 1973
Founding and Former Series Editors:
Gerhard Goos, Juris Hartmanis, and Jan van Leeuwen

Tor Helleseth   Dilip Sarwate
Hong-Yeop Song   Kyeongcheol Yang (Eds.)

Sequences and Their Applications – SETA 2004

Third International Conference
Seoul, Korea, October 24-28, 2004
Revised Selected Papers

 Springer

Volume Editors

Tor Helleseth
University of Bergen
Department of Informatics
Selmer Center, Thormohlensgate 55, 5020 Bergen, Norway
E-mail: tor.helleseth@ii.uib.no

Dilip Sarwate
University of Ilinois at Urbana-Champaign
Department of Electrical and Computer Engineering
1406 West Green Street, Urbana, IL 61801, USA
E-mail: sarwate@uiuc.edu

Hong-Yeop Song
Yonsei University
School of Electronics and Electrical Engineering
Seoul 120-749, Korea
E-mail: hy.song@coding.yonsei.ac.kr

Kyeongcheol Yang
Pohang University of Science and Technology
Department of Electronics and Electrical Engineering
Pohang, Kyungbuk 790-784, Korea
E-mail: kcyang@postech.ac.kr

Library of Congress Control Number: 2005925881

CR Subject Classification (1998): E.4, F.2, I.1, E.3, F.1, G.1

ISSN   0302-9743
ISBN-10   3-540-26084-6 Springer Berlin Heidelberg New York
ISBN-13   978-3-540-26084-4 Springer Berlin Heidelberg New York

Springer is a part of Springer Science+Business Media

springeronline.com

© Springer-Verlag Berlin Heidelberg 2005
Printed in Germany

Typesetting: Camera-ready by author, data conversion by Scientific Publishing Services, Chennai, India
Printed on acid-free paper   SPIN: 11423461   06/3142   5 4 3 2 1 0

Preface

This volume contains the refereed proceedings of the 3rd International Conference on Sequences and Their Applications (SETA 2004), held in Seoul, Korea during October 24–28, 2004. The previous two conferences, SETA 1998 and SETA 2001, were held in Singapore and Bergen, Norway, respectively. These conferences are motivated by the many widespread applications of sequences in modern communication systems. These applications include pseudorandom sequences in spread spectrum systems, code-division multiple-access, stream ciphers in cryptography and several connections to coding theory.

The Technical Program Committee of SETA 2004 received 59 submitted papers, many more than the submissions to previous SETA conferences. The Committee therefore had the difficult task of selecting the 33 papers to be presented at the Conference in addition to four invited papers. The authors of papers presented at the conference were invited to submit full papers that were refereed before appearing in this proceedings.

These proceedings have been edited by the Co-chairs of the Technical Program Committee for SETA 2004: Tor Helleseth of the University of Bergen, Norway, and Dilip Sarwate of the University of Illinois at Urbana-Champaign, USA, and Technical Program Committee members Hong-Yeop Song of Yonsei University, Korea, and Kyeongcheol Yang of Pohang University of Science and Technology, Korea. The editors wish to thank the other members of the Technical Program Committee: Serdar Boztas (Royal Melbourne Institute of Technology, Australia), Claude Carlet (INRIA and University of Paris 8, France), Zongduo Dai (University of Science and Technology of China, Beijing, China), Cunsheng Ding (Hong Kong University of Science and Technology, Hong Kong, China), Hans Dobbertin (Ruhr University Bochum, Germany), Pingzhi Fan (Southwest Jiaotong University, China), Solomon W. Golomb (University of Southern California, USA), Guang Gong (University of Waterloo, Canada), Tom Høholdt (Technical University of Denmark, Denmark), Andrew Klapper (University of Kentucky, USA), P. Vijay Kumar (University of Southern California, USA), Vladimir Levenshtein (Keldysh Institute of Applied Mathematics, Russia), Oscar Moreno (University of Puerto Rico, Puerto Rico), Harald Niederreiter (National University of Singapore, Singapore), Matthew Parker (University of Bergen, Norway), Kenneth G. Paterson (Royal Holloway, University of London, UK), Aleksander Pott (Otto von Guericke University Magdeburg, Germany), Hans Schotten (Qualcomm CDMA Technologies, Nürnberg, Germany), Patrick Sole (CNRS-I3S, ESSI, Sophia Antipolis, France), Naoki Suehiro (University of Tsukuba, Japan) for providing clear, insightful, and prompt reviews of the submitted papers.

In addition to the contributed papers, there were four invited papers. These papers provide an overview of new developments in some important areas related to sequences. The invited papers by Jedwab and by Parker both include an updated overview and some recent results on the constructions of some exciting new families of sequences with a merit factor more than 6.3. Klapper gives an overview of the fascinating topic of feedback with carry shift registers, while Dobbertin and Leander present new and recent results on bent functions.

We wish to thank Jong-Seon No and Habong Chung for their support as General Co-chairs of SETA 2004; Dong-Joon Shin, Wonjin Sung and Jun Heo for their support as members of the Organizing Committee of SETA 2004. Last but not least, we thank all the authors of the papers presented at SETA 2004 for their help in making this conference a resounding success. Finally, we also thank the Korea Research Foundation (KTF) for its financial support.

March, 2005

Tor Helleseth
Dilip Sarwate
Hong-Yeop Song
Keongcheol Yang

Organization

SETA 2004

October 24–28, 2004, Seoul, Korea

General Co-chairs
Jong-Seon No, Seoul National University, Korea
Habong Chung, Hongik University, Korea

Program Co-chairs
Tor Helleseth, University of Bergen, Norway
Dilip Sarwate, University of Illinois at Urbana-Champaign, USA

Secretary and Treasury
Dong-Joon Shin, Hanyang University, Korea

Local Arrangements
Wonjin Sung, Sogang University, Korea

Registration
Jun Heo, Konkuk University, Korea

Publication Co-editors
Hong-Yeop Song, Yonsei University, Korea
Kyeongcheol Yang, POSTECH, Korea

Technical Program Committee for SETA 2004

Program Co-chairs

Tor Helleseth .... University of Bergen, Norway
Dilip Sarwate .... University of Illinois at Urbana-Champaign, USA

Program Committee

Serdar Boztas .... RMIT University, Australia
Claude Carlet .... INRIA and University of Paris 8, France
Zongduo Dai .... University of Science and Technology of China, Beijing, China
Cunsheng Ding ... Hong Kong University of Science and Technology, Hong Kong,
            China
Hans Dobbertin .... Ruhr University Bochum, Germany
Pingzhi Fan .... Southwest Jiaotong University, China
Solomon W. Golomb .... University of Southern California, USA
Guang Gong .... University of Waterloo, Canada
Tom Høholdt .... Technical University of Denmark, Denmark
Andrew Klapper .... University of Kentucky, USA
P. Vijay Kumar .... University of Southern California, USA
Vladimir Levenshtein .... Keldysh Institute of Applied Mathematics, Russia
Oscar Moreno .... University of Puerto Rico, Puerto Rico
Harald Niederreiter .... National University of Singapore, Singapore
Matthew G. Parker .... University of Bergen, Norway
Kenneth G. Paterson .... Royal Holloway, University of London, UK
Alexander Pott .... Otto von Guericke University Magdeburg, Germany
Hans Schotten .... Qualcomm CDMA Technologies, Nürnberg, Germany
Patrick Solé .... CNRS-I3S, ESSI, Sophia Antipolis, France
Hong-Yeop Song .... Yonsei University, Korea
Naoki Suehiro .... University of Tsukuba, Japan
Kyeongcheol Yang .... Pohang University of Science and Technology, Korea

Table of Contents

Perfect Sequences

Sequence Constructions

Sequences over $\mathbb{Z}_m$

Sequence Generator Properties and Applications

Multi-dimensional Sequences

Optics and OFDM Applications

Polynomials and Functions

A Survey of Some Recent Results
on Bent Functions

Hans Dobbertin and Gregor Leander

Department of Mathematics,
Ruhr-University Bochum,
D-44780 Bochum, Germany
{hans.dobbertin, gregor.leander}@ruhr-uni-bochum.de

Abstract. We report about recent results and methods in the study of
bent functions. Here we focus on normality and trace expansions of bent
functions.

1   Introduction

In this paper we present an overview about recent developments in the study
of bent functions. We summarize and cite new results and techniques from the
preprints [4, 10, 11, 15] and the paper [10].

Bent functions are maximally nonlinear Boolean functions with an even number of variables and were introduced by Rothaus [27] in 1976. Because of their
own sake as interesting combinatorial objects, but also because of their relations
to coding theory and applications in cryptography, they have attracted a lot of
research, specially in the last ten years.

Despite their simple and natural definition, bent functions have turned out to
admit a very complicated structure in general. On the other hand many special
explicit constructions are known, primary ones giving bent functions from scratch
and secondary ones building a new bent function from one or several given bent
functions.

Normality of Bent Functions

Basic criteria of Boolean functions on $\mathbb{F}_2^n$, which are relevant to cryptography, are
for instance its algebraic degree and nonlinearity. Another condition in this line
of research is normality (resp. weak normality), i. e. the existence of a subspace
of $\mathbb{F}_2^n$ with dimension $\frac{n}{2}$ such that the restriction of the given function is constant
(resp. affine).

The notion of normality was introduced by the first author [16] in the study
of bent functions and highly nonlinear balanced Boolean functions. While for
increasing dimension n a counting argument can be used to prove that nearly
all Boolean functions are non-normal, the situation for bent functions is more
difficult. Most of the well studied families of bent functions are obviously normal
and furthermore, unlike for arbitrary Boolean functions, normality has strong

T. Helleseth et al. (Eds.): SETA 2004, LNCS 3486, pp. 1–29, 2005.

consequences for the behavior of bent functions. One of the consequences is, that if a bent function f is constant on an $\frac{n}{2}$-dimensional affine subspace, then f is balanced on each proper coset. In other words, a normal bent function can be understood as a collection of balanced functions. The question whether non-normal bent functions exist at all, is therefore important. The interpretation of a normal bent function as a collection of balanced functions was used in [16] to give a framework for constructions of normal bent functions.

Monomial and Binomial Bent Functions

A complete classification of bent functions is elusive and looks hopeless today. In the second part of this paper we focus on traces of power functions, so called *monomial* Boolean functions. The study of trace expansions is well known in related areas, but has not yet been comprehensively studied for bent functions. This approach turns out to be very fruitful for several reasons. The only known non-normal bent functions are monomial bent functions, demonstrating that the study of monomial functions leads to new classes of bent functions. Furthermore one result of our considerations is, that for each of the well studied families of bent function, there is a monomial bent function belonging to these classes. Moreover, carefully studying the proofs for the monomial bent functions all these families can quite easily be rediscovered. In this sense most of the variety of (at least known) bent functions can already be discovered by the investigation of monomial functions.

In Section 5 we take the next step and extend our focus to linear combinations of two power functions. In particular we focus on Niho power functions, i.e. power functions where the restriction to the subfield of index 2 is linear. Using classical results for the Walsh-Spectrum of these functions and techniques recently developed by the first author, we present several new primary constructions of bent functions. These results are based on new techniques to study certain properties of rational functions. More precisely we present a general procedure to prove that certain rational functions induce one-to-one mappings.

These techniques and the Multivariate-Method developed by the first author (see [17]) follow the same line of reasoning. Both approaches are strongly based on properties of mappings, that can be defined in a global way, meaning that these properties are valid for an infinite chain of finite fields. In both situations this results in generic discussion of specific rational functions. These generic discussions are often relatively easy to describe for the conceptual point of view, while the actual inherent computations require the help of computer algebra. One key step is often to find the factorization of (parameterized) polynomials, which usually is not feasible by hand calculations. Nevertheless, once the factorization has been found, verifying the result is much easier and can here in most cases be done by hand.

2   Preliminaries

Throughout let $n = 2k$ be an even number. We recall some definitions and basic properties.

Walsh-Transform and Bent Functions

Given a Boolean function $f : \mathbb{F}_2^n \to \mathbb{F}_2$, the function

$$a \in \mathbb{F}_2^n \mapsto f^{\mathcal{W}}(a) = \sum_{x \in \mathbb{F}_2^n} (-1)^{f(x) + \langle a, x \rangle}$$

is called the *Walsh transform* of f. Moreover, the values $f^{\mathcal{W}}(a), a \in \mathbb{F}_2^n$ are called the Walsh coefficients of f. The set

$$\{f^{\mathcal{W}}(a), a \in \mathbb{F}_2^n\}$$

is called the Walsh-spectrum of f. Note that the Walsh-spectrum is not changed if we replace f by $f \circ H$ where H is a bijective affine or linear mapping, moreover adding an affine mapping does not change the absolute values of the Walsh-spectrum. Thus in most of our discussions we do not distinguish between these *affine* or *linear* equivalent functions.

By looking at the ± 1 valued function $F = (-1)^f$ the Walsh-transform of f corresponds (up to scaling) to the additive Fourier transform of F.

$$\hat{F} = 2^{-k} \sum_{x \in \mathbb{F}_2^n} F(x)(-1)^{\langle y, x \rangle}.$$

Due to the fact, that this transform is effected by the Hadamard matrix

$$\left((-1)^{\langle y, x \rangle} \right)_{x, y \in \mathbb{F}_2^n}$$

it is also sometimes called *Hadamard transform*. There are a few properties of the Hadamard transform, that we like to recall here.

For the operator $F \to \hat{F}$ we have the *involution law*

$$\hat{\hat{F}} = F$$

and with

$$(F * G)(x) = \sum_{u \in \mathbb{F}_2^n} F(u + x)G(x)$$

the *convolution law*

$$\widehat{F * G} = 2^k \hat{F} \hat{G}.$$

If we regard $\mathbb{R}^{\mathbb{F}_2^n}$ as a inner-product space where,

$$\langle F, G \rangle = \sum_{x \in \mathbb{F}_2^n} F(x)G(x)$$

the map $F \to \hat{F}$ is an orthogonal operator on $\mathbb{R}^{\mathbb{F}_2^n}$, i.e.

$$\sum_{x \in \mathbb{F}_2^n} F(x)G(x) = \sum_{y \in \mathbb{F}_2^n} \hat{F}(y)\hat{G}(y).$$

A measure of the linearity of a Boolean function f with respect to the Walsh transform is defined by

$$\mathrm{Lin}(f) = \max_{a \in \mathbb{F}_2^n} \left| f^{\mathcal{W}}(a) \right|.$$

Obviously we have the upper bound

$$2^n \geq \mathrm{Lin}\, f,$$

and it is attained if and only if f is affine.

For n even, f is called *bent* if $\mathrm{Lin}(f) = 2^{n/2}$, which is the minimal value that can occur and we then have $f^{\mathcal{W}}(a) = \pm 2^{n/2}$ for all $a \in \mathbb{F}_2^n$, since

$$\sum_{a \in \mathbb{F}_2^n} f^{\mathcal{W}}(a)^2 = 2^{2n} \quad \text{(Parseval's equation)}.$$

Note that Parseval's equation is a direct consequence of the above mentioned fact that the operator $F \to \hat{F}$ is orthogonal. Another measurement for the linearity of a Boolean function f is the *autocorrelation function*. It is defined by

$$\mathrm{AC}_f(a) = \sum_{x \in \mathbb{F}_2^n} (-1)^{f(x+a)+f(x)}.$$

Bent functions can also be characterized in terms of their Autocorrelation function, which again follows directly from the orthogonality of the Fourier transform.

Proposition 1. *A Boolean function f on $\mathbb{F}_2^n$ is bent if and only if $\mathrm{AC}_f(a) = 0$ for all non-zero $a \in \mathbb{F}_2^n$.*

Bent functions always occur in pairs. In fact, given a bent function $f : \mathbb{F}_2^n \to \mathbb{F}_2^n$, we define the *dual* f^* of f by the equation

$$(-1)^{f^*(a)}\, 2^{n/2} = f^{\mathcal{W}}(a),$$

i.e. we consider the signs of the Walsh-coefficients of f. Due to the involution law the Fourier transform is self-inverse. Thus the dual of a bent function is again a bent function, and we have the rule $f^{**} = f$.

Every Boolean function can be uniquely described by its Algebraic Normal Form (ANF)

$$f(x) = \sum_{u \in \mathbb{F}_2^n} \lambda_u \prod_{i=1}^{n} x_i^{u_i}, \quad \lambda_u \in \mathbb{F}_2.$$

The degree of a Boolean function is the maximal value of $\mathrm{wt}(u)$ such that $\lambda_u \neq 0$. It was already proven by Rothaus, that the degree of a bent function is at most k. Furthermore the dual of a bent function of degree 2 (resp. k) has also degree 2 (resp. k) (see [27]).

Boolean Functions on $\mathbb{F}_{2^n}$

We will often identify the vector space $\mathbb{F}_2^n$ with the Galois field $L = \mathbb{F}_{2^n}$. As the notion of a Walsh transform refers to a scalar product, it is convenient to choose the isomorphism such that the canonical scalar product $\langle \cdot, \cdot \rangle$ in $\mathbb{F}_2^n$ coincides with the canonical scalar product in L, which is the trace of the product:

$$\langle x, y \rangle = \sum_{i=1}^{n} x_i y_i = \operatorname{tr}_L(xy), \quad x, y \in L.$$

Thus the *Walsh transform* of $f : L \to \mathbb{F}_2$ is defined as

$$f^{\mathcal{W}}(c) = \sum_{x \in L} (-1)^{f(x)} \chi_L(cx), \quad c \in L,$$

where

$$\chi_L(x) := (-1)^{\operatorname{tr}_L(x)}$$

is the canonical additive character on L.

We often make use of the following known properties of the trace function: For all $x \in L$

- $\operatorname{tr}_L(x) = \operatorname{tr}_L(x^2)$,
- $\operatorname{tr}_L(x) = 0$ if and only if $x = y^2 + y$ for some $y \in L$.

Throughout let $K = \mathbb{F}_{2^k}$ be the subfield of L with $[L : K] = 2$. When dealing with Boolean functions on L, in particular *monomial functions*, we will use the following notation. The *conjugate* of $x \in L$ over K will be denoted by $\overline{x}$, i.e.

$$\overline{x} = x^{2^k}.$$

We denote the *relative trace* from L onto K by

$$\operatorname{tr}_{L/K}(x) = x + \overline{x}$$

Note that according to the transitivity law for the trace function we have

$$\operatorname{tr}_L = \operatorname{tr}_K \circ \operatorname{tr}_{L/K}.$$

The relative norm with respect to L/K is defined as

$$\operatorname{norm}_{L/K}(x) = x \overline{x}$$

and maps L onto K.

The *unit circle* of L is the set

$$S = \{u \in L : u\overline{u} = 1\}$$

of all elements having relative norm 1. In other words S is the group of $(2^k + 1)$-st roots of unity, and therefore the order of S is $2^k + 1$, since L^* is cyclic and $2^k + 1$ divides $2^n - 1$.

Note that $S \cap K = \{1\}$ and each non-zero element of L has a unique *polar coordinate representation,* i.e.

$$x = \lambda u$$

with $\lambda \in K^*$ and $u \in S$. According to the analogy to $\mathbb{C}/\mathbb{R}$ we write $\lambda = \|x\|$ for the *length* and $u = \varrho(x)$ for the *angle* of x. We have

$$\text{norm}\, x = \sqrt{x\overline{x}}, \tag{1}$$

$$\varrho(x) = \sqrt{x/\overline{x}}.. \tag{2}$$

Where the symbol $\sqrt{X}$ stands for the inverse of the Frobenius mapping $\varphi(X) = X^2$, which makes sense, as we deal with *finite* fields of characteristic 2. Concretely here $\sqrt{z} = z^{2^{k-1}}$ for $z \in K = \mathbb{F}_{2^k}$.

Normality of Boolean Functions
Normality is a property of the restriction of Boolean functions to subspaces or affine translations of subspaces. For simplicity we call a t-dimensional affine subspace a *flat* of dimension t.

Definition 1. *A function $f : \mathbb{F}_2^n \to \mathbb{F}$ is called* normal *if there exists a flat of dimension m, such that f is constant on this flat.*

As bentness is invariant under addition of affine functions, it is natural to consider a generalization of Definition 1.

Definition 2. *A function $f : \mathbb{F}_2^n \to \mathbb{F}$ is called* weakly-normal *if there exists a flat of dimension m, such that the restriction of f to this flat is affine.*

Unlike for arbitrary Boolean functions, for bent functions being normal has strong consequences. The following well known lemmas state some of the most important properties of normal bent functions.

Lemma 1. *Assume that $f : \mathbb{F}_2^n \to \mathbb{F}_2$ is a normal bent function, which is accordingly constant on an affine subspace $V \subseteq \mathbb{F}_2^n$ with $\dim V = k$. Then f is balanced on each proper coset of V.*

Normality of a bent function is also reflected in the dual bent function, as stated in the next lemma

Lemma 2. *Let $f : \mathbb{F}_2^n \to \mathbb{F}_2$ be a bent function and $V \subset \mathbb{F}_2^n$ any $n/2$-dimensional subspace. Then f is constant on $a + V$ for $a \in \mathbb{F}_2^n$ if and only if $f^* + \phi_a$ is constant on $V^\perp$. Where*

$$\phi_a(x) = \langle a, x \rangle$$

$\square$

Another important observation is, that a bent function f on $\mathbb{F}_2^n$ cannot be constant on an affine subspace of dimension greater than k. This is a direct consequence of the following lemma.

Lemma 3. *Let $f : \mathbb{F}_2^n \to \mathbb{F}$ be a Boolean function, $U \subseteq \mathbb{F}_2^n$ a subspace and $g = f|_U$, then $\text{Lin}(g) \leq \text{Lin}(f)$*

3   Non-normal Bent Functions

In [11, 4, 5] it was shown that most of the known constructions of bent functions lead to normal bent functions only. Moreover, using a computer algorithm, it has been verified that the class of bent functions constructed via Kasami exponents contains non-normal bent functions:

Definition 3. *Let $d = 2^{2r} - 2^r + 1$ be a Kasami exponent with $\gcd(r,n) = 1$ and $\alpha \in \mathbb{F}_{2^n}$. Then we call $f_{\alpha,r} : \mathbb{F}_{2^n} \to \mathbb{F}_2$ with $f_{\alpha,r}(x) = \mathrm{tr}(\alpha x^d)$ a DD-function.*

It was proven by Dillon and Dobbertin in [14], that, under certain conditions, these functions are bent (see Theorem 6).

For n divisible by 4 it is easy to see that the DD-functions are always normal.

Lemma 4. *Let $n = 2k$ with k even. Then DD-functions*

$$\begin{aligned} f : \mathbb{F}_{2^n} &\to \quad \mathbb{F}_2 \\ x &\mapsto \mathrm{tr}(\alpha x^d) \end{aligned}$$

are normal.

Proof. First note that $\gcd(d, 2^n - 1) = 3$, i.e.,

$$U = \{x^d \mid x \in \mathbb{F}_{2^n}^*\} = \{x^3 \mid x \in \mathbb{F}_{2^n}^*\}$$

and there exist $\lambda_1, \lambda_2 \notin U$ such that

$$\mathbb{F}_{2^n}^* = U \cup \lambda_1 U \cup \lambda_2 U.$$

In the case where $4 \mid n$, we will show that λ_1, λ_2 can be chosen in $\mathbb{F}_{2^k}$. It is sufficient to show that there exists $x \in \mathbb{F}_{2^k}$ such that $x \notin U$. Let g be a generator of $\mathbb{F}_{2^k}^*$. g is in U if and only if $g^{\frac{2^n-1}{3}} = 1$. But

$$g^{\frac{2^n-1}{3}} = g^{\frac{(2^k-1)(2^k+1)}{3}}$$
$$= g^{(2^k+1)\frac{2^k-1}{3}} \neq 1$$

as $2^k + 1$ is not divisible by 3 if k is even. So we can choose $\lambda_1 = g$ and $\lambda_2 = g^2$. Note that if $\alpha' = \alpha c^d$ for some $c \in \mathbb{F}_{2^n}^*$ then $f_{\alpha,r}(cx) = f_{\alpha',r}(x)$ for all $x \in \mathbb{F}_{2^n}$. Thus, we can assume that α is in $\{1, g, g^2\} \subset \mathbb{F}_{2^k}$. So for $x \in \mathbb{F}_{2^k}$ we get

$$\begin{aligned} f_{\alpha,k}(x) &= \mathrm{tr}(\alpha x^d) \\ &= \mathrm{tr}_{\mathbb{F}_{2^k}/\mathbb{F}_2}(\mathrm{tr}_{\mathbb{F}_{2^n}/\mathbb{F}_{2^k}}(\alpha x^d)) \\ &= \mathrm{tr}_{\mathbb{F}_{2^k}/\mathbb{F}_2}(\alpha x^d \, \mathrm{tr}_{\mathbb{F}_{2^n}/\mathbb{F}_{2^k}}(1)) \\ &= 0 \ . \end{aligned}$$

This proves the lemma.   □

So we can only hope to get non-normal DD-functions for k odd. Furthermore, as all quadratic bent functions are normal, only the case $r \neq 1$ is interesting. As it is known that all bent functions on $\mathbb{F}_2^6$ are normal, the first possibility for DD-function to be non-normal is $n = 10$.

We found out that for $n = 10$ all the DD-functions are normal but by addition of a linear function they can be modified into non-normal functions.

Fact 1. *Let* $\alpha \in \mathbb{F}_4 \setminus \mathbb{F}_2 \subset \mathbb{F}_{2^{10}}$. *Then there exists* $\beta \in \mathbb{F}_{2^{10}}$ *such that the function* $f : \mathbb{F}_{2^{10}} \to \mathbb{F}_2$ *with*

$$f(x) = \mathrm{tr}(\alpha x^{57} + \beta x)$$

is non-normal.

Verification. This can be verified using the algorithm described in [11].   □

Furthermore, we found that for $n = 14$ and $r = 3$ the corresponding DD-functions are non weakly-normal.

Fact 2. *Let* $\alpha \in \mathbb{F}_4 \setminus \mathbb{F}_2 \subset \mathbb{F}_{2^{14}}$. *The function* $f : \mathbb{F}_{2^{14}} \to \mathbb{F}_2$ *with*

$$f(x) = \mathrm{tr}(\alpha x^{57})$$

is non weakly-normal.

Verification. By using the algorithm described in [11].   □

These results are verified with a computer algorithm, proving these results theoretically is still an open problem. We state the following conjecture.

Conjecture 1. All non quadratic DD-functions on $\mathbb{F}_{2^{2k}}$ with k odd and $k \geq 7$ are non weakly-normal.

One interesting corollary is that (at least some of) the DD bent functions do not belong to the Maiorana McFarland ($\mathcal{M}$), the Partial Spread ($\mathcal{PS}$) nor the class of bent functions constructed by the first author (see [16]) ($\mathcal{N}$) class of bent functions. Note that in general it is very difficult to prove that a given function does not belong to these classes.

Corollary 1. *The DD bent function* $f : \mathbb{F}_{2^{14}} \to \mathbb{F}_2$ *defined by*

$$f(x) = \mathrm{tr}(\alpha x^{57})$$

with $\alpha \in \mathbb{F}_4 \setminus \mathbb{F}_2 \subset \mathbb{F}_{2^{14}}$ *and its dual do not belong to*

$$\mathcal{M} \cup \mathcal{PS} \cup \mathcal{N} .$$

3.1   Secondary Constructions of Non-normal Bent Functions

As a theoretical approach to prove non-normality of any function is not available today, and as the algorithms to check non-normality are strictly limited to very small values of n, we need other approaches to create non-normal functions for arbitrary dimensions.

The easiest example of such a *secondary construction* of non-normal function is the following lemma (see [19]).

Lemma 5. *Let $f : \mathbb{F}_2^n \to \mathbb{F}_2$ be a Boolean function. The following properties are equivalent:*

1. *f is (weakly) normal*
2. *The function*

$$g : \mathbb{F}_2^n \times \mathbb{F}_2 \times \mathbb{F}_2 \to \mathbb{F}_2$$
$$(x, y, z) \mapsto f(x) + yz$$

is (weakly) normal

Thus, given a non-normal function f with n variables Lemma 5 can be used to construct a non-normal function with $n + 2$ variables.

According to this procedure applied recursively, if f is a Boolean function on $\mathbb{F}_2^n$ and if f' is a quadratic bent function on $\mathbb{F}_2^{n'}$, then f is (weakly) normal if and only if $g(x, y) = f(x) + f'(y)$ is (weakly) normal. An important observation from our point of view is that, if the function f in the above lemma is bent, then g is also bent.

S. Gangopadhyay, S. Maitra [21] and C. Carlet conjectured that, much more general, the direct sum of a non-normal bent function with any normal one is non-normal (the same statement for weak normality can easily be reduced to the latter.)

In order to proof this statement we introduce the notion of a normal extension, which in some sense is a generalization of the direct sum of a normal and a non-normal bent functions (see [10]).

Definition 4. *Suppose $U \subseteq V$, let $\beta : U \to \mathbb{F}_2$ and $f : V \to \mathbb{F}_2$ be bent functions. Then we say that f is a normal extension of β, in symbols*

$$\beta \preceq f$$

if there is a direct decomposition $V = U \oplus W_1 \oplus W_2$ such that

(i) $\beta(u) = f(u + w_1)$ for all $u \in U$, $w_1 \in W_1$,
(ii) $\dim W_1 = \dim W_2$.

Note that condition (i) can also be written as

$$f|_{U \oplus W_1} = \beta \oplus 0,$$

here 0 denoting the 0-function on W_1. Thus the restriction of f on $U \oplus W_1$ is a "blown-up" bent function.

We will discuss normal extensions in the context of duality, and writing $\beta \preceq f$ (i.e., f is a normal extension of β) we shall, according to Definition 4, assume that $\beta : U \to \mathbb{F}_2$ and $f : V = U \times W \times W \to \mathbb{F}_2$ are bent functions such that $r = \dim W$, $m = \dim U$ and

$$f(x, y, 0) = \beta(x)$$

for all $x \in U$, $y \in W$.

Lemma 6. *Under the hypothesis above, we have* $f^{\mathcal{W}}(a, 0, c) = 2^r \beta^{\mathcal{W}}(a)$, *that is,* $f^*(a, 0, c) = \beta^*(a)$, *for all* $a \in U$ *and* $c \in W$.

Proof. We compute for all $a \in U$:

$$\sum_{c \in W} (-1)^{f^*(a,0,c)} = 2^{-r-m/2} \sum_{c \in W} \sum_{x \in U} \sum_{y,z \in W} (-1)^{f(x,y,z)+\langle a,x \rangle+\langle c,z \rangle}$$

$$= 2^{-r-m/2} \sum_{x \in U} \sum_{y,z \in W} (-1)^{f(x,y,z)+\langle a,x \rangle} \left(\sum_{c \in W} (-1)^{\langle c,z \rangle} \right)$$

$$= 2^{-m/2} \sum_{x \in U} \sum_{y \in W} (-1)^{f(x,y,0)+\langle a,x \rangle}$$

$$= \left(2^{-m/2} \times 2^r \right) \beta^{\mathcal{W}}(a)$$

$$= 2^r (-1)^{\beta^*(a)}.$$

The left hand sum can take on its extremal values $\pm 2^r$ only if $f^*(a, 0, c) = \beta^*(a)$, for every $c \in W$.

In view of Lemma 6 it is easy to see that f is a normal extension of β if and only if the dual of f is a normal extension of the dual of β, i.e.

Proposition 2. $\beta \preceq f$ *if and only if* $\beta^* \preceq f^*$.

Proof. It is enough to prove that $\beta \preceq f$ implies $\beta^* \preceq f^*$. Thus assume that $\beta \preceq f$. For all $a \in U$, $c \in W$, we have $f^*(a, 0, c) = \beta^*(a)$, according to Lemma 6, and therefore $\beta^* \preceq f^*$, where the subsets $W_1 = \{0\} \times W \times \{0\}$ and $W_2 = \{0\} \times \{0\} \times W$ of V interchange their roles (cf. Definition 4).

This observation is the key to proof the following theorem.

Theorem 1. *(see Theorem 14 in [10])*
Let β be a bent function on U and f a bent function on $V = U \times W \times W$. Assume that $\beta \preceq f$. Let

$$\beta' : U \to \mathbb{F}_2$$

be any bent function. Modify f by setting for all $x \in U$, $y \in W$

$$f'(x, y, 0) = \beta'(x),$$

while $f'(x, y, z) = f(x, y, z)$ for all $x \in U$, $y, z \in W$, $z \neq 0$. Then f' keeps to be bent and we have $\beta' \preceq f'$.

Thus, we have seen that the relation $\beta \preceq f$ for bent functions, in view of the Theorem 1, describes a property of the restriction of f to $V \setminus (U \times W \times \{0\})$ (see Definition 4), which has nothing to do with a particular β. Under this aspect, only the *size*, i.e. the number of variables, of β is of importance.

The main result for the secondary construction of non-normal bent functions is the following Theorem.

Theorem 2. *(see Theorem 15 in [10])*
Suppose that $\beta \preceq f$ for bent functions β and f. If f is normal, then also β is normal.

In the next section we use this result to derive various different constructions of non-normal bent functions.

3.2   Examples of Normal Extensions

In this section we will describe explicit constructions of normal extensions of bent functions. Accordingly, let $\beta : U \to \mathbb{F}_2$ and $f : V = U \times W \times W \to \mathbb{F}_2$ be Boolean functions; let $r = \dim W$ and $m = \dim U$. By Definition 4, $\beta \preceq f$ means that

$$f(x, y, 0) = \beta(x)$$

for all $x \in U$, $y \in W$. The easiest example of a normal extension if the case where f is a direct sum of a normal bent function g and an arbitrary bent function β. In this case Theorem 2 can be restated as

$$\text{normal} \oplus \text{non-normal} = \text{non-normal}.$$

Nevertheless, the notion of a normal extension is more general then the trivial construction, as we will demonstrate with the following constructions.

Proposition 3. *Assume that $r = 1$, i.e., $W = \mathbb{F}_2$. Given bent functions β and g on U, we have $\beta \preceq f$ for f defined on $U \times W \times W$ by setting for all $x \in U$:*

$$f(x, 0, 0) = \beta(x),$$
$$f(x, 1, 0) = \beta(x),$$
$$f(x, 0, 1) = g(x),$$
$$f(x, 1, 1) = g(x) + 1.$$

Moreover, f is a bent function.
Conversely in case $r = 1$, $\beta \preceq f$ occurs for bent functions if and only if, up to equivalence, f is of this form.

In the preceding proposition we get a direct sum if and only if $g = \beta$ or $g = \beta + 1$.
Proposition 3 is a special case of the following general setting, see [6].

Theorem 3. *Let $h : \mathbb{F}_2^n \times \mathbb{F}_2^m \to \mathbb{F}_2$ be a Boolean function such that, for every $z \in \mathbb{F}_2^m$, the function on $\mathbb{F}_2^n$:*

$$h_z : x \to h(x, z)$$

is bent. Then h is bent if and only if the function

$$\varphi_a : z \to h_z^*(a)$$

is bent for every $a \in \mathbb{F}_2^n$. And the dual of h is $h^*(a, b) = \varphi_a^*(b)$.

If for every $z \in \mathbb{F}_2^m$, their exists a $\beta_z \preceq h_z$ with the same decomposition $\mathbb{F}_2^n = U \times W_1 \times W_2$, then $\beta \preceq h$, with the decomposition $\mathbb{F}_2^{n+m} = (U \times \mathbb{F}_2^m) \times W_1 \times W_2$ where

$$\beta : U \times \mathbb{F}_2^m \to \mathbb{F}_2$$

is defined by

$$\beta(u, z) = \beta_z(u).$$

However this Proposition is not constructive. We shall give two special constructive cases of normal extensions of bent functions. The first one is the following proposition, which can be easily verified (see also [9]):

Proposition 4. *Let $f_1, f_2 : U \to \mathbb{F}_2$ and $g_1, g_2 : U' \to \mathbb{F}_2$ be bent functions, then*

$$h : U \times U' \to \mathbb{F}_2$$

with

$$h(x, z) = f_1(x) + g_1(z) + (f_1(x) + f_2(x))(g_1(z) + g_2(z))$$

is bent, and the dual of h is given by

$$h^*(a, b) = f_1^*(a) + g_1^*(b) + (f_1^*(a) + f_2^*(a))(g_1^*(b) + g_2^*(b)),$$

$a \in U$, $b \in U'$.

Indeed, for every $z \in U'$, the function $h_z : x \mapsto h(x, z)$ is bent (since it equals f_1 or $f_1 + 1$ for some values of z, and f_2 or $f_2 + 1$ for the other values) and the function $z \mapsto h_z^*(a)$ is bent too, for every $a \in U$, since it equals $f_1^*(a) + g_1(z) + (f_1^*(a) + f_2^*(a))(g_1(z) + g_2(z))$. This implies Proposition 4, according to Theorem 3.

If we assume that g_1, g_2 in Proposition 4 are normal with the same normality subspace, i.e., $U' = W \times W$ with $g_1(y, 0) = g_2(y, 0) = 0$ for all $y \in W$, then we get

$$h(x, y, 0) = f_1(x),$$

which means that $f_1 \preceq h$.

We state another generalization of Proposition 3, in the framework of Theorem 3, based on the Maiorana-McFarland bent functions. The MM-construction always gives a bent function on V with $\dim V$ even. First represent $V = W \times W$, where W is endowed with a scalar product. Ingredients for this construction are any permutation $\pi : W \to W$ and any Boolean function $\tau : W \to \mathbb{F}_2$. Define for all $x, y \in W$

$$M(x, y) = \langle x, \pi(y) \rangle + \tau(y).$$

It is well-known and very easy to see that M is a bent function on V and that its dual is $M^*(a, b) = \langle b, \pi^{-1}(a) \rangle + \tau(\pi^{-1}(a))$.

Now let

$$(f_w)_{w \in W}, \quad f_w : U \to \mathbb{F}_2$$

be an arbitrary collection of bent functions on U. For the sake of conformity with our previous notation, set $\beta = f_0$ and assume $\pi(0) = 0$, $\tau(0) = 0$.

Then, using M, we get the following normal extension h on $W \times W \times U$ of β by setting

$$h(x, y, z) = f_y(z) + M(x, y).$$

Such an h will be called *an extension of MM-Type* of β.

According to Theorem 3, we have:

Proposition 5. *If β is a bent function and h is a normal extension of MM-type of β, then h is also bent.*

Indeed, we have $h_z^*(a, b) = f_{\pi^{-1}(a)}(z) + \langle b, \pi^{-1}(a) \rangle + \tau(\pi^{-1}(a))$. Note that the dual of h is of the same type as h:

$$h^*(a, b, c) = f_{\pi^{-1}(a)}^*(c) + \langle b, \pi^{-1}(a) \rangle + \tau(\pi^{-1}(a)).$$

4 Monomial Bent Functions

In this section we recall what is known about monomial bent functions. More precisely we are interested in Boolean functions of the form

$$f : L \to \mathbb{F}_2$$
$$f(x) = \text{tr}(\alpha x^d)$$

An exponent d (always understood modulo $2^n - 1$) is called a *bent exponent*, if there exists an α such that the Boolean function $\text{tr}(\alpha x^d)$ is bent. Considering all known cases of monomial bent function it turns out that most of the (known) constructions for bent functions can already be discovered when focusing on monomial functions.

Preliminaries

As $\text{tr}(x^d) = \text{tr}(x^{2d})$ for all x, we can always replace d by any exponent in the cyclotomic coset of d.

There are some necessary conditions for d to be a bent exponent. Let $\text{wt}(d)$ denote the binary weight of d. As bent functions have degree at most k and as the degree of $\text{tr}(\alpha x^d)$ is either 0 or $\text{wt}(d)$ (see [8]), it follows that the binary weight of a bent exponent is at most k.

Furthermore a bent exponent d cannot be coprime to $2^n - 1$. Otherwise, if d is coprime to $2^n - 1$ we get

$$\sum_{x \in L} \chi_L(\alpha x^d) = \sum_{x \in L} \chi_L(\alpha x) = 0$$

for every non-zero $\alpha \in L$ in contradiction to the bentness property.

Note that a function $f(x) = \text{tr}(\alpha x^d)$ cannot be bent for every non-zero α. If f would be bent for every $\alpha \in L^*$, this would allow the construction of a vectorial bent function from L to L which is not possible [26].

Denote $\gcd(d, 2^n - 1)$ by s. Moreover let

$$W = \{\gamma \in L \mid \gamma^d = 1\} = \{\gamma \in L \mid \gamma^s = 1\}.$$

Clearly, f is constant on all cosets λW. Consequently we get

$$f^{\mathcal{W}}(0) = 1 + s \sum_{\lambda \in L^*/W} \chi_L(\alpha x^d) \equiv 1 \bmod s.$$

Assume that $f^{\mathcal{W}}(0) = 2^k$ then s divides $2^k - 1$. On the other hand if $f^{\mathcal{W}}(0) = -2^k$ then s divides $2^k + 1$. As $\gcd(2^k - 1, 2^k + 1) = 1$ we see, that d is coprime either to $2^k - 1$ or to $2^k + 1$. Summarizing we get

Lemma 7. *Let d be a bent exponent. Then $\gcd(d, 2^n - 1) \neq 1$. Furthermore let $f(x) = \mathrm{tr}(\alpha x^d)$ be a bent function. Then*

1. $\gcd(d, 2^k - 1) = 1$ *if and only if* $f^{\mathcal{W}}(0) = -2^k$
2. $\gcd(d, 2^k + 1) = 1$ *if and only if* $f^{\mathcal{W}}(0) = 2^k$

$\square$

Let $\alpha, \beta \in L^*$. If $\beta = \alpha \lambda^d$ for some non-zero element $\lambda \in L$, the functions $x \to \mathrm{tr}(\alpha x^d)$ and $x \to \mathrm{tr}(\beta x^d)$ are linear equivalent. It follows that for our considerations we can always replace α by any element in the same coset of U, where

$$U = \{x^d \mid x \in L^*\} = \{x^{\gcd(d, 2^n - 1)} \mid x \in L^*\}.$$

Note that for any $\gamma \in L$ with $\gamma^d = 1$ the Walsh-coefficients of $f(a)$ and $f(\gamma a)$ are equal. Thus we can always replace a by any element in the same coset of W.

4.1   The Gold Case

This case belongs to the class of quadratic bent functions, the easiest and best understood class of bent functions. It is well known, that the dual of a quadratic bent function is again a quadratic bent function. Furthermore all quadratic bent functions are linear or affine equivalent, which can easily be proven, using the theory of quadratic forms. It follows that the dual of any quadratic bent function is equivalent to the function itself. Nevertheless finding the explicit linear or affine mapping is not always trivial. However in the case of the Gold Exponent the linear transformation can easily be computed.

Theorem 4. *Let $\alpha \in \mathbb{F}_{2^n}$, $r \in \mathbb{N}$ and $d = 2^r + 1$. The function*

$$f : L \to \mathbb{F}_2$$

with

$$f(x) = \mathrm{tr}(\alpha x^d),$$

is bent if and only if

$$\alpha \notin \{x^d \mid x \in \mathbb{F}_{2^n}\}$$

If α is not a d-th power, the linear mapping

$$H(\gamma) = \alpha^{2^r} \gamma^{2^{2r}} + \alpha\gamma$$

is bijective and the dual of f is

$$f^{\mathcal{W}}(a) = (-1)^{f(H^{-1}(a))} f^{\mathcal{W}}(0).$$

Remark 1. Note, that not for every r the coefficient α can be chosen, such that the corresponding monomial function is bent. In the case where $\gcd(d, 2^n - 1) = 1$ every $\alpha \in \mathbb{F}_{2^n}$ is a d-th power. In other words a Gold Exponent d is a bent exponent if and only if $x \to x^d$ is not a bijection.

4.2   The Dillon Case

Let $d = 2^k - 1$. We consider the monomial function

$$f : L \to \mathbb{F}_2$$

with

$$f(x) = \mathrm{tr}(\alpha x^d).$$

This exponent was first considered by Dillon [12] as an example of bent functions belonging to the PS-class. Dillon proved that this function is bent if and only if α is a zero of the Kloosterman Sum.

Obviously this function is constant on λK^* for all $\lambda \in L$, where K is the subfield of index 2 in L. So this leads to a bent function if and only if it belongs to the Partial Spread class.

$$f^{\mathcal{W}}(a) = 2^k \chi_L(\alpha a^{-d}) + K(\alpha)$$

and as $2^k d = -d \bmod 2^n - 1$ and $\alpha^{2^k} = \alpha$ we get the following theorem.

Theorem 5. *Let $a \in L$. Then*

$$f^{\mathcal{W}}(a) = 2^k \chi_L(\alpha a^d) + K(\alpha)$$

As a corollary we get

Corollary 2. *The exponent $d = 2^k - 1$ is a bent exponent. $f(x) = \mathrm{tr}(\alpha x^d)$ is bent if and only if $K(\alpha) = 0$. In this case the dual of f is identical to the function itself.*

In [22] it was shown, that such an α exists for every k, i.e. bent functions of this type exist for every n. Furthermore as

$$S(\alpha) = \sum_{u \in S} \chi_L(\alpha u) = \sum_{u \in S} \chi_L(\alpha u^s)$$

for every integer s coprime to $2^k + 1$ we get

Corollary 3. *For every integer s coprime to $2^k + 1$ the function*

$$f'(x) = \mathrm{tr}(\alpha x^{sd})$$

is bent whenever $K(\alpha) = 0$.

4.3   The Dillon-Dobbertin Case

Regarding the question of normality of bent functions, the Kasami exponent $d = 2^{2r} - 2^r + 1$ is the most interesting case of a bent exponent (see also Section 3). The following theorem was conjectured by Hollmann and Xiang in 1999 ([20], Conjecture 4.4)and was proven by Dillon and Dobbertin in [14].

Theorem 6. *Let n be an even integer with $\gcd(3, n) = 1$. Furthermore let $d = 2^{2r} - 2^r + 1$ with $\gcd(r, n) = 1$ and $\alpha \in \mathbb{F}_{2^n}$. The function*

$$f : \mathbb{F}_2^n \to \mathbb{F}_2^n$$

with

$$f(x) = \mathrm{tr}(\alpha x^d)$$

is bent if and only if $\alpha \notin \{x^3 | x \in \mathbb{F}_{2^n}\}$.

Remark 2. We anticipate that the restriction $\gcd(3, n) = 1$ is not necessary.

Despite the strong similarities of this theorem and the Gold Exponent, the proof of Theorem 6 is distinct more complex and requires very sophisticated techniques. Unfortunately it does not give any insight to the structure of the dual function, as it is proven that $f^{\mathcal{W}}(a)^2 = 2^n$ for all $a \in L$.

As a first step to investigate the dual function we used an algorithm [24], to check if the dual is linear equivalent to the function itself. Remarkably, using our algorithm described, it turns out that this is true for $n = 8$ but not for $n = 10, 12$ or 14. A theoretical approach for computing the dual is an interesting open challenge.

4.4   A New Bent Exponent for $n = 4r$

Based on computer experiments, Anne Canteaut ([3]) conjectured that for r odd and $n = 4r$ the exponent $d = (2^r + 1)^2$ is a bent exponent. This was proven in [23] and the proof of this theorem actually shows, that the corresponding Boolean function belongs to the well known class of Maiorana-McFarland bent functions.

Let $n = 4r$ be an even integer where r is odd. $L = \mathbb{F}_{2^n}$ and $E = \mathbb{F}_{2^r}$. As r is odd the polynomial

$$\beta^4 + \beta + 1 \tag{3}$$

is irreducible over E as it is irreducible over $\mathbb{F}_2$. We conclude that $E[\beta] = L$. Note that β is a primitive element in $\mathbb{F}_{16}$. In particular every element $x \in L$ can be represented as

$$x = x_3\beta^3 + x_2\beta^2 + x_1\beta + x_0$$

with x_i in E. The key step in the proof of the following theorem is to express the function on L as a function in the variables x_i.

Theorem 7. *With the notation from above let* $d = (2^r + 1)^2 = 2^{2r} + 2^{r+1} + 1$ *and* $\alpha = \beta^5$. *The function*

$$f : L \to \mathbb{F}_2$$
$$f(x) = \mathrm{tr}_L(\alpha x^d)$$

is bent. In particular d *is a bent exponent.*

5   Niho-Type Bent Functions

In the this chapter we study traces of a linear combination of Niho power functions. Recall that power function on $\mathbb{F}_{2^n}$ is called a Niho power function if its restriction to $\mathbb{F}_{2^k}$ is linear. The considered functions are therefore weakly normal. In this way, under certain conditions, we get as our main results (Theorems 8, 9 and 10) three primary constructions of bent functions as linear combinations of two Niho power functions. Theorem 9 actually belongs to a more general class of bent functions, discussed in [24]. One advantage of focusing on Niho power functions is, that the classical theorem of Niho [25] serves as a starting point for proving our results. Furthermore we make use of new methods to handle Walsh transforms of Niho power functions from [18].

We introduce a new general method to prove that certain rational functions are one-to-one. This technique follows the sprit of the multivariate method introduced by the first author (see [17]), in the sense that the rational functions considered here introduce one-to-one mappings on an infinite chain of finite fields. This is also reflected in the techniques developed here which, like with the multivariate method, mainly manipulate generic properties of the discussed mappings. Another similarity is, that some of these manipulations can only be treated with the help of computer algebra packages.

The preceding theorems were conjectured based on computer experiments worked out by Canteaut, Carlet and Gaborit for $k \leq 6$. Every found example of that exhaustive search is now covered by one of our theorems.

Niho power functions. Recall that we say d (always understood modulo $2^n - 1$) is a *Niho exponent* and x^d is a *Niho power function,* if the restriction of x^d to $\mathbb{F}_{2^k}$ is linear or in other words

$$d \equiv 2^i \quad (\mathrm{mod}\ 2^k - 1)$$

for some $i < n$. Without loss of generality we can assume that d is in the *normalized form* with $i = 0$, and then we have a unique representation

$$d = (2^k - 1)s + 1$$

with $2 \leq s \leq 2^k$, because here s and s' give the same power function d on $\mathbb{F}_{2^n}$ iff $s \equiv s' \pmod{2^k + 1}$.

Let $L = \mathbb{F}_{2^n}$ and $n = 2k$. We consider Boolean functions

$$f(x) = \text{tr}_L(\alpha_1 x^{d_1} + \alpha_2 x^{d_2})$$

on L, for $\alpha_1, \alpha_2 \in L$, where the $d_i = (2^k - 1)s_i + 1$, $i = 1, 2$, are Niho exponents. We anticipate that if f is bent, then necessarily w.l.o.g.

$$d_1 = (2^k - 1)\frac{1}{2} + 1.$$

This conjecture is suggested by computer experiments. In the sequel we require this choice of d_1. Observe that d_1 is cyclotomic equivalent to $2d_1 = 2^k + 1$.

This special choice of d_1 implies that replacing α_1 by α_1' does not change f if (and only if) $\alpha_1 + \alpha_1' \in K$.

For $\alpha_2 = 0$, we get bent functions iff $\alpha_1 \notin K$, which corresponds to the Gold Case (see 4.1). It seems that there are no more bent functions of the form $f(x) = \text{tr}_L(\alpha x^d)$ with Niho exponent d.

For the following theorems we require that

$$\alpha_1 + \overline{\alpha_1} = \text{norm}\, \alpha_2$$

However, this general form can easily be reduced to the case $\alpha_2 = 1$, as we shall see.

Theorem 8. *(see Theorem 1 in [15])*
Define

$$\boxed{d_2 = (2^k - 1)3 + 1.}$$

If $k \equiv 2 \pmod 4$ assume that $\alpha_2 = \beta^5$ for some $\beta \in L^*$. Otherwise, i.e. if $k \not\equiv 2$ (mod 4), $\alpha_2 \in L^*$ is arbitrary. Then f is a bent function with degree k.

From $\omega(d_2) = \omega(2^k + (2^{k-1} - 1)) = 1 + (k - 1) = k$ we conclude that f, as a multi-variate binary function, has in fact degree k, the maximal degree a bent functions can attain.

Theorem 9. *(see Theorem 2 in [15])*
Suppose that k is odd. Define

$$\boxed{d_2 = (2^k - 1)\frac{1}{4} + 1.}$$

Then f is a bent function of degree 3.

Observe that d_2 is cyclotomic equivalent to and can be replaced by

$$4d_2 = 2^k + 3.$$

From $\omega(4d_2) = 3$ we conclude that f has degree 3.

Theorem 10. (see Theorem 3 in [15])
Suppose that k is even. Define

$$d_2 = (2^k - 1)\frac{1}{6} + 1.$$

Then f is a bent function of degree k.

Note that

$$2d_2 = (1 + 4 + 16 + \cdots + 2^{k-2}) + 2$$

and therefore $w(d_2) = k/2 + 1$ and consequently f has actually degree k.
    Niho's theorem [25] is presented below.

Theorem 11. *Assume that*

$$d = (2^k - 1)s + 1$$

is a Niho exponent and

$$f(x) = \mathrm{tr}(x^d).$$

Then $f^{\mathcal{W}}(c) = (N(c) - 1)2^k$, where $N(c)$ is the number of $u \in S$ such that

$$u^{2s-1} + \overline{u}^{2s-1} + cu + \overline{cu} = 0, \tag{4}$$

for each $c \in L = \mathbb{F}_{2^n}$.
    Thus the Walsh spectrum of f is at most $2s$-valued, and the occurring values are among

$$-2^k, \ 0, \ 2^k, \ 2 \cdot 2^k, \ ..., \ (2s - 2)\, 2^k.$$

The same argument shows that more generally if

$$f(x) = \mathrm{tr}_L\left(\sum_{i=1}^{m} \alpha_i x^{d_i}\right)$$

for Niho exponents $d_i = (2^k - 1)s_i + 1$ $(i = 1, ..., m)$, then $N(c)$ is the number of solutions u in S of

$$cu + \overline{cu} + \sum_{i=1}^{m} \alpha_i u^{1-2s_i} + \sum_{i=1}^{m} \overline{\alpha_i}\, \overline{u}^{1-2s_i} = 0,$$

or equivalently by replacing u by $\overline{u}$

$$cu + \overline{cu} + \sum_{i=1}^{m} \alpha_i u^{2s_i-1} + \sum_{i=1}^{m} \overline{\alpha_i}\, \overline{u}^{2s_i-1} = 0.$$

This means for the f in Theorems 8, 9 and 10, where $s_1 = \frac{1}{2}$ that the equation

$$cu + \overline{cu} + \alpha_1 + \overline{\alpha_1} + \alpha_2 u^{2s_2-1} + \overline{\alpha_2}\, \overline{u}^{2s_2-1} = 0 \tag{5}$$

has to be considered. We assume that $\alpha_2 = 1$ and thus $\alpha_1 + \overline{\alpha_1} = 1$. (The assertion of our theorems can easily be reduced to that case, see section 5.1.) Therefore in order to confirm that f is bent, setting $s = s_2$ we have to show that the number of roots u in $\mathcal{S}$ of

$$G_c(u) = u^{2s-1} + \overline{u}^{2s-1} + cu + \overline{c}\overline{u} + 1 = 0 \tag{6}$$

is either 0 or 2.

Parseval's equation obviously implies that it suffices to prove that (6) never has exactly one solutions. Thus in the case where $c \in K$, i.e. $c = \overline{c}$ we can argue as follows: $G_c(1) = 1$, so in this case $u = 1$ is never a solution. Furthermore

$$G_c(u) = u^{2s-1} + \overline{u}^{2s-1} + c(u + \overline{u}) + 1 = G_c(\overline{u}),$$

thus whenever u is a solution of (6) the conjugate $\overline{u}$ is also a solution and exactly one solution is never possible.

Nevertheless for the proof of theorems 8 and 10 we explicitly show that we have either 0 or 2 solutions to demonstrate the power of our general technique.

In [18] the value distribution of the Walsh spectrum of $\mathrm{tr}(x^{d_2})$ for $d_2 = (2^k - 1)\,3 + 1$ of Theorem 8 has been determined for odd k, which requires to analyze the number of solutions of the closely related equation for $s = 3$:

$$u^5 + \overline{u}^5 + cu + \overline{c}\overline{u} = 0.$$

This problem was settled with the development of new approach using Dickson polynomials [18], which will be explained below. It is also the basic tool for proving the results of the present paper.

Given $c \in L \setminus K$ the idea of [18] is to consider c, $\overline{c}$ and the associated equations $G_c(u) = 0$ and $G_{\overline{c}}(u) = 0$ simultaneously:

$$G_c(u)\, G_{\overline{c}}(u) = 0. \tag{7}$$

Then we can change from the parameters $u \in \mathcal{S}$ and $c \in L$ to new parameters β, resp. γ, T and N in the small field K. The advantage of this procedure is that we end up with an equation where we have to count the solutions with a special "trace condition" instead of counting solutions with a "norm condition", which turns out to be much easier.

The twins $c, \overline{c} \in L \setminus K$ are replaced by the coefficients of their (common) minimal polynomial

$$m_{c,\overline{c}} = X^2 + TX + N$$

over K, that is

$$T = \mathrm{tr}_{L/K}(c) = \mathrm{tr}_{L/K}(\overline{c}) = c + \overline{c},$$
$$N = \mathrm{norm}_{L/K}(c) = \mathrm{norm}_{L/K}(\overline{c}) = c\overline{c}.$$

Necessary and sufficient conditions for $T, N \in K$ to represent $c, \overline{c} \in L \setminus K$ in this way are $T \neq 0$ and

$$\mathrm{tr}_K(N/T^2) = 1. \tag{8}$$

We recall the following simple, but very important observation:

Fact. We have $\mathrm{tr}_K(x) = 0$ for $x \in K$ if and only if there exists some $y \in K$ with $x = y^2 + y$.

Thus (8) means that $X^2 + TX + N$ is irreducible over K. Fortunately (8) can be ignored in this context, as it is included in (9) (see below).
   Similarly β stands for $u, \bar{u} \in \mathcal{S} \setminus \{1\}$ in the sense that

$$m_{u,\bar{u}}(X) = X^2 + \frac{1}{\beta}X + 1.$$

or equivalently

$$\beta = \frac{1}{u + \bar{u}}.$$

A necessary and sufficient condition for β to play this role is

$$\mathrm{tr}_K(\beta) = 1.$$

Sometimes it is convenient to make also use of the parameter γ:

$$\gamma = 1/\beta.$$

Changing to the new parameters, $G_c(u) G_{\bar{c}}(u)$ can be transformed as follows, where $D_i(X)$ denotes the i-th Dickson polynomial over $\mathbb{F}_2$:

$$
\begin{aligned}
G_c(u) G_{\bar{c}}(u) &= \left(u^{2s-1} + \bar{u}^{2s-1} + cu + \overline{cu} + 1\right)\left(u^{2s-1} + \bar{u}^{2s-1} + \bar{c}u + c\bar{u} + 1\right) \\
&= \left(u^{2s-1} + \bar{u}^{2s-1} + 1\right)^2 + \left(u^{2s-1} + \bar{u}^{2s-1} + 1\right)(c + \bar{c})(u + \bar{u}) \\
&\quad + (cu + \overline{cu})(\bar{c}u + c\bar{u}) \\
&= (D_{2s-1}(\gamma) + 1)^2 + (D_{2s-1}(\gamma) + 1)\gamma T + T^2 + \gamma^2 N.
\end{aligned}
$$

Dickson polynomials satisfy the functional equation

$$D_i(X + X^{-1}) = X^i + X^{-i},$$

the iteration rule

$$D_i(D_j(X)) = D_{ij}(X)$$

and can be obtained by the recursion

$$D_{i+2}(X) = X D_{i+1}(X) + D_i(X)$$

with $D_0(X) = 0$ and $D(X) = X$.
   Summarizing we have seen that $G_c(u) G_{\bar{c}}(u) = 0$ with $u \in \mathcal{S}$ is equivalent to the following equation in K:

$$\left(\frac{(D_{2s-1}(1/\beta) + 1)\beta}{T}\right)^2 + \frac{(D_{2s-1}(1/\beta) + 1)\beta}{T} + \beta^2 = \frac{N}{T^2}. \qquad (9)$$

Given T and N we have to count the number of solutions β with trace 1 of (9). Now the trick is that we can look at this solution counting problem also in another way. Given any non-zero T and β with trace 1, we can interpret (9) as *definition* of N. This makes sense, because it then follows, as already mentioned above, that $\mathrm{tr}_K(N/T^2) = \mathrm{tr}_K(\beta) = 1$ and therefore T, N represent c, $\bar{c}$ via $m_{c,\bar{c}}(X) = X^2 + TX + N$. We then have to look at the number of solutions of (9) different from the given β (for more details see [18])). The special cases $T = 0$ and $T = 1$ have to be considered separately.

5.1   Proof of Theorem 8

In order to demonstrate how, in prinicipal, Theorems 8,9 and 10 can be deduced we briefly sketch the proof of the first Theorem. We will demonstrate, that the statements of these Theorems are related to the one-to-one property of certain rational functions. For the proof that these rational functions are indeed one-to-one we develop a new technique, as described in Section 5.2.

The general case $\alpha_1 + \overline{\alpha_1} = \mathrm{norm}\, \alpha_2$ for Theorem 8 follows from $\alpha_2 = 1$ and $\alpha_1 + \overline{\alpha_1} = 1$.

Using Niho's theorem (Theorem 11) in order to confirm Theorem 8 we have to prove that, for all $c \in L = \mathbb{F}_{2^n}$, $n = 2k$, the number of $u \in S$ such that

$$G_c(u) = u^5 + \overline{u}^5 + cu + \overline{cu} + 1 = 0$$

is either 0 or 2 (see (6)). Recall that

$$S = \{u \in L : u\overline{u} = 1\},$$

$K = \mathbb{F}_{2^k}$, and $x \in K$ iff $x \in L$ and $x = \overline{x} = x^{2^k}$. We shall apply the approach described in Section 4. Recall that

$$\beta = 1/(u + \overline{u}), \quad \mathrm{tr}(\beta) = 1,$$
$$T = c + \overline{c},$$
$$N = c\overline{c}.$$

Case 1: $T = 0$, *i.e.* $c \in K$. Then $G_c(u) = 0$ iff

$$u^5 + \overline{u}^5 + c(u + \overline{u}) = 1$$

i.e. iff

$$c = D_5(1/\beta)\beta + \beta = 1/\beta^4 + 1/\beta^2 + 1 + \beta,$$

where $D_5(X) = X^5 + X^3 + X$ denotes the 5-th Dickson polynomial. Thus given c we have no or precisely two solutions $u \in S$ of $G_c(u) = 0$ if and only if

$$\beta \mapsto \Phi(\beta) = 1/\beta^4 + 1/\beta^2 + \beta$$

is one-to-one for $\beta \in T_1$, the set of elements in K with trace 1, which is true by Lemma 8. (For further details concerning this approach see [18] in Section 4, Case 1 especially.)

Case 2a: $T = 1$. Note that this case occurs if and only if $u = 1$ is a solution of $G_c(u) = 0$. Then on the other hand $G_c(u)\,G_{\bar{c}}(u) = 0$ with $u \neq 1$ iff

$$c\bar{c} = \Psi(\beta) = 1/\beta^8 + 1/\beta^2 + \beta, \tag{10}$$

where $\beta = 1/(u+\bar{u}) \in K$ and therefore $\mathrm{tr}_K(\beta) = 1$, see (9). Arguing as before in Case 1 we have to show that Ψ is one-to-one on T_1, which is true by Lemma 8. The two solutions of $G_c(u) = 0$ and $G_{\bar{c}}(u) = 0$ are $u = 1$ and $u = u_0$, respectively $u = 1$ and $u = \overline{u_0}$, where $\beta_0 = 1/(u_0 + \overline{u_0})$ is the unique solution of (10) with trace 1.

Case 2b: $T \notin \mathbb{F}_2$. By (9) we have

$$N = T^2\beta^2 + \Phi_1(\beta)\,T + \Phi_1(\beta)^2,$$

with

$$\Phi_1(\beta) := (D_5(1/\beta) + 1)\,\beta = \Phi(\beta) + 1.$$

We have to show that for each $T \notin \mathbb{F}_2$

$$\beta \mapsto T^2\beta^2 + \Phi_1(\beta)\,T + \Phi_1(\beta)^2$$

maps two-to-one for $\beta \in T_1$. (For details concerning this approach we refer again to [18], Section 4, Case 2 in particular.) In other words, since $u = 1$ is impossible (see Case 2a above), given $T \notin \mathbb{F}_2$ and β with $\mathrm{tr}_K(\beta) = 1$ we have to show that there is a unique non-zero Δ with $\mathrm{tr}_K(\Delta) = 0$ and

$$T^2\beta^2 + \Phi_1(\beta)\,T + \Phi_1(\beta)^2 = T^2(\beta + \Delta)^2 + \Phi_1(\beta + \Delta)\,T + \Phi_1(\beta + \Delta)^2, \tag{11}$$

that is

$$\Delta^2 = \left(\Phi_1(\beta + \Delta) + \Phi_1(\beta)\right)/T + \left(\Phi_1(\beta + \Delta) + \Phi_1(\beta)\right)^2/T^2.$$

Setting $\Delta = x^2 + x$, this means that

$$x^2 + \left(\Phi_1(\beta + x^2 + x) + \Phi_1(\beta)\right)/T + \varepsilon = 0,$$

or equivalently

$$T = \frac{\Phi_1(\beta + x^2 + x) + \Phi_1(\beta)}{x^2 + \varepsilon} \tag{12}$$

for an unique set $\{x, x+1\}$ and $\varepsilon \in \mathbb{F}_2$. The pairs (x, ε) and $(x + 1, \varepsilon + 1)$ give the same T. Hence w.l.o.g. we can choose $\varepsilon = 0$. Then the right hand rational function of equation (12) coincides with $R_a(x)$ for $a = \beta^2$, since $\Phi_1(\beta) = \Phi(\beta)+1$, see (13). Thus the existence of an unique non-zero $\Delta = x^2 + x$ for given T and β is guaranteed in view of Lemma 9. This completes the proof that the Boolean function f in Theorem 8 is bent.

5.2   One-to-One Rational Functions

After these preparations, the verification of our main results comes down to the following two lemmas (to be honest, they have been found for that reason).

Remark 3. The technique used here to prove the below Lemmas 8 and 9 is due to Dobbertin and Leander. It is in some sense similar to the *multi-variate method* (see [17], where the multi-variate method is described in its general form), insofar as a "generic" point of view is taken. As for the multi-variate method, also here algebraic computations are applied, which often need Computer Algebra support. Decomposition of multi-variate polynomials (with variables which are considered to be independent) and formal elimination of variables, i.e. for instance computation of resultants, as basic steps.

We briefly describe the method and roughly explain why it works. Suppose that an irreducible multi-variate polynomial $F(a, x_1, ..., x_m)$ is given, and that we have to show that $F(a, x_1, ..., x_m) = 0$ implies that a has trace 0, i.e. we can represent $a = b^2 + b$ in each of the considered fields. If this fact has "generic" reasons then we can represent these "local" b in a "global" way as a *fixed* rational function of a, x_1, ..., x_m:

$$b = R(a, x_1, ..., x_m) = \frac{C(a, x_1, ..., x_m)}{D(a, x_1, ..., x_m)}.$$

Assume that R in fact exists. Then $X = b$ is a zero of the rational function

$$(X + R(X^2 + X, x_1, ..., x_m))(X + 1 + R(X^2 + X, x_1, ..., x_m)).$$

In the generic case we can expect that this rational function is essentially, up to avoiding denominators, the polynomial

$$F(X^2 + X, x_1, ..., x_m),$$

which therefore factorizes in the form

$$Q(X, x_1, ..., x_m)\, Q(X + 1, x_1, ..., x_m).$$

Thus we consider b as unknown, substitute $a = b^2 + b$ in F and decompose F in order to compute Q. We can assume that a occurs in Q with some odd exponent. Using then $b^2 = b + a$ we reduce Q and get the polynomial $C(a, x_1, ..., x_m) + D(a, x_1, ..., x_m)b$, which gives $R = C/D$. Common zeros of C and D need an extra discussion.

Given a concrete field K of characteristic 2, we find $b \in E$ with $a = b^2 + b$ in some extension field E of K. Thus if $F(a, x_1, ..., x_m) = 0$ for $a, x_1, ..., x_m \in K$, then our generic result implies that $b = R(a, x_1, ..., x_m)$ and therefore $b \in K$, i.e. $\mathrm{tr}_K(a) = 0$.

This simple machinery, which works of course for any non-zero characteristic, will turn out to be very powerful and effective.

Define

$$T_\varepsilon = \{x \in K : \mathrm{tr}_K(x) = \varepsilon\}, \quad \varepsilon \in \mathbb{F}_2.$$

Lemma 8. *Let K be any finite field of characteristic 2. Then the rational functions*

$$\Phi(x) = \frac{1}{x^4} + \frac{1}{x^2} + x$$

and

$$\Psi(x) = \frac{1}{x^8} + \frac{1}{x^2} + x,$$

respectively, induce a permutation of $\mathcal{T}_1$.

Proof. The proof is essentially the same for both rational functions. We consider first $\Phi(x) = 1/x^4 + 1/x^2 + x$. Note that

$$\begin{aligned}
\mathrm{tr}(\Phi(x)) &= \mathrm{tr}(1/x^4) + \mathrm{tr}(1/x^2) + \mathrm{tr}(x)\\
&= \mathrm{tr}(1/x) + \mathrm{tr}(1/x) + \mathrm{tr}(x)\\
&= \mathrm{tr}(x).
\end{aligned}$$

Thus Φ maps $\mathcal{T}_\varepsilon$ into itself. It remains to confirm that for $\Delta \neq 0$

$$\Phi(x + \Delta) = \Phi(x)$$

implies $\mathrm{tr}(x) = 0$. We have $\Phi(x) = U(x)/V(x)$ with polynomials $U(x) = x^5 + x^2 + 1$ and $V(x) = x^4$. Substituting $x^2 = y^2 + y$ the idea is to represent y as a rational function of x and Δ as described above[1]. We see that the polynomial

$$(\Phi(x + \Delta) + \Phi(x))\, V(x + \Delta) V(x) = U(x + \Delta) V(x) + U(x) V(x + \Delta)$$

factorizes in the form

$$\Delta\, Q(\Delta, y)\, Q(\Delta, y + 1)$$

with

$$Q(\Delta, y) = y^4 + y^3 + \Delta^2 y^2 + \Delta y + \Delta^2.$$

On the other hand we can write Q uniquely as

$$Q(\Delta, y) = C(\Delta, x^2) + D(\Delta, x^2) y$$

with polynomials C and D. In fact to compute C and D, reduce Q modulo $y^2 = y + x^2$. Here we have

$$\begin{aligned}
C(\Delta, x) &= x^2 + \Delta^2(x + 1),\\
D(\Delta, x) &= x + \Delta^2 + \Delta.
\end{aligned}$$

Summarizing we conclude for $\Delta \neq 0$ that $\Phi(x + \Delta) = \Phi(x)$ implies $Q(\Delta, y) = 0$ w.l.o.g., thus $x^2 = y^2 + y$ for $y = C(\Delta, x^2)/D(\Delta, x^2)$. Hence $y \in K$ and $\mathrm{tr}(x) = 0$. It remains to confirm that $C(\Delta, x)$ and $D(\Delta, x)$ have no common zeros x in $\mathcal{T}_1$, which is trivial in our case, since already $D(\Delta, x) = 0$ implies $\mathrm{tr}(x) = 0$.

---

[1] We take $x^2 = y^2 + y$ instead of $x = y^2 + y$, since here $U(x + \Delta) V(x) + U(x) V(x + \Delta)$ is a polynomial in x^2.

The other rational function $\Psi(x) = 1/x^8 + 1/x^2 + x$ can ge handled in precisely the same way. Here $U(x) = x^9 + x^6 + 1$ and $V(x) = x^8$. This leads to

$$
\begin{aligned}
Q(\Delta, y) &= y^8 + \Delta y^5 + (\Delta^4 + \Delta^2 + 1)y^4 \\
&\quad + (\Delta^3 + \Delta^2 + \Delta)y^3 + \Delta^3 y^2 + \Delta^3 y + \Delta^4 \\
C(\Delta, x) &= x^4 + (\Delta^4 + \Delta^2)x^2 + \Delta^4 x + \Delta^4, \\
D(\Delta, x) &= \Delta(x^2 + (\Delta^2 + \Delta)x + \Delta^3 + \Delta^2).
\end{aligned}
$$

$C(\Delta, x)$ and $D(\Delta, x)$ have a common zero Δ if and only if the resultant $\mathrm{res}(C, D, \Delta)$ of C and D with respect to Δ is zero. In this case we have

$$
\mathrm{res}(C, D, \Delta) = x^{14},
$$

which is non-zero. In general it suffices here to get a contradiction by showing that the zeros of resultant have trace 0. $\qquad\square$

Lemma 9. *Let K be any finite field of characteristic 2 and suppose that $a \in K$ has absolute trace 1. Then the rational functions*

$$
R_a(x) = \frac{(x+1)(ax^4 + x^3 + ax^2 + x + a^2)(ax^4 + x^3 + (a+1)x^2 + a^2)}{x(x^4 + x^2 + a)^2 a^2}
$$

induce a permutation of $K \setminus \mathbb{F}_2$.

Proof. We first consider R_a. Let $U_a(x)$ and $V_a(x)$ denote the nominator and denominator polynomial of $R_a(x)$, respectively. $V_a(x)$ is non-zero for non-zero x, since $\mathrm{tr}(a) = 1$. We note that $R_a(x)$ can be written as

$$
R_a(x) = \frac{\Phi(\sqrt{a} + x^2 + x) + \Phi(\sqrt{a})}{x^2} \tag{13}
$$

with Φ (see Lemma 8) defined as

$$
\Phi(x) = \frac{1}{x^4} + \frac{1}{x^2} + x.
$$

Thus $R_a(x)$ is non-zero for $x \notin \mathbb{F}_2$, since Φ is one-to-one on $\mathcal{T}_1$ by Lemma 8 and $a, a + x^2 + x \in \mathcal{T}_1$.

To confirm that R_a is one-to-one, we argue as before. Suppose on the contrary that $R_a(x) = R_a(y)$ for $x, y \notin \mathbb{F}_2$, $x \neq y$. We have to present $a = b^2 + b$ in K to get a contradiction to $\mathrm{tr}_K(a) = 1$. Substituting $a = b^2 + b$, the polynomial

$$
(R_a(x) + R_a(y))\, V_a(x)V_a(y) = U_a(x)V_a(y) + U_a(y)V_a(x)
$$

factorizes in the form

$$
a^2\, (x + y)\, Q(b, x, y)\, Q(b+1, x, y)
$$

with

$$Q(b, x, y) = b^6 + (x + y)^4 b^4 + xy(x + y)b^3$$
$$+ (xy(x + y))^3 + (x + 1)^4 (y + 1)^4 + x^2 y^2)b^2$$
$$+ x^2 y^2 (x + y)(xy + x + y)b$$
$$+ x^2 (x + 1)^2 y^2 (y + 1)^2.$$

Reducing Q modulo $b^2 = b + a$ we get

$$Q(b, x, y) = C(a, x, y) + D(a, x, y)b$$

with

$$C(a, x, y) = a^3 + (x + y + 1)^4 a^2$$
$$+ xy(xy + x + y)\left(xy(x + y) + (x + 1)^2(y + 1)^2\right)a$$
$$+ x^2 (x + 1)^2 y^2 (y + 1)^2,$$
$$D(a, x, y) = a^2 + xy(x + y)a + xy(x + 1)^2 (y + 1)^2 (xy + x + y).$$

Summarizing we conclude for $x \neq y$ that $R_a(x) = R_a(y)$ implies $Q(b, x, y) = 0$ w.l.o.g., thus $a = b^2 + b$ for $b = C(a, x, y)/D(a, x, y)$. It remains to confirm that $D(a, x, y)$ has no zeros a in T_1. On the contrary, suppose $D(a, x, y) = 0$. Then $C(a, x, y) = 0$ and $\mathrm{res}(C, D, a) = 0$. Here we have

$$\mathrm{res}(C, D, a) = x^2 (x + 1)^6 y^2 (y + 1)^6 (x + y)^2 (x + y + 1)^6.$$

Consequently $x + y + 1 = 0$, because $x, y \notin \mathbb{F}_2$ and $x \neq y$. On the other hand, from $C = D = 0$ we get a as a rational function in x and y, in our case

$$a = \frac{xy(x + 1)^2 (y + 1)^2}{x^2 + xy + y^2 + 1}.$$

A substitution of $y = x + 1$ yields $a = x^4 + x^2$, which implies that $\mathrm{tr}(a) = 0$, a contradiction.

It remains to show that R_a does not attain the value 1. Conversely assuming $R_a(x) = 1$, i.e. $U_a(x) = V_a(x)$ we have to conclude that a has trace 0. To this end we apply the same technique as before and substitute $a = b^2 + b$. Then the polynomial $U_a(x) + V_a(x)$ factorizes

$$U_a(x) + V_a(x) = Q(b, x)Q(b + 1, x)$$

with

$$Q(b, x) = b^4 + (x^4 + x + 1)b^2 + (x^3 + x^2)b + x^3 + x.$$

For C and D satisfying $Q = C + Db$ we compute

$$C(a, x) = a^2 + (x^4 + x)a + x^3 + x,$$
$$D(a, x) = x(x + 1)^3.$$

Now $C = 0$ contradicts our assumption $x \notin \mathbb{F}_2$. $\qquad\square$

Acknowledgement

We like to thank our colleagues Anne Canteaut, Claude Carlet, Magnus Daum and Patrick Felke for many interesting discussions.

References

1. A. Braeken, C. Wolf, B. Preneel, *A Randomised Algorithm for Checking the Normality of Cryptographic Boolean Functions* , 3rd International Conference on Theoretical Computer Science 2004, J. Levy, E. W. Mayr, and J. C. Mitchell (eds.), Kluwer, pp. 51-66, 2004.
2. A. Braeken, M. Daum, G. Leander, C. Wolf *An Algorithm for Checking Total Non-Normality*, Preprint 2004.
3. A. Canteaut, private communication, June 2004.
4. A. Canteaut, M. Daum, H. Dobbertin, G. Leander, *Normal and Non-Normal Bent Functions,* Proceedings of the Workshop on Coding and Cryptography (WCC 2003), Versailles, France, March 2003, pp. 91-100.
5. A. Canteaut, M. Daum, H. Dobbertin, G. Leander, *Finding Non-Normal Bent Functions*, special issue on Coding and Cryptography, Discrete Applied Mathematics, to appear.
6. C. Carlet, *A construction of bent functions*. Finite Fields and Applications, London Mathematical Society, Lecture Series 233, Cambridge University Press, pp. 47-58, 1996.
7. C. Carlet, *On cryptographic complexity of boolean functions*, Finite Fields with Applications to Coding Theory, Cryptography and Related Areas (Proceedings of Fq6), pages 53–69, Springer-Verlag, 2002.
8. C.Carlet *Codes de Reed-Muller, codes de Kerdock et de Preparata*. PhD thesis, Université Paris 6, 1990.
9. C. Carlet. *On the secondary constructions of resilient and bent functions*. Proceedings of the Workshop on Coding, Cryptography and Combinatorics 2003, published by BirkHauser Verlag, 2004.
10. C.Carlet, H.Dobbertin, G.Leander, *Normal Extensions of Bent Functions*, IEEE Transactions on Information Theory, vol. 50, no. 11, November 2004
11. M. Daum, H. Dobbertin, G. Leander, *An algorithm for checking normality of Boolean functions,* Proceedings of the Workshop on Coding and Cryptography (WCC 2003), Versailles, France, March 2003, pp. 133-142.
12. J.F. Dillon, *Elementary Hadamard Difference sets*, PhD thesis, University of Maryland, 1974.
13. Dillon, J. F., *Elementary Hadamard difference sets,* Proceedings of the Sixth Southeastern Conference on Combinatorics, Graph Theory and Computing, Boca Raton, Florida, Congressus Numerantium No. XIV, Utilitas Math., Winnipeg, Manitoba, 1975, 237-249.
14. J.F. Dillon and H. Dobbertin, *New Cyclic Difference Sets with Singer Parameters*, Finite Fields And Applications, 2004, pp. 342-389.
15. H. Dobbertin, G. Leander, A. Canteaut, C. Carlet, P. Felke, P. Gaborit, *Construction of Bent Functions via Niho Power Functions*, submitted.
16. H. Dobbertin, *Construction of bent functions and highly nonlinear balanced Boolean functions* , in "Fast Software Encryption", Lecture Notes on Computer Science, vol. 1008, Bart Preneel (ed.), Springer Verlag, 1995, 61-74.

17. H. Dobbertin, *Uniformly representable permutation polynomials,* in the Proceedings of "Sequences and their applications–SETA '01", T. Helleseth, P.V. Kumar and K. Yang (eds.), Springer Verlag, 2002, 1-22.

18. H. Dobbertin, P. Felke, T. Helleseth and P. Rosendahl, *Niho type cross-correlation functions via Dickson polynomials and Kloosterman sums,* IEEE Transactions on Information Theory, submitted.

19. S. Dubuc-Camus, *Etude des fonctions Booléennes dégénérées et sans corrélation,* PhD thesis, Université de Caen, 1998.

20. H. Hollmann, Q. Xiang, *On binary cyclic codes with few weights,* in the Proceedings of the Fifth Conference on Finite Fields and Their Applications, Augsburg, Germany, 1999, D. Jungnickel, H. Niederreiter (eds.), Springer Verlag, 251-275.

21. S. Gangopadhyay, S. Maitra, personal communication.

22. G. Lachaud, J. Wolfmann *The Weights of the Orthogonals of the Extended Quadratic Binary Goppa,* IEEE Transactions on Information Theory, Vol. 36, No 3, pages 686-692, May 1990

23. G. Leander, *Monomial Bent Functions,* submitted, 2004.

24. G. Leander, *Non Normal Bent Functions, Monomial and Binomial Bent Functions,* Ph.D. Thesis, Ruhr University Bochum, 2004.

25. Y. Niho, *Multivalued cross-correlation functions between two maximal linear recursive sequences,* Ph.D. Thesis, University of Southern California (1972).

26. K. Nyberg, *Perfect non-linear S-boxes,* in "Advances in Cryptology – Eurocrypt '91", vol. 547, Lecture Notes in Computer Science, Springer Verlag, 1991, 378-386.

27. O.S. Rothaus, *On "bent" functions,* Journal of Combinatorial Theory, Ser. A, 20 (1976), 300-305.

A Survey of the Merit Factor Problem
for Binary Sequences

Jonathan Jedwab

Department of Mathematics, Simon Fraser University,
Burnaby, BC, Canada V5A 1S6
jed@sfu.ca

Abstract. A classical problem of digital sequence design, first studied
in the 1950s but still not well understood, is to determine those binary sequences whose aperiodic autocorrelations are collectively small according
to some suitable measure. The merit factor is an important such measure,
and the problem of determining the best value of the merit factor of long
binary sequences has resisted decades of attack by mathematicians and
communications engineers. In equivalent guise, the determination of the
best asymptotic merit factor is an unsolved problem in complex analysis
proposed by Littlewood in the 1960s that until recently was studied along
largely independent lines. The same problem is also studied in theoretical
physics and theoretical chemistry as a notoriously difficult combinatorial
optimisation problem. The best known value for the asymptotic merit
factor has remained unchanged since 1988. However recent experimental and theoretical results strongly suggest a possible improvement. This
survey describes the development of our understanding of the merit factor problem by bringing together results from several disciplines, and
places the recent results within their historical and scientific framework.

1 Introduction

A *binary sequence*. A of length n is an n-tuple $(a_0, a_1, \ldots, a_{n-1})$ where each a_i
takes the value -1 or 1. The *aperiodic autocorrelation* of the binary sequence A
at shift u is given by

$$C_A(u) := \sum_{i=0}^{n-u-1} a_i a_{i+u} \qquad \text{for } u = 0, 1, \ldots, n-1. \tag{1}$$

Since the 1950s, digital communications engineers have sought binary sequences
whose aperiodic autocorrelations are collectively small according to some suitable
measure of "goodness" (see Section 2.1). This survey deals with an important
such measure, defined by Golay [30] in 1972: the *merit factor* of a binary sequence
A of length n is given by

$$F(A) := \frac{n^2}{2\sum_{u=1}^{n-1}[C_A(u)]^2}, \tag{2}$$

and the best binary sequences are those with the largest merit factor.

T. Helleseth et al. (Eds.): SETA 2004, LNCS 3486, pp. 30–55, 2005.

Let $\mathcal{A}_n$ be the set of all binary sequences of length n. We define F_n to be the optimal value of the merit factor for sequences of length n:

$$F_n := \max_{A \in \mathcal{A}_n} F(A).$$

The principal problem in the study of the merit factor is to determine the asymptotic behaviour of F_n:

The Merit Factor Problem. Determine the value of $\limsup_{n \longrightarrow \infty} F_n$.

Golay's publications reveal a fascination with the Merit Factor Problem spanning a period of nearly twenty years [30], [31], [32], [33], [34], [35]; the closing words of [35], published after Golay's death, refer to the study of the merit factor as "...this challenging and charming problem".

Prior to Golay's definition of merit factor in 1972, Littlewood [54] and other analysts studied questions concerning the norms of polynomials with ± 1 coefficients on the unit circle of the complex plane. As we describe in Section 2.2, the Merit Factor Problem is precisely equivalent to a natural such question involving the L_4 norm. This survey traces the historical development of the two (mostly independent) streams of investigation: the merit factor of binary sequences, and the L_4 norm of complex-valued polynomials with ± 1 coefficients on the unit circle.

A benchmark result on the asymptotic behaviour of the merit factor was given by Newman and Byrnes [62] in 1990:

Proposition 1. *The mean value of $1/F$, taken over all sequences of length n, is* $\dfrac{n-1}{n}$.

Proposition 1 shows that the asymptotic mean value of $1/F$ over all sequences of length n is 1. We cannot follow [38] in concluding that the asymptotic mean value of F itself over all sequences of length n is 1 [68], but we expect that "good" sequences will have an asymptotic value of F greater than 1. Indeed, the best known asymptotic results to date are given by explicitly constructed families of sequences whose merit factor tends to 6 (see Theorems 10, 15 and 16). The current state of knowledge regarding the Merit Factor Problem can therefore be summarised as:

$$6 \leq \limsup_{n \longrightarrow \infty} F_n \leq \infty. \tag{3}$$

Both of the extreme values in (3) have been conjectured to be the true value of the lim sup:

Conjecture 2 (Høholdt and Jensen, 1988 [39]). $\limsup_{n \longrightarrow \infty} F_n = 6$.

Conjecture 3 (Littlewood, 1966 [53–§6]). $\limsup_{n \longrightarrow \infty} F_n = \infty$.

Littlewood [53] also proposed stronger versions of Conjecture 3:

(i) $\lim_{n\longrightarrow\infty} F_n = \infty$
(ii) $1/F_n = O(1/\sqrt{n})$ for infinitely many n
(iii) $1/F_n = O(1/\sqrt{n})$ for all n.

My impression is that most researchers are reluctant to take seriously even the weakest of Littlewood's proposals, Conjecture 3, perhaps because the identity of their originator does not seem to be widely known.

Considerable computational evidence has been amassed regarding the value of F_n for specific values of n, in order to shed light on the Merit Factor Problem. Where computationally feasible, the actual value of F_n has been calculated; for larger values of n we have lower bounds on F_n via the identification of good, though not necessarily optimal, sequences (see Section 3). Indeed, Conjectures 2 and 3 are both based at least partially on numerical data. Conjecture 2 was made in light of Theorem 10 and its proof in [39], together with an examination of large values of F found in [4] for sequences of odd length between 100 and 200. Conjecture 3 and the stronger versions (i), (ii) and (iii) listed above were based primarily on calculation of F_n for $7 \leq n \leq 19$; Littlewood [53] asserted that "the evidence seems definitely in favour of [these conjectures]", and reiterated in 1968 [54] that "the numerical evidence for [these conjectures] is very strong". In terms of the computational power readily available nowadays, the range of Littlewood's calculations from the 1960s appears woefully inadequate! By 1996 Mertens [58] had calculated the value of F_n for $n \leq 48$ and reached the "tentative conclusion" that $\lim_{n\longrightarrow\infty} F_n > 9$. Currently the value of F_n has been calculated [59] for $n \leq 60$, and large values of F are known [45] for $61 \leq n \leq 271$ (see Figure 1).

Some authors seem to have conjectured that $\limsup_{n\longrightarrow\infty} F_n$ is given by the largest merit factor value known to be consistently achievable for long sequences at the time of writing. For example, Newman and Byrnes [62] incorrectly conjectured in 1990 that $\lim_{n\longrightarrow\infty} F_n = 5$, "... based on extensive numerical evidence employing the Bose-Einstein statistics methodology of statistical mechanics". Likewise, as noted above, Høholdt and Jensen [39] based Conjecture 2 in part on the best known merit factors reported in 1985 in [4] for sequences of odd length between 100 and 200, which they described as "either strictly smaller than or suspiciously close to 6"; however the current data underlying Figure 1 shows that the best merit factor is actually greater than 8 for all of these odd sequence lengths. In contrast, Golay [33] proposed in 1982 that $\limsup_{n\longrightarrow\infty} F_n \simeq 12.32$ (see Section 4.7) and yet in 1983 wrote that [34] "... the eventuality must be considered that no systematic synthesis will ever be found which will yield higher merit factors [than 6]"!

Recent work of Borwein, Choi and Jedwab [13] provides numerical evidence, from sequences up to millions of elements in length, that $\limsup_{n\longrightarrow\infty} F_n > 6.34$ (see Section 5). This conclusion, which would increase the best known asymptotic merit factor for the first time since 1988, is implied by Conjecture 20 on the behaviour of a specified infinite family of sequences.

The remainder of this survey is organised as follows. Section 2 gives a detailed practical motivation for the Merit Factor Problem from digital sequence design, together with a theoretical motivation from complex analysis. Section 3 describes various experimental computational approaches that have been used to gather numerical data, including exhaustive search and stochastic algorithms. Section 4 explains the main theoretical approaches that have been used to analyse the Merit Factor Problem. Section 5 outlines the method and results of [13], which suggest a new lower bound for $\limsup_{n \longrightarrow \infty} F_n$. Section 6 is a selection of challenges for future study.

This survey is concerned only with binary sequences, although the definition of merit factor has been extended to real-valued sequences (see for example [1]) as well as to binary arrays of dimension larger than 1 (see for example [8]). I have found the earlier surveys of Jensen and Høholdt [41] and Høholdt [38] to be helpful in preparing this paper, particularly when writing Section 4.

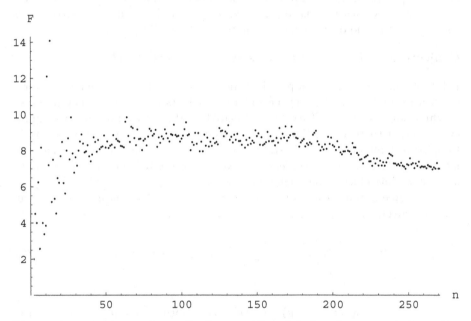

Fig. 1. The optimal merit factor (for $2 \leq n \leq 60$) and the best known merit factor (for $61 \leq n \leq 271$) for binary sequences of length n

2   Practical and Theoretical Motivation

This section shows how the Merit Factor Problem arises independently in digital sequence design and complex analysis.

2.1   Digital Sequence Design

Since the 1950s, digital communications engineers have sought to identify those binary sequences whose aperiodic autocorrelations are collectively as small as possible, for application in synchronisation, pulse compression and especially radar [74]. This classical problem of digital sequence design remains largely unsolved. In 1953 Barker [2] proposed that an ideal binary sequence of length n is one for which

$$C(u) = -1 \text{ or } 0 \text{ for } 0 < u < n,$$

but could find examples only for lengths $n = 3$, 7 and 11. Subsequent authors relaxed Barker's condition to:

$$|C(u)| = 1 \text{ or } 0 \text{ for } 0 < u < n, \tag{4}$$

and binary sequences satisfying (4) became known as *Barker sequences*. By a parity argument, no binary sequence can have a smaller value of $|C(u)|$ than a Barker sequence for any u. However the only non-trivial lengths for which Barker sequences are known to exist are 2, 3, 4, 5, 7, 11 and 13, and it has long been conjectured that no other sequence lengths n are possible:

Conjecture 4. *There is no Barker sequence of length $n > 13$.*

I do not know who first proposed Conjecture 4 but it is implied by Ryser's Conjecture [67] of 1963 on cyclic difference sets (see [48] for recent progress on this conjecture), and Turyn [74] declared in 1968: "There is overwhelming evidence that there are no Barker sequences [with $n > 13$]". A weaker version of Conjecture 4, alluded to in [74], states that there are only finitely many lengths n for which a Barker sequence of length n exists. In order to continue the historical account we introduce some further definitions.

The *periodic autocorrelation* of a binary sequence $A = (a_0, a_1, \ldots, a_{n-1})$ at shift u is given by

$$R_A(u) := \sum_{i=0}^{n-1} a_i a_{(i+u) \bmod n} \qquad \text{for } u = 0, 1, \ldots, n-1, \tag{5}$$

so that

$$R_A(u) = C_A(u) + C_A(n-u) \qquad \text{for } 0 < u < n. \tag{6}$$

A (v, k, λ) *cyclic difference set* is a k-element subset D of the cyclic group $\mathbb{Z}_v$ for which the multiset of differences $\{d_1 - d_2 : d_1, d_2 \in D, d_1 \neq d_2\}$ contains each non-zero element of $\mathbb{Z}_v$ exactly λ times (see [6] for background on difference sets, including generalisation to non-cyclic groups). The following is well-known (see for example [3]):

Proposition 5. *A (v, k, λ) cyclic difference set D is equivalent to a binary sequence $A = (a_0, a_1, \ldots, a_{v-1})$ having k elements -1 and constant periodic autocorrelation $R_A(u) = v - 4(k - \lambda)$ for $0 < u < v$, via the relationship*

$$i \in D \text{ if and only if } a_i = -1, \text{ for } 0 \leq i < v.$$

Turyn and Storer [72] showed in 1961 that Conjecture 4 is true for odd n, and that if a Barker sequence A of even length $n > 2$ exists then it satisfies $R_A(u) = 0$ for $0 < u < n$. Therefore, by Proposition 5, if there is a Barker sequence of length $v > 13$ then there is a cyclic difference set in $\mathbb{Z}_v$ satisfying $v = 4(k - \lambda)$. Difference sets satisfying this condition are known as *Hadamard* difference sets, and [73] must satisfy

$$(v, k, \lambda) = (4N^2, 2N^2 - N, N^2 - N) \text{ for integer } N. \tag{7}$$

(See [20] for a survey of difference sets with parameters (7); note that the difference set parameters (v, k, λ) given in (17) are unfortunately *also* called Hadamard.)

In his celebrated 1965 paper [73], Turyn showed that if there is a cyclic difference set with parameters (7) in $\mathbb{Z}_{4N^2}$ having $N \neq 1$, then $N \geq 55$. The paper [73] established the systematic use of algebraic number theory in the study of difference sets, which is now a standard and much-used technique. (See [6] for an overview of this technique, and some precursors to [73]; see [49] for dramatic improvements to the smallest open case for a Barker sequence and for a cyclic difference set with parameters (7).) R. Turyn has confirmed [personal communication, May 2003] that the chain of reasoning presented here, beginning with the search for binary sequences with small aperiodic autocorrelations, was the principal motivation behind [73]. I find it noteworthy that there has been a striking expansion of knowledge regarding difference sets since the publication of [73] and yet we still have not reached a comparably deep understanding of the original motivating problem.

Once it became apparent that the ideal behaviour given by a Barker sequence is unlikely to be achieved beyond length 13, attention turned [69], [74] to two measures of how closely the aperiodic autocorrelations of a binary sequence A of length n can collectively approach the ideal behaviour. These two measures are:

$$\sum_{u=1}^{n-1} [C_A(u)]^2 \tag{8}$$

and

$$M(A) := \max_{0 < u < n} |C_A(u)|. \tag{9}$$

The first measure (8) is simply $n^2/(2F(A))$, which was actually used by communications engineers as a measure of the "goodness" of a binary sequence several years before Golay [30] defined the merit factor in 1972 (see Section 3.2 for mention of Lunelli's work [55] of 1965 in this context).

The second measure (9) has been less well studied. By analogy with F_n, define

$$M_n := \min_{A \in \mathcal{A}_n} M(A)$$

to be the optimal value of $M(A)$ for sequences of length n. By exhaustive search, Turyn [74] showed that $M_n \leq 2$ for $n \leq 21$ in 1968 and Lindner [51] determined M_n for $n \leq 40$ in 1975 using specialised hardware. In 1990 Cohen, Fox and

Baden [18] found M_n for $n \leq 48$ by fixing sequence elements one pair at a time, working from the endpoints towards the centre and retaining only sequences with a prescribed maximum value of $|C(u)|$. Further calculations along similar lines by Coxson, Hirschel and Cohen [19] in 2001 found M_n for $n \leq 69$. From [18] and [19] we have $M_n \leq 3$ for all $n \leq 48$ and $M_n \leq 4$ for all $n \leq 69$; the value of M_n broadly increases with n, but not monotonically. In 1968 Moon and Moser [61] used elementary counting arguments to establish an (apparently weak) upper bound on M_n:

Theorem 6. *For any fixed $\epsilon > 0$, $M_n \leq (2 + \epsilon)(n \log n)^{1/2}$.*

I can see three possible explanations for the relative popularity of the measure (8) over (9). The first is that we have a more developed theoretical framework for studying the merit factor (see Section 4). The second is that the merit factor is a natural measure of the energy efficiency of a binary sequence used for physical transmission of information (see (13) and the comments following it). The third was offered by Turyn [74–p. 199] in 1968: "Intuitively one would expect [determination of $\limsup_{n \longrightarrow \infty} F_n$] to be easier [than determination of $\liminf_{n \longrightarrow \infty} M_n$]"!

2.2   Complex Analysis

We now describe an equivalent formulation of the Merit Factor Problem from complex analysis. Let $P_A(z) := \sum_{i=0}^{n-1} a_i z^i$ be the complex-valued polynomial whose coefficients are the elements of the binary sequence $A = (a_0, a_1, \ldots, a_{n-1})$ of length n. The L_4 *norm* of the polynomial $P_A(z)$ on the unit circle of the complex plane is defined to be

$$\|P_A(z)\|_4 := \left(\int_0^1 |P_A(\exp(2\pi\theta\sqrt{-1}))|^4 d\theta \right)^{1/4}, \qquad (10)$$

and it is straightforward to show [53] that

$$\|P_A(z)\|_4^4 = n^2 + 2\sum_{u=1}^{n-1} [C_A(u)]^2. \qquad (11)$$

Therefore the merit factor of the sequence A is related to the L_4 norm of the corresponding polynomial of degree $n - 1$ by

$$\|P_A(z)\|_4^4 = n^2 \left(1 + \frac{1}{F(A)} \right), \qquad (12)$$

and a large merit factor corresponds to a small L_4 norm. (See [17] for a survey of extremal problems involving the L_4 norm and other norms of complex-valued polynomials with ± 1 coefficients.)

Statements such as Proposition 1 and Conjecture 3 were originally made in terms of the L_4 norm of ± 1 polynomials, but have been expressed here in terms of the merit factor using (12). (Results that are recast in terms of the merit factor in this way often appear initially to be different from their original L_4 formulation because complex analysis usually considers polynomials of degree n, whereas (11) and (12) involve a polynomial of degree $n-1$.) However the results from the sequence design literature have not always been known to the complex analysis community, and vice-versa. For example, Littlewood [53] stated the correspondence (11) in 1966 without naming the aperiodic autocorrelation coefficients, and highlighted the lengths $n = 7, 11$ and 13 as supporting binary sequences satisfying (4) without mentioning Barker sequences (see Section 2.1). If Littlewood had been aware of the nonexistence results for Barker sequences already found by Turyn and Storer [72] in 1961 and by Turyn [73] in 1965, I do not believe he would have proposed Conjecture 3 in 1966 [53], repeated it in 1968 [54–Problem 19], and on both occasions cited as strong evidence the calculated value of F_n for $7 \leq n \leq 19$. Likewise, in 1988 Høholdt and Jensen [39] restated the correspondence (11) and explicitly linked the merit factor of binary sequences to complex-valued polynomials with ± 1 coefficients on the unit circle (to my knowledge, for the first time). But they then declared: "Unfortunately, there have been no results on [integrals of the type (10)], which can give new information on the behavior of the merit factor", whereas Littlewood [54] had given Theorem 11 in 1968.

Since $\int_0^1 |P_A(\exp(2\pi\theta\sqrt{-1}))|^2 d\theta = \sum_{i=0}^{n-1} a_i^2 = n$, we deduce from (2), (10) and (11) that

$$\int_0^1 \left\{ \left| P_A(\exp(2\pi\theta\sqrt{-1})) \right|^2 - n \right\}^2 d\theta = \frac{n^2}{F(A)}. \tag{13}$$

The left-hand side of (13) measures, in terms of power, how much the amplitude spectrum of the continuous-time signal corresponding to the sequence A deviates from its mean value n [4]. Therefore a larger merit factor corresponds to a more uniform distribution of the signal energy over the frequency range, which is of particular importance in spread-spectrum radio communications.

We have seen in Proposition 5 that a cyclic difference set is equivalent to a binary sequence A having constant periodic autocorrelation at all non-zero shifts. In terms of the corresponding polynomial $P_A(z)$, it is equivalent to the value $|P_A(z)|$ being constant at all complex nth roots of unity except 1.

Conjecture 3 is related to another old conjecture from complex analysis involving the supremum norm of ± 1 polynomials on the unit circle:

Conjecture 7 (Erdős, 1957 [24–Problem 22]). *There exists a constant $c > 0$ such that, for all n and for all binary sequences $A = (a_0, a_1, \ldots a_{n-1})$ of length n,*

$$\sup_{|z|=1} |P_A(z)| \geq (1+c)\sqrt{n}$$

where $P_A(z) := \sum_{i=0}^{n-1} a_i z^i$.

(Conjecture 7 was posed in [24] as a question as to whether a suitable $c > 0$ exists for complex-valued sequences satisfying $|a_i| = 1$ for all i, and restated

[25] in 1962 as a conjecture. Kahane [43] showed that the conjecture is false for complex-valued sequences, but the restriction of the question to binary sequences remains open.) Since $\sup_{|z|=1} |P_A(z)| \geq ||P_A(z)||_4$, we deduce from (12) that if Conjecture 3 is false then Conjecture 7 is true.

Furthermore, by (2) and (4), a Barker sequence of even length n must have merit factor n, so if Conjecture 3 is false then the weaker version of Conjecture 4 (that there are only finitely many Barker sequences) is also true. So determining whether $\limsup_{n \to \infty} F_n$ is unbounded would be of great significance: if so then a 1966 conjecture due to Littlewood is established; and if not then both a 1957 conjecture due to Erdös and a forty-year-old conjecture on the finiteness of the number of Barker sequences are true!

3   Computational Approaches

An experimental approach to mathematics has long provided "... a compelling way to generate understanding and insight; to generate and confirm or confront conjectures; and generally to make mathematics more tangible, lively and fun..." [9]. The view of mathematics as an experimental science has become more prominent as computers of steadily increasing power have become widely accessible to perform the role of "laboratory". In this spirit, considerable computational evidence has been collected regarding the value of F_n for specific values of n, in order to better understand the Merit Factor Problem.

The value of F_n has been calculated for $n \leq 60$ using exhaustive computation (see Section 3.2). A lower bound on F_n has been found for $61 \leq n \leq 271$ using stochastic search algorithms to identify good, though not necessarily optimal, sequences (see Section 3.3). The available evidence for $n \leq 271$ is summarised in Figure 1. The two largest known values of F_n are $F_{13} \simeq 14.1$ and $F_{11} = 12.1$, both of which arise from Barker sequences (see Section 2.1); no other values of $F_n \geq 10$ are known. The known lower bounds for F_n suffer a reduction for values of n beyond about 200. However one would expect a reduction of this sort owing to the increased computational burden for larger n and the large number of local optima in the search landscape (see Section 3.3). Indeed, previous versions of Figure 1, representing less extensive computational effort, have exhibited a similar phenomenon but at smaller values of n.

3.1   Skew-Symmetric Sequences

A common strategy for extending the reach of merit factor computations (both exhaustive and stochastic) is to impose restrictions on the structure of the sequence. The most popular of these historically has been the restriction to a *skew-symmetric* binary sequence, defined by Golay [30] in 1972 as a binary sequence $(a_0, a_1, \ldots, a_{2m})$ of odd length $2m + 1$ for which

$$a_{m+i} = (-1)^i a_{m-i} \text{ for } i = 1, 2, \ldots, m. \tag{14}$$

(Condition (14) had also been noted by Littlewood [53], [54] in relation to a question involving the supremum norm of ± 1 polynomials on the unit circle).

Skew-symmetric sequences are known to attain the optimal merit factor value F_n for the following odd values of $n < 60$: 3, 5, 7, 9, 11, 13, 15, 17, 21, 27, 29, 39, 41, 43, 45, 47, 49, 51, 53, 55, 57 and 59. Indeed, Golay [32] used the observation that all odd length Barker sequences are skew-symmetric to propose the skew-symmetric property as a sieve in searching for sequences with large merit factor. The computational advantage of this sieve is that it roughly doubles the sequence length that can be searched with given computational resources. Furthermore, as noted by Golay [30], half the aperiodic autocorrelations of a skew-symmetric sequence are 0:

Proposition 8. *A skew-symmetric binary sequence A of odd length has $C_A(u) = 0$ for all odd u.*

Golay [32] proposed that the asymptotic optimal merit factor of the set of skew-symmetric sequences is equal to $\limsup_{n \to \infty} F_n$, so that nothing is lost by restricting attention to this set. Although Golay's argument was heuristic and relied on the unproven "Postulate of Mathematical Ergodicity" (see Section 4.7), we know [10–p. 33] that the asymptotic value of the *mean* merit factor does not change when we restrict to skew-symmetric sequences (by comparison with Proposition 1):

Proposition 9. *The mean value of $1/F$, taken over all skew-symmetric sequences of odd length n, is $\dfrac{(n-1)(n-2)}{n^2}$.*

The optimal merit factor over all skew-symmetric sequences of odd length n was calculated by Golay [32] for $n \leq 59$ in 1977. It was then calculated independently by Golay and Harris [35] for $n \leq 69$ in 1990 and by de Groot, Würtz and Hoffmann [22] for $n \leq 71$ in 1992. It would be feasible, using the methods of Section 3.2, to extend these results to lengths up to around 119 (involving 60 arbitrary sequence elements), although it is not clear to me that this would represent a useful investment of computational resources.

In 1990 Golay and Harris [35] found good skew-symmetric sequences for odd lengths n in the range $71 \leq n \leq 117$ by regarding a skew-symmetric sequence as the interleaving of two constrained sequences: one symmetric and the other anti-symmetric. They formed candidate sets S_1 of symmetric sequences and S_2 of anti-symmetric sequences, each of whose members had large merit factor relative to other sequences of the same type, and then found the largest merit factor over all interleavings of a sequence from S_1 with a sequence from S_2.

3.2 Exhaustive Computation

The value of F_n has been calculated:

(i) for "small n" in 1965 by Lunelli [55], as referenced in [74] (and expressed in terms of minimising (8))
(ii) for $7 \leq n \leq 19$ by Swinnerton-Dyer, as presented by Littlewood [54] in 1966 (and expressed in terms of minimising (11))

(iii) for $n \leq 32$ by Turyn, as presented by Golay [33] in 1982
(iv) for $n \leq 48$ in 1996 by Mertens [58]
(v) for $n \leq 60$ (the current record) by Mertens and Bauke [59].

The merit factor of a binary sequence of length n can be calculated from (1) and (2) in $O(n^2)$ operations. The computations for a given sequence can be re-used to calculate the merit factor of other sequences of the same length by changing one sequence element at a time and updating the aperiodic autocorrelations in $O(n)$ operations for each such change. The determination of F_n by calculating the merit factor in this way for all 2^n sequences of length n, for example by using a Gray code, requires $O(n2^n)$ operations. However the algorithm of [58] and [59] reduces the exponential term of the complexity of determining F_n from 2^n to roughly 1.85^n by means of a branch-and-bound algorithm. The principle is to fix the sequence elements one pair at a time, working from the endpoints towards the centre and pruning the search tree by bounding $\sum_{u>0}[C(u)]^2$. (See Section 2.1 for a a similar idea applied to the calculation of M_n in [18] and [19], reducing the exponential term of the complexity of that calculation from 2^n to roughly 1.4^n.)

3.3   Stochastic Search Algorithms

For a given sequence length n, the search for a good lower bound for F_n can be viewed as a combinatorial optimisation problem over the space of 2^n binary sequences. This problem, often referred to as the "low autocorrelated binary string problem", has been studied in theoretical physics in connection with quantum models of magnetism as well as in theoretical chemistry. Early results [4], [5], [22] from the application of simulated annealing and evolutionary algorithms to the optimisation problem were rather disappointing, finding merit factor values no larger than about 6 for sequence lengths of around 200 and often failing to find previously known large merit factor values. Bernasconi [5] predicted from computational experiments that "...stochastic search procedures will not yield merit factors higher than about $F = 5$ for long sequences" (referring to lengths greater than about 200), and the problem was declared [22] to be "...amongst the most difficult optimization problems".

Several authors [5], [58], [60] suggested that the combinatorial landscape of the search space exhibits "golf-hole" behaviour, in the sense that the sequences attaining F_n are extremely isolated within the landscape (see [65] for an overview of combinatorial landscapes). This suggestion appears to have originated with an unfavourable comparison between empirically obtained merit factor values and the value of approximately $12.32\ldots$ conjectured by Golay [33] for $\limsup_{n \to \infty} F_n$, even though Golay's value depends on an unproven hypothesis (see Section 4.7). But, while the landscape has an exceptionally large number of local optima [23], after detailed analysis Ferreira, Fontanari and Stadler [27] found no evidence of "golf-hole" behaviour and suggested that the difference in difficulty between this and other problems of combinatorial optimisation is quantitative rather than qualitative.

As recognised in [22], the performance of stochastic search algorithms can vary significantly according to the care with which the algorithm parameters are tuned. In 1998 Militzer, Zamparelli and Beule [60] used an evolutionary algorithm to obtain more encouraging numerical results for sequence lengths up to about 200 — although they still considered that the search for a binary sequence attaining the optimal value F_n "...resembles the search for a needle in the haystack"! The best currently known stochastic search results, on which Figure 1 is based, are due to Borwein, Ferguson and Knauer [14] (with possible updates listed at [45]). These results rely on a combination of algorithmic improvements and extended use of considerable computational resources.

The basic method underlying many of the stochastic search algorithms is to move through the search space of sequences by changing only one, or sometimes two, sequence elements at a time. This method was suggested by Golay [31] as early as 1975, in relation to skew-symmetric sequences. The merit factor of any close neighbour sequence can be calculated in $O(n)$ operations from knowledge of the aperiodic autocorrelations of the current sequence. The search algorithm specifies when it is acceptable to move to a neighbour sequence, for example when it has merit factor no smaller than the current sequence, or when it has the largest merit factor amongst all close neighbours not previously visited. The search algorithm must also specify how to choose a new sequence when no acceptable neighbour sequence can be found. The method of [14] augments this search strategy to allow the addition or removal of one outer sequence element at a time.

Many authors [4], [22], [32], [35] [60] have applied stochastic search algorithms only to skew-symmetric sequences in order to obtain results for lengths that would otherwise be out of computational reach (see Section 3.1). The results of [14] for all lengths $n \geq 103$ (both odd and even) are based on searches for which the initial sequence is skew-symmetric.

Despite recent improvements, no stochastic search algorithm has yet been found that reliably produces binary sequences with merit factor greater than 6 in reasonable time for large n. Therefore such algorithms cannot yet shed light on whether the known range for $\limsup_{n \to \infty} F_n$ given in (3) can be narrowed.

4 Theoretical Approaches

In this section we consider theoretical approaches to the Merit Factor Problem, based mostly on infinite families of binary sequences with specified structure.

4.1 Legendre Sequences

We begin with the strongest proven asymptotic result. The *Legendre sequence* $X = (x_0, x_1, \ldots, x_{n-1})$ of prime length n is defined by:

$$x_i := \left(\frac{i}{n} \right) \text{ for } 0 \leq i < n,$$

where $\left(\frac{i}{n} \right)$ is the Legendre symbol (which takes the value 1 if i is a quadratic residue modulo n and the value -1 if not; we choose the convention that $\left(\frac{i}{n} \right) := 1$

if $i = 0$). Given a sequence $A = (a_0, a_1, \ldots, a_{n-1})$ of length n and a real number r, we write A_r for the sequence $(b_0, b_1, \ldots, b_{n-1})$ obtained by *rotating* (equivalently, cyclically shifting) the sequence A by a multiple r of its length:

$$b_i := a_{(i+\lfloor rn \rfloor) \bmod n} \text{ for } 0 \leq i < n. \tag{15}$$

In 1981 Turyn calculated the merit factor of the rotation X_r of the Legendre sequence X for sequence lengths up to 10,000, as reported in [34]. Based on Turyn's work Golay [34] gave a derivation of the asymptotic value of this merit factor, which accorded with Turyn's calculations but relied on heuristic arguments as well as the "Postulate of Mathematical Ergodicity" (see Section 4.7). In 1988 Høholdt and Jensen [39] proved that the expression derived by Golay is in fact correct:

Theorem 10. *Let X be a Legendre sequence of prime length n. Then*

$$\frac{1}{\lim\limits_{n \longrightarrow \infty} F(X_r)} = \begin{cases} \frac{1}{6} + 8(r - \frac{1}{4})^2 & \text{for } 0 \leq r \leq \frac{1}{2} \\ \frac{1}{6} + 8(r - \frac{3}{4})^2 & \text{for } \frac{1}{2} \leq r \leq 1. \end{cases} \tag{16}$$

Therefore the asymptotic merit factor of the optimal rotation of a Legendre sequence is 6 and occurs for $r = 1/4$ and $r = 3/4$. Borwein and Choi [12] subsequently determined the exact, rather than the asymptotic, value of $F(X_r)$ for all r. In the optimal cases $r = 1/4$ and $r = 3/4$, this exact value involves the class number of the imaginary quadratic field $\mathbb{Q}(\sqrt{-n})$.

The analytical method used by Høholdt and Jensen [39], and its refinement in [12], applies only to odd-length sequences and depends crucially on the relationship of the sequence to a cyclic difference set, in this case belonging to the parameter class

$$(v, k, \lambda) = (n, (n-1)/2, (n-3)/4) \text{ for integer } n \equiv 3 \pmod 4. \tag{17}$$

(Many constructions for difference sets with parameters (17) are known; see [75] for a survey and [37] for an important recent result. The parameter class (17) is referred to as *Hadamard*, but unfortunately so is another parameter class (7).)

It is well known (see for example [6]) that, for $n \equiv 3 \pmod 4$, a Legendre sequence X of length n is equivalent to a cyclic difference set in $\mathbb{Z}_n$ with parameters from the class (17), known as a *quadratic residue* or *Paley* difference set. By Proposition 5 this is equivalent to X having constant periodic autocorrelation at all non-zero shifts, and this property is retained under all rotations of the sequence:

$$R_{X_r}(u) = -1 \text{ for } 0 \leq r \leq 1 \text{ and } 0 < u < n. \tag{18}$$

Therefore, from (6), every rotation of a Legendre sequence of length $n \equiv 3 \pmod 4$ has the property that its aperiodic autocorrelations sum in pairs to -1. Of course this does not imply that the individual aperiodic autocorrelations will themselves have small magnitude but one might hope that, for some rotation, the full set of aperiodic autocorrelations will have a small sum of squares. Indeed, R. Turyn has indicated [personal communication, May 2003] that this was exactly his rationale for investigating these sequences. (Similar reasoning was used by Boehmer [7] in seeking binary sequences A with a small value of $M(A)$, as defined in (9), from sequences having small periodic autocorrelation at all non-zero shifts.) For $n \equiv 1 \pmod 4$, all rotations of a Legendre sequence are equivalent to a *partial difference set* in $\mathbb{Z}_n$ [56–Theorem 2.1] and can be dealt with in a similar manner to the case $n \equiv 3 \pmod 4$, for example [13] by slight modification to the sequence.

An asymptotic merit factor of 6, as given by the family of Legendre sequences in Theorem 10, is the largest so far proven. We shall see in Section 5 that there is strong evidence, although not a proof, that an asymptotic merit factor greater than 6.34 can be achieved via a related family of sequences. I find it remarkable that both of these constructions, as well as others described in this section, rely directly on special *periodic* autocorrelation properties of the sequences. In my opinion this is a reflection of the current paucity of powerful tools for analysing aperiodic autocorrelations independently of their periodic properties. Indeed, while periodic behaviour lends itself readily to mathematical investigation via techniques from algebra and analysis based on an underlying cyclic group or finite field, it remains the case [26–p. 269] that "... the aperiodic correlation properties of sequences are notoriously difficult to analyse"; see [64] for further discussion of this point. Given the appropriate mathematical tools, I believe we might uncover asymptotic merit factors significantly greater than those suggested by the results of Section 5.

The method introduced by Høholdt and Jensen [39] to calculate the asymptotic merit factor of rotated Legendre sequences was subsequently applied to further families of odd-length binary sequences corresponding to cyclic difference sets with parameters in the class (17) (see Sections 4.4 and 4.6).

4.2   Rudin-Shapiro Sequences

We next consider the earliest asymptotic merit factor result of which I am aware. Given sequences $A = (a_0, a_1, \ldots, a_{n-1})$ of length n and $A' = (a'_0, a'_1, \ldots, a'_{n'-1})$ of length n' we write $A; A'$ for the sequence $(b_0, b_1, \ldots, b_{n+n'-1})$ given by *appending* A' to A:

$$b_i := \begin{cases} a_i & \text{for } 0 \le i < n \\ a'_{i-n} & \text{for } n \le i < n + n'. \end{cases} \tag{19}$$

The *Rudin-Shapiro sequence pair* $X^{(m)}, Y^{(m)}$ of length 2^m is defined recursively [66], [71] by:

$$X^{(m)} := X^{(m-1)}; Y^{(m-1)}, \tag{20}$$
$$Y^{(m)} := X^{(m-1)}; -Y^{(m-1)}. \tag{21}$$

where $X^{(0)} = Y^{(0)} := [1]$. In 1968 Littlewood [54–p. 28] proved (in the language of complex-valued polynomials — see Section 2.2):

Theorem 11. *The merit factor of both sequences* $X^{(m)}, Y^{(m)}$ *of a Rudin-Shapiro pair of length* 2^m *is* $\dfrac{3}{(1 - (-1/2)^m)}.$

Therefore the asymptotic merit factor of both sequences of a Rudin-Shapiro pair is 3. To my knowledge, Theorem 11 is the first explicit construction of an infinite family of binary sequences, each with known merit factor, whose asymptotic merit factor is non-zero. It is not surprising that such constructions exist, because by Proposition 1 the expected asymptotic value of $1/F$ for a *randomly-chosen* binary sequence is 1. Nonetheless such a construction did not appear in the digital sequence design literature until Theorem 11 was rediscovered in generalised form as Theorem 12 in 1985. (In fact Littlewood [52–p. 334] performed the "straightforward calculations" leading to Theorem 11 as early as 1961, but the stated values in [52] are incorrect.)

Rudin-Shapiro sequence pairs are a special case of binary Golay complementary sequence pairs. (H. Shapiro suggests [70] that, in terms of historical precedence, a more suitable name than "Rudin-Shapiro" would be "Golay-Shapiro"; the confusion seems to have arisen from several mistaken citations of [71] as having been published in 1957, only two years prior to [66], rather than 1951.) Golay complementary pairs were introduced by Golay [28], [29] in 1949 in connection with a problem in infrared multislit spectroscopy and have seen repeated practical application since then, most recently in multicarrier wireless transmission (see [21] for details and recent results).

4.3   Generalisations of the Rudin-Shapiro Sequences

We now describe two generalisations of the Rudin-Shapiro sequences. Unfortunately neither improves on the asymptotic merit factor of 3 achieved in Theorem 4.2.

A first generalisation involves binary sequences $X^{(m)} = (x_0, x_1, \ldots, x_{2^m-1})$ of length 2^m that are defined recursively via:

$$x_{2^i+j} := (-1)^{j+f(i)} x_{2^i-j-1} \text{ for } 0 \le j < 2^i \text{ and } 0 \le i < m, \qquad (22)$$

where $x_0 := 1$ and f is any function from $\mathbb{N}$ to $\{0,1\}$. If we take

$$f(i) = \begin{cases} 0 \text{ if } i = 0 \text{ or } i \text{ is odd} \\ 1 \text{ if } i > 0 \text{ is even} \end{cases}$$

then the resulting sequence $X^{(m)}$ satisfies (20), and if we take the same function f but switch the value of $f(m-1)$ then the resulting sequence $Y^{(m)}$ satisfies (21); so the sequences of a Rudin-Shapiro pair are special cases of (22).

In 1985 Høholdt, Jensen and Justesen [40] established:

Theorem 12. *The merit factor of the sequence $X^{(m)}$ defined in (22) is*

$$\frac{3}{(1 - (-1/2)^m)}$$

for any function f.

Therefore the asymptotic merit factor of this family of sequences is 3.

Using polynomial notation, we can further generalise the Rudin-Shapiro sequences by regarding (22) as the special case $X^{(0)} = 1$ of the recursive construction

$$P_{X^{(m)}}(z) := P_{X^{(m-1)}}(z) \pm z^{2^{m-1}} P^*_{X^{(m-1)}}(-z), \tag{23}$$

where $P^*(z)$ is defined to be $z^{n-1}P(1/z)$ for a polynomial $P(z)$ of degree $n-1$. In 2000 Borwein and Mossinghoff [15] considered choices for the initial polynomial other than $X^{(0)} = 1$, having unrestricted degree, but concluded that the asymptotic merit factor achievable from (23) in this way is never more than 3.

The asymptotic merit factor results of Sections 4.2 and 4.3 are unusual in that they do not rely on special periodic autocorrelation properties. Instead, the merit factor is calculated directly from the defining recurrence relations.

4.4   Maximal Length Shift Register Sequences

A *maximal length shift register sequence* (often abbreviated to an ML-sequence or *m*-sequence) $X = (x_0, x_1, \dots, x_{2^m-2})$ of length $2^m - 1$ is defined by:

$$x_i := (-1)^{\mathrm{tr}(\beta\alpha^i)} \text{ for } 0 \le i < 2^m - 1,$$

where α is a primitive element of the field $\mathbb{F}_{2^m}$, β is a fixed element of the same field, and tr() is the trace function from $\mathbb{F}_{2^m}$ to $\mathbb{F}_2$. The name given to these sequences arises from an alternative definition involving a linear recurrence relation of period $2^m - 1$ that can be physically implemented using a shift register with m stages [36]. A maximal length shift register sequence is equivalent to a type of cyclic difference set with parameters from the class (17), known as a *Singer difference set*.

Sarwate [68] showed in 1984 that:

Theorem 13. *The mean value of $1/F$, taken over all n rotations of a maximal length shift register sequence of length $n = 2^m - 1$, is $\dfrac{(n-1)(n+4)}{3n^2}$.*

([40] points out that Theorem 13 could be derived from much earlier results due to Lindholm [50].) Theorem 13 implies that for any length $n = 2^m - 1$ there is some rotation of a maximal length shift register sequence of length n with merit factor of *at least* $3n^2/((n-1)(n+4))$, which asymptotically equals 3. This suggests the possibility of achieving an asymptotic merit factor greater than 3 by choosing a suitable rotation of a maximal length shift register sequence, but in 1989 Jensen and Høholdt [41] used the method introduced in [39] to show that this is not possible:

Theorem 14. *The asymptotic merit factor of any rotation of a maximal length shift register sequence is 3.*

4.5   Jacobi Sequences

A *Jacobi sequence* $X = (x_0, x_1, \dots, x_{n-1})$ of length $n = p_1 p_2 \dots p_r$, where $p_1 < p_2 < \dots < p_r$ and each p_j is prime, is defined by:

$$x_i := \left(\frac{i}{p_1}\right)\left(\frac{i}{p_2}\right)\cdots\left(\frac{i}{p_r}\right) \text{ for } 0 \le i < n.$$

We can regard Jacobi sequences as the "product" of r Legendre sequences; for $r > 1$ such sequences do not correspond to difference sets.

In 2001 Borwein and Choi [11] proved:

Theorem 15. *Let X be a Jacobi sequence of length $n = p_1 p_2 \ldots p_r$, where $p_1 < p_2 < \ldots p_r$ and each p_j is prime. Then, provided that $n^\epsilon/p_1 \longrightarrow 0$ for any fixed $\epsilon > 0$ as $n \longrightarrow \infty$, $1/\lim_{n \to \infty} F(X_r)$ is given by (16).*

Therefore, provided that (roughly speaking) p_1 does not grow significantly more slowly than n, the asymptotic merit factor of the optimal rotation of a Jacobi sequence is 6. The case $r = 2$ of Theorem 15, subject to a more restrictive condition on the growth of p_1, was given earlier by Jensen, Jensen and Høholdt [42].

4.6   Modified Jacobi Sequences

We next consider a modification of the Jacobi sequences of Section 4.5 for the case $r = 2$, as introduced by Jensen, Jensen and Høholdt [42]. A *modified Jacobi sequence* $X = (x_0, x_1, \ldots, x_{n-1})$ of length $n = pq$, where p and q are distinct primes, is defined by:

$$x_i := \begin{cases} 1 & \text{for } i \equiv 0 \pmod{q} \\ -1 & \text{for } i > 0 \text{ and } i \equiv 0 \pmod{p} \\ \left(\dfrac{i}{p}\right)\left(\dfrac{i}{q}\right) & \text{for all other } i \text{ for which } 0 \le i < n. \end{cases}$$

In the case $q = p+2$, a modified Jacobi sequence is called a *Twin Prime sequence* and corresponds to a type of cyclic difference set with parameters from the class (17), known as a *Twin Prime difference set*.

In 1991 Jensen, Jensen and Høholdt [42] used the method introduced in [39] to prove:

Theorem 16. *Let X be a modified Jacobi sequence of length pq, where p and q are distinct primes. Then, provided that $((p+q)^5 \log^4(n))/n^3 \longrightarrow 0$ as $n \longrightarrow \infty$, $1/\lim_{n \to \infty} F(X_r)$ is given by (16).*

Therefore, provided that p grows roughly as fast as q, the asymptotic merit factor of the optimal rotation of a modified Jacobi sequence (and in particular a Twin Prime sequence) is 6.

4.7   Golay's "Postulate of Mathematical Ergodicity"

The aperiodic autocorrelations $C(1), C(2), \ldots, C(n-1)$ of a sequence that is chosen at random from the 2^n binary sequences of length n are clearly dependent random variables. However in 1977 Golay [32] proposed, with an appeal to intuition and by analogy with statistical mechanics, a "Postulate of Mathematical Ergodicity" that states roughly:

The correct value of $\limsup_{n \longrightarrow \infty} F_n$ can be found by treating $C(1), C(2), \ldots,$ $C(n-1)$ as *independent* random variables for large n.

(The statement of the Postulate in [32] and its restatement in [33] are not entirely precise; indeed, Massey disclosed [57] that he was asked to mediate a dispute between Golay and the referees over the level of rigour of [33].)

Assuming the Postulate, Golay argued in [32] that $\lim_{n \longrightarrow \infty} F_n = 2e^2 \simeq 14.78$. In 1982 Golay [33] identified a "convenient, but faulty, approximation" in [32] and, by refining its heuristic arguments, concluded instead that the Postulate implies:

Conjecture 17 (Golay, 1982 [33]). $\lim_{n \longrightarrow \infty} F_n = 12.32 \ldots$

Golay [33] also argued that:

(i) restriction to skew-symmetric sequences (see Section 3.1) does not change the asymptotic optimal merit factor
(ii) it is "most likely" that $F_n \leq 12.32 \ldots$ for all $n \neq 13$.

Bernasconi [5] gave a more transparent derivation of the value $12.32 \ldots$ as an estimate for the asymptotic optimal merit factor, based on "an uncontrolled approximation for the partition function". Although the underlying assumption in [5] and [33] is clearly identified as unproven, its derived consequences are sometimes quoted as fact (for example, conclusion (ii) above and Conjecture 17 are treated in [4] and [16] respectively as proven results). Massey [57] wrote that he "would not want to bet on the contrary [to Conjecture 17]".

Golay [34] used the Postulate to predict correctly the asymptotic merit factor of a rotated Legendre sequence (see Section 4.1). Further evidence in support of the Postulate was given by Ferreira, Fontanari and Stadler [27], who found unexpectedly good agreement between experimentally determined parameters of the combinatorial search landscape (see Section 3.3) and those predicted by the Postulate. Nonetheless I am sceptical about its use: I do not find the arguments proposed in its favour in [33] to be convincing, and it seems not to be falsifiable except by direct disproof of Conjecture 17 or conclusion (i) above.

5   Periodic Appending

This section contains an overview of recent results of Borwein, Choi and Jedwab [13] that strongly suggest that $\limsup_{n \longrightarrow \infty} F_n > 6$. These results were motivated by the discoveries of A. Kirilusha and G. Narayanaswamy in 1999, working as summer students under the supervision of J. Davis at the University of Richmond.

We shall make use of the definition of rotation and appending of sequences as given in (15) and (19). Given a sequence $A = (a_0, a_1, \ldots, a_{n-1})$ of length n and a real number t satisfying $0 \leq t \leq 1$, we write A^t for the sequence $(b_0, b_1, \ldots, b_{\lfloor tn \rfloor - 1})$ obtained by *truncating* A to a fraction t of its length:

$$b_i := a_i \text{ for } 0 \leq i < \lfloor tn \rfloor.$$

Let X be a Legendre sequence of prime length n. We know from Theorem 10 that $\lim_{n \to \infty} F(X_{\frac{1}{4}}) = 6$. Kirilusha and Narayanaswamy [44] investigated how the merit factor of $X_{\frac{1}{4}}$ changes as sequence elements are successively appended. They observed:

Proposition 18. *Let $\{A_n\}$ and $\{B_n\}$ be sets of binary sequences, where each A_n has length n and each B_n has length $o(\sqrt{n})$. Then*

$$\frac{1}{F(A_n; B_n)} = \frac{1}{F(A_n)} + o(1).$$

It follows that up to $o(\sqrt{n})$ *arbitrary* sequence elements ± 1 can be appended to $X_{\frac{1}{4}}$ without changing the asymptotic merit factor of 6. Kirilusha and Narayana-swamy [44] then asked which choice of *specific* sequence elements yields the best merit factor when appended to $X_{\frac{1}{4}}$. To considerable surprise, they found that when the appended sequence elements are identical to some truncation of $X_{\frac{1}{4}}$, the merit factor appears to increase to a value consistently greater than 6.2!

This phenomenon was studied in detail in [13]. A key realisation was that the number of appended elements should take the form $\lfloor tn \rfloor$ for some fixed t, rather than the form $\lfloor n^\alpha \rfloor$ for fixed $\alpha < 1$ as suggested in [44]. Figure 2 shows the variation of $F(X_r; (X_r)^t)$ with r for the optimal value of t, for a large fixed length $n = 259499$. Extensive numerical evidence was presented in [13] to suggest that:

(i) for large n, the merit factor of the appended sequence $X_{\frac{1}{4}}; (X_{\frac{1}{4}})^t$ is greater than 6.2 when $t \simeq 0.03$

(ii) for large n, the merit factor of the appended sequence $X_r; (X_r)^t$ is greater than 6.34 for $r \simeq 0.22$ and $r \simeq 0.72$, when $t \simeq 0.06$.

I do not have a complete explanation for these apparent properties but they appear to rely on X_r having small periodic autocorrelation at all non-zero shifts, as given by (18). It seems that this causes the aperiodic autocorrelations of the appended sequence $X_r; (X_r)^t$ to be collectively small (for some values of $r \neq 0$ and for an appropriate value of t). In fact the process of successively appending initial elements of the sequence to itself would give a progressively larger merit factor if not for the single shift $u = n$. At this shift, the initial $\lfloor tn \rfloor$ elements of X_r are mapped onto copies of themselves and the resulting contribution of $(\lfloor tn \rfloor)^2$ to $\sum [C(u)]^2$ cannot be allowed to grow too large.

This intuition was formalised in [13], leading to an asymptotic relationship between the merit factor of the appended sequence $X_r; (X_r)^t$ and the merit factor of two truncated sequences $(X_r)^t$ and $(X_{r+t})^{1-t}$:

Theorem 19 ([13–Theorem 6.4 and equation (20)]). *Let X be a Legendre sequence of prime length n and let r, t satisfy $0 \le r \le 1$ and $0 < t \le 1$. Then, for large n,*

$$\frac{1}{F(X_r; (X_r)^t)} \sim \begin{cases} 2\left(\frac{t}{1+t}\right)^2 \left(\frac{1}{F((X_r)^t)} + 1\right) + \left(\frac{1-t}{1+t}\right)^2 \left(\frac{1}{F((X_{r+t})^{1-t})}\right) & \text{for } t < 1 \\ \frac{1}{2}\left(\frac{1}{F(X_r)} + 1\right). & \text{for } t = 1 \end{cases}$$

$$\tag{24}$$

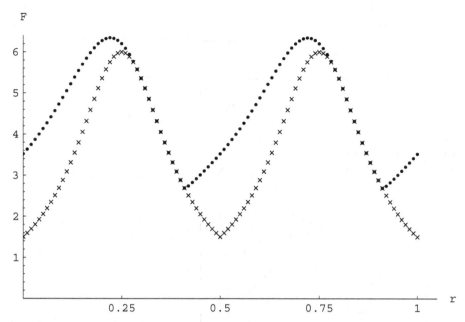

Fig. 2. The merit factor of the r-rotated Legendre sequence of length 259499 before ($\times$) and after ($\bullet$) appending of the optimal number of its own initial elements, for varying r

(The single shift $u = n$ is responsible for a contribution of $2\left(\frac{t}{1+t}\right)^2$ to the right-hand side of (24) for $t \leq 1$; if this contribution were zero for $t = 1$ we would have $F(X_r; X_r) = 2F(X_r)$.) Given Theorem 19, it is sufficient to determine an asymptotic form for the function $t^2/F((X_r)^t)$ for any (r, t) satisfying $0 \leq r \leq 1$ and $0 < t \leq 1$; this asymptotic form is already known for $t = 1$ from Theorem 10. Numerical evidence from sequences up to millions of elements in length leads to:

Conjecture 20 ([13–Conjecture 7.5]). *Let X be a Legendre sequence of prime length n. Then*

$$g(r,t) := \begin{cases} \lim\limits_{n \longrightarrow \infty} \left(\dfrac{t^2}{F((X_r)^t)}\right) & \text{for } 0 < t \leq 1 \\ 0 & \text{for } t = 0 \end{cases} \tag{25}$$

is well-defined for any $r, t \in [0, 1]$ and is given by

$$g(r,t) = t^2(1 - \tfrac{4}{3}t) + h(r,t),$$

where

$$h(r + \tfrac{1}{2}, t) := h(r, t) \text{ for } 0 \leq r \leq \tfrac{1}{2} \text{ and } 0 \leq t \leq 1$$

and $h(r, t)$ is defined for $0 \leq r \leq 1/2$ and $0 \leq t \leq 1$ in Figure 3.

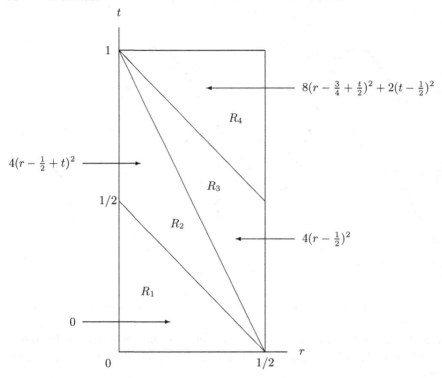

Fig. 3. The function $h(r,t)$ for the range $0 \le r \le 1/2$ and $0 \le t \le 1$, in regions R_1, R_2, R_3 and R_4

By Proposition 7.6 of [13], the definition of $h(r,t)$ given in Figure 3 is needed for only one of the regions R_2 and R_3, because if the definition holds in either one then it holds in the other.

Support for Conjecture 20 is given by calculations [13] showing that for

$$(r,t) \in G := \{0, 1/64, 2/64, \ldots, 1\} \times \{1/64, 2/64, \ldots, 1\},$$

the maximum discrepancy between the conjectured and actual value of $\frac{t^2}{F((X_r)^t)}$ is

$$\max_{(r,t) \in G} \left| \frac{t^2}{F((X_r)^t)} - g(r,t) \right| = \begin{cases} 0.00484 \text{ for } n = 22783 \\ 0.00122 \text{ for } n = 259499 \\ 0.00025 \text{ for } n = 4433701. \end{cases} \tag{26}$$

Subject to Conjecture 20, Theorem 19 implies that the maximum value of $\lim_{n \to \infty} F(X_r; (X_r)^t)$ over $r, t \in [0, 1]$ is approximately 6.3421, occurring at $r \simeq 0.2211$ and 0.7211 and $t \simeq 0.0578$, and the maximum value of the limit $\lim_{n \to \infty} F(X_{\frac{1}{4}}; (X_{\frac{1}{4}})^t)$ over $t \in [0, 1]$ is approximately 6.2018, occurring at $t \simeq 0.0338$ [13]. These values are in excellent agreement with calculated data.

Furthermore, experimental results [13] suggest that, provided p and q grow roughly as fast as each other, appending the initial elements of a modified Jacobi sequence of length $n = pq$ (see Section 4.6) to itself produces the same asymptotic behaviour as for Legendre sequences. Likewise, M. Parker [personal communication, June 2003] has found numerical evidence that the same is true for a family of sequences described in [63], provided that the sign of sequence elements is reversed under rotation and under appending.

Independently of [13], Kristiansen [46] presented sequences of length up to 20,000 having merit factor greater than 6.3, also inspired by Kirilusha and Narayanaswamy [44]. Each of the sequences in [46] is obtained by searching over a set of sequences derived from a Legendre sequence. [46] gives an approximate value for the total number of sequence elements resulting from the search but does not contain a theoretical explanation of the merit factor properties of the sequences. In response to a preprint of [13], Kristiansen and Parker [47] recognised that the sequences in [46] could more easily be viewed as an appending of a rotated Legendre sequence.

6   Challenges

I conclude with a personal selection of challenges concerning the Merit Factor Problem, arranged in order of increasing significance.

1. **Find a binary sequence X of length $n > 13$ for which $F(X) \geq 10$.**
   Such a sequence would give the largest known merit factor, with the exception of Barker sequences of length 11 and 13 (see Section 3).
2. **Find a binary sequence X for which $F(X) > 14.1$.**
   Regarding such a possibility, Massey [57] wrote in 1990: "Golay always regarded the length 13 Barker Sequences, whose merit factor is 14.08... as a singularity of nature whose goodness would never again be attained". Attractive though such a result would be, I have not placed it any higher on the list of challenges because I believe the study of the merit factor is fundamentally concerned with asymptotic behaviour, not the identification of a particular sequence with an unusually large value of F.
3. **Prove that Conjecture 2 is false.**
   This might be achieved, for example, by determining the asymptotic value of $t^2/F((X_r)^t)$ for a Legendre sequence X for appropriate r and t (see Conjecture 20). A disproof of Conjecture 2 would give a proven new lower bound on $\limsup_{n \to \infty} F_n$ for the first time since 1988.
4. **Find a binary sequence family X for which $\lim_{n \to \infty} F(X) > 6.3421\ldots$**
   The apparent lower bound of 6.3421... implied by Conjecture 20 arises by reference to periodic properties of Legendre sequences. I believe that better bounds might be found from a direct analysis of aperiodic behaviour (see Section 4.1).
5. **Find a binary sequence family X for which $\lim_{n \to \infty} F(X)$ is an integer greater than 6.**
   Although I do not have a satisfying explanation, I find it remarkable that the Legendre, Rudin-Shapiro, generalised Rudin-Shapiro (22), maximal length shift register, Jacobi, and modified Jacobi sequences all have an asymptotic

merit factor that takes an integer value (see Section 4). One might expect that some other infinite families of sequences behave similarly.

6. **Determine whether** $\limsup_{n\longrightarrow\infty} F_n$ **is finite and, if so, determine its value.**

This is a restatement of the Merit Factor Problem. If $\limsup_{n\longrightarrow\infty} F_n$ is infinite then Conjecture 3 from 1966 is true, whereas if it is finite then Conjecture 7 from 1957 is true and furthermore there are only finitely many Barker sequences (see Section 2.2). If $\limsup_{n\longrightarrow\infty} F_n$ takes any value other than 12.32... then Golay's "Postulate of Mathematical Ergodicity" is false (see Section 4.7).

Acknowledgements

I am grateful to R. Turyn for generously sharing insights into the historical background of the Merit Factor Problem, and to L. Goddyn for suggesting the use of Figure 3 to represent the function $h(r,t)$.

This work was supported by NSERC of Canada via Discovery Grant # 31-611394.

References

1. M. Antweiler and L. Bömer. Merit factor of Chu and Frank sequences. *Electron. Lett.*, **26**:2068–2070, 1990.
2. R.H. Barker. Group synchronizing of binary digital systems. In W. Jackson, editor, *Communication Theory*, pages 273–287. Academic Press, New York, 1953.
3. L.D. Baumert. *Cyclic Difference Sets*. Lecture Notes in Mathematics 182. Springer-Verlag, New York, 1971.
4. G.F.M. Beenker, T.A.C.M. Claasen, and P.W.C. Hermens. Binary sequences with a maximally flat amplitude spectrum. *Philips J. Res.*, **40**:289–304, 1985.
5. J. Bernasconi. Low autocorrelation binary sequences: statistical mechanics and configuration state analysis. *J. Physique*, **48**:559–567, 1987.
6. T. Beth, D. Jungnickel, and H. Lenz. *Design Theory*. Cambridge University Press, Cambridge, 2nd edition, 1999. Volumes I and II.
7. A.M. Boehmer. Binary pulse compression codes. *IEEE Trans. Inform. Theory*, **IT-13**:156–167, 1967.
8. L. Bömer and M. Antweiler. Optimizing the aperiodic merit factor of binary arrays. *Signal Processing*, **30**:1–13, 1993.
9. J. Borwein and D. Bailey. *Mathematics by Experiment: Plausible Reasoning in the 21st Century*. A.K. Peters, Natick, 2004.
10. P. Borwein. *Computational Excursions in Analysis and Number Theory*. CMS Books in Mathematics. Springer-Verlag, New York, 2002.
11. P. Borwein and K.-K.S. Choi. Merit factors of polynomials formed by Jacobi symbols. *Canad. J. Math.*, **53**:33–50, 2001.
12. P. Borwein and K.-K.S. Choi. Explicit merit factor formulae for Fekete and Turyn polynomials. *Trans. Amer. Math. Soc.*, **354**:219–234, 2002.
13. P. Borwein, K.-K.S. Choi, and J. Jedwab. Binary sequences with merit factor greater than 6.34. *IEEE Trans. Inform. Theory*, **50**:3234–3249, 2004.

14. P. Borwein, R. Ferguson, and J. Knauer. The merit factor of binary sequences. In preparation.

15. P. Borwein and M. Mossinghoff. Rudin-Shapiro-like polynomials in L_4. *Math. of Computation*, **69**:1157–1166, 2000.

16. F. Brglez, X.Y. Li, M.F. Stallman, and B. Militzer. Evolutionary and alternative algorithms: reliable cost predictions for finding optimal solutions to the LABS problem. *Information Sciences*, 2004. To appear.

17. K.-K.S. Choi. Extremal problems about norms of Littlewood polynomials. 2004. Preprint.

18. M.N. Cohen, M.R. Fox, and J.M. Baden. Minimum peak sidelobe pulse compression codes. In *IEEE International Radar Conference*, pages 633–638. IEEE, 1990.

19. G.E. Coxson, A. Hirschel, and M.N. Cohen. New results on minimum-PSL binary codes. In *IEEE Radar Conference*, pages 153–156. IEEE, 2001.

20. J.A. Davis and J. Jedwab. A survey of Hadamard difference sets. In K.T. Arasu et al., editors, *Groups, Difference Sets and the Monster*, pages 145–156. de Gruyter, Berlin-New York, 1996.

21. J.A. Davis and J. Jedwab. Peak-to-mean power control in OFDM, Golay complementary sequences, and Reed-Muller codes. *IEEE Trans. Inform. Theory*, **45**:2397–2417, 1999.

22. C. de Groot, D. Würtz, and K.H. Hoffmann. Low autocorrelation binary sequences: exact enumeration and optimization by evolutionary strategies. *Optimization*, **23**:369–384, 1992.

23. V.M. de Oliveira, J.F. Fontanari, and P.F. Stadler. Metastable states in high order short-range spin glasses. *J. Phys. A: Math. Gen.*, **32**:8793–8802, 1999.

24. P. Erdös. Some unsolved problems. *Mich. Math. J.*, **4**:291–300, 1957.

25. P. Erdös. An inequality for the maximum of trigonometric polynomials. *Ann. Polon. Math.*, **12**:151–154, 1962.

26. P. Fan and M. Darnell. *Sequence Design for Communications Applications.* Communications Systems, Techniques and Applications. Research Studies Press, Taunton, 1996.

27. F.F. Ferreira, J.F. Fontanari, and P.F. Stadler. Landscape statistics of the low autocorrelated binary string problem. *J. Phys. A: Math. Gen.*, **33**:8635–8647, 2000.

28. M.J.E. Golay. Multislit spectroscopy. *J. Opt. Soc. Amer.*, **39**:437–444, 1949.

29. M.J.E. Golay. Static multislit spectrometry and its application to the panoramic display of infrared spectra. *J. Opt. Soc. Amer.*, **41**:468–472, 1951.

30. M.J.E. Golay. A class of finite binary sequences with alternate autocorrelation values equal to zero. *IEEE Trans. Inform. Theory*, **IT-18**:449–450, 1972.

31. M.J.E. Golay. Hybrid low autocorrelation sequences. *IEEE Trans. Inform. Theory*, **IT-21**:460–462, 1975.

32. M.J.E. Golay. Sieves for low autocorrelation binary sequences. *IEEE Trans. Inform. Theory*, **IT-23**:43–51, 1977.

33. M.J.E. Golay. The merit factor of long low autocorrelation binary sequences. *IEEE Trans. Inform. Theory*, **IT-28**:543–549, 1982.

34. M.J.E. Golay. The merit factor of Legendre sequences. *IEEE Trans. Inform. Theory*, **IT-29**:934–936, 1983.

35. M.J.E. Golay and D.B. Harris. A new search for skewsymmetric binary sequences with optimal merit factors. *IEEE Trans. Inform. Theory*, **36**:1163–1166, 1990.

36. S.W. Golomb. *Shift Register Sequences.* Aegean Park Press, California, revised edition, 1982.

37. T. Helleseth, P.V. Kumar, and H. Martinsen. A new family of ternary sequences with ideal two-level autocorrelation function. *Designs, Codes and Cryptography*, **23**:157–166, 2001.
38. T. Høholdt. The merit factor of binary sequences. In A. Pott et al., editors, *Difference Sets, Sequences and Their Correlation Properties*, volume 542 of *NATO Science Series C*, pages 227–237. Kluwer Academic Publishers, Dordrecht, 1999.
39. T. Høholdt and H.E. Jensen. Determination of the merit factor of Legendre sequences. *IEEE Trans. Inform. Theory*, **34**:161–164, 1988.
40. T. Høholdt, H.E. Jensen, and J. Justesen. Aperiodic correlations and the merit factor of a class of binary sequences. *IEEE Trans. Inform. Theory*, **IT-31**:549–552, 1985.
41. H.E. Jensen and T. Høholdt. Binary sequences with good correlation properties. In L. Huguet and A. Poli, editors, *Applied Algebra, Algebraic Algorithms and Error-Correcting Codes, AAECC-5 Proceedings*, volume 356 of *Lecture Notes in Computer Science*, pages 306–320. Springer-Verlag, Berlin, 1989.
42. J.M. Jensen, H.E. Jensen, and T. Høholdt. The merit factor of binary sequences related to difference sets. *IEEE Trans. Inform. Theory*, **37**:617–626, 1991.
43. J.-P. Kahane. Sur les polynômes á coefficients unimodulaires. *Bull. London Math. Soc.*, **12**:321–342, 1980.
44. A. Kirilusha and G. Narayanaswamy. Construction of new asymptotic classes of binary sequences based on existing asymptotic classes. Summer Science Program Technical Report, Dept. Math. Comput. Science, University of Richmond, July 1999.
45. J. Knauer. Merit Factor Records. Online. Available: <http://www.cecm.sfu.ca/~jknauer/labs/records.html>, November 2004.
46. R.A. Kristiansen. On the Aperiodic Autocorrelation of Binary Sequences. Master's thesis, University of Bergen, March 2003.
47. R.A. Kristiansen and M.G. Parker. Binary sequences with merit factor > 6.3. *IEEE Trans. Inform. Theory*, **50**:3385–3389, 2004.
48. K.H. Leung, S.L. Ma, and B. Schmidt. Nonexistence of abelian difference sets: Lander's conjecture for prime power orders. *Trans. Amer. Math. Soc.*, **356**:4343–4358, 2004.
49. K.H. Leung and B. Schmidt. The field descent method. *Designs, Codes and Cryptography*. To appear.
50. J.H. Lindholm. An analysis of the pseudo-randomness properties of subsequences of long m-sequences. *IEEE Trans. Inform. Theory*, **IT-14**:569–576, 1968.
51. J. Lindner. Binary sequences up to length 40 with best possible autocorrelation function. *Electron. Lett.*, **11**:507, 1975.
52. J.E. Littlewood. On the mean values of certain trigonometrical polynomials. *J. London Math. Soc.*, **36**:307–334, 1961.
53. J.E. Littlewood. On polynomials $\sum^n \pm z^m$, $\sum^n e^{\alpha_m i} z^m$, $z = e^{\theta i}$. *J. London Math. Soc.*, **41**:367–376, 1966.
54. J.E. Littlewood. *Some Problems in Real and Complex Analysis*. Heath Mathematical Monographs. D.C. Heath and Company, Massachusetts, 1968.
55. L. Lunelli. Tabelli di sequenze $(+1, -1)$ con autocorrelazione troncata non maggiore di 2. Politecnico di Milano, 1965.
56. S.L. Ma. A survey of partial difference sets. *Designs, Codes and Cryptography*, **4**:221–261, 1994.
57. J.L. Massey. Marcel J.E. Golay (1902-1989). Obituary, in: *IEEE Information Theory Society Newsletter*, June 1990.

58. S. Mertens. Exhaustive search for low-autocorrelation binary sequences. *J. Phys. A: Math. Gen.*, **29**:L473–L481, 1996.
59. S. Mertens and H. Bauke. Ground States of the Bernasconi Model with Open Boundary Conditions. Online. Available: <http://odysseus.nat.uni-magdeburg.de/~mertens/bernasconi/open.dat>, November 2004.
60. B. Militzer, M. Zamparelli, and D. Beule. Evolutionary search for low autocorrelated binary sequences. *IEEE Trans. Evol. Comput.*, **2**:34–39, 1998.
61. J.W. Moon and L. Moser. On the correlation function of random binary sequences. *SIAM J. Appl. Math.*, **16**:340–343, 1968.
62. D.J. Newman and J.S. Byrnes. The L^4 norm of a polynomial with coefficients ± 1. *Amer. Math. Monthly*, **97**:42–45, 1990.
63. M.G. Parker. Even length binary sequence families with low negaperiodic autocorrelation. In S. Boztas and I. E. Shparlinski, editors, *Applied Algebra, Algebraic Algorithms and Error-Correcting Codes, AAECC-14 Proceedings*, volume 2227 of *Lecture Notes in Computer Science*, pages 200–210. Springer-Verlag, 2001.
64. K.G. Paterson. Applications of exponential sums in communications theory. In M. Walker, editor, *Cryptography and Coding*, volume 1746 of *Lecture Notes in Computer Science*, pages 1–24. Springer-Verlag, Berlin, 1999.
65. C.M. Reidys and P.F. Stadler. Combinatorial landscapes. *SIAM Review*, **44**:3–54, 2002.
66. W. Rudin. Some theorems on Fourier coefficients. *Proc. Amer. Math. Soc.*, **10**:855–859, 1959.
67. H.J. Ryser. *Combinatorial Mathematics*. Carus Mathematical Monographs No. 14. Mathematical Association of America, Washington, DC, 1963.
68. D.V. Sarwate. Mean-square correlation of shift-register sequences. *IEE Proceedings Part F*, **131**:101–106, 1984.
69. M.R. Schroeder. Synthesis of low peak-factor signals and binary sequences with low autocorrelation. *IEEE Trans. Inform. Theory*, **IT-16**:85–89, 1970.
70. H. Shapiro. Harold Shapiro's Research Interests. Online. Available: <http://www.math.kth.se/~shapiro/profile.html>, November 2004.
71. H.S. Shapiro. Extremal Problems for Polynomials and Power Series. Master's thesis, Mass. Inst. of Technology, 1951.
72. R. Turyn and J. Storer. On binary sequences. *Proc. Amer. Math. Soc.*, **12**:394–399, 1961.
73. R.J. Turyn. Character sums and difference sets. *Pacific J. Math.*, **15**:319–346, 1965.
74. R.J. Turyn. Sequences with small correlation. In H.B. Mann, editor, *Error Correcting Codes*, pages 195–228. Wiley, New York, 1968.
75. Q. Xiang. Recent results on difference sets with classical parameters. In A. Pott et al., editors, *Difference Sets, Sequences and Their Correlation Properties*, volume 542 of *NATO Science Series C*, pages 419–437. Kluwer Academic Publishers, Dordrecht, 1999.

A Survey of Feedback with Carry Shift Registers

Andrew Klapper

University of Kentucky, Department of Computer Science,
779 A Anderson Hall, Lexington, KY 40506-0046, USA
klapper@cs.uky.edu
http://www.cs.uky.edu/~klapper/

Abstract. Feedback with carry shift registers (FCSRs) are arithmetic analogs of linear feedback shift registers (LFSRs). In this paper we survey some of the basic properties of FCSRs. For comparison, we first review some basic facts about LFSRs. We then define FCSRs and discuss their relation to the N-adic numbers. This leads to the analysis of periodicity of FCSR sequences, their exponential representation, and a description of maximal period FCSR sequences. We also discuss an arithmetic analog of cross-correlations, the FCSR register synthesis problem, and how FCSRs can be efficiently implemented in parallel architecture.

1 Introduction

Fast generators of sequences with good statistical properties play a role in many application areas such as keystream generation for stream ciphers, spreading sequences for CDMA communications systems, sequences for radar ranging, error correcting codes, and pseudorandom number generation for quasi-Monte Carlo integration. In many cases appropriate sequences can be generated by linear feedback shift registers (LFSRs) or by generators constructed from LFSRs.

In 1994 a new class of sequence generators, *feedback with carry shift registers* (FCSRs), was invented. FCSRs share many desirable properties with LFSRs. This paper surveys the basic properties of FCSRs. For comparison, we first review basic facts about LFSRs. We then define FCSRs and discuss their relation to the N-adic numbers. We show that each FCSR sequence has an associated N-adic number. This number is in fact a rational number, and its rational representation gives information about the sequence and the FCSR. This representation leads to an analysis of the periodicity of FCSR sequences, their exponential representation, and a description of maximal period FCSR sequences. We also discuss an arithmetic analog of cross-correlations and see that maximal period FCSR sequences have remarkable arithmetic correlation properties. Next we consider the register synthesis problem — for a given sequence prefix, find the smallest generator in a given class that outputs the sequence. For LFSRs the register synthesis problem is solved by the Berlekamp-Massey algorithm. For FCSRs we see that the register synthesis problem can be solved similarly. Finally, we show how FCSRs can be efficiently implemented in parallel architecture.

T. Helleseth et al. (Eds.): SETA 2004, LNCS 3486, pp. 56–71, 2005.

Throughout the paper, we indicate general references for individual sections in the section headers.

2   Linear Feedback Shift Registers [14–Chapter 8]

Linear feedback shift registers (LFSRs) are simple mechanisms that can be used to efficiently generate large period sequences over finite fields. They can be thought of either as finite state devices with output or, more mathematically, as linear recurrences over fields. They can be pictured as finite state device as in Figure 1. In this figure q_i and a_j are elements of a finite field F.

The LFSR changes states by shifting its contents right one position, putting $q_1 a_{k-1} + \cdots + q_k a_0$ in the leftmost cell, and outputting a_0. Equivalently, the LFSR generates an infinite sequence $A = a_0, a_1, \cdots, a_{k-1}, a_k, a_{k+1}, \cdots$ from the initial values $a_0, \cdots, a_{k-1}$ by the linear recurrence

$$a_k = q_1 a_{k-1} + \cdots + q_k a_0.$$

LFSRs are analyzed using various associated algebraic structures. The *generating function* $\alpha(A, x) = \sum_{i=0}^{\infty} a_i x^i$ is a power series associated with the output sequence. The *connection polynomial* is a polynomial $q(x) = \sum_{i=1}^{k} q_i x^i - 1$ associated with the coefficients of the linear recurrence. The Fundamental Theorem of Linear Feedback Shift Registers relates the generating function and connection polynomial.

Theorem 1. *If $\alpha(A, x)$ is the generating function of the output from a LFSR with a particular initial state, and $q(x)$ is the connection polynomial of the LFSR, then*

$$\alpha(A, x) = \frac{u(x)}{q(x)},$$

for some polynomial $u(x) \in F[x]$.

The third algebraic structure that is useful in analyzing LFSRs relates them to the multiplicative structure of an extension field. We describe here only a simple case. More general cases involve sums of terms such as the one in the following theorem.

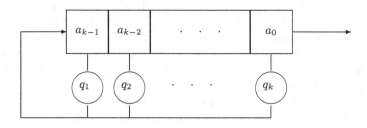

Fig. 1. A Linear Feedback Shift Register

Theorem 2. *Let $q \in F[x]$ be irreducible, let $E = F[x]/(q)$, and let α be a root of q in E. Then E is a field and*

$$a_i = Tr_F^E(c\alpha^i).$$

It is immediate from the definition that the output A from a LFSR is eventually periodic since the LFSR has finitely many possible states.

Theorem 3. *A is (purely) periodic if and only if $\deg(u) < \deg(q)$. In general the eventual period divides the least N such that $q(x)|x^N - 1$, the multiplicative order of x mod $q(x)$. It is at most $|F|^k - 1$, and divides $|(F[x]/(q))^*|$.*

This leads to a definition of the maximal period LFSR sequences, arguably the most important such sequences.

Definition 1. *A is an m-sequence if its period equals $|F|^k - 1$. This is equivalent to saying $q(x)$ is a primitive polynomial.*

M-sequences are important in part because they have excellent distributional properties. Let $A = a_0, a_1, \cdots$ be periodic with period L. Then an *occurrence* of $B = b_0, b_1, \cdots, b_{r-1}$ in A is an index i, $0 \leq i \leq L - 1$ so that

$$a_i = b_0, a_{i+1} = b_1, \cdots, a_{i+r-1} = b_{r-1}.$$

Theorem 4. *If A is an m-sequence with period $|F|^k - 1$ and $1 \leq r \leq k$, and b is the number of occurrences of B in A, then*

$$b = \begin{cases} |F|^{k-r} & \text{if } B \neq (0, 0, \cdots, 0); \\ |F|^{k-r} - 1 & \text{if } B = (0, 0, \cdots, 0). \end{cases}$$

Proof: An occurrence of B corresponds to a state $(*, *, \cdots, *, b_{r-1}, \cdots, b_0)$. $\quad\square$

The LFSR Synthesis Problem is the following:

Instance: A prefix $a_0, a_1, \cdots, a_r$ of A
Find: The shortest LFSR with output A.

Suppose that $\alpha(A, x) = u(x)/q(x)$ with $u(x)$ and $q(x)$ relatively prime. Let $\lambda = \lambda(u, q) = \max\{\deg(u)+1, \deg(q)\}$. Then the LFSR synthesis problem can be solved by the Berlekamp-Massey algorithm [17], which computes $u(x)$ and $q(x)$ given 2λ consecutive symbols of A. To understand what comes later in this paper, it is helpful to have some idea of how the Berlekamp-Massey algorithm works. It processes one symbol at a time. At the ith stage, polynomials u_i and q_i are known so that u_i/q_i approximates u/q up to degree i. If it fails to approximate u/q up to degree $i + 1$ (we say a *discrepancy* occurs), the approximation is updated by adding a multiple of an earlier (and carefully chosen) approximation:

$$u_{i+1} = u_i + dx^{i-k}u_k \text{ and } q_{i+1} = q_i + dx^{i-k}q_k \text{ where } d \in F.$$

A critical fact that makes the proof of correctness work is that $\lambda(u_{i+1}, q_{i+1})$ can be described in terms of of $\lambda(u_i, q_i)$. This is possible because the degree of a sum of polynomials is at most the maximum of their degrees.

M-sequences have also been studied for their *cross-correlation* properties. For period L binary sequences the cross-correlations are defined as follows.

$$\Theta_{A,B}(t) = \sum_{i=0}^{L-1} (-1)^{a_i - b_{i+t}}$$
$$= |\{i : a_i = b_{i+t}\}| - |\{i : a_i \neq b_{i+t}\}|$$
$$= \#0\text{s} - \#1\text{s in one period of } \alpha(A, x) - \alpha(B_t, x), \quad (1)$$

where $B_t = b_t, b_{t+1}, \cdots$ is the shift of B by t places. When $A = B$, the cross-correlation is called the *autocorrelation* of A. In some applications, it is desirable to have sequences whose autocorrelations with $t \neq 0$ are all as close to zero as possible. In other applications, it is desirable to have large families of sequences whose pairwise cross-correlations are as low as possible.

For m-sequences with period $L = |F|^k - 1$ it is known that $\Theta_{A,A}(t) = -1$ if $t \neq L$. However, for two distinct m-sequence of the same period, $\Theta_{A,B}(t)$ is known only in special cases. Moreover, it follows from Parseval's identity that $\max_t(\Theta_{A,B}(t)) \geq |F|^{k/2} - 1$ if A is an m-sequence and B is any sequence. This puts a fundamental limit on what we can hope to achieve.

3 FCSR/MWC Sequences [10]

Feedback with carry shift registers are similar to LFSRs, but with the addition of an "extra memory" that retains a carry from one stage to the next. It is not immediately clear, but it is a finite state device (see Section 7). A diagram of an FCSR is given in Figure 2.

The FCSR changes states by computing the linear combination

$$\sigma = q_1 a_{k-1} + \cdots + q_k a_0 + m,$$

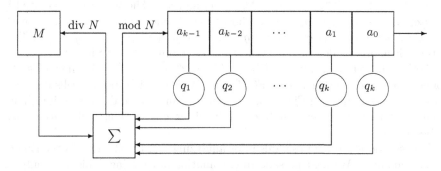

Fig. 2. A Feedback with Carry Shift Register

M	a_3	a_2	a_1	a_0
0	1	1	1	1
1	1	1	1	1
2	0	1	1	1
2	1	0	1	1
2	0	1	0	1
2	0	0	1	0
1	1	0	0	1
1	0	1	0	0
1	0	0	1	0
1	0	0	0	1

M	a_3	a_2	a_1	a_0
1	0	0	0	0
0	1	0	0	0
0	0	1	0	0
0	1	0	1	0
0	1	1	0	1
1	0	1	1	0
1	1	0	1	1
1	1	1	0	1
1	1	1	1	0
1	1	1	1	1

Fig. 3. The states of an FCSR with connection number $q = 27$

then setting $a_k = \sigma \bmod N$ and $m' = \lfloor \sigma/N \rfloor$. The main part of the new state is $(a_k, \cdots, a_1)$ while the new value of the extra memory is m'. It appears to be possible that the extra memory grows unboundedly during some infinite execution of the FCSR. We shall see in Section 7 that this is not the case, and will give explicit bounds on the size of the extra memory in terms of the q_i.

FCSRs were invented by Goresky and Klapper [10], and independently by Marsaglia and Zaman [16] and Couture and L'Ecuyer [4]. The view point of FCSRs is due to Goresky and Klapper and was motivated by the cryptanalysis of the summation combiner [18, 21]. Marsaglia, Zaman, Couture, and L'Ecuyer, on the other hand, took the point of view of recurrences with carry and were motivated by the desire for good random number generators for such applications as quasi-Monte Carlo simulation. Marsaglia and Zaman treated only a special case, so-called Add with Carry recurrences, and Couture and L'Cuyer generalized these recurrences to Multiply with Carry (MWC) recurrences. MWC recurrences are equivalent to FCSRs. With $a_i, q_i \in \{0, 1, \cdots, N-1\}$ for $i = 0, \cdots, k-1$ and $m_{k-1} \in \mathbf{Z}$, a MWC recurrence defines two sequences $a_i \in \{0, 1, \cdots, N-1\}$, $m_i \in \mathbf{Z}$ for $i = k, k+1, \cdots$ by the equation

$$a_n + N m_n = q_1 a_{n-1} + \cdots + q_k a_{n-k} + m_{n-1}$$

for $i = k, k+1, \cdots$. This uniquely defines a_n and m_n. The sequence generated is $A = a_0, a_1, \cdots, a_{k-1}, a_k, a_{k+1}, \cdots$. The first algebraic structure associated with an FCSR is the *connection number* $q = q_k N^k + q_{k-1} N^{k-1} + \cdots + q_1 N - 1$. This is the arithmetic analog of the connection polynomial of a LFSR.

Let's consider an example. If $q = 27 = 2^4 + 2^3 + 2^2 - 1$, then q is the connection number of a length 4 FCSR with $N = 2$ and "taps" on cells $0, 1, 2$. A series of states is given in Figure 3. The output can be read down the right hand columns. Note that the eventual period in this case is 18. This will be explained in Section 4.

The second algebraic mechanism for analyzing FCSR sequences is the ring of N-adic numbers. We next present an explanation of this ring. This is a complex subject that has been studied by mathematicians since at least the early 1900s.

Intuitively, we want a ring that is an arithmetic or "with carry" analog of the ring of power series over $\mathbf{Z}/(N)$. Let

$$\mathbf{Z}_N = \{\sum_{i=0}^{\infty} a_i N^i, a_i \in \{0, 1, \cdots, N-1\}\}.$$

We can define a ring structure on $\mathbf{Z}_N$ by defining addition and multiplication of N-adic numbers as follows.

Suppose we are given N-adic numbers $\alpha = \sum_{i=0}^{\infty} a_i N^i$ and $\beta = \sum_{i=0}^{\infty} b_i N^i$.

1. Addition: The sum $\alpha + \beta = \sum_{i=0}^{\infty} c_i N^i$ is the N-adic number determined by the infinite system of equations

$$a_0 + b_0 = c_0 + N d_0$$
$$a_1 + b_1 + d_0 = c_1 + N d_1$$

$$\cdot$$
$$\cdot$$
$$\cdot$$

2. Multiplication: The product $\alpha\beta = \sum_{i=0}^{\infty} c_i N^i$ is the N-adic number determined by the infinite system of equations

$$a_0 b_0 = c_0 + N d_0$$
$$a_1 b_0 + a_0 b_1 + d_0 = c_1 + N d_1$$

$$\cdot$$
$$\cdot$$
$$\cdot$$

Theorem 5. $\mathbf{Z}_N$ *is a ring.*

It may help the reader get a feel for the N-adic numbers to notice that

$$-1 = (N-1) + (N-1)N + (N-1)N^2 + \cdots. \tag{2}$$

That is, if we add 1 to the N-adic number on the right hand side of equation (2), the carries propagate infinitely and we get 0.

Next notice that the ordinary integers are a subset of the N-adic numbers. Indeed, the positive integers are just the N-adic numbers with finitely many nonzero terms. The negative integers are obtained by multiplying the positive integers by -1.

Moreover, many N-adic numbers are invertible. Let $\alpha = \sum_{i=0}^{\infty} a_i N^i$. To invert α, we want an N-adic number $\gamma = \sum_{i=0}^{\infty} c_i N^i$ so that $1 = \alpha\gamma$. This is accomplished if the infinite system of equations

$$1 = a_0 c_0 + N d_0$$
$$0 = a_1 c_0 + a_0 c_1 + N d_1$$

$$\cdot$$
$$\cdot$$
$$\cdot$$

has a solution. It can be seen that this is the case if and only if $\gcd(a_0, N) = 1$. Consequently, we have

$$\{\frac{u}{q} : \gcd(N, q) = 1, u, q \in \mathbf{Z}\} \subseteq \mathbf{Z}_N.$$

Now suppose that $A = a_0, a_1, \cdots$ is the output from an FCSR with connection number $q = \sum_{i=1}^{k} q_i N^i - 1$. Then $\alpha(A, N) = \sum_{i=0}^{\infty} a_i N^i$ is the N-adic number associated with A.

Theorem 6. *For some $u \in \mathbf{Z}$ we have*

$$\alpha(A, N) = \frac{u}{q} \in \mathbf{Z}_N.$$

Proof: For $j \geq k$ we have

$$a_j + N M_{j+1} = q_1 a_{j-1} + \cdots q_{j-k} a_0 + M_j,$$

so

$$a_j = q_1 a_{j-1} + \cdots q_{j-k} a_0 + M_j - N M_{j+1}.$$

We can multiply the jth equation by N^j and sum from k to infinity. Thus

$$\sum_{j=k}^{\infty} a_j N^j = \sum_{r=1}^{k} q_r \sum_{j=k}^{\infty} a_{j-r} N^j + N^k M_k$$

$$= \sum_{r=1}^{k} q_r N^r \sum_{j=k}^{\infty} a_{j-r} N^{j-r} + N^k M_k.$$

This implies that

$$\alpha(A, N) - \sum_{j=0}^{k-1} a_j N^j = \sum_{r=1}^{k} q_r N^r (\alpha(A, N) - \sum_{j=0}^{k-r-1} a_{j-r} N^{j-r}) + N^k M_k.$$

Now we can solve for $\alpha(A, N)$. We see that

$$\alpha(A, N) = \frac{\sum_{r=0}^{k} q_r N^r \sum_{j=0}^{k-r-1} a_{j-r} N^{j-r} - N^k M_k}{q}.$$

$\square$

We can characterize the periodic sequences in terms of this representation.

Theorem 7. *A is periodic if and only if $-q \leq u \leq 0$.*

Proof: If t is the eventual period of A, then $u/q = v/(1 - N^t)$ for some integer v. A is purely periodic if and only if

$$\frac{u}{q} = v_0 + v_1 N + \cdots + v_{t-1} N^{t-1} + v_0 N^t + v_1 N^{t+1} + \cdots + v_{t-1} N^{2t-1} + \cdots$$

$$= \frac{v}{1 - N^t},$$

so $v = v_0 + v_1 N + \cdots + v_{t-1} N^{t-1}$. Thus A is purely periodic if and only if $0 \le v \le N^t - 1$, which is equivalent to $-q \le u \le 0$.   □

There is an analog of the trace representation of LFSR sequences, Theorem 2. We call this the *exponential representation*.

Theorem 8. *If A is a periodic sequence generated by a FCSR with connection number q, and A is not all $(N-1)s$, then $a_i = (cN^{-i} \bmod q) \bmod N$.*

This unusual combination of operations means the following. First find $\gamma = N^{-1}$ in $\mathbf{Z}/(q)$. Next find $c\gamma^i$ in $\mathbf{Z}/(q)$. Now lift this to an integer in the set $\{0, 1, \cdots, q-1\}$ whose reduction modulo q is $c\gamma^i$. Reduce this element modulo modulo N to an element of S.

Proof: Let u/q and v/q be the N-adic numbers corresponding to the outputs from consecutive states of the FCSR. Then

$$\frac{u}{q} = a_i + \frac{v}{q} N$$

so

$$u = q a_i + N v.$$

If we reduce this equation modulo q then we see that $v = N^{-1} u$. If we reduce this equation modulo N then we see that $a_i = q^{-1} u \equiv -u$. Thus the series of numerators of the rational representations of the successive tails of A are obtained by successive multiplication by N^{-1} modulo q. The successive outputs are obtained by reducing these numerators modulo N.   □

4 ℓ-Sequences [10]

As with LFSRs, we are most interested in the FCSR sequences of largest period. We take q as a measure of the size of the FCSR (or perhaps more properly $\lfloor \log(q+1) \rfloor$).

Theorem 9. *The eventual period of an FCSR sequence A with connection number q is a divisor of $\mathrm{ord}_q(N)$, the multiplicative order of N modulo q.*

The largest possible value of $\mathrm{ord}_q(N)$ is $\phi(q)$. In the extreme case, q is prime and $\phi(q) = q - 1$.

Definition 2. *A is an ℓ-sequence if it is generated by a FCSR with connection number q and the period of A is $\phi(q)$.*

This is equivalent to saying that N is primitive modulo q, and implies that q is a power of a prime. For example, the FCSR with connection number $q = 27$, whose states are given in Figure 3, outputs an ℓ-sequence if we start it in a periodic state. Indeed, we saw that the period is $18 = \phi(27)$.

It is important to know how likely it is to find ℓ-sequences. Unfortunately, it is not even known whether there are infinitely many ℓ-sequences.

Conjecture 1. (Artin, Heilbronn [8]) The number of primes q with n bits for which N is a primitive root is asymptotically $cn/\ln(n)$, where c is a constant.

In fact, as was pointed out to the author by Sol Golomb, this conjecture is known to be true for all but at most two primes q, although it is not known to be true for all but any particular two primes. A similar statement holds for prime powers q and all but at most three prime powers. However, it is also known that if N is primitive modulo a prime q and modulo q^2, then it is primitive modulo q^r for every r. This provides us with infinite families of ℓ-sequences.

One of the reasons we are interested in ℓ-sequences is that they have excellent distribution properties. This is what makes them useful for pseudo-Monte Carlo simulation.

Theorem 10. *Suppose q is prime and let A be an ℓ-sequence generated by an FCSR with connection number q. Let $1 \leq r \leq k$. The number of occurrences of any block $B = b_0, \cdots, b_{r-1}$ in A is N^{k-r} or $N^{k-r} - 1$.*

Proof: Let $\beta = \sum_{i=0}^{r-1} b_i N^i \in \mathbf{Z}$ and let $q = N^k + e$ for some integer e. The number of occurrences of B in a period of A equals the number of shifts of A that start with B. This in turn equals the number of integers u with $-u/q \equiv \beta \bmod N^r$ and $0 < u < q$. Since q is invertible mod N, this equals the number of integers u with $u \equiv -q\beta \bmod N^r$ and $0 < u < q$. If $v = q\beta \bmod N$ with $0 \leq v < N$ then the set of $u \equiv -q\beta \bmod N^r$ such that $0 < u < q$ is

$$\begin{cases} v, v + N^r, \cdots, v + N^{k-r} N^r, & \text{if } v < e \\ v, v + N^r, \cdots, v + (N^{k-r} - 1) N^r, & \text{if } v \geq e. \end{cases}$$

$\square$

A slightly weaker statement holds when q is a power of a prime.

5   Arithmetic Correlations [6, 11]

It appears that the ordinary correlation properties of ℓ-sequences are not very good. We would not expect them to be since the linearity that is intrinsic to cross-correlations is absent from FCSRs. However, we can look for an arithmetic analog of cross-correlations. We treat only the binary case, $N = 2$, here. We look to our third interpretation of cross-correlations, equation (1), for inspiration. For binary sequences with period L we define the *arithmetic cross-correlations* [11] as

$$\Delta_{A,B}(t) = \#\ 0s - \#\ 1s \text{ in one period of } \alpha(A, 2) - \alpha(B_t, 2).$$

Some care must be taken in how we interpret this definition. By $\alpha(A, 2) - \alpha(B_t, 2)$ we mean compute the difference as 2-adic numbers. Even if the original 2-adic numbers have periodic coefficient sequences, the difference may not be purely periodic. The difference will, however, be eventually periodic (in fact it can be shown that the difference is periodic at least from the Lth term on). In the definition of arithmetic cross-correlations we mean that the numbers of zeros and ones should be counted only in a legitimate period of the periodic part.

For ℓ-sequences we have $L = q-1$. In this case we have identically zero shifted arithmetic autocorrelations: $\Theta_{A,A}(t) = 0$ if $t \neq L$. Essentially this fact has been used previously in arithmetic coding. Moreover, suppose that A is an ℓ sequence with connection number q. By a *decimation* of A we mean a sequence formed by taking every dth symbol of A for some positive integer d. A decimation is *proper* if d is relatively prime to the period of A, so that the decimation has the same period as A. Suppose that B and C are two proper decimations of A and that B and C are cyclically distinct. Then it can be shown that $\Theta_{A,B}(t) = 0$. This remarkable fact is in stark contrast to the situation with ordinary cross-correlations. We do not, however, know any applications, and pose this as a major open problem.

It is also interesting to ask how many cyclically distinct sequences we obtain by decimating an ℓ-sequence.

Conjecture 2. If $q > 13$ is prime, then all proper decimations of an ℓ-sequence with connection number q are cyclically distinct.

It would follow from this conjecture that we have a family of $\phi(q-1)$ cyclically distinct sequences with identically zero arithmetic cross-correlations. Note that the conjecture is equivalent to saying that A is cyclically distinct from all its proper decimations. In some cases we know that the conjecture is true or nearly true [6].

1. Exhaustive search has shown that the conjecture is true for $13 < q < 2,000,000$.
2. The $d = (q - 2)$-fold decimation is cyclically distinct from A. (Proof of this fact uses an analysis of the bit patterns that can occur in ℓ-sequences.)
3. If $q \equiv 1 \bmod 4$, then the $d = ((q+1)/2)$-fold decimation is cyclically distinct from A. (Proof of this fact uses elementary number theory.)
4. Goresky, Klapper, Murty, Shparlinski showed that for any $\varepsilon > 0$ there is a constant $c > 0$, so that there are $\leq cq^{2/3+\varepsilon}$ decimations of A that are cyclic permutations of A. (Proof of this fact uses exponential sums.)
5. Conjecture 2 is true if $q = 2p+1 = 8r+3$ with q, p, r prime and 2 a primitive root mod q. (This is a consequence of the previous result.)

We also observe that Conjecture 2 is equivalent to the following number theoretic conjecture.

Conjecture 3. Suppose $q > 13$ is prime. Suppose that c and d are integers with $\gcd(c, q) = 1$ and $\gcd(d, q-1) = 1$, such that for $0 \leq x < q$, x is even if and only if $cx^d \bmod q$ is even. Then $c = d = 1$.

6   Register Synthesis [13]

The problem solved by the Berlekamp-Massey algorithm, that of finding a smallest LFSR that outputs a given sequence, is one instance of a general class of problems. We consider a fixed class of sequence generators $\mathcal{G}$. We assume that there is some notion of the size of a generator in $\mathcal{G}$, typically approximately the log (with base equal to the alphabet size) of the number of states of the generator. Then the *register synthesis problem for $\mathcal{G}$* is

Instance: A prefix $a_0, a_1, \cdots, a_{r-1}$ of A

Problem: Find the smallest $G \in \mathcal{G}$ that outputs A

A solution will only successfully output a generator of A if sufficiently many symbols of A are available. Thus the effectiveness of such an algorithm depends on the number, K, of symbols that are needed for success and the time complexity of algorithm. K is typically expressed as a function of λ.

If there is an efficient solution to this problem, then there is an associated security measure, the $\mathcal{G}$-complexity of A, defined to be the size of the smallest $G \in \mathcal{G}$ that outputs A. The $\mathcal{G}$-complexity of A is denoted by $\lambda_{\mathcal{G}}(A)$ (or simply $\lambda(A)$ if $\mathcal{G}$ is clear from the context). If there is an efficient algorithm solving the register synthesis problem for $\mathcal{G}$, then every keystream must have large $\lambda_{\mathcal{G}}(A)$.

The register synthesis problem has been solved in several cases.

LFSRs: The Berlekamp-Massey algorithm solves the LFSR synthesis problem for sequences over a field with $K = 2\lambda(A)$ and time complexity in $O(K^2)$ [17]. Reeds and Sloane extended the Berlekamp-Massey algorithm to sequences over $\mathbf{Z}/(N)$ [20].

Quadratic FSRs: A quadratic feedback shift register (QFSR) is similar to a LFSR but the new symbol is computed as a polynomial of the previous state of degree at most two. Chan and Games described an algorithm that solves the register synthesis problem for QFSRs with $K \in O(\lambda(A)^2)$ and time complexity in $O(\lambda(A)^6 \log(\lambda(A))) \subseteq O(K^3 \log(K))$ [3].

FCSRs: Goresky and Klapper [10] gave a solution to the FCSR-synthesis problem when N is prime with $K = 2(\lambda(A) + \log(\lambda(A)))$ and time complexity in $O(K^2 \log(K))$. It is based on the lattice-theoretic analysis of N-adic numbers due to Mahler [15] and De Weger [22].

Xu and Klapper [13] gave a solution to the FCSR-synthesis problem when N is arbitrary with $K = 6\lambda(A) + 16$ and time complexity in $O(K^2 + KN^{3/2} \log(N))$. The details of this algorithm are given below.

Arnault, Berger, and Necer showed that the Euclidean algorithm could also be used to solve the FCSR-synthesis problem for arbitrary N [2].

The Berlekamp-Massey algorithm proceeds by computing successively better rational approximations to a given power series. It refines the approximations when the current approximation incorrectly predicts the next symbol (i.e., when a *discrepancy* occurs) by adding a multiple of a previous approximation to the current one, as described in Section 2. It is natural to try to adapt this algorithm to the FCSR synthesis problem by simply replacing power series by N-adic numbers and polynomials by integers. Unfortunately, the resulting algorithm fails to find an FCSR that outputs A when a relatively small number of symbols of A are available. The problem is that when we produce a new rational approximation by combining with an earlier approximation, the size of the resulting rational approximation may grow too much, due to the carry from addition of integers.

Xu and Klapper found a way to overcome this difficulty [13]. The idea is to produce a new rational approximation that is correct for not just one, but three additional symbols. To make this work we need two mathematical objects. The first is an *index function* $\phi : \mathbf{Z} \to \mathbf{R}^{\geq 0} \cup \{-\infty\}$, defined by

$$\phi(x) = \begin{cases} \lfloor \log_N |x| \rfloor & \text{if } x \neq 0 \\ -\infty & \text{if } x = 0. \end{cases}$$

The second is an *interpolation set* $P = [-\lfloor N^2/2 \rfloor, \lfloor N^2/2 \rfloor] \cap \mathbf{Z}$. Central to the operation and correctness of the algorithm is the following lemma.

Lemma 1. *For all $x, y \in \mathbf{Z}$ the following hold.*

1. $\phi(xy) \leq \phi(x) + \phi(y) + 1$.
2. $\phi(x \pm y) \leq \max\{\phi(x), \phi(y)\} + 1$.
3. $\phi(Nx) = \phi(x) + 1, \ k \in \mathbf{N}$.
4. $\exists s, t \in P: N^3 | (sx + ty)$.
5. $\forall s, t, \ \phi(sx + ty) \leq \max(\phi(x), \phi(y)) + 2$.

The goal of the algorithm is to find, for every i, a pair of integers q_i, u_i so that q_i is odd and $\alpha \equiv u_i/q_i \bmod N^i$. That is, $N^i | (u_i - q_i \alpha)$. When a discrepancy occurs, we let

$$(u_{i+1}, q_{i+1}) = s(u_i, q_i) + tN^{i-m}(u_m, q_m),$$

with s, t chosen from P so that

$$N^{i+3} | (u_{i+1} - q_{i+1} \alpha).$$

The full algorithm is given in Figure 4, where we let $\Phi(x, y) = \max(\phi(x), \phi(y))$.

It follows that all secure keystreams must have large N-adic complexity. This observation leads to an attack on Rueppel and Massey's summation combiner [18]. This is a stream cipher constructed from (in the simplest case) two maximal period binary LFSR sequences A and B of period $2^r - 1$ and $2^s - 1$, respectively. The sequences are added term by term modulo N, but the carry is saved and added in at the next stage. It has been shown that the result has linear complexity $\sim 2^{r+s}$ [21]. Unfortunately, the addition with carry just amounts to addition of N-adic numbers, so the N-adic complexity is at most one plus the sum of the N-adic complexities of A and B. These have N-adic complexities bounded by their periods, so the output of the summation combiner has N-adic complexity at most $2^r + 2^s - 1$. In specific cases it may in fact be much smaller than this. FCSRs were invented specifically to cryptanalyze the summation combiner.

Meier and Staffelbach [19] and others have also found correlation attacks on the summation combiner.

Rational_Approximation {
 input $A = \{a_i \in S, 0 \le i \le k\}$
 $\alpha \longleftarrow 1 + N \sum_{i=0}^{k} a_i N^i$
 $(u_0, q_0) \longleftarrow (0, 1)$
 $(u_1, q_1) \longleftarrow (1 + a_0 N + a_1 N^2, 1 + N^4)$
 $m \longleftarrow 0$
 for $(i = m + 1 \text{ to } k - 1)$
  if $((u_i - q_i \alpha) \not\equiv 0 \bmod N^{i+1})$ {
   if ($\exists s \ne 0 \in P$ with $(N^{i+3} \mid s(u_i - q_i \alpha)))$
    $(u_{i+1}, q_{i+1}) \longleftarrow s(u_i, q_i)$
   else {
    Find $s, t \in P$, not both zero, with
     $N^{i+3} \mid s(u_i - q_i \alpha) + t N^{i-m}(u_m - q_m \alpha)$
     $(u_{i+1}, q_{i+1}) \longleftarrow s(u_i, q_i) + t N^{i-m}(u_m, q_m)$
   }
   if $(\Phi(u_{i+1}, q_{i+1}) > \Phi(u_i, q_i) \wedge$
    $\Phi(u_i, q_i) \le i - m + \Phi(u_m, q_m) \wedge t \ne 0)$
    $m \longleftarrow i$
  }
 output u, q with $1 + N(u/q) = u_k/q_k$
}

Fig. 4. Xu and Klapper's Algorithm for FCSR Synthesis

7 Implementation [5, 10]

In this section we discuss some of the details of the implementation of FCSRs. We begin by showing that an FCSR is a finite state device [10]. Let $w = \sum_{i=1}^{k} |q_i|$.

Theorem 11. *If an FCSR is in a periodic state (meaning that the output sequence from this state is purely periodic) then the memory is in the range $0 \le m < w$. If the initial memory m_{k-1} is greater than or equal to w, then it will monotonically decrease within $\lfloor \log_N(m_{k-1} - w) \rfloor + k$ steps until it is in the range $0 \le m < w$. If the initial memory is less than 0, then it will monotonically increase so that within $\lceil \log_N(|m_{k-1}|) \rceil + k$ steps it is in the range $0 \le m < w$.*

It follows that an FCSR with connection number $q = \sum_{i=1}^{k} q_i N^i - 1$ and periodic output requires at most $k + \log_N(w) \le k + \log_N(k) + 1$ N-ary cells.

Next we describe an alternate architecture for FCSRs that allows fast parallel implementation. There is a so-called Galois architecture for LFSRs, depicted in Figure 5.

To change states, a constant multiple of the rightmost cell is added to the ith cell and the results are put in the $(i - 1)$st cell. That is, the new value of a_{i-1} is $a_i + q_i a_0 \bmod N$. It can be seen that a periodic sequence is generated by this device if and only if it is generated by a LFSR with connection polynomial $\sum_{i=1}^{k} q_i x^i - 1$. The advantage is that all the state changes of the individual

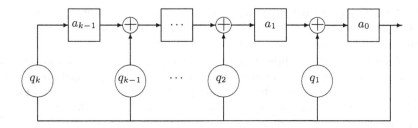

Fig. 5. Galois mode implementation of LFSRs

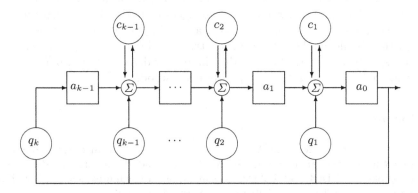

Fig. 6. Galois mode implementation of FCSRs

cells can be performed in parallel so the time required for a state change is independent of k.

Similarly, there is a Galois architecture for FCSRs, depicted in Figure 6 [5]. We need an additional carry cell between each pair of ordinary cells. To change states, we compute $\sigma_i = a_i + q_i a_0 + c_i$. The new value of a_{i-1} is $\sigma_i \bmod N$, and the new value of c_i is $\lfloor \sigma_i / N \rfloor$. It can be seen that a periodic sequence is generated by this device if and only if it is generated by a FCSR with connection number $\sum_{i=1}^{k} q_i N^i - 1$. Again, the advantage is that all the state changes of the individual cells can be performed in parallel, so the time required for a state change is independent of k.

8   Summary

We have described the structure and basic properties of feedback with carry shift registers and the sequences they output. We have seen that many of these properties can be analyzed with the help of the N-adic numbers. This includes the period of the sequences and the distribution of sub-blocks in the sequences. It also leads to an arithmetic version of the cross-correlation, and we have seen that we can produce families of sequences with vanishing pairwise arithmetic cross-correlations. We have also seen that there is an efficient solution to the FCSR synthesis problem. Its quality is within a small constant factor of the quality of the Berlekamp-Massey algorithm.

We have mentioned a couple of open problems related to FCSRs, and there are many more. Of a very general nature is the search for more applications of these lovely devices. It may be possible, for example, to find specific classes of FCSRs that have even better statistical properties for quasi-Monte Carlo integration. It should also be possible to use FCSRs as building blocks in the design of stream ciphers. Arnault, Berger, and Necer have designed such a stream cipher that is based on a combination of LFSRs and FCSRs [1].

In other papers we have described a number of generalizations of both LFSRs and FCSRs. The most general form is the algebraic feedback shift registers [12]. The essential idea is that whenever you have a ring like the power series over a field or the N-adic numbers, you can construct algebraic feedback shift registers. Various specific cases have been studied in greater detail [7, 9, 13], but there are still many instances that are only poorly understood.

References

1. F. Arnault, T. Berger, and A. Necer: A new class of stream ciphers combining LFSR and FCSR architectures. In A. Menezes and P. Sarkar, ed.: Progress in Cryptology - INDOCRYPT 2002. Lecture Notes in Computer Science, Vol. 2551. Springer-Verlag, Berlin Heidelberg New York (2002) 22-33.
2. F. Arnault, T. Berger, and A. Necer: Feedback with carry shift registers synthesis with the Euclidean algorithm. IEEE Trans. Info. Thy. **50** (2004) 910- 917.
3. A. Chan and R. Games: On the quadratic span of de Bruijn sequences. IEEE Trans. Info. Thy. **36** (1990) 822-829.
4. R. Couture and P. L'Ecuyer: On the lattice structure of certain linear congruential sequences related to AWC/SWB generators. Math. Comp. **62** (1994) 799–808.
5. M. Goresky and A. Klapper: Fibonacci and Galois Mode Implementation of Feedback with Carry Shift Registers. IEEE Trans. Info. Thy. **48** (2002) 2826-2836.
6. M. Goresky, A. Klapper. R. Murty, and I. Shparlinski: On Decimations of ℓ-sequences. SIAM J. Disc. Math **18** (2004) 130-140.
7. M. Goresky and A. Klapper: Polynomial pseudo-noise sequences based on algebraic feedback shift registers, under review.
8. C. Hooley: On Artin's conjecture. J. Reine Angew. Math. **22** (1967) 209-220.
9. A. Klapper: Distributional properties of d-FCSR sequences. J. Complexity **20** (2004) 305-317.
10. A. Klapper and M. Goresky: Feedback Shift Registers, Combiners with Memory, and 2-Adic Span. J. Cryptology **10** (1997) 111-147.
11. A. Klapper and M. Goresky: Arithmetic Cross-Correlations of FCSR Sequences. IEEE Trans. Info. Thy. **43** (1997) 1342-1346.
12. A. Klapper and J. Xu: Algebraic feedback shift registers. Theoretical Comp. Sci. **226** (1999) 61-93.
13. A. Klapper and J. Xu: Register synthesis for algebraic feedback shift registers based on non-primes. Designs, Codes, and Crypt. **31** (2004) 227-25.
14. R. Lidl and H. Niederreiter: Finite Fields, Encycl. Math. Appl. **20**. Addision Wesley, Reading, MA (1983).

15. K. Mahler: On a geometrical representation of p-adic numbers. Ann. Math. **41** (1940) 8-56.
16. G. Marsaglia and A. Zaman: A new class of random number generators. Annals of Appl. Prob. **1** (1991) 462-480.
17. J. Massey: Shift-register synthesis and BCH decoding. IEEE Trans. Info. Thy. **IT-15** (1969) 122-127.
18. J. Massey and R. Rueppel: Method of, and Apparatus for, Transforming a Digital Data Sequence into an Encoded Form, U.S. Patent No. 4,797,922 (1989).
19. W. Meier and O. Staffelbach: Correlation properties of combiners with memory in stream ciphers. J. Cryptology **5** (1992) 67-86.
20. J. Reeds and N. Sloane: Shift-register synthesis (modulo m). SIAM J. Comp. **14** (1985) 505-513.
21. R. Rueppel: Analysis and Design of Stream Ciphers. Springer-Verlag, Berlin Heidelberg New York (1986).
22. B. M. de Weger: Approximation lattices of p-adic numbers. J. Num. Th. **24** (1986) 70-88.

Univariate and Multivariate Merit Factors

Matthew G. Parker

The Selmer Center, Department of Informatics, University of Bergen,
PB 7800, N-5020 Bergen, Norway
matthew@ii.uib.no
http://www.ii.uib.no/~matthew

Abstract. Merit factor of a binary sequence is reviewed, and constructions are described that appear to satisfy an asymptotic merit factor of 6.3421... Multivariate merit factor is characterised and recursive Boolean constructions are presented which satisfy a non-vanishing asymptote in multivariate merit factor. Clifford merit factor is characterised as a generalisation of multivariate merit factor and as a type of quantum merit factor. Recursive Boolean constructions are presented which, however, only satisfy an asymptotic Clifford merit factor of zero. It is demonstrated that Boolean functions obtained via quantum error correcting codes tend to maximise Clifford merit factor. Results are presented as to the distribution of the above merit factors over the set of binary sequences and Boolean functions.

1 Introduction

This paper reviews spectral properties of binary sequences and Boolean functions. It deals with aperiodic and continuous spectral properties of the sequence or function, as quantified by *merit factor* and aperiodic *sum-of-squares*, from which merit factor is derived. The *sum-of-squares* can be computed in two ways, firstly by the sum-of-squares of the autocorrelation coefficients and, secondly, by the sum of the fourth powers of the magnitudes of the spectral values. Merit factor quantifies the continuous mean-square deviation from the average power spectrum of the sequence or Boolean function. Therefore it quantifies the degree of uniformity of spectral energy distribution for the sequence or Boolean function. It is an attractive metric because it computes a continuous (infinite) property of the sequence or function by using a relatively small amount of discrete finite computation. We demonstrate constructions for binary sequences and for Boolean functions such that the associated merit factors asymptote to constant values for large sizes. These asymptotes result from convenient number-theoretic relationships for the sum-of-squares of the associated aperiodic autocorrelation coefficients.

The univariate merit factor ($\mathcal{MF}$) of a $(1, -1)$ binary sequence has been relatively well-studied [1, 2, 3, 4, 5, 6, 7, 8, 9] resulting in a few well-known constructions based on quadratic residues which have tried to maximise *asymptotic merit factor* $\mathcal{F}$ [10, 6, 8, 7]. Until recently there was a longstanding conjecture

T. Helleseth et al. (Eds.): SETA 2004, LNCS 3486, pp. 72–100, 2005.
© Springer-Verlag Berlin Heidelberg 2005

[10, 6] that the maximum $\mathcal{F}$ achievable by an infinite binary construction is 6.0. In Section 2 we report on a recent construction by Kristiansen and Parker [11, 12], independently obtained by Borwein, Choi, and Jedwab [13], that satisfies $\mathcal{F} > 6.3$. This result is discussed in detail elsewhere in this proceedings [14]. We also report on a variant of this construction which appears to achieve the same asymptote.

In comparison to the univariate case, merit factor for the multivariate case remains largely unstudied, apart from some activity with respect to aperiodic binary (two-dimensional) arrays (e.g. see [15]), and with respect to the periodic sum-of-squares metric for a Boolean function [16]. In Section 3 we consider the extreme multivariate case where each dimension is of size 2 and the alphabet is $(1, -1)$ binary. These multivariate 'arrays' are conveniently specified by Boolean functions. We are, therefore, interested here in merit factors of sequences described by Boolean functions [11, 17]. We demonstrate that, as with the one-dimensional case, the *multivariate merit factor* $(\mathcal{MMF})$ for infinite constructions often asymptotes to a constant value, at least for recursive quadratic constructions. These constructions exhibit linear recursive formulae for both univariate and multivariate sum-of-squares and, in these cases, asymptotic merit factors are easily computed. This research was initially inspired by a previous result by Høholdt, Jensen, and Justesen [5] who proved the recursion $\gamma_n = 2\gamma_{n-1} + 8\gamma_{n-2}$ for the univariate sum-of-squares of the *Golay-Rudin-Shapiro sequence* of length 2^n [18, 19, 20].

In Section 4 we discuss our aim to characterise and evaluate *quantum merit factor* $(\mathcal{QMF})$ of a Boolean function where, in this context, the Boolean function of n binary variables is actually interpreted as a pure quantum multipartite state of n quantum bits (qubits) [21, 22]. $\mathcal{QMF}$ quantifies the degree of uniformity of energy distribution for the state with respect to the set of transform spectra resulting from the infinite set of transforms comprising all n-fold tensor products of 2×2 unitary matrices. High $\mathcal{QMF}$ indicates a high degree of uncertainty as to the joint value obtained by observing the n qubits in any local measurement basis and is a measure of entanglement of the n qubits [23, 24, 25]. Using brute-force algorithms on classical computers, it is not possible to compute $\mathcal{QMF}$ beyond about $n = 4$ qubits to any reasonable accuracy. It is therefore desirable to find faster algorithms to evaluate $\mathcal{QMF}$, and to recursively construct 'graphical' quantum states (*quantum graphs*) [26, 21, 27, 28, 29, 30] such that their $\mathcal{QMF}$ is computed precisely via simple recursive relationships for their quantum sum-of-squares. This paper achieves both these goals. The second goal is motivated, in part, by the recent proposal for *measurement-driven quantum computation* based on the idea of pre-entangling an array of qubits, where quantum computation is then undertaken by a series of well-chosen quantum measurements [26, 31, 23, 24]. The form of inter-qubit pre-entanglement chosen for the array can be modelled, precisely, by a quadratic Boolean function of n variables, as shown by Parker and Rijmen at SETA01 [21]. Moreover, *stabilizer quantum error-correcting codes* (QECCs) are exactly described using quadratic Boolean functions [27, 28, 29, 30]. So the metric of $\mathcal{QMF}$ can be used to evaluate *entanglement* of a graph-based multipartite quantum state (QECC), where large $\mathcal{QMF}$ indicates high entanglement [1] .

In Section 5 we back off somewhat from the problem of $\mathcal{QMF}$ to consider the evaluation of something we call *Clifford merit factor* $(\mathcal{CMF})$. Instead of computing merit factor with respect to the infinite set of transforms formed from tensor products of all 2×2 unitary transforms we, instead, compute merit factor with respect to the finite set of transforms formed from tensor products of members of the *Local Clifford Group* [33, 34, 35, 30]. $\mathcal{CMF}$ is a natural generalisation of $\mathcal{MMF}$ as it is computed via a collection of *fixed-multivariate* aperiodic autocorrelations over the set of all possible fixings [36], and gives a good indication of the quantum energy distribution for the associated quantum state. We further show that $\mathcal{CMF}$ is a measure of quantum entanglement of the associated multipartite state as it remains invariant with respect to local unitary transform of the state. We also, quite unexpectedly [2], arrive at the conclusion that $\mathcal{CMF}$ is precisely equal to $\mathcal{QMF}$. We also find that, for recursively constructed graphs, $\mathcal{CMF}$ can, once again, be exactly computed via sum-of-squares recursions. $\mathcal{CMF}$ is typically maximised over quadratic Boolean functions which describe zero-dimension QECCs with maximum distance [33, 35, 36, 37, 38, 22, 39]. Graphs constructed from adjacency matrices of a bordered-quadratic residue form tend to maximise $\mathcal{CMF}$ [30]. This nicely mirrors the univariate situation where quadratic residue constructions are central to the optimisation of $\mathcal{MF}$.

We conclude by listing some interesting open problems that this research suggests.

1.1   Key to Notation

We introduce some of the notation and fundamental spectral concepts that we use. All of the metrics discussed can be viewed as arising from the output spectra with respect to unitary transforms over complex space (i.e. the result of a set of matrix-vector products).

Consider matrix-vector products, Ts, in complex space, where T is a $2^n \times 2^n$ unitary matrix, and s is a $2^n \times 1$ vector, where both matrix and vector have entries from $\mathbb{C}$. 'Unitary' means that $TT^\dagger = I$, where '$\dagger$' means conjugate-transpose and I is the identity matrix. Transform T is constructed using the following unitary primitives:

$$U(\theta, \phi) = \begin{pmatrix} \cos\theta & \sin\theta e^{i\phi} \\ \sin\theta & -\cos\theta e^{i\phi} \end{pmatrix}, \qquad 0 \le \theta < \frac{\pi}{2}, \quad 0 \le \phi < \pi, \tag{1}$$

where $i^2 = -1$.

---

[1] QMF satisfies the requirement for an entanglement metric that it is invariant with respect to local unitary transform of the associated state [32, 21].

[2] It was not the author's original intention to establish the equivalence of $\mathcal{CMF}$ and $\mathcal{QMF}$ but it appears that they are equivalent.

Define $I, H, N \in \{U\}$, where

$$I = \begin{pmatrix} 1 & 0 \\ 0 & 1 \end{pmatrix}, H = \frac{1}{\sqrt{2}} \begin{pmatrix} 1 & 1 \\ 1 & -1 \end{pmatrix}, \text{ and } N = \frac{1}{\sqrt{2}} \begin{pmatrix} 1 & i \\ 1 & -i \end{pmatrix}.$$

Define the *tensor product* (or *Kronecker product*) as:

$$A \otimes B = \begin{pmatrix} a_{00}B & a_{01}B & \cdots \\ a_{10}B & a_{11}B & \cdots \\ \vdots & \vdots & \ddots \end{pmatrix}.$$

We introduce the notion of a set of identically-dimensioned unitary matrices, $\{A_0, A_1, \ldots, A_{k-1}\}$, such that an associated set, $\{A_0, A_1, \ldots, A_{k-1}\}^n$, comprises all n-fold tensor products of members of $\{A_0, A_1, \ldots, A_{k-1}\}$, giving a total of k^n unitary matrices, each of size $2^n \times 2^n$.

Example: $\{H\}^n = H \otimes H \otimes \ldots \otimes H$ defines a set of one $2^n \times 2^n$ unitary matrix, which implements the *Walsh-Hadamard* transform.

Example: $\{I, H\}^n = \{I \otimes \ldots \otimes I \otimes I, \quad I \otimes \ldots \otimes I \otimes H, \quad I \otimes \ldots \otimes H \otimes I,$
$I \otimes \ldots \otimes H \otimes H, \quad \ldots \quad H \otimes \ldots \otimes H \otimes H\}$ defines a set of 2^n distinct unitary matrices of size $2^n \times 2^n$ which implement the so-called $\{I, H\}^n$-*transform*.

Let $\mathcal{D}$ be the set of all diagonal and antidiagonal 2×2 unitary matrices. Thus

$$\mathcal{D} = \{\begin{pmatrix} a & 0 \\ 0 & b \end{pmatrix}, \begin{pmatrix} 0 & c \\ d & 0 \end{pmatrix}, \} \tag{2}$$

$\forall a, b, c, d$ such that $|a| = |b| = |c| = |d| = 1$.

We use '$\simeq$' to indicate that two $2^n \times 2^n$ matrices A and B are $\mathcal{D}$-*equivalent*, where,

$$A \simeq B \quad \Rightarrow \quad A = \Delta B \quad \text{for some } \Delta \in \mathcal{D}^n. \tag{3}$$

Then $\mathcal{D}^n\{U\}^n$ comprises all $2^n \times 2^n$ local unitary transforms.

We further define $\{V\} = \{U\}_{\theta=\pi/4}$, i.e. V is the subset of U where all matrix entries have the same magnitude. We also define the infinite transform sets $\{W\} \simeq \{V\}N$ and $\{X\} \simeq \{W\}N$. We can partition $\{V\}$ into matrix pairs, F_α and F'_α, where,

$$\{V\} = \{F_\alpha, F'_\alpha \mid \forall \alpha \in \mathcal{C}, |\alpha| = 1, \quad 0 \leq \angle\alpha < \frac{\pi}{2}\}, \tag{4}$$

where $F_\alpha = \frac{1}{\sqrt{2}} \begin{pmatrix} 1 & \alpha \\ 1 & -\alpha \end{pmatrix}$, and $F'_\alpha = \frac{1}{\sqrt{2}} \begin{pmatrix} 1 & i\alpha \\ 1 & -i\alpha \end{pmatrix}$. The rows of F_α relate to the residue system, mod $(x-\alpha)(x+\alpha) = (x^2 - \alpha^2)$, as left-multiplication of a vector, s, by F_α can be interpreted as evaluating the residues of $s(x) = s_0 + s_1 x$ mod $(x - \alpha)$ and mod $(x + \alpha)$. Similarly, the rows of F'_α relate to a residue system, mod $(x - i\alpha)(x + i\alpha) = (x^2 + \alpha^2)$. The combined rows of F_α and F'_α therefore relate to a residue system, mod $(x^4 - \alpha^4)$.

1.2   Useful Example

Here is an example of the spectral computations underlying $\mathcal{MF}$, $\mathcal{MMF}$, and $\mathcal{CMF}$. Let $p(\mathbf{x}) : Z_2^n \rightarrow Z_2$ be the Boolean function $p(\mathbf{x}) = x_0 x_1$, where $n = 2$. From p we create a 4×1 bipolar vector, $s = (s_{00}, s_{01}, s_{10}, s_{11})^T$, where $s_{ab} = (-1)^{p(x_0=a, x_1=b)}$. Thus $s = (-1)^{p(\mathbf{x})} = (1, 1, 1, -1)^T$. One computes the merit factor by first computing the *sum-of-squares* metric. This, in turn, can be computed directly by computing the sum-of-squares of the out-of-phase autocorrelation coefficient magnitudes, but here we, equivalently, sum the fourth powers of spectral magnitudes, whilst retaining the nomenclature 'sum-of-squares' for the resultant *sum-of-squares* metric.

To compute $\mathcal{MF}$ for p we proceed as follows, where $N = 2^n = 4$:

- $S = \frac{1}{2} \begin{pmatrix} 1 & 1 & 1 & 1 \\ 1 & i & -1 & -i \\ 1 & -1 & 1 & -1 \\ 1 & -i & -1 & i \end{pmatrix} s = (1, i, 1, -i)^T.$

- $S' = \frac{1}{2} \begin{pmatrix} 1 & \omega & i & \omega^3 \\ 1 & \omega^3 & -i & \omega \\ 1 & \omega^5 & i & \omega^7 \\ 1 & \omega^7 & -i & \omega^5 \end{pmatrix} s = \frac{1}{\sqrt{2}}(1 + \omega, -1 + \omega^7, -1 + \omega, 1 + \omega^7)^T,$

    where $\omega = \sqrt{i}$.
- The univariate sum-of-squares, γ, is
    $\gamma = \frac{1}{2}(\frac{N}{2}(\sum_k |S_k|^4 + \sum_k |S'_k|^4) - N^2) = \frac{1}{2}(\frac{4}{2}(4 + 6) - 16) = 2.$
- $\mathcal{MF} = \frac{N^2}{2\gamma} = 4.0.$

To compute $\mathcal{MMF}$ for p we proceed as follows, where $n = 2$:

- $S^{00} = (H \otimes H)s = (1, 1, 1, -1)^T.$
- $S^{01} = (H \otimes N)s = (1, 1, i, -i)^T.$
- $S^{10} = (N \otimes H)s = (1, i, 1, -i)^T.$
- $S^{11} = (N \otimes N)s = (1 + i, 0, 0, 1 - i)^T.$
- The multivariate sum-of-squares, σ, is
    $\sigma = \frac{1}{2}((\sum_{\mathbf{r} \in \{0,1\}^n} \sum_{\mathbf{k} \in \{0,1\}^n} |S_{\mathbf{k}}^{\mathbf{r}}|^4) - 4^n) = \frac{1}{2}(4 + 4 + 4 + 8 - 16) = 2.$
- $\mathcal{MMF} = \frac{4^n}{2\sigma} = 4.0.$

To compute $\mathcal{CMF}$ for p we proceed as follows, where $n = 2$:

- $S^{00} = (I \otimes I)s = (1, 1, 1, -1)^T.$
- $S^{01} = (I \otimes H)s = (\sqrt{2}, 0, 0, \sqrt{2})^T.$
- $S^{02} = (I \otimes N)s = (\omega, \omega^7, \omega^7, \omega)^T.$
- $S^{10} = (H \otimes I)s = (\sqrt{2}, 0, 0, \sqrt{2})^T.$
- $S^{11} = (H \otimes H)s = (1, 1, 1, -1)^T.$
- $S^{12} = (H \otimes N)s = (1, 1, i, -i)^T.$
- $S^{20} = (N \otimes I)s = (\omega, \omega^7, \omega^7, \omega)^T.$
- $S^{21} = (N \otimes H)s = (1, i, 1, -i)^T.$
- $S^{22} = (N \otimes N)s = (1 + i, 0, 0, 1 - i)^T.$
- The fixed-multivariate sum-of-squares, $\mathcal{E}$, is
    $\mathcal{E} = \frac{1}{2}((\sum_{\mathbf{r} \in \{0,1,2\}^n} \sum_{\mathbf{k} \in \{0,1\}^n} |S_{\mathbf{k}}^{\mathbf{r}}|^4) - 6^n) = \frac{1}{2}(4 + 8 + 4 + 8 + 4 + 4 + 4 + 4 + 8 - 36) = 6.$
- $\mathcal{CMF} = \frac{6^n}{2\mathcal{E}} = 3.0.$

1.3   The Rough Guide to Transform Spectra

We also provide a "map" (Fig. 1) that indicates the types of spectra we will be dealing with and how they relate to each other. For simplicity, the map only deals with input sequences, s, of length 2^n. The map represents sets of spectral outputs, S, by their associated transforms, T, from which S is computed where $S = Ts$, and the different forms of T are indicated on the map. All three metrics, $\mathcal{MF}$, $\mathcal{MMF}$, and $\mathcal{CMF}$, describe a property of an infinite spectral set - indicated on the map by an infinite set of transforms - but, as just shown in the example, each of the three metrics can be computed using only a finite set of spectral points. For sequences of length $N = 2^n$, the spectral outputs with respect to (w.r.t.) the univariate continuous Fourier transform occur as a strict subset of the spectral set $\{S\} = \{V\}^n s$ [40]. For example the univariate spectral points generated by rows of the matrix:

$$\frac{1}{2} \begin{pmatrix} 1 & 1 & 1 & 1 \\ 1 & i & -1 & -i \\ 1 & -1 & 1 & -1 \\ 1 & -i & -1 & i \end{pmatrix}$$

can be found as a subset of the rows of the matrices $H \otimes H$ and $N \otimes H$. The matrix multisets $\{V^I\}$, $\{W^N\}$, and $\{X^H\}$, are defined in definitions 9 and 16.

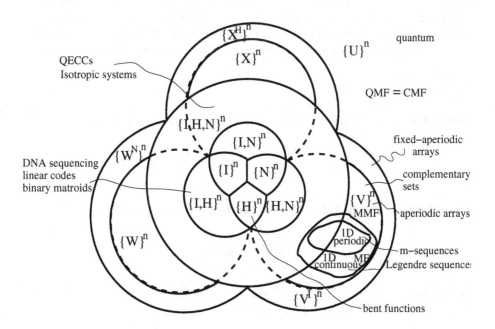

Fig. 1. Map of spectral outputs described by their associated transforms

2   Univariate Merit Factor

The univariate aperiodic autocorrelation of s is given by,

$$u_k = \sum_{j=0}^{N-1} s_j s_{j+k}^*, \qquad -N < k < N, \tag{5}$$

where $s_j = 0$, $N \geq j < 0$.

Alternatively, by representing s as a polynomial $s(z) = s_0 + s_1 z + \ldots + s_{N-1} z^{N-1}$, we can express u as $u(z) = u_{1-N} z^{1-N} + u_{2-N} z^{2-N} + \ldots + u_{N-2} z^{N-2} + u_{N-1} z^{N-1}$, where

$$u(z) = s(z)s(z^{-1})^*. \tag{6}$$

Using the polynomial form for s, define the *univariate continuous fourier transform* of s by,

$$S_k(L,c) = \frac{1}{\sqrt{N}} s(e^{\frac{\pi i(2tk+c)}{L}}), \qquad 0 \leq k < N, \tag{7}$$

where $L = tN$, $t \in \{1, 2, \ldots, \infty\}$, with $c \in \{\{1, 2, \ldots, 2t - 1\} | \gcd(c, 2t) = 1\} + \{0, t\} \bmod 2t$ if t odd, and $c \in \{\{1, 2, \ldots, 2t - 1\} | \gcd(c, 2t) = 1\}$ if t even. Each (L, c) pair defines a different $N \times N$ unitary transform, $T(L, c)$, such that $S(L, c) = T(L, c)s$, where $T_{kj}(L, c) = \frac{1}{\sqrt{N}} e^{\frac{\pi i j(2tk+c)}{L}}$ and the infinite set of spectral points, $\{S_k(L, c)\}$, for valid triples (k, L, c) approximates the continuous univariate Fourier transform spectra infinitely closely. Therefore (7) evaluates $s(z)$ at all points on the unit circle. We symbolically represent this evaluation as

 .

Definition 1. *The* sum-of-squares, γ, *of the sequence,* s, *is given by,*

$$\gamma = \frac{1}{2}\left(\left(\sum_{k=1-N}^{N-1} |u_k|^2\right) - N^2\right) = \sum_{k=1}^{N-1} |u_k|^2. \tag{8}$$

Definition 2. *The* merit factor, $\mathcal{MF}$, *of the sequence,* s, *is given by,*

$$\mathcal{MF} = \frac{N^2}{2\gamma}. \tag{9}$$

Definition 3. *Let* $s_{\mathcal{A}}$ *be a length N sequence generated by construction $\mathcal{A}$. Then the* asymptotic merit factor *of* $s_{\mathcal{A}}$ *is given by,*

$$\mathcal{F} = \lim_{N \to \infty} \mathcal{MF}(s_{\mathcal{A}}).$$

2.1   Transform $\Leftrightarrow$ Autocorrelation Duality

The *Wiener-Kinchine theorem* states that (5) and (7) are related by

$$\sum_{k=1-N}^{N-1} |u_k|^2 = \frac{N}{2}(\sum_{k=0}^{N-1} |S_k(L,c)|^4 + |S_k(L,c')|^4), \tag{10}$$

where $c' = c + t \bmod 2t$, $c = 0$ if $t = 1$, $c \in \{\{1, 2, \ldots, 2t - 1\}| \gcd(c, 2t) = 1\}$ if t odd, and $c \in \{\{1, 2, \ldots, t - 1\}| \gcd(c, 2t) = 1\}$ if t even. The reason for the choice of pairings (c, c') becomes clear when we consider an embedding of the non-modular polynomial multiplication (6) in a polynomial modulus (i.e. we realise an aperiodic autocorrelation using a constaperiodic autocorrelation). Specifically, let

$$u'(z) = s(z)s(z^{-1})^* \bmod (z^{2N} - \epsilon), \tag{11}$$

where ϵ is a complex root of one of order t, $t \in \{1, 2, \ldots, \infty\}$. Then,

$$\begin{aligned} u'_k &= u_k, & 0 \le k < N, \\ u'_k &= \epsilon^{-1} u_{k-2N}, & N < k < 2N, \\ u'_k &= u'_N = 0 & \text{otherwise.} \end{aligned}$$

In particular,

$$\sum_{k=0}^{2N-1} |u'_k|^v = \sum_{k=1-N}^{N-1} |u_k|^v, \qquad \forall v. \tag{12}$$

So, from (8), we can use (11) instead of (6) to compute γ. (10) follows directly from (11) and (12) because we can factorise (11) into two residue computations mod $(z^N - \eta)$ and mod $(z^N + \eta)$, where η is a complex root of one of order $2t$ such that $\eta^2 = \epsilon$. Then $s(z)s(z^{-1})^* \bmod (z^N - \eta)$ and $s(z)s(z^{-1})^* \bmod (z^N + \eta)$ can be computed by evaluating $s(z)s(z^{-1})^*$ at the N residues $z \in \{e^{\frac{\pi i(2tk+c)}{L}} | 0 \le k < N\}$, and at the N residues $z \in \{e^{\frac{\pi i(2tk+c')}{L}} | 0 \le k < N\}$, respectively. In particular $u'(e^{\frac{\pi i(2tk+d)}{L}}) = |S_k(L,d)|^2$, $d \in \{c, c'\}$. One then obtains (10) by Parseval (or the Chinese Remainder Theorem). The main point here is that we obtain (10) and exactly the same value of γ for any choice of complex root, ϵ, of order $2t$. By considering all such ϵ, $|S_k(L,d)|$ ranges over the continuous fourier magnitude spectrum and, therefore, as γ is independent of ϵ, γ evaluates a property of the continuous fourier spectra, namely the mean-square deviation from the flat continuous fourier power spectrum. Specifically,

$$\gamma = \frac{1}{4\pi} \int_0^{2\pi} (|s(e^{i\omega})|^2 - N)^2 d\omega.$$

In this paper we choose to compute γ by selecting $\epsilon = 1$, leading to $L = N$ and $(c, c') = (0, 1)$, and allowing us to abbreviate $S_k(L, d)$ to S_k and S'_k for $d = c$ and $d = c'$, respectively, as done in Section 1.2.

2.2   Expected Values and Constructions

The maximum merit factor known is for the length $N = 13$ sequence, 0101001100000, for which $\mathcal{MF} = 14.083$, although there is no proof that this is the true maximum over all N. $\mathcal{F}$ exists for many infinite sequence constructions. Experimental results suggest that, for a random binary sequence, $\mathcal{F} = 1.0$, as indicated by the following graphs of random samplings for (from left to right) $N = 16, 64, 512,$ and 1024, with merit factor and # sequences on x and y-axes, respectively, with x-axes ranging linearly from 0 to 4, and where the highest peak is centred around $\mathcal{MF} = 1.0$ ever more tightly as N increases (we leave the graph axes unmarked as we simply wish to indicate the general trend as N increases):

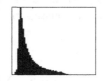

Although binary sequences with merit factors around 8.0 or 9.0 have been found up to lengths $N = 250$, the maximum known asymptotic merit factor was, until recently, $\mathcal{F} = 6.0$. This asymptote is satisfied by the *Legendre* construction [10, 6, 41], the *Jacobi* and *modified Jacobi* construction [8, 42], and is conjectured to be satisfied by a negaperiodic construction of Parker [43]. In his recent master's thesis [11], Kristiansen describes a construction based on an extended Legendre sequence which satisfies, experimentally, $\mathcal{F} > 6.3$. Independently, Borwein, Choi and Jedwab [13] proposed a construction which satisfies, experimentally, $\mathcal{F} = 6.3421\ldots$. These two constructions generate essentially the same sequence, although only [13] discovered the periodic form of the extension and provided theoretical arguments as to the precise values of the asymptote and construction parameters. Detailed descriptions of the constructions can be found in [12] and [13], and elsewhere in this publication [14]. Both Kristiansen and Parker, and Borwein, Choi and Jedwab were influenced by prior work of two master's students of Jim Davis, Kirilusha and Narayanaswamy [44], who first developed the essential form of the construction by realising that extending a $\frac{1}{4}$-rotated Legendre sequence by up to $O(\sqrt{N})$ elements does not change $\mathcal{F}$ from 6.0. Moreover they noticed that if the extension was periodic they could even increase $\mathcal{F}$ above 6.0, although they did not uncover an asymptote. A summary of some of the constructions with large $\mathcal{F}$ is now given.

Legendre Construction [10]

- Select a prime integer, m.
- Construct the $\{1, -1\}$ sequence, $l = (l_0, l_1, \ldots, l_{m-1})^T$, of length m, such that $l_j = 1$ if $\exists k$ such that $k^2 = j \bmod m$, (in which case j is called a *quadratic residue*, mod m). Otherwise $l_j = -1$. By convention, $l_0 = 1$.
- Construct s as the periodic rotation of l by $\frac{1}{4}$ of its length:

$$s_j = l_{j + \lfloor \frac{m}{4} \rfloor} \bmod m.$$

l is the *Legendre sequence* and satisfies $\mathcal{F}(l) = 1.5$.

Theorem 1. *For s a $\frac{1}{4}$-rotated Legendre sequence:*

$$\mathcal{F}(s) = 6.0.$$

Construction – Borwein, Choi and Jedwab [13]

- Construct a Legendre sequence, l, using a prime, m.
- Construct l^r to be a periodic rotation of l by $0.2211m$ (or by $0.7211m$):

$$l^r_j = l_{j+\lfloor rm \rfloor} \bmod m, \qquad r \in \{0.2211, 0.7211\}.$$

- Construct the length $1.0578m$ "BCJ-sequence", s, as the periodic extension of l^r by $0.0578m$:

$$s_j = l^r_j, \qquad 0 \le j < m, \qquad s_{j+m} = l^r_j, \qquad 0 \le j \le \lfloor 0.0578m \rfloor.$$

Conjecture 1. *For s a "BCJ-sequence":*

$$\mathcal{F}(s) = 6.3421\ldots.$$

Construction – Kristiansen [11, 12]

- Construct a Legendre sequence, l, using a prime, m.
- Assign $k = 0$ and $l^k = l$.
- Step A: Construct l^+ and l^- as the periodic and negaperiodic rotations of l^k by one element:

$$l^+_j = l^k_{j+1} \bmod m, \qquad l^-_j = (-1)^{\lfloor \frac{j+1}{m} \rfloor} l^k_{j+1} \bmod m, \qquad 0 \le j < m.$$

- If $\mathcal{MF}(l^+) \ge \mathcal{MF}(l^-)$ then assign $l^{k+1} = l^+$ else assign $l^{k+1} = l^-$.
- Assign $k = k + 1$.
- If $k < 0.31m$ then loop back to step A.
- Construct the sequence, $\mathcal{T}$, such that,

$$\mathcal{T} = l | l^1_{m-1} | l^2_{m-1} | \cdots | l^k_{m-1},$$

where '$a|b$' means concatenate b onto the end of a.
- Construct the "K-sequence", s^r, of length $\lfloor 1.059m \rfloor$, such that,

$$s^r_j = \mathcal{T}_{j+rm}, \qquad 0 \le j < \lfloor 1.059m \rfloor,$$

where $r < 0.242m$.

Conjecture 2. *For s_r a "K-sequence", $\exists r$ such that*

$$\mathcal{F}(s^r) > 6.3\ldots.$$

The intuition behind the construction of Kristiansen and Parker is that a sequence with high merit factor should contain subsequences with moderately high merit factor. After becoming aware of the preprint [13], Kristiansen and Parker realised that, in all but four small-length cases, $\mathcal{MF}(l^+)$ appears to be always greater than $\mathcal{MF}(l^-)$ for each k-iteration. It follows that, to within some inaccuracies in periodic extension and rotation length, the optimal "K-sequence" is the "BCJ-sequence".

Construction – Parker

Empirical evidence indicates that the asymptote of $\mathcal{F}(s) = 6.3421\ldots$ also holds true for a periodic rotation and extension of the (modified)-Jacobi construction. We here summarise yet another construction that appears to satisfy the same asymptote, namely the negaperiodic rotation and extension of the negaperiodic construction of [43].

- Construct a Legendre sequence, l, using a prime, m.
- Construct $\mathcal{N}$ such that

$$\mathcal{N} = (l|l) \odot (1, 1, -1, -1, 1, 1, -1, -1, \ldots)^T,$$

  where '$a|b$' is the concatenation of vectors, and $(w) = (u) \odot (v)$ implies $w_i = u_i v_i$.
- Construct $\mathcal{N}^r$ as the negaperiodic rotation of l by $0.4705(2m)$ (or by $0.9705(2m)$):

$$\mathcal{N}_j^r = (-1)^{\frac{h}{2m}} \mathcal{N}_h \bmod 2m,$$

  where $h = j + \lfloor r(2m) \rfloor$, $0 \leq j < 2m$, and $r \in \{0.4705, 0.9705\}$.
- Construct the length $1.0578(2m)$ "P-sequence", s, as the negaperiodic extension of $\mathcal{N}^r$ by $0.0578(2m)$:

$$s_j = \mathcal{N}_j^r, \qquad 0 \leq j < 2m, \qquad s_{j+2m} = -\mathcal{N}_j^r, \qquad 0 \leq j \leq \lfloor 0.0578(2m) \rfloor.$$

Conjecture 3.

$$\mathcal{F}(\mathcal{N}) = 6.0.$$

Conjecture 4. *For s a "P-sequence":*

$$\mathcal{F}(s) = 6.34\ldots.$$

An alternative periodic version of the same construction is as follows.

- Construct a Legendre sequence, l, using a prime, m.
- Construct $\mathcal{L}$ such that

$$\mathcal{L} = (l|l)$$

- Construct $\mathcal{L}^r$ as the periodic rotation of l by $0.4705(2m)$ (or by $0.9705(2m)$):

$$\mathcal{L}_j^r = \mathcal{L}_{j+\lfloor r(2m) \rfloor} \bmod 2m,$$

  where $0 \leq j < 2m$ and $r \in \{0.4705, 0.9705\}$.
- Construct the length $1.0578(2m)$ sequence, s', as the periodic extension of $\mathcal{L}^r$ by $0.0578(2m)$:

$$s_j' = \mathcal{L}_j^r, \qquad 0 \leq j < 2m, \qquad s_{j+2m}' = \mathcal{L}_j^r, \qquad 0 \leq j \leq \lfloor 0.0578(2m) \rfloor.$$

- Construct the length $1.0578(2m)$ "P-sequence", s, such that

$$s = s' \odot (1, 1, -1, -1, 1, 1, -1, -1, \ldots)^T.$$

The Golay-Rudin-Shapiro Construction

Both the *m-sequence* and *Golay-Rudin-Shapiro sequence* [18, 19, 20] satisfy $\mathcal{F} = 3.0$. The latter construction can be described using Boolean functions as shown by Davis and Jedwab [45]. Define $p(\mathbf{x}) : \mathcal{Z}_2^n \rightarrow \mathcal{Z}_2$ as

$$p(\mathbf{x}) = (\sum_{i=0}^{n-2} x_{\pi(i)} x_{\pi(i+1)}) + (\sum_{i=0}^{n-1} c_i x_i) + d, \qquad (13)$$

where $\pi : \mathcal{Z}_n \rightarrow \mathcal{Z}_n$ is a permutation of the integers, mod n, $c_i, d \in \mathcal{Z}_2$.

- Construct the length 2^n sequence, s^π, such that,

$$s_j^\pi = (-1)^{p(x_i = j_i)}, \qquad (14)$$

where $j = \sum_{i=0}^{n-1} 2^{j_i}$ and $j_i \in \{0, 1\}$, $\forall i$.

Theorem 2. *[1, 5] When π is the identity permutation, then $s = s^\pi$ is the Golay-Rudin-Shapiro sequence and*

$$\mathcal{F}(s) = 3.0.$$

Proof. Let γ_n be the sum-of-squares for s constructed from p over n binary variables. It can be shown that [5],

$$\gamma_n = 2\gamma_{n-1} + 8\gamma_{n-2}.$$

In closed form, $\gamma_n = \frac{4^n}{6} - \frac{(-2)^n}{6}$. The asymptote follows from (9) as $n \rightarrow \infty$. $\square$

Remark: It is currently unclear whether $\mathcal{F}(s^\pi) = 3.0$ over the complete set of permutations, π, or whether asymptotes above and below 3.0 can be obtained by suitable choice of permutation [11]. For instance, for $n = 8$, $2.27 \leq \mathcal{MF}(s^\pi) \leq 4.49$.

Table 1. Conjectures on $\mathcal{F}$ for certain graphical constructions

graph	$p(\mathbf{x})$	$\mathcal{F}(s)$	γ_n - recursion
circle	$(\sum_{i=0}^{n-2} x_i x_{i+1}) + x_{n-1} x_0$	1	$4\gamma_{n-1} + 12\gamma_{n-2} - 64\gamma_{n-3} + 256\gamma_{n-5}$
complete	$\sum_{i<j, 1 \leq j < n} x_i x_j$	0	$\gamma_n = 10\gamma_{n-1} - 36\gamma_{n-2} + 88\gamma_{n-3}$
			$-96\gamma_{n-4} - 512\gamma_{n-5} + 1024\gamma_{n-6}$
star	$x_0(x_1 + x_2 + \ldots + x_{n-1})$	0	$\gamma_n = 16\gamma_{n-1} - 68\gamma_{n-2} - 48\gamma_{n-3}$
			$+768\gamma_{n-4} - 1024\gamma_{n-5}$

Other Graphical Constructions

Quadratic Boolean functions, $p(\mathbf{x})$, have a natural interpretation as graphs where, for $p(\mathbf{x}) = \sum_{i<j} a_{ij} x_i x_j$, the adjacency matrix, Γ, of the associated graph satisfies $\Gamma_{ij} = \Gamma_{ji} = a_{ij}$ for $i < j$ and $\Gamma_{ii} = 0$. Thus one can view the Golay-Rudin-Shapiro sequence as the *path graph* with a particular ordering of the vertices. Table 1 summarises conjectures, first presented in [17], as to the value of $\mathcal{MF}$ for a few other simple recursive graph constructions.

3   Multivariate Merit Factor

The multivariate merit factor $\mathcal{MMF}$ was first investigated by Gulliver and Parker in [17], as a modification of the metric first introduced by Kristiansen in [11]. Define the multivariate sequence, s, with each dimension of s of length 2, such that,

$$s = (s_{0...00}, s_{0...01}, s_{0...10}, \ldots, s_{1...11})^T$$
$$s_\mathbf{j} \in \{1, -1\}, \qquad\qquad \mathbf{j} \in \{0,1\}^n$$
$$s_\mathbf{j} = 0, \qquad\qquad \text{otherwise.}$$

The multivariate sequence, s, is always, in this paper, constructed via its associated Boolean function, p, such that,

$$s = (-1)^{p(\mathbf{x})},$$

where $s_\mathbf{j} = (-1)^{p(\mathbf{x}=\mathbf{j})}$, and $\mathbf{x}, \mathbf{j} \in \mathbb{Z}_2^n$.

The *multivariate aperiodic autocorrelation* of s is given by,

$$u_\mathbf{k} = \sum_{\mathbf{j} \in \{0,1\}^n} s_\mathbf{j} s_{\mathbf{j}+\mathbf{k}}^*, \qquad \mathbf{k} \in \{-1,0,1\}^n. \tag{15}$$

Alternatively, by representing s as a multivariate polynomial,

$$s(z_0, z_1, \ldots, z_{n-1}) = s_{0...00} + s_{0...01} z_0 + s_{0...10} z_1 + \ldots + s_{1...11} z_{n-1} \cdots z_1 z_0,$$

we can compute u where,

$$u(z_0, z_1, \ldots, z_{n-1}) = s(z_0, z_1, \ldots, z_{n-1}) s(z_0^{-1}, z_1^{-1}, \ldots, z_{n-1}^{-1})^*. \tag{16}$$

Define the *multivariate continuous fourier transform* of s by,

$$S_\mathbf{k}(\mathbf{L}, \mathbf{c}) = 2^{-\frac{n}{2}} s(z_j = e^{\frac{\pi i(2t_j k_j + c_j)}{L_j}} \mid 0 \le j < n), \qquad \mathbf{k} \in \{0,1\}^n, \tag{17}$$

where $\mathbf{L} = 2t$, $t_j \in \{1, 2, \ldots, \infty\}$, with $c_j \in \{\{1, 2, \ldots, 2t_j - 1\} \mid \gcd(c_j, 2t_j) = 1\} + \{0, t_j\} \bmod 2t_j$ if t_j odd, and $c_j \in \{\{1, 2, \ldots, 2t_j - 1\} \mid \gcd(c_j, 2t_j) = 1\}$ if t_j even. Each $(\mathbf{L}, \mathbf{c})$ pair defines a different $2^n \times 2^n$ unitary transform, $T(\mathbf{L}, \mathbf{c})$, such that $S(\mathbf{L}, \mathbf{c}) = T(\mathbf{L}, \mathbf{c})s$, where

$$T(\mathbf{L}, \mathbf{c}) = 2^{-\frac{n}{2}} \bigotimes_{j=0}^{n-1} \begin{pmatrix} 1 & e^{\frac{\pi i c_j}{L_j}} \\ 1 & -e^{\frac{\pi i c_j}{L_j}} \end{pmatrix},$$

and the infinite set of spectral points, $\{S_\mathbf{k}(\mathbf{L}, \mathbf{c})\}$, for valid vector triples $(\mathbf{k}, \mathbf{L}, \mathbf{c})$ approximates the continuous multivariate Fourier transform spectra infinitely closely. From section 1.1 it is apparent that

$$\{T(\mathbf{L}, \mathbf{c})\} = \{V\}.$$

(17) evaluates $s(z_0, z_1, \ldots, z_{n-1})$ at all points on the multi-unit circle. We symbolically represent this evaluation as .

Definition 4. *The* multivariate sum-of-squares, σ, *of the sequence,* s, *is given by,*

$$\sigma = \frac{1}{2}\left(\left(\sum_{\mathbf{k}\in\{-1,0,1\}^n} |u_{\mathbf{k}}|^2\right) - 4^n\right) = \frac{1}{2}\sum_{\mathbf{k}\in\{-1,0,1\}^n, \mathbf{k}\neq\mathbf{0}} |u_{\mathbf{k}}|^2. \quad (18)$$

Definition 5. *The* multivariate merit factor, $\mathcal{MMF}$, *of the sequence,* s, *is given by,*

$$\mathcal{MMF} = \frac{4^n}{2\sigma}. \quad (19)$$

Definition 6. *Let* $s_{\mathcal{A}}$ *be a length* 2^n *multivariate sequence generated by construction* $\mathcal{A}$. *Then the* asymptotic multivariate merit factor *of* $s_{\mathcal{A}}$ *is given by,*

$$\mathcal{F}^{\mathcal{M}} = \lim_{n\to\infty} \mathcal{MMF}(s_{\mathcal{A}}).$$

$\mathcal{MMF}$ Symmetries

Lemma 1. *Let* $s = (-1)^{p(\mathbf{x})}$, *where* p *is a Boolean function of* n *variables. Let* $s' = (-1)^{p'(\mathbf{x})}$, *where*

$$p'(\mathbf{x}) = p(\tilde{x}_{\pi(0)}, \tilde{x}_{\pi(1)}, \ldots, \tilde{x}_{\pi(n-1)}) + \left(\sum_{i=0}^{n-1} c_i x_i\right) + d,$$

where $\tilde{x} \in \{x, x+1\}$, $\pi : \mathcal{Z}_n \to \mathcal{Z}_n$ *is a permutation of the integers, mod* n, *and* $c_i, d \in \mathcal{Z}_2$. *Then,*

$$\mathcal{MMF}(s') = \mathcal{MMF}(s).$$

3.1 Transform $\Leftrightarrow$ Autocorrelation Duality

Let $\mathbf{r} \in \{0,1\}^n$ and define $\mathbf{d}(\mathbf{r}) = (d(r)_0, d(r)_1, \ldots, d(r)_{n-1})$ such that

$$d(r)_j = c_j + r_j t_j, \text{ mod } 2t_j,$$

where $c_j \in \{\{1, 2, \ldots, 2t_j - 1\}| \gcd(c_j, 2t_j) = 1\}$ if t_j odd, and $c_j \in \{\{1, 2, \ldots, t_j - 1\}| \gcd(c_j, 2t_j) = 1\}$ if t_j even, $0 \leq j < n$. A multivariate version of the *Wiener-Kinchine theorem* states that (15) and (17) are related by

$$\sum_{\mathbf{k}\in\{-1,0,1\}^n} |u_{\mathbf{k}}|^2 = \sum_{\mathbf{k},\mathbf{r}\in\{0,1\}^n} |S_{\mathbf{k}}(\mathbf{L}, \mathbf{d}(\mathbf{r}))|^4. \quad (20)$$

We realise the aperiodic autocorrelation by embedding the non-modular polynomial multiplication (16) in a polynomial modulus: Let

$$u'(z_0, z_1, \ldots, z_{n-1}) = s(z_0, z_1, \ldots, z_{n-1})s(z_0^{-1}, z_1^{-1}, \ldots, z_{n-1}^{-1})^* \text{ mod } \prod_{j=0}^{n-1}(z_j^4 - \epsilon_j) \quad (21)$$

where ϵ_j is a complex root of one of order t_j, $t_j \in \{1, 2, \ldots, \infty\}$, $0 \le j < n$. Then, with $\mathbf{k} \in \{-1, 0, 1\}^n$ and $\mathbf{k}' \in \{0, 1, 3\}^n$,

$$u'_{\mathbf{k}'} = \left(\prod_{j=0}^{n-1} \epsilon_j^{-\lfloor \frac{k'_j}{2} \rfloor} \right) u_{\mathbf{k}}, \qquad k'_j = k_j \bmod 4.$$

In particular,

$$\sum_{\mathbf{k}' \in \{0,1,3\}^n} |u'_{\mathbf{k}'}|^v = \sum_{\mathbf{k} \in \{-1,0,1\}^n} |u_{\mathbf{k}}|^v, \qquad \forall v. \tag{22}$$

So, from (18), we can use u' instead of u to compute σ. (20) follows directly from (21) and (22) because we can factorise (21) into two residue computations per dimension, mod $(z_j^2 - \eta_j)$ and mod $(z_j^2 + \eta_j)$, where η_j is a complex root of one of order $2t_j$ such that $\eta_j^2 = \epsilon_j$. The two residue computations per dimension are realised by left-multiplication by matrices, F_{α_j} and F'_{α_j} (see (4)), where $\eta_j = \alpha_j^2$. Then, for each $\mathbf{r} \in \{0,1\}^n$, we compute $s(z_0, z_1, \ldots, z_{n-1}) s(z_0^{-1}, z_1^{-1}, \ldots, z_{n-1}^{-1})^* \bmod \prod_{j=0}^{n-1} (z_j^2 - (-1)^{r_j} \eta_j)$ by evaluating $s(z)s(z^{-1})^*$ at the 2^n residues, $z_j \in \{e^{\frac{\pi i (2t_j k_j + d(r)_j)}{L_j}} | \mathbf{k} \in \{0,1\}^n\}$. In particular $u'(z_j = e^{\frac{\pi i (2t_j k_j + d(r)_j)}{L_j}} \mid 0 \le j < n) = |S_{\mathbf{k}}(\mathbf{L}, \mathbf{d}(\mathbf{r}))|^2$. One then obtains (20) by Parseval (or the Chinese Remainder Theorem). We obtain (20) and exactly the same value of σ for any choice of vector of complex roots, $\bar{\epsilon} = (\epsilon_0, \epsilon_1, \ldots, \epsilon_{n-1})$, where ϵ_j has order t_j. The infinite set of transforms $\{V\}^n$ is obtained by considering all such $\bar{\epsilon}$, so that $|S_{\mathbf{k}}(\mathbf{L}, \mathbf{d}(\mathbf{r}))|$ ranges over the continuous multivariate fourier spectrum. Therefore, σ evaluates the mean-square deviation from the flat continuous multivariate fourier power spectrum. Specifically,

$$\sigma = \frac{1}{4\pi} \int_0^{2\pi} \int_0^{2\pi} \cdots \int_0^{2\pi} (|s(e^{i\omega_0}, e^{i\omega_1}, \ldots, e^{i\omega_{n-1}})|^2 - 2^n)^2 d\bar{\omega},$$

where $\bar{\omega} = (\omega_0, \omega_1, \ldots, \omega_{n-1})$.

In this paper we choose to compute σ by selecting $\bar{\eta} = (1, 1, \ldots, 1)$, leading to $L_j = 2$ and $(c_j, c'_j) = (0, 1)$, $\forall j$. Therefore $T(\mathbf{L}, \mathbf{c}) = \{H, N\}^n$ for H and N as defined in Section 1.1, and $S_{\mathbf{k}}(\mathbf{L}, \mathbf{d}(\mathbf{r}))$ can be abbreviated to $S_{\mathbf{k}}^{\mathbf{r}}$, as done in Section 1.2. $\{S_{\mathbf{k}}^{\mathbf{r}}\}$ is a set of 4^n spectral points.

3.2   Expected Values and Constructions

Maximising the $\mathcal{MMF}$ of a Boolean function indicates a minimum mean-square deviation from the flat continuous multivariate fourier power spectrum. Unlike the univariate case, the Boolean multivariate problem does not appear to have been investigated before [17]. Initial investigations suggest that the maximum $\mathcal{MMF}$ may be for the $n = 2$ variable sequence 0001, for which $\mathcal{MMF} = 4$. Table 2 shows the equivalence classes for Boolean functions of $n = 2$ to 5 variables, where the set of inequivalent functions is obtained from [46].

Table 2. Complete set of multivariate merit factors for $n = 2$ to $n = 5$

n	# inequivalent functions	# equivalence classes with list of $\mathcal{MMF}$s
2	2	2 classes **4.000**, 0.8
3	5	3 classes **2.667**, 1.143, 0.421
4	39	18 classes **3.200**, 1.778, 1.600, 1.455, 1.333, 1.231, 1.143, 1.067, 1.000, 0.941, 0.842, 0.800, 0.727, 0.696, 0.640, 0.552, 0.400, 0.246
5	22442	80 classes **2.909** − 0.152

Experimental results suggest that, for a random Boolean function of n variables, $\mathcal{F}^{\mathcal{M}} = 1.0$, as indicated here by the random samplings for (from left to right) $n = 4, 6, 9$ and 10, where $\mathcal{MMF}$ and #sequences are x and y-axes, respectively, with x-axes ranging linearly from $\mathcal{MMF} = 0$ to 4, and where the highest peak approaches $\mathcal{F}^{\mathcal{M}} = 1.0$ as n increases.

In comparison, a sampling of just quadratic Boolean functions for $n = 4, 6, 9$ and 10 indicates a wider range of $\mathcal{MMF}$s for a given n than for the full space of Boolean functions although, once again, the highest peak appears to approach $\mathcal{F}^{\mathcal{M}} = 1.0$ as n gets large. Once again, the x-axis ranges linearly from $\mathcal{MMF} = 0$ to 4 and the y-axis indicates #sequences.

 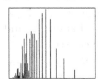

Conjecture 5. *A random Boolean function satisfies,*

$$\mathcal{F}^{\mathcal{M}} = 1.0.$$

Definition 7. *Define* $\mathcal{Q}$ *to be the complete set of homogeneous quadratic Boolean functions over n variables, i.e.* $q \in \mathcal{Q}$ *iff* $q = \sum_{j<k} c_{jk} x_j x_k$, $c_{jk} \in \mathcal{Z}_2$.

Definition 8. *Let* $\mathcal{S}$ *be an arbitrary subset of n-variable Boolean functions. Define* $\mathcal{S}_{\mathcal{Q}} = \{s + q \mid \forall s \in \mathcal{S}, q \in \mathcal{Q}\}$.

Theorem 3. *The average value of* $\frac{1}{\mathcal{MMF}}$ *with respect to any set* $\mathcal{S}_{\mathcal{Q}}$ *is,*

$$average\ _{\mathcal{S}_{\mathcal{Q}}}\left(\frac{1}{\mathcal{MMF}}\right) = \frac{2^n - 1}{2^n}.$$

Proof. Using arguments similar to [9], observe, from (16) and (18), that,

$$2\sigma + 4^n = \sum_{j+k=l+m} s_j s_k s_l s_m, \tag{23}$$

where $\mathbf{j}, \mathbf{k}, \mathbf{l}, \mathbf{m} \in \{0,1\}^n$ and the '+' for the subscript of the summation is not mod 2. Now $p(\mathbf{x}) = 0$ if $p(\mathbf{x})$ is a homogeneous quadratic and $\mathrm{wt}(\mathbf{x}) \le 1$, where $\mathrm{wt}(\mathbf{y})$ means the number of non-zero components of $\mathbf{y}$. We partition the summation (23) as follows:

- $\mathrm{wt}(\mathbf{j}), \mathrm{wt}(\mathbf{k}), \mathrm{wt}(\mathbf{l}), \mathrm{wt}(\mathbf{m}) \le 1$:
  - $\mathbf{j} = \mathbf{k} = \mathbf{l} = \mathbf{m} \to$ this case contributes 2^n to the summation.
  - $\mathbf{j} = \mathbf{l}, \mathbf{k} = \mathbf{m}$, or $\mathbf{j} = \mathbf{m}, \mathbf{k} = \mathbf{l}$. $\to$ there are $\frac{2^n(2^n-1)}{2}$ pairs in 4 configurations each, contributing a total of $4\frac{2^n(2^n-1)}{2}$ to the summation.
- Otherwise there are one or more of $\mathbf{j}, \mathbf{k}, \mathbf{l}$ and $\mathbf{m}$ with weight > 1. W.l.o.g. assume that $\mathbf{j}$ has weight 2 or greater. In particular, assume that $\mathbf{j}$ is 1 in positions j_a and j_b. We are summing over $|\mathcal{S}|$ copies of each of the homogeneous quadratics. Exactly half of these quadratics will contain the monomial $x_a x_b$. Therefore the contribution to the summation in this case is zero.

Therefore (23) evaluates to $2^n + 4\frac{2^n(2^n-1)}{2}$ and Theorem 3 follows. $\square$

Corollary 1. *The set of n-variable Boolean functions of degree d or less satisfies, average $(\frac{1}{\mathcal{MMF}}) = \frac{2^n-1}{2^n}$ for any d, $2 \le d \le n$, and, consequently, average $(\frac{1}{\mathcal{MMF}}) \to 1.0$ as $n \to \infty$.*

Remark: Theorem 3 is similar to a theorem by Newman and Byrnes [9] for the univariate case which states that, for a random binary sequence of length N, average $(\frac{1}{\mathcal{MF}}) = \frac{N-1}{N}$.

Table 3 is taken from [17] and summarises constructions, described by Boolean functions, $p(\mathbf{x})$, where $s = (-1)^{p(\mathbf{x})}$. The constructions represent a larger class of $\mathcal{MMF}$-invariant sequences, as generated by Lemma 1, and the recursions have all been proven. σ_n is the value of σ for the construction over n variables.

Remark: From Theorems 2 and Table 3 the values for univariate and multivariate sum-of-squares for the path are the same if π is the identity permutation.

Table 3. $\mathcal{F}^{\mathcal{M}}$ for certain graphical constructions [17]

graph	$p(\mathbf{x})$	$\mathcal{F}^{\mathcal{M}}(s)$	σ_n - recursion	σ_n - closed form
path	$\sum_{i=0}^{n-2} x_i x_{i+1}$	3	$2\sigma_{n-1} + 8\sigma_{n-2}$	$\frac{4^n}{6} - \frac{(-2)^n}{6}$
circle	$(\sum_{i=0}^{n-2} x_i x_{i+1}) + x_{n-1}x_0$	1	$2\sigma_{n-1} + 8\sigma_{n-2}$	$\frac{4^n}{2} - \frac{(-2)^n}{2}$
complete	$\sum_{i<j, 1 \le j < n} x_i x_j$	0	$10\sigma_{n-1} - 20\sigma_{n-2}$ $-40\sigma_{n-3} + 96\sigma_{n-4}$	$\frac{6^n}{4} - \frac{4^n}{2} + \frac{2^n}{2} - \frac{(-2)^n}{4}$
star	$x_0(x_1 + x_2 + \ldots + x_{n-1})$	0	$12\sigma_{n-1} - 44\sigma_{n-2}$ $+48\sigma_{n-3}$	$2^n - \frac{4^n}{2} + \frac{6^n}{6}$

Conjecture 6. *The maximum $\mathcal{MMF}$ is always obtained by the path graph.*

It is expected that a much larger class of Boolean functions which generalises the path graph, as described in [40], will generate a large set of multivariate sequences with maximal or near-maximal $\mathcal{MMF}$. This set can also be seen as arising from the union of certain *Golay complementary sets* of length 2^n [18, 47] and satisfies a tight upper-bound on the *peak-to-average power ratio* of the spectra with respect to $\{V\}^n$ - for this reason the sequences should have high $\mathcal{MMF}$.

4   Towards a Quantum Merit Factor

Section 3 has established that $\mathcal{MMF}$ quantifies a spectral property with respect to the infinite set of $2^n \times 2^n$ local unitary transforms, $\{V\}^n$. In contrast, one quantifies the spectral properties of a *pure quantum state* of n qubits with respect to the infinite set of all possible $2^n \times 2^n$ local unitary transforms, $\mathcal{D}^n\{U\}^n$, (see (1) and (2)), where $\{V\} \subset \{U\}$ [48, 21, 32, 25]. This leads to the idea of a *quantum merit factor* ($\mathcal{QMF}$), derived from a *quantum sum-of-squares* metric.

Lemma 2. *Let T and T' be two $2^n \times 2^n$ matrices such that $T' \simeq T$ (see (3)). Let $S = Ts$ and $S' = T's$. Then,*

$$\sum_{\mathbf{k}} |S'_{\mathbf{k}}|^v = \sum_{\mathbf{k}} |S_{\mathbf{k}}|^v, \qquad v \geq 0.$$

We wish to compute $\mathcal{QMF}$ by summing the fourth powers of spectral magnitudes with respect to $\mathcal{D}^n\{U\}^n$ but it follows from Lemma 2 that we need only sum over the spectra with respect to $\{U\}^n$ to compute $\mathcal{QMF}$. Symbolically, for $n = 1$, we view this as summarising the fourth powers over the complete sphere (otherwise known as the *Bloch Sphere* [48]):

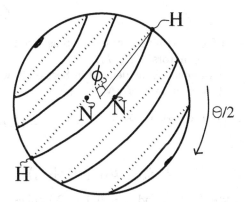

Fig. 2. The Bloch Sphere with points on the sphere described by $U(\theta, \phi)$ (see (1))

where H and N are indicated on the 'equator' of each sphere. For $n > 1$ this becomes a summation over the joint n-sphere:

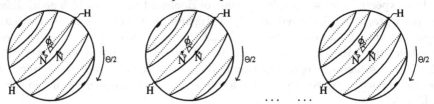

$\mathcal{MMF}$ only quantifies merit factor with respect to the equators of the n-sphere. In section 5 we characterise and investigate the *Clifford merit factor* ($\mathcal{CMF}$), which we show quantifies merit factor with respect to three pole/equator pairs. We then generalise to show that $\mathcal{CMF}$ quantifies merit factor with respect to the complete n-sphere, and is therefore equal to $\mathcal{QMF}$.

5   Clifford Merit Factor

Definition 9. *For F_α, F'_α as defined in (4), let $\{V^I\}$ be an infinite multi-set of transforms, where $\frac{1}{3}$ of all elements in $\{V^I\}$ are the 2×2 identity, I, and where,*

$$\{V^I\} = \{\{I, F_\alpha, F'_\alpha\} \mid \forall \alpha \in \mathcal{C}, |\alpha| = 1\}.$$

Note that $|\{V^I\}| = \frac{3}{2}|\{V\}|$. Just as it is sufficient to compute merit factor with respect to $\{V\}^n$ by summing fourth powers of spectral magnitudes with respect to $\{H, N\}^n$ so, for the merit factor with respect to $\{V^I\}^n$, it is sufficient to sum up fourth powers of spectral magnitudes with respect to $\{I, H, N\}^n$. We represent this transform set visually as,

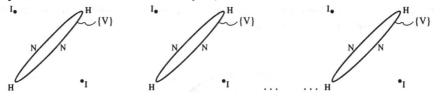

which evaluates a merit factor with respect to any tensor combination of discrete 'poles', I, and continuous 'equators', $\{H, N\}$.

For the multivariate sequence, s, as defined in Section 3, we evaluate the set of 3^n spectra, $\{S\}$, with respect to $\{I, H, N\}^n$, where

$$S = \{S^{\mathbf{r}}\} = \{S^{00\ldots0}, S^{00\ldots1}, S^{00\ldots2} \ldots, S^{22\ldots2}\} = \{I, H, N\}^n s,$$

and

$$S^{\mathbf{r}} = \{S^{\mathbf{r}}_{\mathbf{k}}\} = (S^{\mathbf{r}}_{00\ldots0}, S^{\mathbf{r}}_{00\ldots1} \cdots S^{\mathbf{r}}_{11\ldots1})^T,$$

where $\mathbf{r} \in \{0, 1, 2\}^n$, $\mathbf{k} \in \{0, 1\}^n$, and $r_i = 0, 1$ or 2 implies I, H or N, respectively, in tensor position i. $\{S^{\mathbf{r}}_{\mathbf{k}}\}$ is a set of 6^n spectral points.

Definition 10. *The* Clifford sum-of-squares, $\mathcal{E}$, *of the sequence, s, is given by,*

$$\mathcal{E} = \frac{1}{2}((\sum_{\substack{\mathbf{k} \,\in\, \{0,1\}^n \\ \mathbf{r} \,\in\, \{0,1,2\}^n}} |S_\mathbf{k}^\mathbf{r}|^4) - 6^n). \qquad (24)$$

Definition 11. *The* Clifford merit factor, $\mathcal{CMF}$, *of the sequence, s, is given by,*

$$\mathcal{CMF} = \frac{6^n}{2\mathcal{E}}. \qquad (25)$$

Definition 12. *Let s_A be a length 2^n multivariate sequence generated by construction $\mathcal{A}$. Then the* asymptotic Clifford merit factor *of s_A is given by,*

$$\mathcal{F}^C = \lim_{n \to \infty} \mathcal{CMF}(s_A).$$

Let $\mathbf{b}, \mathbf{e} \in \mathcal{Z}_2^n$. Let wt($\mathbf{b}$) be the binary weight of vector $\mathbf{b}$. Let $p(\mathbf{x}_{\mathbf{b},\mathbf{e}})$: $\mathcal{Z}_2^{n-\text{wt}(\mathbf{e})} \to \mathcal{Z}_2$, be the restriction of p to $n - \text{wt}(\mathbf{e})$ variables, where $x_i = b_i$ if $e_i = 1$, where $\mathbf{b} \preceq \mathbf{e}$, and '$\preceq$' means that $b_i \le e_i, \forall i$.

Define $s_{\mathbf{e},\mathbf{b}} = (-1)^{p(\mathbf{x}_{\mathbf{b},\mathbf{e}})}$, where $s_{\mathbf{j},\mathbf{e},\mathbf{b}} = (-1)^{p(\mathbf{x}_{\mathbf{b},\mathbf{e}}=\mathbf{j})}$, $\mathbf{j} \in \mathcal{Z}_2^{n-\text{wt}(\mathbf{e})}$, $s_{\mathbf{j},\mathbf{e},\mathbf{b}} = 0$ otherwise. The *fixed-aperiodic autocorrelation* [36] of s is given by,

$$u_{\mathbf{k},\mathbf{b},\mathbf{e}} = \sum_{\mathbf{j} \in \{0,1\}^{n-\text{wt}(\mathbf{e})}} s_{\mathbf{j},\mathbf{b},\mathbf{e}} s_{\mathbf{j}+\mathbf{k},\mathbf{b},\mathbf{e}}^*, \quad \mathbf{k} \in \{-1,0,1\}^{n-\text{wt}(\mathbf{e})}. \qquad (26)$$

An alternative to Definition 10 for $\mathcal{E}$ is given by Definition 13.

Definition 13. *The* Clifford sum-of-squares, $\mathcal{E}$, *of the sequence, s, is given by,*

$$\begin{aligned}
\mathcal{E} &= \tfrac{1}{2}((\sum_{\mathbf{e} \in \{0,1\}^n} \sum_{\mathbf{b} \in \{0,1\}^n, \mathbf{b} \preceq \mathbf{e}} \sum_{\mathbf{k} \in \{-1,0,1\}^{n-\text{wt}(\mathbf{e})}} |u_{\mathbf{k},\mathbf{b},\mathbf{e}}|^2) - 6^n) \\
&= \tfrac{1}{2} \sum_{\mathbf{e} \in \{0,1\}^n} \sum_{\mathbf{b} \in \{0,1\}^n, \mathbf{b} \preceq \mathbf{e}} \sum_{\substack{\mathbf{k} \in \{-1,0,1\}^{n-\text{wt}(\mathbf{e})} \\ \mathbf{k} \ne \{0\}^{n-\text{wt}(\mathbf{e})}}} |u_{\mathbf{k},\mathbf{b},\mathbf{e}}|^2.
\end{aligned} \qquad (27)$$

We refer to these metrics as "Clifford" because the unitary matrix set, $\{I, H, N\}$, generates the *Local Clifford Group* [34, 24, 49]. This means that $\{I, H, N\}$ stabilize the *Pauli matrices*, I, $\left(\begin{smallmatrix} 0 & 1 \\ 1 & 0 \end{smallmatrix}\right)$, $\left(\begin{smallmatrix} 1 & 0 \\ 0 & -1 \end{smallmatrix}\right)$, and $i\left(\begin{smallmatrix} 0 & -1 \\ 1 & 0 \end{smallmatrix}\right)$.

$\mathcal{CMF}$ Symmetries

If $|S_\mathbf{k}^\mathbf{r}| = |S_\mathbf{j}^\mathbf{r}|, \forall \mathbf{j}, \mathbf{k} \in \mathcal{Z}_2^n$, then we call $S^\mathbf{r}$ a *flat* spectra. In such a case we express $S^\mathbf{r}$ as

$$S^\mathbf{r} = \omega^{4p^\mathbf{r}(\mathbf{x}) + a(\mathbf{x})},$$

where ω is a complex root of one of order 8 and $p^\mathbf{r}(\mathbf{x}) : \mathcal{Z}_2^n \to \mathcal{Z}_2$ is a Boolean function. Let $s^\mathbf{r} = (-1)^{p^\mathbf{r}(\mathbf{x})}$.

Definition 14. *Define the* IHN-orbit, s_{orb}, *of s, by*

$$s_{orb} = \{s^\mathbf{r} \mid \forall \mathbf{r} \text{ such that } S^\mathbf{r} \text{ is flat and } \deg(a(\mathbf{x})) \le 1\}.$$

Let s' and $p'(\mathbf{x})$ be as defined in Lemma 1. Then,

Lemma 3. *For* $s^{\mathbf{r}} \in s_{orb}$,

$$\mathcal{CMF}(s'^{\mathbf{r}}) = \mathcal{CMF}(s).$$

The IHN-orbit is largest in size for $p(\mathbf{x})$ quadratic where the symmetry reduces to a graphical symmetry called *local complementation* [50, 51, 37], also referred to as *vertex-neighbour-complementation* [30].

5.1   Transform $\Leftrightarrow$ Autocorrelation Duality

From (24) and (27),

$$\sum_{\mathbf{e}\in\{0,1\}^n} \sum_{\mathbf{b}\in\{0,1\}^n, \mathbf{b}\preceq\mathbf{e}} \sum_{\mathbf{k}\in\{-1,0,1\}^{n-\text{wt}(\mathbf{e})}} |u_{\mathbf{k},\mathbf{b},\mathbf{e}}|^2 = \sum_{\substack{\mathbf{k}\in\{0,1\}^n \\ \mathbf{r}\in\{0,1,2\}^n}} |S_{\mathbf{k}}^{\mathbf{r}}|^4. \qquad (28)$$

Proof. The autocorrelation of (26) is the union of a set of multivariate aperiodic autocorrelations where, for fixed $\mathbf{e}$ and $\mathbf{b}$, each such autocorrelation is of the form of (15) and is computed over $n - \text{wt}(\mathbf{e})$ variables, after having fixed $\text{wt}(\mathbf{e})$ variables, x_i, to b_i, if $e_i = 1$. This fixing is mirrored in the spectral domain by assigning $r_i = 0$ iff $e_i = 1$. In other words, matrix I occurs in the ith tensor position of the transform $T \in \{I, H, N\}^n$ iff $e_i = 1$, where the first and second rows of I reflect $x_i = b_i = 0$ and $x_i = b_i = 1$, respectively. (28) follows by summing instances of (20) for each choice of $\mathbf{e}$ and $\mathbf{b}$. $\qquad \square$

5.2   Clifford Merit Factor Is Quantum Merit Factor

Definition 15. *The* normalised quantum sum-of-squares *with respect to the transform set,* $\{A\}^n$, *is given by,*

$$\mathcal{E}_{\{A\}^n} = \frac{3^n}{2}\left(\frac{||S||^4_{\{A\}^n}}{|\{A\}|^n} - 2^n\right), \qquad (29)$$

where $||S||^4_{\{A\}^n}$ *is the sum of the fourth powers of the spectral magnitudes with respect to the transform set* $\{A\}^n$.

We recover definition 10 from (29) by assigning $\{A\} = \{I, H, N\}$. For $\{A\} = \{V^I\}$ we obtain $\mathcal{E}_{\{V^I\}^n} = 2^{n-1}\left(\frac{||S||^4_{\{V^I\}^n}}{|\{V\}|^n} - 1\right)$, by substituting $|\{V^I\}| = \frac{3|\{V\}|}{2}$.

Lemma 4.

$$\mathcal{E}_{\{I,H,N\}^n Z} = \mathcal{E}_{\{V^I\}^n Z}, \qquad \forall Z \in \mathcal{D}^n\{U\}^n.$$

Recalling section 1.1, (3), (4), and definition 9,

Definition 16. *Define* $\{W^N\} \simeq \{V^I\}N$, *and* $\{X^H\} \simeq \{W^N\}N$.

$\{V^I\}, \{W^N\}$, and $\{X^H\}$ describe the following pole/equator pairs, respectively:

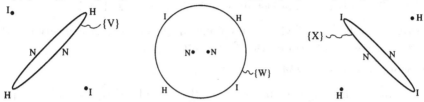

Proposition 1. $\mathcal{E}_{\{A\}^n} = \mathcal{E}$ *for* $\{A\} = \{\{V^I\}, \{W^N\}, \{X^H\}\}$, *and* $\{A\} = \{\{V\}, \{W\}, \{X\}\}$, *which we visualise as:*

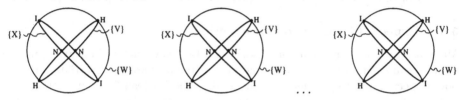

Proof. We know that $\mathcal{E}_{\{V^I\}} = \mathcal{E}$. As $\{W^N\} \simeq \{V^I\}N$ it follows, from Lemma 4, that $\mathcal{E}_{\{W^N\}} = \mathcal{E}_{\{I,H,N\}N} = \mathcal{E}_{\{N,I,H\}} = \mathcal{E}$. Likewise, as $\{X^H\} \simeq \{V^I\}N^2$ it follows, from Lemma 4, that $\mathcal{E}_{\{X^H\}} = \mathcal{E}_{\{I,H,N\}N^2} = \mathcal{E}_{\{H,N,I\}} = \mathcal{E}$. The simplification to $\{\{V\}, \{W\}, \{X\}\}$ occurs because we can remove surplus element triples, $\{I, H, N\}$, from $\{\{V^I\}, \{W^N\}, \{X^H\}\}$ without changing the normalised spectral sum. The argument extends to any tensor combination of the three pole/equator pairs when $n > 1$. □

Let $F_{\tilde{\alpha}} = F_{\alpha_0} \otimes F_{\alpha_1} \otimes \ldots \otimes F_{\alpha_{n-1}}$, where F_α was defined in (4). We now generalise $\{A\}$ in proposition 1 by assigning $\{A\} = \{V^I\}^n \{I^n, F_{\tilde{\alpha}}, F_{\tilde{\alpha}} F_{\tilde{\beta}}\}$:

Theorem 4.

$$\mathcal{E}_{\{V^I\}^n \{I^n, F_{\tilde{\alpha}}, F_{\tilde{\alpha}} F_{\tilde{\beta}}\}} = \mathcal{E}, \qquad \forall F_{\tilde{\alpha}}, F_{\tilde{\beta}} \in \{V\}^n.$$

Proof. We see that $\{I, H, N\}F_\alpha \simeq \{F_\alpha, I, F'_\alpha\}$, and we already know that $\mathcal{E}_{\{F_\alpha, I, F'_\alpha\}} = \mathcal{E}$. Therefore, from Lemma 4, $\mathcal{E}_{\{V^I\}F_\alpha} = \mathcal{E}$. Using Lemma 4 repeatedly, $\mathcal{E}_{\{V^I\}F_\alpha F_\beta} = \mathcal{E}_{\{I,H,N\}F_\alpha F_\beta} = \mathcal{E}_{\{F_\alpha,I,F'_\alpha\}F_\beta} = \mathcal{E}_{\{V^I\}F_\beta} = \mathcal{E}$. The argument extends to any tensor combination when $n > 1$. □

Lemma 5.

$$\{U\} \simeq \{V\}\{V\}.$$

Proof. $F_\alpha F_\beta = \frac{(1+\alpha)}{2} \begin{pmatrix} 1 & \frac{(1-\alpha)}{(1+\alpha)}\beta \\ \frac{(1-\alpha)}{(1+\alpha)} & \beta \end{pmatrix} = \mu \begin{pmatrix} 1 & 0 \\ 0 & -i \end{pmatrix} \frac{(1+\alpha)}{2\mu} \begin{pmatrix} 1 & \frac{(1-\alpha)}{(1+\alpha)}\beta \\ i\frac{(1-\alpha)}{(1+\alpha)} & i\beta \end{pmatrix},$

$\forall F_\alpha, F_\beta \in \{V\}$, where $\mu = \sqrt{\alpha}$. The lemma follows by assigning $\beta = e^{i\phi}$, $\cos\theta = \frac{(1+\alpha)}{2\mu}$, and $\sin\theta = i\frac{(1-\alpha)}{2\mu}$. □

Theorem 5.

$$\mathcal{CMF}(Ts) = \mathcal{CMF}(s), \qquad \forall T \in \mathcal{D}^n\{U\}^n.$$

Proof. Follows directly from Definition 11, and lemmas 2, 4 and 5.   □

Theorem 5 implies that $\mathcal{CMF}$ is an entanglement metric as it is invariant with respect to local unitary transform of the state, s [25].

Theorem 6. $\mathcal{CMF}$ *is* $\mathcal{QMF}$.

Proof. From figure 2, and lemma 5, we see that α and β specify θ and ϕ, respectively. Over all $\alpha, \beta \in \mathcal{C}$, and $d \in \mathcal{D}$, we have, from theorem 5 that, for $n = 1$, $\mathcal{E}_{\mathcal{D}\{U\}} = \mathcal{E}_{I,H,N} = \mathcal{E}$ is invariant, where each spectral point is counted the same number of times. The argument extends to tensor products when $n > 1$.   □

5.3   Expected Values and Constructions

Maximising Clifford merit factor ($\mathcal{CMF}$) of a Boolean function indicates a minimum mean-square deviation from the joint flat continuous multivariate fourier power spectrum of the sequences associated to the Boolean function and all its subspace fixings. Table 4 shows equivalence classes for Boolean functions of $n = 2$ to 5 variables, where sets of inequivalent functions are obtained from [46].

Table 4. complete set of Clifford merit factors for $n = 2$ to $n = 5$

n	# inequivalent functions	# equivalence classes with list of $\mathcal{CMF}$s
2	2	2 classes **3.0**, 1.286
3	4	4 classes **2.077**, 1.421, 1.286, 0.730
4	34	18 classes **1.723**, 1.588, 1.473, 1.446, 1.373, 1.286, 1.266, 1.209, 1.141, 1.125, 1.080, 1.025, 0.976, 0.920, 0.786, 0.730, 0.675, 0.463
5	22050	193 classes **1.723** − 0.311

Experimental results suggest that, for a random Boolean function of n variables, $\mathcal{F}^C = 1.0$, as indicated here by the random samplings for (from left to right) $n = 4, 6, 9$ and 10, where $\mathcal{CMF}$ and #sequences are x and y-axes, respectively, with x-axes ranging from $\mathcal{CMF} = 0$ to 4, and where the highest peak approaches $\mathcal{F}^C = 1.0$ as n increases.

In comparison, a sampling of just the quadratic Boolean functions for $n = 4, 6, 9$ and 10 indicates a wider range of $\mathcal{CMF}$s for a given n than for the full space of Boolean functions although, once again, the highest peak appears to approach $\mathcal{F}^C = 1.0$ as n gets large. Once again, the x-axes range linearly from $\mathcal{CMF} = 0$ to 4, and the y-axes indicate #sequences.

 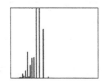 

Conjecture 7. *A random Boolean function satisfies,*

$$\mathcal{F}^C = 1.0.$$

Theorem 7. *The average value of $\frac{1}{\mathcal{CMF}}$ with respect to any set $\mathcal{S}_Q$ (see definition 8) is,*

$$average \ _{\mathcal{S}_Q}\left(\frac{1}{\mathcal{CMF}}\right) = \frac{3^n - 2^n}{3^n}.$$

Proof. It follows directly from Theorem 3, by summing up the multivariate sum-of-squares over every fixed-subspace of $\mathcal{S}_Q$. Each member of each coset of Q is represented the same number of times over each subspace.   □

Corollary 2. *The set of n-variable Boolean functions of degree d or less satisfies,*

$$average \ \left(\frac{1}{\mathcal{CMF}}\right) = \frac{3^n - 2^n}{3^n}$$

for any d, $2 \leq d \leq n$, and, consequently, average $\left(\frac{1}{\mathcal{CMF}}\right) \to 1.0$ as $n \to \infty$.

Table 5 summarises constructions described by Boolean functions, $p(\mathbf{x})$, where $s = (-1)^{p(\mathbf{x})}$. The associated recursions originate from [52]. The constructions represent a larger class of $\mathcal{CMF}$-invariant sequences, as generated by Lemma 3, and the recursions have all been proven using the results of [39]. The star and complete graph are in the same IHN-orbit. None of the constructions in Table 5 satisfy a non-vanishing value for $\mathcal{F}^C$. We appear to obtain maximum values of $\mathcal{CMF}$ for s constructed from quadratic Boolean functions which describe optimal QECCs [33, 30, 53, 46, 36, 39, 37]. Table 6 shows maximal values of $\mathcal{CMF}$ for $n = 2$ to 5, and highest found values of $\mathcal{CMF}$ for $n = 6$ to 9, and all represent QECCs with optimal distance. The associated QECC is obtained from an additive $[n, 2^n, \text{distance}]$ code over GF(4) where the associated generator matrix, G, satisfies $G = \omega \mathcal{I} + \Gamma$, where Γ is the adjacency matrix of the graph associated with the quadratic Boolean function, $p(\mathbf{x})$, $\mathcal{I}$ is the $n \times n$ identity matrix, and

Table 5. $\mathcal{F}^C$ for certain graphical constructions

graph	$\mathcal{F}^C(s)$	$\mathcal{E}_n$ - recursion	$\mathcal{E}_n$ - closed form
path	0	$10\mathcal{E}_{n-1} - 8\mathcal{E}_{n-2} - 96\mathcal{E}_{n-3}$	$\frac{6^n}{2} + (\frac{5-3\sqrt{5}}{20})(2-2\sqrt{5})^n$ $+(\frac{5+3\sqrt{5}}{20})(2+2\sqrt{5})^n$
circle	0	$14\mathcal{E}_{n-1} - 48\mathcal{E}_{n-2} - 64\mathcal{E}_{n-3} + 38\mathcal{E}_{n-4}$	$\frac{4^n}{2} + \frac{(2-2\sqrt{5})^n}{2} + \frac{(2+2\sqrt{5})^n}{2} - \frac{6^n}{2}$
complete $\equiv$ star	0	$18\mathcal{E}_{n-1} - 104\mathcal{E}_{n-2} - 192\mathcal{E}_{n-3}$	$\frac{8^n}{4} - \frac{6^n}{2} + \frac{4^n}{2}$

Table 6. Boolean functions with maximal $\mathcal{CMF}$ for $n = 2$ to 5 and large (possibly maximal) $\mathcal{CMF}$ for $n = 6$ to 9 with their associated QECC distances

n	$p(\mathbf{x})$	$\mathcal{CMF}(s)$	QECC distance
2	01	**3.0**	2
3	$01, 02$	**2.08**	2
4	$01, 12, 23$	**1.72**	2
5	$01, 02, 13, 24, 34$	**1.72**	3
6	$01, 02, 03, 04, 05, 12, 23, 34, 45, 51$	1.72	4
7	$03, 06, 14, 16, 25, 26, 34, 35, 45$	1.43	3
8	$02, 03, 04, 12, 13, 15, 26, 37, 46, 47, 56, 57, 67$	1.40	4
	$05, 06, 07, 13, 15, 17, 24, 25, 27, 36, 37, 46, 47, 67$	1.40	4
9	$02, 04, 08, 13, 15, 18, 26, 28, 37, 38, 47, 48, 56, 58, 67, 68, 78$	1.30	4
	$04, 07, 08, 13, 14, 18, 24, 25, 28, 36, 37, 56, 57, 58, 67, 68$	1.30	4
	$06, 07, 08, 14, 16, 18, 25, 26, 28, 34, 35, 37, 38, 47, 48, 57, 58, 68$	1.30	4
	$04, 07, 08, 14, 16, 18, 25, 26, 28, 34, 35, 37, 57, 58, 67, 68$	1.30	4
	$01, 07, 08, 14, 18, 23, 25, 28, 36, 37, 45, 46, 57, 58, 67, 68$	1.30	4

$\omega^2 + \omega + 1 = 0$ over GF(4). The results are only exhaustive for $n = 2$ to 5. In the table, expressions of the form $ab, cd, \ldots$ are short for $x_a x_b + x_c x_d + \ldots$.

A few cubics and quartics have recently been found which equal the $\mathcal{CMF}$ values in Table 6 [54], but none have been found yet with greater $\mathcal{CMF}$.

Many high-distance QECCs are of (bordered) quadratic-residue [30]. Let l be a Legendre sequence of prime length m, where $m = 4k + 1$, as described in Section 2. Construct $p(\mathbf{x})$ over $n = m$ variables such that,

$$p(\mathbf{x}) = \sum_{j=0}^{n-1} l_j \sum_{i=0}^{n-j-1} x_i x_{i+j}.$$

For the bordered version, construct $p(\mathbf{x})$ over $n = m + 1$ variables such that,

$$p(\mathbf{x}) = \sum_{j=1}^{n-1} x_0 x_i + \sum_{j=1}^{n-1} l_j \sum_{i=1}^{n-j-1} x_i x_{i+j}.$$

Then, for both non-bordered and bordered versions, $s = (-1)^{p(\mathbf{x})}$ has a relatively high and sometimes optimal $\mathcal{CMF}$. The examples in Table 6 for $n = 5$ and 6 are equivalent, by Lemma 3, to (bordered)-quadratic residue constructions.

There is a connection with recent results in graph theory. Aigner and van der Holst have defined an *interlace polynomial*, $Q(z)$, which summarises various spectral properties of a graph [55], this being a generalisation of an interlace polynomial, $q(z)$, defined by Arratia, Bollobas, and Sorkin [56], where both polynomials are variants of *Tutte* and *Tutte-Martin* polynomials as defined by Bouchet [57]. Moreover a further interlace polynomial, $Q_{HN}(z)$ has recently been defined in [58], and it is shown there that, for sequences constructed from quadratic Boolean functions, $\sigma_n = 2^{n-1}(Q_{HN,n}(4) - 2^n)$ and $\mathcal{E}_n = 2^{n-1}(Q_n(4) - 3^n)$.

In contrast to $\mathcal{CMF}$, most entanglement measures are computationally infeasible beyond about 4 qubits. So $\mathcal{CMF}$ is a useful measure in a quantum context as it is (currently) computationally viable up to about $n = 12$ qubits. Moreover, for graph states and, in particular, recursively constructed graph states, $\mathcal{CMF}$ gives us an entanglement measure of a pure multipartite system for large n.

6   Conclusion

The univariate, multivariate, and Clifford merit factors ($\mathcal{MF}$, $\mathcal{MMF}$, and $\mathcal{CMF}$, resp.) have been reviewed. Constructions which achieve the best-known asymptotic merit factor, $\mathcal{F}$, have been described. $\mathcal{MMF}$ has been characterised for the extreme case where each dimension is of length 2. The associated multivariate sequences therefore have a natural description via Boolean functions. The average value for $\frac{1}{\mathcal{MMF}}$ was established. We presented 'graphical' constructions for which recursions in multivariate sum-of-squares exist leading, in some cases, to non-vanishing asymptotic multivariate merit factor $\mathcal{F}^{\mathcal{M}}$. We conjectured that maximal $\mathcal{F}^{\mathcal{M}}$ is satisfied by the path graph. We characterised $\mathcal{CMF}$ as a generalisation of $\mathcal{MMF}$ and proved it is invariant to local unitary transform and is, moreover, a quantum merit factor, $\mathcal{QMF}$. The average value for $\frac{1}{\mathcal{CMF}}$ was established. We presented 'graphical' constructions for which recursions in Clifford sum-of-squares exist, although all associated asymptotic Clifford merit factors, $\mathcal{F}^{\mathcal{C}}$, are zero. We demonstrated that sequences constructed from quantum error correcting codes appear to maximise $\mathcal{CMF}$.

We finish with a list of open problems suggested by this paper:

- Establish whether $\mathcal{MF} = 14.083$ is maximal over all binary sequences.
- Prove $\mathcal{F} = 6.3421\ldots$ for the relevant constructions of Section 2.
- Establish the range of $\mathcal{F}$ for the univariate sequence constructed via the path graph under all possible index permutations.
- Prove the recursions in univariate sum-of-squares for the recursive graphical constructions of Section 2.
- Prove that $\mathcal{F} = \mathcal{F}^{\mathcal{M}} = \mathcal{F}^{\mathcal{C}} = 1.0$ for a random univariate or multivariate sequence of length N, 2^n, or 2^n, respectively.
- Establish whether the maximum $\mathcal{MMF}$ and $\mathcal{CMF}$ over all multivariate binary sequences are 4.00 and 3.00, respectively for the sequence constructed from $p(\mathbf{x}) = x_0 x_1$.

- Prove that the maximal $\mathcal{MMF}$ over n variables and, therefore, the maximal $\mathcal{F}^{\mathcal{M}}$ is always obtained by the path graph.
- Find an infinite multivariate sequence construction such that $\mathcal{F}^{\mathcal{C}} > 0$.
- Prove whether, for n variables, $\mathcal{CMF}$ is always optimised by quadratic Boolean functions, or give a counter-example.

References

1. Littlewood, J.E.: Some Problems in Real and Complex Analysis. Heath Mathematical Monographs (1968)
2. Golay, M.J.E.: A class of finite binary sequences with alternate autocorrelation values equal to zero. IEEE Trans. Inform. Theory **IT-18** (1972) pp. 449–450
3. Golay, M.J.E.: The merit factor of long low autocorrelation binary sequences. IEEE Trans. Inform. Theory **28** (1982) pp. 543–549
4. Golay, M.J.E.: A new search for skewsymmetric binary sequences with optimal merit factors. IEEE Trans. Inform. Theory **36** (1990) pp. 1163–1166
5. Høholdt, T., Jensen, H.E., Justesen, J.: Aperiodic correlations and the merit factor of a class of binary sequences. IEEE Trans. Inform. Theory **31** (1985) pp. 549–552
6. Høholdt, T., Jensen, H.E.: Determination of the merit factor of Legendre sequences. IEEE Trans. Inform. Theory **IT-34** (1988) pp. 161–164
7. Høholdt, T.: The merit factor of binary sequences. In Pott, A., Kumar, P.V., Helleseth, T., Jungnickel, D., eds.: Difference Sets, Sequences and their Correlation Properties, Bad Winsheim, 2–14 August 1998. Series C: Mathematical and Physical Sciences, Kluwer Academic Publishers (1999) pp. 227–237 Long version: http://arxiv.org/quant-ph/0107106.
8. Jensen, J.M., Jensen, H.E., Høholdt, T.: The merit factor of binary sequences related to difference sets. IEEE Trans. Inform. Theory **37** (1991) pp. 617–626
9. Newman, D.J., Byrnes, J.S.: The l^4 norm of a polynomial with coefficients ± 1. Amer. Math. Monthly **97** (1990) pp. 42–45
10. Golay, M.J.E.: The merit factor of legendre sequences. IEEE Trans. Inform. Theory **29** (1983) pp. 934–936
11. Kristiansen, R.A.: On the aperiodic autocorrelation of binary sequences. Master's thesis, Selmer Centre, Inst. for Informatics, University of Bergen, Norway (2003) http://www.ii.uib.no/~matthew/Masters/notes.ps.
12. Kristiansen, R.A., Parker, M.G.: Binary sequences with merit factor > 6.3. IEEE Trans. Inform. Theory **50** (2004) pp. 3385–3389
13. Borwein, P., Choi, K.K.S., Jedwab, J.: Binary sequences with merit factor greater than 6.34. IEEE Trans. Inform. Theory **50** (2004) pp. 3234–3249
14. Jedwab, J.: A survey of the merit factor problem for binary sequences. In: Proc. of SETA04. Lecture Notes in Computer Science. Springer-Verlag (2005) this issue.
15. Ramakrishna, G.S., Mow, W.H.: A new search for optimal binary arrays with minimum peak sidelobe levels. In: Proc. of SETA04. Lecture Notes in Computer Science, Springer-Verlag (2005) This issue.
16. Zhang, X.M., Zheng, Y.: Gac - the criterion for global avalanche characteristics of cryptographic functions. J. Universal Computer Science **1** (1995) pp. 320–337
17. Gulliver, T.A., Parker, M.G.: The multi-dimensional aperiodic merit factor of binary sequences. preprint, http://www.ii.uib.no/~matthew/ISITRecursions.pdf (2003)

18. Golay, M.J.E.: Multislit spectroscopy. J. Opt. Soc. Amer. **39** (1949) pp. 437–444
19. Shapiro, H.S.: Extremal problems for polynomials. Master's thesis, M.I.T. (1951)
20. Rudin, W.: Some theorems on fourier coefficients. Proc. Amer. Math. Soc. **10** (1959)
21. Parker, M.G., Rijmen, V.: The quantum entanglement of binary and bipolar sequences. In Helleseth, T., Kumar, P.V., Yang, K., eds.: Sequences and Their Applications, SETA'01. Discrete Mathematics and Theoretical Computer Science Series, Springer (2001) Long version: http://arxiv.org/quant-ph/0107106.
22. Riera, C., Parker, M.G.: Generalised bent criteria for boolean functions (i). preprint, http://www.ii.uib.no/~matthew/LCPartIf.pdf (2004)
23. Briegel, H.J., Raussendorf, R.: Persistent entanglement in arrays of interacting particles. Physical Review Letters **86** (2001) pp. 910–913
24. Hein, M., Eisert, J., Briegel, H.J.: Multi-party entanglement in graph states. Phys. Rev. A **69** (2004) http://arxiv.org/quant-ph/0307130.
25. Verstraete, F.: A Study of Entanglement in Quantum Information Theory. PhD thesis, Dept. Elektrotechniek, Katholieke Universiteit, Leuven, Belgium (2002)
26. Parker, M.G.: Quantum factor graphs. Annals of Telecom. (2001) pp. 472–483 http://arxiv.org/quant-ph/0010043.
27. Schlingemann, D., Werner, R.F.: Quantum error-correcting codes associated with graphs. Phys. Rev. A **65** (2002) http://arxiv.org/quant-ph/0012111.
28. Glynn, D.G.: On self-dual quantum codes and graphs. Submitted to Elect. J. Combinatorics. http://homepage.mac.com/dglynn/.cv/dglynn/Public/SD-G3.pdf-link.pdf (2002)
29. Grassl, M., Klappenecker, A., Rotteler, M.: Graphs, quadratic forms, and quantum codes. In: Proc. IEEE Int. Symp. Inform. Theory. (2002) p. 45
30. Glynn, D.G., Gulliver, T.A., Maks, J.G., Gupta, M.K.: The Geometry of Additive Quantum Codes. Springer-Verlag (2004)
31. Raussendorf, R., Briegel, H.J.: A one-way quantum computer. Phys. Rev. Lett. **86** (2001) pp. 5188–5191
32. Zyczkowski, K., Bengtsson, I.: Relativity of pure states entanglement. Ann. Phys. **295** (2002)
33. Calderbank, A.R., Rains, E.M., Shor, P.M., Sloane, N.J.A.: Quantum error correction via codes over GF(4). IEEE Trans. Inform. Theory **44** (1998) pp. 1369–1387 http://arxiv.org/quant-ph/9608006.
34. Klappenecker, A., Rotteler, M.: Clifford codes. In Brylinski, R., Chen, G., eds.: Mathematics of Quantum Computation. CRC Press (2002)
35. Gottesman, D.: Stabilizer Codes and Quantum Error Correction. PhD thesis, California Institute of Technology (1997) http://arxiv.org/quant-ph/9705052.
36. Danielsen, L.E., Gulliver, T.A., Parker, M.G.: Aperiodic propagation criteria for Boolean functions. ECRYPT Internal Document, STVL-UiB-1-APC-1.0. http://www.ii.uib.no/~matthew/GenDiff4.pdf (2004)
37. Danielsen, L.E., Parker, M.G.: Spectral orbits and peak-to-average power ratio of boolean functions with respect to the $\{I, H, N\}^n$ transform. In: Proc. of SETA04. Lecture Notes in Computer Science, Springer-Verlag (2005) this issue, http://www.ii.uib.no/~matthew/seta04-parihn.pdf.
38. Grassl, M.: Bounds on d_{min} for additive $[[n, k, d]]$ QECC. Web page (2003) http://iaks-www.ira.uka.de/home/grassl/QECC/TableIII.html.
39. Riera, C., Petrides, G., Parker, M.G.: Generalised bent criteria for boolean functions (ii). preprint, http://www.ii.uib.no/~matthew/LCPartIIf.pdf (2004)

40. Parker, M.G., Tellambura, C.: A construction for binary sequence sets with low peak-to-average power ratio. Technical Report 242, Dept. of Informatics, University of Bergen, Norway (2003) http://www.ii.uib.no/publikasjoner/texrap/pdf/2003-242.pdf, update at:http://www.ii.uib.no/~matthew/Construct04.pdf.

41. Borwein, P., Choi, K.K.S.: Explicit merit factor formulae for fekete and turyn polynomials. Trans. Amer. Math. Soc. **354** (2002) pp. 219–234

42. Borwein, P., Choi, K.K.S.: Merit factors of polynomials formed by jacobi symbols. Canad. J. Math. **53** (2001) pp. 33–50

43. Parker, M.G.: Even length binary sequence families with low negaperiodic autocorrelation. In: Applied Algebra, Algebraic Algorithms and Error-Correcting Codes, AAECC-14 Proceedings. Number 2227 in Lecture Notes in Computer Science. Springer-Verlag (2001) p. 200–209

44. Kirilusha, A., Narayanaswamy, G.: Construction of new asymptotic classes of binary sequences based on existing asymptotic classes. Technical report, Dept. Math. and Comput. Science, Univ. of Richmond (1999) http://www.mathcs.richmond.edu/~jad/summer.html.

45. Davis, J.A., Jedwab, J.: Peak-to-mean power control in OFDM, Golay complementary sequences and Reed-Muller codes. IEEE Trans. Inform. Theory **45** (1999) pp. 2397–2417

46. Danielsen, L.E.: Database of self-dual quantum codes. Web page (2004) http://www.ii.uib.no/~larsed/vncorbits/.

47. Golay, M.J.E.: Complementary series. IRE Trans. Inform. Theory **IT-7** (1961) pp. 82–87

48. Nielsen, M., Chuang, I.: Quantum Computation and Quantum Information. Cambridge University Press (2000)

49. Van den Nest, M., Dehaene, J., De Moor, B.: Graphical description of the action of local Clifford transformations on graph states. Phys. Rev. A **69** (2004) http://arxiv.org/quant-ph/0308151.

50. Bouchet, A.: Isotropic systems. European J. Combin. **8** (1987) pp. 231–244

51. Bouchet, A.: Recognizing locally equivalent graphs. Discrete Math. **114** (1993) pp. 75–86

52. Storøy, D.: Master's thesis - in preparation. Selmer Centre, Inst. for Informatics, University of Bergen, Bergen, Norway (2005)

53. Gulliver, T.A., Kim, J.L.: Circulant based extremal additive self-dual codes over GF(4). IEEE Trans. Inform. Theory **50** (2004) pp. 359–366

54. Danielsen, L.E.: Master's thesis - in preparation. Selmer Centre, Inst. for Informatics, University of Bergen, Bergen, Norway (2005)

55. Aigner, M., van der Holst, H.: Interlace polynomials. Linear Algebra and its Applications **377** (2004) pp. 11–30

56. Arratia, R., Bollobas, B., Sorkin, G.B.: The interlace polynomial of a graph. J. Combin. Theory Ser. B **92** (2004) pp. 199–233 http://arxiv.org/abs/math/0209045.

57. Bouchet, A.: Tutte-martin polynomials and orienting vectors of isotropic systems. Graphs Combin. **7** (1991) pp. 235–252

58. Riera, C., Parker, M.G.: Spectral interpretations of the interlace polynomial. preprint, http://www.ii.uib.no/~matthew/WCC4.pdf (2004)

Discrete Fourier Transform, Joint Linear Complexity and Generalized Joint Linear Complexity of Multisequences

Wilfried Meidl

Johann Radon Institute for Computational and Applied Mathematics,
Austrian Academy of Sciences, Altenbergerstrasse 69, A-4040 Linz, Austria
and
Temasek Laboratories, National University of Singapore,
5 Sports Drive 2, Singapore 117508
wilfried.meidl@oeaw.ac.at

Abstract. Let $S_1, S_2, \ldots, S_t$ be t N-periodic sequences over $\mathbb{F}_q$. The *joint linear complexity* $L(S_1, S_2, \ldots, S_t)$ is the least order of a linear recurrence relation that $S_1, S_2, \ldots, S_t$ satisfy simultaneously. Since the $\mathbb{F}_q$-linear spaces $\mathbb{F}_q^t$ and $\mathbb{F}_{q^t}$ are isomorphic, a multisequence can also be identified with a single sequence $\mathcal{S}$ having its terms in the extension field $\mathbb{F}_{q^t}$. The linear complexity $L(\mathcal{S})$ of $\mathcal{S}$, i.e. the length of the shortest recurrence relation with coefficients in $\mathbb{F}_{q^t}$ that $\mathcal{S}$ satisfies, may be significantly smaller than $L(S_1, S_2, \ldots, S_t)$. We investigate relations between $L(\mathcal{S})$ and $L(S_1, S_2, \ldots, S_t)$, in particular we establish lower bounds on $L(\mathcal{S})$ expressed in terms of $L(S_1, S_2, \ldots, S_t)$.

Keywords: Multisequences; Linear complexity; Generalized discrete Fourier transform; Stream ciphers.

1 Introduction

Let $S = s_0, s_1, \ldots$ be a sequence with terms in the finite field $\mathbb{F}_q$ (or shortly over $\mathbb{F}_q$). If for a positive integer N the terms of S satisfy $s_{i+N} = s_i$ for all $i \geq 0$, then we say that S is N-periodic. The *linear complexity* $L(S)$ of the N-periodic sequence S is the smallest nonnegative integer c for which there exist coefficients $d_1, d_2, \ldots, d_c \in \mathbb{F}_q$ such that

$$s_j + d_1 s_{j-1} + \cdots + d_c s_{j-c} = 0 \quad \text{for all } j \geq c. \tag{1}$$

Trivially, the linear complexity of an N-periodic sequence can be at most N.

The concept of the linear complexity is very useful in the study of the security of stream ciphers (cf. [10], [11]). Recent developments point towards an interest in word-based stream ciphers. The theory of such stream ciphers requires the study of parallel streams of finitely many sequences, i.e. of multisequences. For previous work on the synthesis of multisequences we refer to [2–Appendix A], [9], [12], [13], [14] and [15].

T. Helleseth et al. (Eds.): SETA 2004, LNCS 3486, pp. 101–112, 2005.
© Springer-Verlag Berlin Heidelberg 2005

Consider t periodic sequences $S_1, S_2, \ldots, S_t$ over $\mathbb{F}_q$. Without loss of generality we assume that they have common period N. The *joint linear complexity* $L(S_1, S_2, \ldots, S_t)$ is the least order of a linear recurrence relation that $S_1, S_2, \ldots, S_t$ satisfy simultaneously.

Since the $\mathbb{F}_q$-linear spaces $\mathbb{F}_q^t$ and $\mathbb{F}_{q^t}$ are isomorphic, a multisequence can also be identified with a single sequence $\mathcal{S}$ having its terms in the extension field $\mathbb{F}_{q^t}$. If $s_i^{(r)} \in \mathbb{F}_q$ denotes the ith term of the rth sequence S_r and $\{\beta_1, \beta_2, \ldots, \beta_t\}$ is a basis of $\mathbb{F}_{q^t}$ over $\mathbb{F}_q$, then the ith term of $\mathcal{S}$ is given by $\sigma_i = \sum_{r=1}^t \beta_r s_i^{(r)}$.

In some cases the linear complexity of $\mathcal{S}$ is significantly smaller than the joint linear complexity of the t corresponding sequences over $\mathbb{F}_q$, which is clearly not desirable for vectorized stream ciphers.

Example. The binary 7-periodic sequences $S_1 = 1, 0, 1, 0, 1, 0, 1, \ldots$, $S_2 = 0, 1, 0, 0, 0, 0, 0, \ldots$, $S_3 = 0, 1, 1, 1, 0, 0, 0, \ldots$ have the maximal possible joint linear complexity 7. The corresponding sequence $\mathcal{S} = 1, \alpha^2 + \alpha, \alpha^2 + 1, \alpha^2, 1, 0, 1, \ldots$ over $\mathbb{F}_2/(x^3 + x + 1) \simeq \mathbb{F}_8$ with basis $1, \alpha, \alpha^2$ over $\mathbb{F}_2$, where α is a root of $x^3 + x + 1$, satisfies the recurrence relation $s_j + (\alpha^2 + \alpha)s_{j-1} + (\alpha^2 + \alpha + 1)s_{j-3} = 0$ for all $j \geq 3$. In fact, the linear complexity of $\mathcal{S}$ is 3.

The joint linear complexity of t N-periodic sequences over $\mathbb{F}_q$ can also be interpreted as the $\mathbb{F}_q$-linear complexity of the corresponding N-periodic sequence $\mathcal{S}$ over $\mathbb{F}_{q^t}$, which is the least order of a linear recurrence relation in $\mathbb{F}_q$ that $\mathcal{S}$ satisfies (cf. [1–pp. 27], [2–pp. 83–85]). Thus all results can also be seen as results on the relationship between the usual linear complexity and the $\mathbb{F}_q$-linear complexity of a sequence $\mathcal{S}$ over $\mathbb{F}_{q^t}$. Having the applications in vectorized stream ciphers in mind we will formulate most results in terms of the joint linear complexity of t parallel sequences. Following the notation in [1], where the $\mathbb{F}_q$-linear complexity of a sequence $\mathcal{S}$ over $\mathbb{F}_{q^t}$ has been called the generalized linear complexity of $\mathcal{S}$ we will call the linear complexity $L(\mathcal{S})$ of the sequence $\mathcal{S}$ over $\mathbb{F}_{q^t}$ associated with the multisequence $S_1, S_2, \ldots, S_t$ the *generalized joint linear complexity* of the multisequence. Strictly speaking we should call it the generalized joint linear complexity with respect to the given basis $\{\beta_1, \beta_2, \ldots, \beta_t\}$, since we obtain a different sequence $\bar{\mathcal{S}}$ over $\mathbb{F}_{q^t}$ if we choose a different basis $\{\bar{\beta}_1, \bar{\beta}_2, \ldots, \bar{\beta}_t\}$ and the linear complexities of the sequences $\mathcal{S}$ and $\bar{\mathcal{S}}$ may be different. But this distinction does not play a role for the results of this article.

After collecting some preliminary results in Section 2 we will establish a sharp lower bound on the generalized joint linear complexity of t N-periodic sequences $S_1, S_2, \ldots, S_t$ over $\mathbb{F}_q$ expressed in terms of the joint linear complexity of $S_1, S_2, \ldots, S_t$. This lower bound estimates the possible decrease of the linear complexity if the calculations are shifted to a proper extension field. We also get conditions under which the decrease of the linear complexity will always be small or always zero. We will provide an exact formula for number of t-tuples of N-periodic sequences $S_1, S_2, \ldots, S_t$ over $\mathbb{F}_q$ with joint linear complexity N and minimal possible generalized joint linear complexity, and an exact formula for the expected value of the generalized joint linear complexity of a multisequence over $\mathbb{F}_q$ with joint linear complexity N. As a tool we choose interesting relations between linear complexity and certain discrete Fourier transforms. The results are a further contribution to the theory of linear complexity and they are also of some practical interest for the security of vectorized stream ciphers.

2   Preliminaries

The coefficients of the linear recurrence relation (1) for the sequence S give rise to the *minimal feedback polynomial* $f(x) = 1 + d_1 x + \cdots + d_c x^c \in \mathbb{F}_q[x]$. Thus, the linear complexity $L(S)$ is equal to the degree of $f(x)$. For an N-periodic sequence $S = s_0, s_1, \ldots$ the minimal feedback polynomial $f(x)$ is, up to a nonzero multiplicative constant, given by

$$f(x) = \frac{x^N - 1}{\gcd(x^N - 1, S^N(x))}, \tag{2}$$

where $S^N(x) := s_0 + s_1 x + \cdots + s_{N-1} x^{N-1}$. We remark that if S is a sequence over $\mathbb{F}_{q^t}$ then we can take the greatest common divisor over $\mathbb{F}_q[x]$ in order to obtain a feedback polynomial and thus a recurrence relation with coefficients in $\mathbb{F}_q$. The degree of $f(x)$ is then the $\mathbb{F}_q$-linear complexity of S. For further background on linear recurrence sequences we refer to [5–Chapter 8].

If $N = p^v n$ with $\gcd(n, p) = 1$, where p is the characteristic of $\mathbb{F}_q$ (we will use the notation $p = char\mathbb{F}_q$), and α is a primitive nth root of unity in some extension field of $\mathbb{F}_q$, then it follows from (2) that the degree of $f(x)$ is given by $N - \sum_{i=0}^{n-1} \nu_i$, where ν_i is the minimum of p^v and the multiplicity of α^i as a root of $S^N(x)$. This correspondence is summarized in the Günther-Blahut Theorem (cf. [6]). To present the Günther-Blahut Theorem, we need the following definition.

Definition 1. *The Günther weight of a matrix of t-tuples is the number of its t-tuples that are nonzero or that lie below a nonzero t-tuple.*

For $t = 1$, the Günther weight has been defined in [6]. If the matrix consists of one row, then the Günther weight is identical with the Hamming weight.

Theorem 1 (Günther–Blahut Theorem). *Let S be an N-periodic sequence with terms in the finite field $\mathbb{F}_q$ of characteristic p, $N = p^v n$, $\gcd(n, p) = 1$. Then the linear complexity of S is the Günther weight of the $p^v \times n$ matrix*

$$\begin{pmatrix} S^N(1) & S^N(\alpha) & \cdots & S^N(\alpha^{n-1}) \\ (S^N)^{[1]}(1) & (S^N)^{[1]}(\alpha) & \cdots & (S_N)^{[1]}(\alpha^{n-1}) \\ \vdots & & & \\ (S^N)^{[p^v-1]}(1) & (S^N)^{[p^v-1]}(\alpha) & \cdots & (S^N)^{[p^v-1]}(\alpha^{n-1}) \end{pmatrix}, \tag{3}$$

where $S^N(x)$ is the polynomial corresponding to the sequence S, $(S^N)^{[k]}(x) = \sum_{i=k}^{N-1} \binom{i}{k} s_i x^{i-k}$ is the kth Hasse derivative (cf. [3]) of $S^N(x)$, and α is any primitive nth root of unity in some extension field of $\mathbb{F}_q$.

Remark 1. If $\gcd(N,p) = 1$ then the matrix (3) reduces to the N-tuple $(S^N(1), S^N(\alpha), \ldots, S^N(\alpha^{N-1}))$ known as the *discrete Fourier transform* of the N-tuple representing the period of the sequence S. In this case Theorem 1 is known as Blahuts Theorem (see [4–Sect. 6.8], [11]).

In [9] Theorem 1 has been generalized to multisequences.

Proposition 1 ([9], Proposition 3). *The joint linear complexity* $L(S_1, S_2, \ldots, S_t)$ *of the* t *N-periodic sequences* $S_1, S_2, \ldots, S_t$ *with terms in the finite field* $\mathbb{F}_q$ *of characteristic* p, *where* $N = p^v n$ *and* $\gcd(n,p) = 1$, *is the Günther weight of the* $p^v \times n$ *matrix*

$$
\begin{pmatrix}
\mathbf{S}^N(1) & \mathbf{S}^N(\alpha) & \cdots & \mathbf{S}^N(\alpha^{n-1}) \\
(\mathbf{S}^N)^{[1]}(1) & (\mathbf{S}^N)^{[1]}(\alpha) & \cdots & (\mathbf{S}_N)^{[1]}(\alpha^{n-1}) \\
\vdots & & & \\
(\mathbf{S}^N)^{[p^v-1]}(1) & (\mathbf{S}^N)^{[p^v-1]}(\alpha) & \cdots & (\mathbf{S}^N)^{[p^v-1]}(\alpha^{n-1})
\end{pmatrix}
\tag{4}
$$

of t-*tuples* $(\mathbf{S}^N)^{[k]}(\alpha^j) := ((S_1^N)^{[k]}(\alpha^j), \ldots, (S_t^N)^{[k]}(\alpha^j))$ *of Hasse derivatives, where* α *is any primitive* nth *root of unity in some extension field of* $\mathbb{F}_q$ *and* $S_r^N(x)$ *is the polynomial corresponding to the sequence* S_r, $1 \le r \le t$.

In all further considerations the matrices (3) and (4) will be called the *generalized discrete Fourier transform* (GDFT) for the sequence S and the multisequence $S_1, S_2, \ldots, S_t$, respectively. Let $S_1, S_2, \ldots, S_t$ be t N-periodic sequences with terms in $\mathbb{F}_q$ and let $\{\beta_1, \beta_2, \ldots, \beta_t\}$ be a basis of $\mathbb{F}_{q^t}$ over $\mathbb{F}_q$. Then the GDFT for the multisequence with terms in $\mathbb{F}_q$ can easily be transformed into a GDFT for the sequence S with terms in $\mathbb{F}_{q^t}$ that can be identified with the multisequence.

Lemma 1. *Let* $S_1, S_2, \ldots, S_t$ *be* t *N-periodic sequences with terms in the finite field* $\mathbb{F}_q$ *of characteristic* p, *and let* $\{\beta_1, \beta_2, \ldots, \beta_t\}$ *be a basis of* $\mathbb{F}_{q^t}$ *over* $\mathbb{F}_q$. *Then the matrix*

$$
\begin{pmatrix}
\mathbf{S}^N(1)\beta^T & \mathbf{S}^N(\alpha)\beta^T & \cdots & \mathbf{S}^N(\alpha^{n-1})\beta^T \\
(\mathbf{S}^N)^{[1]}(1)\beta^T & (\mathbf{S}^N)^{[1]}(\alpha)\beta^T & \cdots & (\mathbf{S}_N)^{[1]}(\alpha^{n-1})\beta^T \\
\vdots & & & \\
(\mathbf{S}^N)^{[p^v-1]}(1)\beta^T & (\mathbf{S}^N)^{[p^v-1]}(\alpha)\beta^T & \cdots & (\mathbf{S}^N)^{[p^v-1]}(\alpha^{n-1})\beta^T
\end{pmatrix}
\tag{5}
$$

where $\beta := (\beta_1, \beta_2, \ldots, \beta_t)$, *is a generalized discrete Fourier transform for the* N-*periodic sequence* S *with terms in* $\mathbb{F}_{q^t}$ *that can be identified with* $S_1, S_2, \ldots, S_t$.

Proof. For $0 \le j \le n-1$, $0 \le k \le p^v - 1$ we have

$$
(S^N)^{[k]}(\alpha^j) = \sum_{i=k}^{N-1} \binom{i}{k} \sigma_i(\alpha^j)^{i-k} = \sum_{i=k}^{N-1} \binom{i}{k} \sum_{r=1}^{t} \beta_r s_i^{(r)}(\alpha^j)^{i-k}
$$

$$
= \sum_{r=1}^{t} \beta_r \sum_{i=k}^{N-1} \binom{i}{k} s_i^{(r)}(\alpha^j)^{i-k} = \sum_{r=1}^{t} \beta_r (S_r^N)^{[k]}(\alpha^j). \qquad \square
$$

Let $C_1, C_2, \ldots, C_h$ be the cyclotomic cosets modulo n relative to powers of q and let $l_1, l_2, \ldots, l_h$ be their cardinalities. If $d \equiv jq^b \bmod n$, $b \geq 0$, is an element of the cyclotomic coset of j, then we have

$$(S_r^N)^{[k]}(\alpha^d) = ((S_r^N)^{[k]}(\alpha^j))^{q^b}, \quad 1 \leq r \leq t, 0 \leq k \leq p^n - 1.$$

Consequently the matrices (4), (5) are of a specific form, which we call the *GDFT* form. We collect some properties of the GDFT (see [7, 8, 9]).

1. If we have h different cyclotomic cosets modulo n, then the GDFT is uniquely determined by h columns, one column for each cyclotomic coset.
2. If l_r is the cardinality of the cyclotomic coset C_r of j_r, then the entries in the column of α^{j_r} in the matrix (4) are elements of $\mathbb{F}_{q^{l_r}}^t$.
3. We have a bijective correspondence between the set of matrices in GDFT form and the set of (t) N-periodic sequences over $\mathbb{F}_q$.
4. The Günther weight $g(M)$ of a matrix M in GDFT form can be written in the form

$$g(M) = \sum_{r=1}^{h} \rho_r l_r, \quad 0 \leq \rho_r \leq p^v.$$

We will need some properties of cyclotomic cosets.

Lemma 2. *Let j_r be an element of the cyclotomic coset C_r modulo n relative to powers of q, let l_r be the cardinality of C_r, and let $m_r = \gcd(t, l_r)$.*

(i) *Then C_r splits into m_r cyclotomic cosets modulo n relative to powers of q^t, each of cardinality l_r/m_r.*
(ii) *The l_r/m_r elements $j_r q^{cm_r}$, $0 \leq c \leq l_r/m_r - 1$, form the cyclotomic coset of j_r modulo n relative to powers of q^t.*
(iii) *For each integer u the set $\{j_r q^u, j_r q^{u+1}, \ldots, j_r q^{u+m_r-1}\}$ is a set of representatives of the m_r cyclotomic cosets modulo n relative to powers of q^t in which C_r splits.*

Proof. (i) The cardinality of the cyclotomic coset of j_r modulo n relative to powers of q^t is the smallest integer b_1 for which we have $j_r(q^t)^{b_1} \equiv j_r \bmod n$. Since $j_r q^b \equiv j_r \bmod n$ if and only if b is a multiple of l_r we get $b_1 = \mathrm{lcm}(t, l_r)/t$ or equivalently $b_1 = l_r/\gcd(t, l_r)$.

(ii) Since $m_r = \gcd(t, l_r)$ we have $cm_r = et - fl_r$ for suitable integers e, f. Thus $j_r q^{cm_r} \equiv j_r(q^t)^e \bmod n$, i.e. $j_r q^{cm_r}$ is in the cyclotomic coset of j_r modulo n relative to powers of q^t. Moreover we have $j_r q^{c_1 m_r} \not\equiv j_r q^{c_2 m_r} \bmod n$ if $0 \leq c_1 < c_2 \leq l_r/m_r - 1$.

(iii) Suppose that for $0 \leq i_1 < i_2 \leq m_r - 1$ the elements $j_r q^{u+i_1}$ and $j_r q^{u+i_2}$ are in the same cyclotomic coset modulo n relative to powers of q^t. Then we have $u + i_2 + et = u + i_1 + fl_r$ for suitable integers e, f, or equivalently $et - fl_r = i_1 - i_2$ which is a contradiction to $m_r = \gcd(t, l_r)$. □

Finally we remark that for several classes of integers n the cardinalities of the cyclotomic cosets modulo n relative to powers of q are known. We refer to [7, 8] where for several classes of integers n the cardinalities of the cyclotomic cosets modulo n relative to powers of q have been used to obtain closed formulas for the expected linear complexity of random N-periodic sequences over $\mathbb{F}_q$, where $N = np^v$, $p = char\mathbb{F}_q$.

3   Main Results

The first result is an immediate consequence of Lemma 2, the considerations at the beginning of Section 2 and the fact that for $N = p^v n$ we have $x^N - 1 = (x^n - 1)^{p^v}$ and the canonical factorization of $x^n - 1$ in $\mathbb{F}_q[x]$ is given by

$$x^n - 1 = \prod_{i=1}^{h} f_i(x), \text{ where } f_i(x) := \prod_{j \in C_i} (x - \alpha^j),$$

α is a primitive nth root of unity in an extension field of $\mathbb{F}_q$ and C_i, $1 \leq i \leq h$, are the cyclotomic cosets modulo n relative to powers of q. We state this result in view of the concept of the $\mathbb{F}_q$-linear complexity.

Corollary 1. *Let $N = p^v n$, $\gcd(n, p) = 1$, $p = char\mathbb{F}_q$, let l be the order of q modulo n and let t_0 and t_1 be two divisors of $t \geq 1$. Then the $\mathbb{F}_{q^{t_0}}$-linear complexity and the $\mathbb{F}_{q^{t_1}}$-linear complexity of any N-periodic sequence S over $\mathbb{F}_{q^t}$ are the same if and only if $\gcd(l, t_0) = \gcd(l, t_1)$. In particular $m = \gcd(l, t)$ is the smallest integer for which the $\mathbb{F}_{q^m}$-linear complexity of any N-periodic sequence S over $\mathbb{F}_{q^t}$ is the same as its linear complexity.*

From the multisequence point of view one can see from Corollary 1 that the joint linear complexity and the generalized joint linear complexity will always be the same if and only if $\gcd(l, t) = 1$ (see also [9–Proposition 2]).

Let $S_1, S_2, \ldots, S_t$ be t N-periodic sequences with terms in $\mathbb{F}_q$ and let $\{\beta_1, \beta_2, \ldots, \beta_t\}$ be a basis of $\mathbb{F}_{q^t}$ over $\mathbb{F}_q$. By the above discussion the joint linear complexity of $S_1, S_2, \ldots, S_t$ is given by the Günther weight of the corresponding GDFT (4), and the generalized joint linear complexity of $S_1, S_2, \ldots, S_t$ is given by the Günther weight of the GDFT (5) for $\mathcal{S}$, where $\mathcal{S}$ is the N-periodic sequence over $\mathbb{F}_{q^t}$ associated with $S_1, S_2, \ldots, S_t$. Hence in order to obtain relations between the joint linear complexity and the generalized joint linear complexity we may compare the Günther weights of the matrices (4) and (5). Certainly both matrices are in GDFT form. The important amendment is that for the joint linear complexity the cyclotomic cosets modulo n are considered relative to powers of q, while for the generalized joint linear complexity we have to consider the cyclotomic cosets modulo n relative to powers of q^t. The following theorem establishes a lower bound for the generalized joint linear complexity for a multisequence with given joint linear complexity.

Theorem 2. *Let $N = np^v$, $\gcd(n, p) = 1$, $p = char\mathbb{F}_q$, let $S_1, \ldots, S_t$ be t N-periodic sequences over $\mathbb{F}_q$ and let*

$$L(S_1, S_2, \ldots, S_t) = \sum_{r=1}^{h} \rho_r l_r, \quad 0 \leq \rho_r \leq p^v,$$

be their joint linear complexity written as the sum of cardinalities of cyclotomic cosets modulo n relative to powers of q. Then the generalized joint linear complexity $L(\mathcal{S})$ of $S_1, S_2, \ldots, S_t$ is lower bounded by

$$L(\mathcal{S}) \geq \sum_{r=1}^{h} \rho_r \frac{l_r}{m_r}, \quad where \quad m_r = \gcd(l_r, t), 1 \leq r \leq h. \tag{6}$$

Moreover, this lower bound will always be attained by certain multisequences.

Proof. We show the first part with a quite simple argument.

Evidently the Günther weight of the GDFT of the form (5) can be written as $L(\mathcal{S}) = \sum_{r=1}^{h} \rho_r \omega_r \frac{l_r}{m_r}$, with $0 \leq \omega_r \leq m_r$, if $\sum_{r=1}^{h} \rho_r l_r$ is the Günther weight of the GDFT of the form (4).

We will show that $\omega_r \geq 1$, i.e. that $(\mathbf{S}^N)^{[k]}(\alpha^{j_r q^i})\beta^T$ cannot be the zero tuple for all elements $j_r q^i$ in the cyclotomic coset C_r of j_r modulo n relative to powers of q, provided that $(\mathbf{S}^N)^{[k]}(\alpha^{j_r})$ is not the zero tuple. Because of the one to one correspondence between t-tuples of N-periodic sequences with terms in $\mathbb{F}_q$ and GDFTs of the form (4) there exist sequences $\bar{S}_1, \bar{S}_2, \ldots, \bar{S}_t$ (with joint linear complexity $(p^v - k)l_r$) for which the entries in their GDFT are all zero, except $(\mathbf{S}^N)^{[k]}(\alpha^{j_r q^i})$, $0 \leq i \leq l_r - 1$. If $(\mathbf{S}^N)^{[k]}(\alpha^{j_r q^i})\beta^T$ is zero for all i, $0 \leq i \leq l_r - 1$, then the associated sequence $\mathcal{S}$ with terms in $\mathbb{F}_{q^t}$ has linear complexity 0, i.e. $\mathcal{S}$ is the zero sequence which is a contradiction.

We will show the second part as an easy consequence of subsequent further considerations. We will need these considerations thereafter in order to obtain some enumeration results. □

The representation of a given linear complexity as a linear combination of the cardinalities of the cyclotomic cosets may not be unique. Since the lower bound (6) depends on the exact representation, in certain cases we may have different lower bounds (6) for the same absolute value for the joint linear complexity. For certain period lengths N or linear complexities $L(S)$ this representation is unique. We give two examples.

1. $N = q^v n$, n prime, $p^v < l$, $L(S)$ arbitrary,
2. $L(S) = N$, N arbitrary.

For arbitrary linear complexities and period lengths $N = p^v n$ we can use the fact that l divides $\varphi(n)$, where φ is Euler's totient function, and l_r divides l for $1 \leq r \leq h$ to establish a general lower bound on the generalized joint linear complexity.

Corollary 2. *Suppose that $N = p^v n$, $\gcd(n,p) = 1$, $p = \mathrm{char}\mathbb{F}_q$, and l is the multiplicative order of q modulo n. Let $S_1, S_2, \ldots, S_t$ be t N-periodic sequences over $\mathbb{F}_q$ with joint linear complexity $L(S_1, S_2, \ldots, S_t)$. Then the generalized joint linear complexity $L(\mathcal{S})$ of $S_1, S_2, \ldots, S_t$ satisfies*

$$L(S_1, S_2, \ldots, S_t) \geq L(\mathcal{S}) \geq \frac{L(S_1, S_2, \ldots, S_t)}{\gcd(l,t)} \geq \frac{L(S_1, S_2, \ldots, S_t)}{\gcd(\varphi(n), t)},$$

where φ is Euler's totient function.

We now explicitly describe the contribution of a cyclotomic C_r coset modulo n relative to powers of q to the generalized joint linear complexity of a multisequence $S_1, S_2, \ldots, S_t$ over $\mathbb{F}_q$.

Let j_r be a representative of C_r, let l_r be the cardinality of C_r, and suppose that $\gcd(l_r, t) = m_r$. Then by Lemma 2(iii) the set $\{j_r q^t, j_r q^{t-1}, \ldots, j_r q^{t-m_r+1}\}$ is a set of representatives of the m_r cyclotomic cosets modulo n in which C_r splits relative to powers of q^t. For $0 \le u \le m_r - 1$ the element in the $(k+1)$st row $0 \le k \le p^v - 1$, and the column corresponding to $\alpha^{j_r q^{t-u}}$ in the matrix (5) is given by

$$(\mathbf{S}^N)^{[k]}(\alpha^{j_r q^{t-u}})\beta^T = (S_1^N)^{[k]}(\alpha^{j_r q^{t-u}})\beta_1 + \ldots + (S_t^N)^{[k]}(\alpha^{j_r q^{t-u}})\beta_t$$
$$= ((S_1^N)^{[k]}(\alpha^{j_r}))^{q^{t-u}}(\beta_1^{q^u})^{q^{t-u}} + \ldots + ((S_t^N)^{[k]}(\alpha^{j_r}))^{q^{t-u}}(\beta_t^{q^u})^{q^{t-u}}$$
$$= \left((S_1^N)^{[k]}(\alpha^{j_r})\beta_1^{q^u} + \ldots + (S_t^N)^{[k]}(\alpha^{j_r})\beta_t^{q^u}\right)^{q^{t-u}} =: (a_u^{[k]})^{q^{t-u}} \in \mathbb{F}_{q^{tl_r/m_r}}.$$

Note that $(\mathbf{S}^N)^{[k]}(\alpha^{j_r q^{t-u}})\beta^T = 0$ if and only if $a_u^{[k]} = 0$. Consider the vector

$$
\begin{pmatrix} a_0^{[k]} \\ a_1^{[k]} \\ \vdots \\ a_{m_r-1}^{[k]} \end{pmatrix}
=
\begin{pmatrix}
\beta_1 & \beta_2 & \cdots & \beta_t \\
\beta_1^q & \beta_2^q & \cdots & \beta_t^q \\
\vdots & \vdots & & \vdots \\
\beta_1^{q^{m_r-1}} & \beta_2^{q^{m_r-1}} & \cdots & \beta_t^{q^{m_r-1}}
\end{pmatrix}
\begin{pmatrix} (S_1^N)^{[k]}(\alpha^{j_r}) \\ (S_2^N)^{[k]}(\alpha^{j_r}) \\ \vdots \\ (S_t^N)^{[k]}(\alpha^{j_r}) \end{pmatrix},
\qquad (7)
$$

and define the $p^v \times m_r$ matrix A_r over $\mathbb{F}_{q^{tl_r/m_r}}$ by

$$
A_r :=
\begin{pmatrix}
a_0^{[0]} & a_1^{[0]} & \cdots & a_{m_r-1}^{[0]} \\
a_0^{[1]} & a_1^{[1]} & \cdots & a_{m_r-1}^{[1]} \\
\vdots & \vdots & & \vdots \\
a_0^{[p^v-1]} & a_1^{[p^v-1]} & \cdots & a_{m_r-1}^{[p^v-1]}
\end{pmatrix}.
\qquad (8)
$$

Observe that the $(u+1)$st column of A_r has exactly the same nonzero positions as the column in the GDFT (5) corresponding to $\alpha^{j_r q^{t-u}}$, $0 \le u \le m_r - 1$. Since the cardinality of the cyclotomic coset of $j_r q^{t-u}$ modulo n relative to powers of q^t is l_r/m_r, each column of A_r represents l_r/m_r columns in (5). Thus the contribution of the cyclotomic coset C_r modulo n relative to powers of q to the generalized joint linear complexity of $S_1, S_2, \ldots, S_t$ is the Günther weight of A_r multiplied by l_r/m_r.

From the first part of the proof we already know that $(a_0^{[k]}, a_1^{[k]}, \ldots, a_{m_r-1}^{[k]})^T$ in (7) is the zero vector if and only if $(\mathbf{S}^N)^{[k]}(\alpha^{j_r})$ is the zero vector in $\mathbb{F}_{q^{l_r}}^t$. Consequently the linear transformation (7) defines a bijection from the set of t-tuples over $\mathbb{F}_{q^{l_r}}$ into the set of m_r-tuples over $\mathbb{F}_{q^{tl_r/m_r}}$.

We remark that the number of rows of A_r that are nonzero or that lie below a nonzero row multiplied by l_r yields the contribution of C_r to the joint linear complexity. Thus with the choice of A_r we can control the contribution of C_r to the joint linear complexity and to the generalized joint linear complexity.

We now complete the proof of Theorem 2 and show that the lower bound (6) is attained. For each cyclotomic coset C_r we choose A_r such that the first $p^v - \rho_r$ rows are zero rows, the $(p^v - \rho_r + 1)$st row is not the zero row, and A_r has Günther weight ρ_r. For instance we can choose $a_0^{[p^v - \rho_r]} \neq 0$, and 0 for all other entries of A_r. The corresponding multisequence has the required properties. $\square$

The majority of t-tuples of N-periodic sequences with terms in a finite field $\mathbb{F}_q$ with characteristic p, $N = np^v, \gcd(n, p) = 1$, have (maximal possible) joint linear complexity N. By Proposition 1 the exact number coincides with the number $\mathcal{N}_N^t(N)$ of matrices of the form (4) and Günther weight N. Taking into account the specific structure of the GDFT, with combinatorial arguments we get

$$\mathcal{N}_N^t(N) = \prod_{r=1}^h (q^{l_r t} - 1)(q^{l_r t})^{p^v - 1},$$

where h is the number of the cyclotomic cosets modulo n and l_r is the cardinality of the cyclotomic coset C_r, $1 \leq r \leq h$. Hence it is reasonable to investigate the behavior of multisequences with joint linear complexity N. In the following corollary we establish an exact formula for the number of t-tuples of N-periodic sequences over $\mathbb{F}_q$ with joint linear complexity N and minimal generalized joint linear complexity.

Corollary 3. *Let $N = np^v, \gcd(n, p) = 1$, $C_1, C_2, \ldots, C_h$ be the cyclotomic cosets modulo n relative to powers of q, let $l_1, l_2, \ldots, l_h$ be their cardinalities, and let $m_r = \gcd(l_r, t)$ for $1 \leq r \leq h$. Then there exist exactly*

$$\prod_{r=1}^h m_r (q^{\mathrm{lcm}(l_r, t)} - 1)(q^{\mathrm{lcm}(l_r, t)})^{p^v - 1}$$

t-tuples of N-periodic sequences with terms in the finite field $\mathbb{F}_q$ of characteristic p, joint linear complexity N and minimal possible generalized joint linear complexity

$$L(\mathcal{S}) = \sum_{r=1}^h p^v l_r / m_r. \tag{9}$$

Proof. For each cyclotomic coset C_r we have to choose a $p^v \times m_r$ matrix A_r over $\mathbb{F}_{q^{l_r t/m_r}}$ with first row different from the zero row and with Günther weight p^v. In other words for $1 \leq r \leq h$ the matrix A_r must have exactly one column different from the zero column and furthermore the first entry of this nonzero column must not be 0. With simple combinatorial arguments we obtain the required formula. $\square$

Corollary 3 indicates that there is a non-negligible number of t-tuples of N-periodic sequences with the maximal joint linear complexity N and comparatively small generalized joint linear complexity. For instance, if N is a prime different from p, and l is the multiplicative order of q modulo N, then we have

$(N-1)/l$ cyclotomic cosets of cardinality l and one cyclotomic coset $C_1 = \{0\}$ of cardinality 1. In this case (9) reduces to $1 + (N-1)/\gcd(l,t)$.

In the articles [7, 8, 9] it has been shown that the expected value of the linear complexity of an N-periodic sequence as well as of the joint linear complexity of t N-periodic sequences is close to N. Thus, a significant difference between the joint linear complexity and the generalized joint linear complexity of t N-periodic sequences with terms in $\mathbb{F}_q$ does not seem to be likely. Using similar techniques as in [8, 9] we are able to derive an exact formula for the expected value of the generalized joint linear complexity of t N-periodic sequences with terms in $\mathbb{F}_q$ and joint linear complexity N.

Theorem 3. *Let $N = np^v$ with $p = char\,\mathbb{F}_q$, and $\gcd(n,p) = 1$. Let $l_1, l_2, \ldots, l_h$ be the cardinalities of the cyclotomic cosets modulo n relative to powers of q. Then the expected value E of the generalized joint linear complexity of random t N-periodic sequences with terms in $\mathbb{F}_q$ and joint linear complexity N is given by*

$$E = N - \sum_{r=1}^{h} \frac{l_r q^{l_r t}}{q^{l_r t} - 1} \frac{(1 - q^{\mathrm{lcm}(l_r,t) - l_r t})(1 - q^{-\mathrm{lcm}(l_r,t)p^v})}{q^{\mathrm{lcm}(l_r,t)} - 1}.$$

Proof. The expected contribution of a cyclotomic coset C_r with representative j_r to the Günther weight of the GDFT (5) is given by $l_r/\gcd(l_r,t)$ multiplied by the expected Günther weight of the matrix A_r (8). Since we consider t N-periodic sequences over $\mathbb{F}_q$ with joint linear complexity N, and hence $\mathbf{S}^N(\alpha^{j_r})$ cannot be the zero tuple, we have to presuppose that the first row of A_r is not the zero row.

We first calculate the sum $W(d,s,c)$ of the Günther weights of all $d \times s$ matrices with terms in $\mathbb{F}_{q^c}$. Let $\mathcal{M}$ denote the set of $d \times s$ matrices M with columns $\mathbf{k}_1, \ldots, \mathbf{k}_s$ in $\mathbb{F}_{q^c}^d$. For a nonzero column $\mathbf{k}_b$ we let $u(\mathbf{k}_b)$ denote the least positive integer u such that the uth coordinate of $\mathbf{k}_b$ is nonzero. Then for the Günther weight $g(M)$ of M we have

$$g(M) = \sum_{\substack{b=1 \\ \mathbf{k}_b \neq 0}}^{s} (d - u(\mathbf{k}_b) + 1).$$

Hence

$$W(d,s,c) = \sum_{M \in \mathcal{M}} \sum_{\substack{b=1 \\ \mathbf{k}_b \neq 0}}^{s} (d - u(\mathbf{k}_b) + 1) = d \sum_{b=1}^{s} \sum_{\substack{M \in \mathcal{M} \\ \mathbf{k}_b \neq 0}} 1 - \sum_{b=1}^{s} \sum_{\substack{M \in \mathcal{M} \\ \mathbf{k}_b \neq 0}} (u(\mathbf{k}_b) - 1)$$

$$=: T_1 - T_2.$$

For the first term T_1 we get

$$T_1 = d \sum_{b=1}^{s} (q^{cd} - 1)(q^{cd})^{s-1} = ds(q^{cds} - q^{cd(s-1)}).$$

For the second term T_2 we have

$$T_2 = \sum_{b=1}^{s}\sum_{u=1}^{d}(u-1)\sum_{\substack{M\in\mathcal{M}\\u(\mathbf{k}_b)=u}} 1 = \sum_{b=1}^{s}\sum_{u=1}^{d}(u-1)(q^c-1)q^{c(d-u)}q^{cd(s-1)}$$

$$= s\left(q^{cds}(1-q^{-c})\sum_{u=0}^{d-1}u(q^{-c})^u\right).$$

With the identity

$$\sum_{u=0}^{d-1}uz^u = \frac{z-dz^d+(d-1)z^{d+1}}{(z-1)^2}$$

for any real number $z\neq 1$, we obtain

$$T_2 = sq^{cds}(1-q^{-c})\frac{q^{-c}-d(q^{-c})^d+(d-1)(q^{-c})^{d+1}}{(q^{-c}-1)^2}$$

$$= s\frac{q^{cds}}{q^c-1}(1-dq^{-c(d-1)}+(d-1)q^{-cd}) = s\frac{q^{cds}}{q^c-1}(1-q^{-cd})-dsq^{cd(s-1)}.$$

Combining the formulas for T_1 and T_2 we get

$$W(d,s,c) = dsq^{cds}-s\frac{q^{cds}}{q^c-1}(1-q^{-cd}) = sq^{cds}\left(d-\frac{1-q^{-cd}}{q^c-1}\right). \qquad (10)$$

We need the sum of the Günther weights of all $p^v\times m_r$ matrices with terms in $\mathbb{F}_{q^{l_rt/m_r}}$ with the additional property that the first row is not the zero row. This is exactly given by $W(p^v,m_r,c)-W(p^v-1,m_r,c):=W$ with $c=l_rt/m_r$. With (10) we have

$$W = m_rq^{cm_r(p^v-1)}\left(q^{cm_r}p^v-\frac{q^{cm_r}(1-q^{-cp^v})}{q^c-1}-p^v+1+\frac{1-q^{-c(p^v-1)}}{q^c-1}\right)$$

$$= m_rq^{cm_r(p^v-1)}\left(p^v(q^{cm_r}-1)+\frac{(q^c-q^{cm_r})(1-q^{-cp^v})}{q^c-1}\right).$$

To obtain the expected contribution $E(C_r)$ of the cyclotomic coset C_r to the Günther weight of (5), W has to be multiplied by l_r/m_r and divided by $(q^{l_rt}-1)q^{l_rt(p^v-1)}$, the number of $p^v\times m_r$ matrices over $\mathbb{F}_{q^{l_rt/m_r}}$ for which the first row is not the zero row. Writing l_rt/m_r for c, this yields

$$E(C_r) = l_rp^v+\frac{l_r}{q^{l_rt}-1}\frac{(q^{l_rt/m_r}-q^{l_rt})(1-q^{-l_rt/m_rp^v})}{q^{l_rt/m_r}-1}$$

$$= l_rp^v+\frac{l_rq^{l_rt}}{q^{l_rt}-1}\frac{(1-q^{-l_rt/m_r(m_r-1)})(1-q^{-l_rt/m_rp^v})}{q^{l_rt/m_r}-1}.$$

With $E=\sum_{r=1}^{h}E(C_r)$ and $\sum_{r=1}^{h}l_r=n$ we get the assertion. $\qquad\square$

Remark 2. If $v = 0$, i.e. $\gcd(N, q) = 1$, then the formula in Theorem 3 reduces to

$$E = N - \sum_{r=1}^{h} \frac{l_r\left(q^{l_r t - \mathrm{lcm}(l_r, t)} - 1\right)}{q^{l_r t} - 1}.$$

Acknowledgment

The research was supported by DSTA research grant R-394-000-011-422.

References

1. Cusick, T., Ding, C., Renvall, A.: Stream Ciphers and Number Theory (Revised Edition), North-Holland, Amsterdam (2004)
2. C. Ding, G. Xiao, W. Shan, The Stability Theory of Stream Ciphers, Lecture Notes in Computer Science, Vol. 561, Springer, Berlin, 1991.
3. H. Hasse, Theorie der höheren Differentiale in einem algebraischen Funktionenkörper mit vollkommenem Konstantenkörper bei beliebiger Charakteristik, J. Reine Angew. Math. 175 (1936) 50–54.
4. D. Jungnickel, Finite Fields: Structure and Arithmetics, Bibliographisches Institut, Mannheim, 1993.
5. R. Lidl and H. Niederreiter, Finite Fields, Addison-Wesley, Reading, MA, 1983.
6. J. L. Massey, S. Serconek, Linear complexity of periodic sequences: a general theory, in: N. Koblitz (Ed.), Advances in Cryptology–CRYPTO '96, Lecture Notes in Computer Science, Vol. 1109, Springer, Berlin, 1996, pp. 358–371.
7. W. Meidl, H. Niederreiter, Linear complexity, k-error linear complexity, and the discrete Fourier transform, J. Complexity 18 (2002) 87–103.
8. W. Meidl, H. Niederreiter, On the expected value of the linear complexity and the k-error linear complexity of periodic sequences, IEEE Trans. Inform. Theory 48 (2002) 2817–2825.
9. W. Meidl, H. Niederreiter, The expected value of the joint linear complexity of periodic multisequences, J. Complexity 19 (2003) 61–72.
10. H. Niederreiter, Linear complexity and related complexity measures for sequences, Progress in Cryptology - INDOCRYPT 2003 (T. Johansson, S.Maitra, Eds.), Lecture Notes in Computer Science 2904, 1–17, Springer-Verlag 2003.
11. R. A. Rueppel, Stream ciphers, in: G. J. Simmons (Ed.), Contemporary Cryptology: The Science of Information Integrity, IEEE Press, New York, 1992, pp. 65–134.
12. S. Sakata, Extension of the Berlekamp-Massey algorithm to N dimensions, Information and Computation 84 (1990), 207–239.
13. L.P. Wang, H. Niederreiter, Enumeration results on the joint linear complexity of multisequences, Finite Fields Appl., to appear.
14. L.P. Wang, Y.F. Zhu, $F[x]$-lattice basis reduction algorithm and multisequence synthesis, Sci. China Ser. F 44 (2001) 321–328.
15. C. P. Xing, Multi-sequences with almost perfect linear complexity profile and function fields over finite fields, J. Complexity 16 (2000), 661–675.

Expected Value of the Linear Complexity of Two-Dimensional Binary Sequences*

Xiutao Feng and Zongduo Dai

State Key Laboratory of Information Security,
(Graduate School of Chinese Academy of Sciences), Beijing, 100049
fengxt@mails.gscas.ac.cn, yangdai@public.bta.net.cn

Abstract. In this work, based on the technique of multi-continued fractions [6, 7, 8], we study the normalized expected value $e(2, n)$ of the linear complexity of binary sequences of dimension 2. As a result, $e(2, n)$ is determined, and moreover, it is found that $e(2, n) \rightarrow \frac{2}{3}$ as n goes into infinity.

1   Introduction

In stream ciphers, the linear complexity of sequences of elements in the binary field is a fundamental concept for measuring the randomness and unpredictability of key streams. A lot of research have been done on the linear complexity and related complexity measures for sequences; see [1] for a recent survey. However, these works have mainly concentrated on single sequences. Recent developments in stream ciphers show strong interests in word-based (or vectorized) stream ciphers. The study of such stream ciphers requires the study of the (joint) linear complexity of multi-sequences.

Let F be the binary field, m and n be two positive integers. By a multi-sequence $\underline{r}$ of dimension m and length n, we mean an m-tuple of sequences:

$$\underline{r} = \begin{pmatrix} r_1 \\ \vdots \\ r_j \\ \vdots \\ r_m \end{pmatrix} = \begin{pmatrix} r_{1,1} & r_{1,2} & \cdots & r_{1,n} \\ \vdots & \vdots & & \vdots \\ r_{j,1} & r_{j,2} & \cdots & r_{j,n} \\ \vdots & \vdots & & \vdots \\ r_{m,1} & r_{m,2} & \cdots & r_{m,n} \end{pmatrix}, \ r_{j,i} \in F, \tag{1}$$

where each component r_j is a sequence over F. The *linear complexity* (or *joint linear complexity* in some paper) of a multi-sequence $\underline{r}$, denoted by $L_{\underline{r}}$, is defined as the shortest length of the linear feedback shift register capable of generating all its component sequences simultaneously. Assume that multi-sequences of dimension m are distributed evenly, then the expected value of the linear complexity

---

* This work is partly supported by NSFC (Grant No. 60173016), and the National 973 Project (Grant No. 1999035804).

T. Helleseth et al. (Eds.): SETA 2004, LNCS 3486, pp. 113–128, 2005.

of multi-sequences of dimension m and length n, denoted by $E(m, n)$, can be expressed as:

$$E(m, n) = \sum_{\underline{r}} \frac{L_{\underline{r}}}{2^{mn}} = \frac{1}{2^{mn}} \sum_{d=1}^{n} d \cdot |R_{n,d}(m)|, \tag{2}$$

where $\underline{r}$ runs over all possible multi-sequences of dimension m and length n, 2^{mn} is the total number of such sequences, $R_{n,d}(m)$ is the set of all multi-sequences $\underline{r}$ of dimension m, length n and linear complexity d, and $|R_{n,d}(m)|$ means the size of $R_{n,d}(m)$. We call $\frac{E(m,n)}{n}$ the normalized expected value of the linear complexity of those sequences, denoted by $e(m, n)$.

As for single sequences, the normalized expected value of the linear complexity of sequences of length n is obtained in [2] as shown below:

$$e(1, n) = \frac{1}{2} + \frac{4 + \epsilon_n}{18n} + O(2^{-n}/n), \tag{3}$$

where $\epsilon_n \equiv n \pmod 2$.

Later the normalized expected value of the linear complexity of periodically repeated random sequences of period n, denoted by $e(1, \overline{n})$, is provided in [3]:

$$e(1, \overline{n}) = 1 - \sum_{d|n_1} \frac{\phi(d)(1 - 2^{-2^r \cdot n_d})}{n(2^{n_d} - 1)}, \tag{4}$$

where $n = 2^r \cdot n_1$, $\gcd(2, n_1) = 1$, $\phi(d)$ is the Euler's function and n_d is the order of 2 modulo d.

Recently, the normalized expected value of the linear complexity of periodic multi-sequences of dimension m and period n, denoted by $e(m, \overline{n})$, is obtain in [4]. It can be restated as follows:

$$e(m, \overline{n}) = 1 - \sum_{d|n_1} \frac{\phi(d)(1 - 2^{-m2^r \cdot n_d})}{n(2^{mn_d} - 1)}, \tag{5}$$

where the notations made in (4) are kept in (5), which is the generalization of (4).

As mentioned above, $e(1, n)$ was known, but $e(m, n)$ with $m > 1$ was undetermined. However, both H. Niederreiter and C.S. Ding had the following conjecture:

Conjecture [5] on $e(m, n)$. *The normalized expected value $e(m, n)$ goes to $\frac{m}{m+1}$ when n goes to infinity.*

Based on the technique of multi-continued fractions [6, 7, 8], the above conjecture is solved for the case $m = 2$ in this work. The approach in solving this case should work for the general case.

Main Theorem. *Let $n \geq 1$. Then*

$$e(2, n) = \frac{2}{3} + \frac{46 + 3\epsilon_n}{147n} - (\frac{3}{14} + \frac{\omega_n}{882n}) \cdot 2^{-n} - (\frac{1}{7} + \frac{38}{441n}) \cdot 2^{-2n}, \tag{6}$$

where $0 \leq \epsilon_n \leq 2$, $\epsilon_n = n \pmod 3$, $\omega_n = 200$ if $n \equiv 0 \pmod 2$ and $\omega_n = 151$ if $n \equiv 1 \pmod 2$. As a consequence,

$$\lim_{n \to \infty} e(2, n) = \frac{2}{3}.$$

2   Preliminaries: $R_{n,d}(m)$ and $C_{n,d}(m)$

The problem of evaluating $e(m, n)$ is essentially the problem of evaluating the size of $R_{n,d}(m)$. It is shown in [8] that there exists a 1-1 correspondence between elements in $R_{n,d}(m)$ and elements in a set $C_{n,d}(m)$. Hence it is obtained

$$|R_{n,d}(m)| = |C_{n,d}(m)|, \tag{7}$$

where $C_{n,d}(m)$ is a set of certain multi-continued fractions and defined in [8]. Below, according to our requirement, we are going to restate the set $C_{n,d}(m)$. We start with HT-pairs.

2.1 HT-Pairs

In this subsection, we fix a positive integer m. Let both $\underline{h}$ and $\underline{t}$ be μ-tuples of integers, where μ is a positive integer. We call the pair $(\underline{h}, \underline{t})$ a *pre-HT-pair* of dimension m and length μ if it satisfies the following two conditions:

Condition H: $1 \leq h_k \leq m$, $1 \leq k \leq \mu$;
Condition T: $t_k \geq 1$, $1 \leq k \leq \mu$.

Associated to the pre-HT-pair $(\underline{h}, \underline{t})$ of length μ, we define the following parameters:

$$v_{k,j} = \sum_{1 \leq i \leq k,\ h_i = j} t_i, \quad v_{0,j} = 0, \quad v_k = v_{k,h_k},$$
$$d_k = \sum_{1 \leq i \leq k} t_i,$$
$$n_k = d_k + v_{k-1,h_k}$$

for $1 \leq k \leq \mu$ and $1 \leq j \leq m$. We call d_μ the d-value, denoted by $d(\underline{h}, \underline{t})$, and n_μ the n-value, denoted by $n(\underline{h}, \underline{t})$.

Definition 1. *Let $(\underline{h}, \underline{t})$ be a pre-HT-pair of length μ. Then it is called an HT-pair if it satisfies the following additional condition:*

Condition HT: $(h_k, v_{k-1,h_k}) < (h_{k+1}, v_{k+1}), \quad 1 \leq k < \mu,$

where the order is defined as below: for any (h, v) and (h', v'), define $(h, v) < (h', v')$ if and only if $v < v'$ or $v = v'$ but $h < h'$.

2.2   The Set $\mathcal{C}_{n,d}(m)$

Given any *expansion* C of the following form

$$C = \begin{bmatrix} h_1, h_2, \cdots, h_k, \cdots, h_\mu \\ 0, \underline{a}_1, \underline{a}_2, \cdots, \underline{a}_k, \cdots, \underline{a}_\mu \end{bmatrix}, \tag{8}$$

where μ is a positive integer, $1 \le h_k \le m$ and $\underline{a}_k = (a_{k,1}, a_{k,2}, \cdots, a_{k,m})$ is an m-tuple of polynomials over F such that $\deg(a_{k,h_k}) \ge 1$ for $1 \le k \le \mu$, where $\deg(a_{k,h_k})$ denotes the degree of the polynomial a_{k,h_k}, we can get a pre-HT-pair $(\underline{h}, \underline{t})$, where

$$\underline{h} = (h_1, h_2, \cdots, h_k, \cdots, h_\mu),$$
$$\underline{t} = (t_1, t_2, \cdots, t_k, \cdots, t_\mu), \quad t_k = \deg(a_{k,h_k}),$$

which is called the pre-HT-pair of C, denoted by $HT(C)$.

Definition 2. *The expansion C of the form (8) is called a multi-strict continued fraction of dimension m if it satisfies the following two conditions:*

C-1: *$HT(C)$ is an HT-pair of dimension m;*
C-2: *Denote by $[0,p]$ the set made of integers x such that $0 \le x \le p$. Then C matches its HT-pair in the sense that each component $a_{k,j}$ of the polynomial tuple $\underline{a}_k$ has the form:*

$$a_{k,j} = z^{t_k} \delta_j + \sum_{x \in B_{k,j} \cap [0, v_{k,j}]} a_{k,j,x} z^{v_{k,j} - x}, \quad a_{k,j,x} \in F, \ 1 \le j \le m,$$

where $\delta_j = 0$ if $j \ne h_k$ and $\delta_{h_k} = 1$, and $B_{k,j} \ (= B_{k,j}(C))$ is the solution set of the equation $E_{k,j} \ (= E_{k,j}(C))$ on integer x:

$$E_{k,j}: \quad \begin{cases} (h_k, v_{k-1,h_k}) < (j, x) < (h_{k+1}, v_{k+1}), \ 1 \le k < \mu \\ (h_\mu, v_{\mu-1,h_\mu}) < (j, x), \hspace{3.2cm} k = \mu \end{cases}. \tag{9}$$

We call

$$A_{\mu,j} = A_{\mu,j}(C) = \{ x \in B_{\mu,j} \cap [0, v_{\mu,j}] \mid a_{\mu,j,x} = 1 \}, \ 1 \le j \le m,$$

the supporting sets of C. Denote by $\mathcal{C}(m)$ the set of all possible multi-strict continued fractions of dimension m. Then the set $\mathcal{C}_{n,d}(m)$ can be defined as:

$$\mathcal{C}_{n,d}(m) = \left\{ C \in \mathcal{C}(m) \ \middle| \ \begin{array}{l} n(HT(C)) \le n, \ d(HT(C)) = d, \\ A_{\mu,j}(C) \subseteq [0, n-d], 1 \le j \le m \end{array} \right\}.$$

3   Partition of the Set $\mathcal{C}_{n,d}(2)$

From now we only consider the case $m = 2$, and simply write $\mathcal{C}_{n,d} = \mathcal{C}_{n,d}(2)$ and $\mathcal{C} = \mathcal{C}(2)$. For any HT-pair $(\underline{h}, \underline{t})$ of length μ, denote

$$v = \min\{v_{\mu,1}, v_{\mu,2}\},$$

which is called the v-value of $(\underline{h}, \underline{t})$ and denoted by $v(\underline{h}, \underline{t})$.

By means of the v-value of HT-pairs of elements in the set $\mathcal{C}_{n,d}$, we can make a partition of the set $\mathcal{C}_{n,d}$. Denote

$$\mathcal{C}_{n,d,v} = \{C \in \mathcal{C}_{n,d} \mid v(HT(C)) = v\}.$$

It is clear that

$$\mathcal{C}_{n,d} = \bigcup_{0 \le v \le d/2} \mathcal{C}_{n,d,v} \quad \text{and} \quad |\mathcal{C}_{n,d}| = \sum_{0 \le v \le d/2} |\mathcal{C}_{n,d,v}|. \tag{10}$$

Likewise, we can also make a partition of the set of HT-pairs itself. Let HT be the set of all possible HT-pairs of dimension 2. Denote

$$HT_{n,d,v} = \{(\underline{h}, \underline{t}) \in HT \mid n(\underline{h}, \underline{t}) \le n, d(\underline{h}, \underline{t}) = d, v(\underline{h}, \underline{t}) = v\}.$$

Clearly, the set $\mathcal{C}_{n,d,v}$ is also defined equivalently as

$$\mathcal{C}_{n,d,v} = \{C \in \mathcal{C} \mid HT(C) \in HT_{n,d,v}, A_{\mu,j} \subseteq [0, n-d], 1 \le j \le 2\}.$$

Fixing d and v, by definitions, it is easy to see that both $HT_{n,d,v}$ and $\mathcal{C}_{n,d,v}$ are the monotonically increased sets on n for the relation " $\subseteq$ ", i.e., for $\forall n \ge 1$,

$$HT_{n,d,v} \subseteq HT_{n+1,d,v},$$
$$\mathcal{C}_{n,d,v} \subseteq \mathcal{C}_{n+1,d,v}.$$

Lemma 1. *Denote* $\overline{v} = d - v$. *Let* $n \ge d + \overline{v}$. *Then*

1. $HT_{n,d,v} = HT_{d+\overline{v}-1,d,v}$.
2. $\mathcal{C}_{n,d,v} = \mathcal{C}_{d+\overline{v},d,v}$. *As a consequence,* $|\mathcal{C}_{n,d,v}| = |\mathcal{C}_{d+\overline{v},d,v}|$.

Proof. 1. We only need to prove that $HT_{n,d,v} \subseteq HT_{d+\overline{v}-1,d,v}$. For any $(\underline{h}, \underline{t}) \in HT_{n,d,v}$ of length μ, note that $\overline{v} = d - v = \max\{v_{\mu,1}, v_{\mu,2}\}$, by $n(\underline{h}, \underline{t}) = d_\mu + v_{\mu-1,h_\mu} = d + v_\mu - t_\mu \le d + \overline{v} - 1$, we have $(\underline{h}, \underline{t}) \in HT_{2d-v-1,d,v}$. Thus $HT_{n,d,v} \subseteq HT_{d+\overline{v}-1,d,v}$.

2. Let $C \in \mathcal{C}_{n,d,v}$. First, by Item 1, we have $HT(C) \in HT_{n,d,v} = HT_{d+\overline{v},d,v}$. Second, by definition, $A_{\mu,j}(C) \subseteq [0, v_{\mu,j}] \subseteq [0, \overline{v}] = [0, (d+\overline{v}) - d]$. Thus $C \in \mathcal{C}_{d+\overline{v},d,v}$. It follows that $\mathcal{C}_{n,d,v} \subseteq \mathcal{C}_{d+\overline{v},d,v}$. ∎

From the above lemma, it is enough to consider the size of $\mathcal{C}_{n,d,v}$ for $n \le d + \overline{v}$. As for the set $\mathcal{C}_{n,d,v}$, elements in it can be classified further according to their HT-pairs. For any given HT-pair $(\underline{h}, \underline{t}) \in HT_{n,d,v}$, denote

$$\mathcal{C}_n(\underline{h}, \underline{t}) = \{C \in \mathcal{C}_{n,d,v} \mid HT(C) = (\underline{h}, \underline{t})\}.$$

Then we have

$$\mathcal{C}_{n,d,v} = \bigcup_{(\underline{h}, \underline{t}) \in HT_{n,d,v}} \mathcal{C}_n(\underline{h}, \underline{t}) \quad \text{and} \quad |\mathcal{C}_{n,d,v}| = \sum_{(\underline{h}, \underline{t}) \in HT_{n,d,v}} |\mathcal{C}_n(\underline{h}, \underline{t})|. \tag{11}$$

The following lemma formulates the size of $\mathcal{C}_n(\underline{h}, \underline{t})$, which will be in use later.

Lemma 2. *Let $(\underline{h}, \underline{t}) \in HT_{n,d,v}$. Denote*

$$X_{k,j} = \begin{cases} B_{k,j} \cap [0, v_{k,j}], & 1 \le k < \mu, j = 1, 2 \\ B_{\mu,j} \cap [0, v_{\mu,j}] \cap [0, n - d], & k = \mu, j = 1, 2 \end{cases}$$

and $e_{k,j} = |X_{k,j}|$. Then

$$|\mathcal{C}_n(\underline{h}, \underline{t})| = 2^e, \quad e = \sum_{1 \le k \le \mu} \sum_{1 \le j \le m} e_{k,j}.$$

Proof. Let $C \in \mathcal{C}$. Keep the notations made for C. Denote

$$A_{k,j}(C) = \{ x \mid a_{k,j,x} = 1 \}, 1 \le k \le \mu, j = 1, 2.$$

It is easy to see C is determined uniquely by $HT(C)$ and the set $\{A_{k,j}(C) | 1 \le k \le \mu, j = 1, 2\}$. By definition, $C \in \mathcal{C}_n(\underline{h}, \underline{t})$ if and only if $A_{k,j}(C) \subseteq X_{k,j}$. Therefore, to make $C \in \mathcal{C}_n(\underline{h}, \underline{t})$, the coefficient $a_{k,j,x}$ has two possible choices 0 and 1 for each $x \in X_{k,j}$, hence $a_{k,j}$ has totally $2^{e_{k,j}}$ possible choices. So

$$|\mathcal{C}_n(\underline{h}, \underline{t})| = 2^e, \quad e = \sum_{1 \le k \le \mu} \sum_{1 \le j \le m} e_{k,j}.$$

$\blacksquare$

Hence the problem of evaluating $e(2, n)$ is reduced further to the problem of evaluating $|\mathcal{C}_{n,d,v}|$ or $|\mathcal{C}_n(\underline{h}, \underline{t})|$ for $1 \le d \le n \le d + \bar{v}$ and $0 \le v \le d/2$.

4   Structure of the Set $HT_{n,d,v}$

In this section we study the structure of $HT_{n,d,v}$, as the set $HT_{n,d,v}$ is critical to determine the set $\mathcal{C}_{n,d,v}$, based on (11). Set $l = d - 2v$ and $t = n - d - v$, and we will rewrite the triple (n, d, v) in the following form, based on the three parameters v, l and t, as below:

$$(n, d, v) = (3v + l + t, 2v + l, v),$$

where $0 \le v \le d/2$, $0 \le l \le d$ and $-v \le t \le l - 1$, (note: By Item 1 of Lemma 1, it is sound that we suppose that $t \le l - 1$ when we study the set $HT_{3v+l+t,2v+l,v}$). Given a $(\underline{h}, \underline{t}) \in HT_{3v+l+t,2v+l,v}$, denote

$$h = \min \{ j \mid v_{\mu,j} = v, 1 \le j \le 2 \},$$
$$c = n_\mu - d_\mu - v,$$
$$b = l(\mu, h),$$
$$\bar{b} = l(\mu, \bar{h}),$$

where $\bar{h} \in \{1, 2\}$, $\bar{h} \ne h$, and $l(\mu, h)$ is equal to the maximal index k no more than μ such that $h_k = h$ if such k exists, and $l(\mu, h) = 0$ otherwise. Clearly, we have the following facts: 1) $v_b = v$; 2) $v_{\overline{b}} = v + l$; 3) If $b < \mu$, then $v_\mu = v + l$; 4) If $\overline{b} < \mu$, then $v_\mu = v$.

Throughout the rest of this paper, when given an HT-pair $(\underline{h}, t)$, for simplification, if without special declaration, we always keep all notations defined as above for $(\underline{h}, t)$, such as h_k, t_k, μ, $v_{k,j}$, d_k, n_k, v, l, h, $\overline{h}$, c, b, $\overline{b}$ and so on.

Lemma 3. *Let* $(\underline{h}, t) \in HT$, $d(\underline{h}, t) = 2v + l$ *and* $v(\underline{h}, t) = v$. *Then*

1.

$$c = \begin{cases} l - t_\mu, & \text{if } \overline{b} = \mu \\ -t_\mu, & \text{if } \overline{b} < \mu \end{cases}.$$

As a consequence, $c \leq l - 1$.

2. *If* $\overline{b} < \mu$, *let* $i = \mu - \overline{b}$. *Then*

$$l + i - 1 \begin{cases} \leq t_{\overline{b}}, & \text{if } \overline{h} = 1 \\ < t_{\overline{b}}, & \text{if } \overline{h} = 2 \end{cases}.$$

3. $(\underline{h}, t) \in HT_{3v+l+t, 2v+l, v}$ *if and only if* $c \leq t$.

Proof. 1. Note that

$$v_\mu = \begin{cases} v + l, & \overline{b} = \mu \\ v, & \overline{b} < \mu \end{cases},$$

we can obtain

$$c = n_\mu - d_\mu - v = v_{\mu-1, h_\mu} - v = (v_\mu - v) - t_\mu = \begin{cases} l - t_\mu, & \overline{b} = \mu \\ -t_\mu, & \overline{b} < \mu \end{cases}.$$

2. Note that

$$(\overline{h}, v + l - t_{\overline{b}}) = (h_{\overline{b}}, v_{\overline{b}-1, h_{\overline{b}}}) < (h_{\overline{b}+1}, v_{\overline{b}+1}) = (h, v_{\mu-i+1}) \leq (h, v - i + 1).$$

If $\overline{h} = 2$, then $v + l - t_{\overline{b}} < v - i + 1$. Thus $l + i - 1 < t_{\overline{b}}$. If $\overline{h} = 1$, then $v + l - t_{\overline{b}} \leq v - i + 1$. Thus $l + i - 1 \leq t_{\overline{b}}$.

3. Since $c = n_\mu - d_\mu - v$ and $t = n - d_\mu - v$, thus $c \leq t \Leftrightarrow n_\mu \leq n$. ■

By the above lemma we can obtain the following conclusion.

Corollary 1. *Let* $t \leq l - 1$. *Denote by* $HT_{n,d,v}(t_{\overline{b}} \geq 2)$ *the subset of the set* $HT_{n,d,v}$ *made of all HT-pairs* $(\underline{h}, t)$ *in* $HT_{n,d,v}$ *with* $t_{\overline{b}} \geq 2$. *Similarly, we have the sets* $HT_{n,d,v}(t_{\overline{b}} = 1)$, $HT_{n,d,v}(t_{\overline{b}} = 1, \overline{b} < \mu)$, *and so on. Then for* $(\underline{h}, t) \in HT_{3v+l+t, 2v+l, v}$, *we have*

1. *When* $l \geq 2$, *if* $\overline{b} < \mu$ *or* $t \leq l - 2$, *then* $t_{\overline{b}} \geq 2$. *As a consequence, if* $l \geq 2$ *and* $t_{\overline{b}} = 1$, *then* $t = l - 1$ *and* $\overline{b} = \mu$. *In other words, we have*

$$HT_{3v+l+t, 2v+l, v} = HT_{3v+l+t, 2v+l, v}(t_{\overline{b}} \geq 2), \ t \leq l - 2,$$
$$HT_{3v+2l-1, 2v+l, v}(t_{\overline{b}} = 1) = HT_{3v+2l-1, 2v+l, v}(t_{\overline{b}} = 1, \overline{b} = \mu).$$

2. When $l = 1$, if $t_{\bar{b}} = 1$, $t < 0$ or $t = 0$ and $\bar{b} < \mu$, then $\bar{b} = \mu - 1$ and $\bar{h} = 1$. In other words, we have

$$HT_{3v+t+1,2v+1,v}(t_{\bar{b}} = 1) = HT_{3v+t+1,2v+1,v}(t_{\bar{b}} = 1, \bar{b} = \mu - 1, \bar{h} = 1), t < 0,$$
$$HT_{3v+1,2v+1,v}(t_{\bar{b}} = 1, \bar{b} < \mu) = HT_{3v+1,2v+1,v}(t_{\bar{b}} = 1, \bar{b} = \mu - 1, \bar{h} = 1).$$

3. When $l = 0$, we have $t_\mu = -c \geq -t$. In other words, we have $HT_{3v+t,2v,v} = \dot{\bigcup}_{-t \leq i \leq v} HT_{3v+t,2v,v}(t_\mu = i)$. In particular, we have

$$HT_{3v+t,2v,v} = HT_{3v+t,2v,v}(t_\mu = -t) \dot{\cup} HT_{3v+t,2v,v}(t_\mu \geq 1 - t), \; -v < t < 0,$$
$$HT_{3v+t,2v,v}(t_\mu \geq 1 - t) = HT_{3v+t-1,2v,v}, \; -v < t < 0,$$
$$HT_{2v,2v,v} = HT_{2v,2v,v}(t_\mu = v),$$

where $\dot{\cup}$ means the disjoint union of sets.

Proof. 1. If $\bar{b} < \mu$, by Item 2 of Lemma 3, $t_{\bar{b}} \geq l + (\mu - \bar{b}) - 1 \geq 2$. If $t \leq l - 2$ and $\bar{b} = \mu$, by Items 1 and 3 of Lemma 3, $t_{\bar{b}} = t_\mu = l - c \geq l - t \geq 2$.
2. If $t < 0$, then $c \leq t < 0 = l - t_{\bar{b}}$. By Item 1 of Lemma 3, we have $\bar{b} < \mu$. Further by Item 2 of Lemma 3, we have $1 \leq \mu - \bar{b} = l + (\mu - \bar{b}) - 1 \leq t_{\bar{b}} = 1$. It implies that $\bar{b} = \mu - 1$ and $\bar{h} = 1$.
3. It follows directly from Item 1 of Lemma 3. ∎

5 Reduction of the $HT_{3v+l+t,2v+l,v}$

In this section we make reductions for the sets $HT_{3v+l+t,2v+l,v}$, based on some constructed mappings, which are defined as follows.

Mappings. Let $t \leq l - 1$. For any $(\underline{h}, \underline{t}) \in HT_{3v+t+l,2v+l,v}$, we define the mappings ξ, η, ϱ, χ, $\bar{\xi}$, $\bar{\eta}$, $\bar{\zeta}$ and ϑ as below:

1. When $l \geq 0$, define $\bar{\xi}$ and $\bar{\eta}$:

$$\varphi(\underline{h}, \underline{t}) = \begin{cases} (\underline{h}, (\cdots, t_{\bar{b}} + 1, \cdots)), & \varphi = \bar{\xi} \\ ((\underline{h}, \bar{h}), (\underline{t}, 1)), & \varphi = \bar{\eta} \end{cases};$$

2. When $l \geq 1$, define ϱ:

$$\varrho(\underline{h}, \underline{t}) = ((\underline{h}, h), (\underline{t}, l));$$

3. When $l \geq 1$, define $\bar{\zeta}$:

$$\bar{\zeta}(\underline{h}, \underline{t}) = \begin{cases} (\underline{h}, (\cdots, t_{\bar{b}} - 1, \cdots)), & t_{\bar{b}} > 1 \\ ((h_1, \cdots, h_{\bar{b}-1}, h_{\bar{b}+1}, \cdots, h_\mu), (t_1, \cdots, t_{\bar{b}-1}, t_{\bar{b}+1}, \cdots, t_\mu)), & t_{\bar{b}} = 1 \end{cases};$$

4. When $l = 0$, define ξ, η and χ:

$$\varphi(\underline{h}, \underline{t}) = \begin{cases} (\underline{h}, (\cdots, t_b + 1, \cdots)), & \varphi = \xi \\ ((\underline{h}, 1), (\underline{t}, 1)), & \varphi = \eta \\ ((h_1, \cdots, h_{\mu-1}, 1, h_\mu), (t_1, \cdots, t_{\mu-1}, 1, t_\mu)), & \varphi = \chi \end{cases}$$

5. When $l = 0$, let $(\underline{h}, \underline{t}) \in HT_{3v+t,2v,v}(t_\mu = -t)$. Define:

$$\vartheta(\underline{h}, \underline{t}) = (\underline{h}', \underline{t}'),$$

where $\underline{h}' = (h_1, \cdots, h_{\mu-1})$ and $\underline{t}' = (t_1, \cdots, t_{\mu-1})$.

Lemma 4. *Let $t \leq l - 1$. Denote*

$$S = HT_{3v+l+t,2v+l,v}$$
$$S' = HT_{3v+l+t-1,2v+l-1,v}.$$

Then

1. *If $l \geq 2$, then $\overline{\xi}$ is an injective from S' onto $S(t_{\overline{b}} \geq 2)$, and $\overline{\zeta}$ is its inverse.*
2. *If $l \geq 2$ and $t = l - 1$, then $\overline{\eta}$ is an injective from S' onto $S(t_{\overline{b}} = 1, \overline{b} = \mu)$, and $\overline{\zeta}$ is its inverse.*
3. *If $l = 1$, then*
   (a) *$\overline{\xi}$ is an injective from S' onto $S(t_{\overline{b}} \geq 2, \overline{h} = 2)$, and $\overline{\zeta}$ is its inverse.*
   (b) *ξ is an injective from S' onto $S(t_{\overline{b}} \geq 2, \overline{h} = 1)$, and $\overline{\zeta}$ is its inverse.*
   (c) *χ is an injective from $S'(h_\mu = 2)$ onto $S(t_{\overline{b}} = 1, \overline{b} = \mu - 1, \overline{h} = 1)$, and $\overline{\zeta}$ is its inverse.*
4. *If $l = 1$ and $t = 0$, then*
   (a) *$\overline{\eta}$ is an injective from S' onto $S(t_{\overline{b}} = 1, \overline{b} = \mu, \overline{h} = 2)$, and $\overline{\zeta}$ is its inverse.*
   (b) *η is an injective from S' onto $S(t_{\overline{b}} = 1, \overline{b} = \mu, \overline{h} = 1)$, and $\overline{\zeta}$ is its inverse.*
5. *If $l = 0$, denote $v' = v + t$ and $S'' = HT_{3v'-2t,2v'-t,v'}$. Then ϱ is an injective from S'' onto $HT_{3v+t,2v,v}(t_\mu = -t)$, and ϑ is its inverse.*

Proof. Here we will exemplify Item 1 to prove that the conclusion is correct, and since the proofs of others are very similar to the proof of Item 1, we will omit them. Below, we will prove Item 1 by three steps. For any given $(\underline{h}, \underline{t}) \in S'$, let $\overline{\xi}(\underline{h}, \underline{t}) = (\underline{h}', \underline{t}')$.

First, we argue that $(\underline{h}', \underline{t}') \in S(t_{\overline{b} \geq 2})$. By the definition of $\overline{\xi}$, $(\underline{h}', \underline{t}')$ satisfies both Condition H and Condition T. Thus $(\underline{h}', \underline{t}')$ is a pre-HT-pair. For simplification, we make convention that all parameters with apostrophe are associated with $(\underline{h}', \underline{t}')$ and all without apostrophe are associated with $(\underline{h}, \underline{t})$. Compare some parameters associated with $(\underline{h}, \underline{t})$ and $(\underline{h}', \underline{t}')$, and we obtain

$$\mu' = \mu, h'_k = h_k,$$

$$t'_k = \begin{cases} t_k, & k \neq \overline{b} \\ t_{\overline{b}} + 1, & k = \overline{b} \end{cases},$$

$$v'_{k,j} = \begin{cases} v_{k,j}, & 1 \leq k < \overline{b} \text{ or } j = h \\ v_{k,j} + 1, & \overline{b} \leq k \leq \mu, j = \overline{h} \end{cases}, v'_{k-1,h_k} = v_{k-1,h_k},$$

$$v'_k = \begin{cases} v_k, & k \neq \overline{b} \\ v_{\overline{b}} + 1, & k = \overline{b} \end{cases},$$

$$n'_\mu = n_\mu + 1, d'_\mu = d_\mu + 1,$$

$$v' = v, l' = l - 1, t' = t, c' = c.$$

One can check that for $1 \leq k < \mu$

$$(h'_k, v'_{k-1,h'_k}) = (h_k, v_{k-1,h_k}) < (h_{k+1}, v_{k+1}) \leq (h'_{k+1}, v'_{k+1})$$

and

$$c' = n'_\mu - d'_\mu - v' = n_\mu - d_\mu - v = c \leq t = t'.$$

The former inequalities show that $(\underline{h}', \underline{t}')$ satisfies Condition HT, too. Thus it is an HT-pair. The latter shows further that $(\underline{h}', \underline{t}') \in S$. Note that $t'_{\overline{5}} = t_{\overline{5}} + 1 \geq 2$, thus the argument follows.

Second, for any $(\underline{h}, \underline{t}) \in S(t_{\overline{5} \geq 2})$, let $\overline{\zeta}(\underline{h}, \underline{t}) = (\underline{h}', \underline{t}')$. Similarly to the first step, one can check $(\underline{h}', \underline{t}') \in S'$ as well.

Finally, by the definitions of $\overline{\xi}$ and $\overline{\zeta}$, it is easy to check that

$$\overline{\xi} \cdot \overline{\zeta}|_{S(t_{\overline{5} \geq 2})} = Id \quad \text{and} \quad \overline{\zeta} \cdot \overline{\xi}|_{S'} = Id,$$

where Id denotes the identity mapping. Thus $\overline{\xi}$ is an injective from the set S' onto the set $S(t_{\overline{5}} \geq 2)$, and $\overline{\zeta}$ is its inverse. ∎

6   The Size of $\mathcal{C}_{3v+l+t,2v+l,v}$

By Item 2 of Lemma 1, it gives the following immediate result:

$$|\mathcal{C}_{3v+l+t,2v+l,v}| = |\mathcal{C}_{3v+2l,2v+l,v}|, \tag{12}$$

where $t \geq l$. Thus below we shall only consider the case $t \leq l$.

Lemma 5. *Let* $(\underline{h}, \underline{t}) \in HT_{3v+l+t-1,2v+l,v}$, *where* $t \leq l$. *Then*

$$|\mathcal{C}_{3v+l+t}(\underline{h}, \underline{t})| = \begin{cases} 2\,|\mathcal{C}_{3v+l+t-1}(\underline{h}, \underline{t})|, \, l \geq t \geq 1 \\ 4\,|\mathcal{C}_{3v+l+t-1}(\underline{h}, \underline{t})|, \, l = 0 \text{ or } t \leq 0 \end{cases}.$$

As a consequence, we have

$$|\mathcal{C}_{3v+2l,2v+l,v}| = \begin{cases} 2\,|\mathcal{C}_{3v+2l-1,2v+l,v}|, \, l \geq 1 \\ 4\,|\mathcal{C}_{3v-1,2v,v}|, \quad\quad l = 0 \end{cases} \tag{13}$$

and

$$|\mathcal{C}_{3v+t,2v,v}(t_\mu \geq 1-t)| = 4\,|\mathcal{C}_{3v+t-1,2v,v}|, \, -v < t < 0. \tag{14}$$

where $\mathcal{C}_{3v+t,2v,v}(t_\mu \geq 1-t)$ *is a subset of* $\mathcal{C}_{3v+t,2v,v}$ *made of all possible* C *in* $\mathcal{C}_{3v+t,2v,v}$ *such that* $t_\mu \geq 1-t$, *and* t_μ *is the parameter associated with* $HT(C)$.

Proof. Let $E_{k,j}$, $B_{k,j}$, $X_{k,j}$, $e_{k,j}$, etc., and $E'_{k,j}$, $B'_{k,j}$, $X'_{k,j}$, $e'_{k,j}$, etc., be parameters associated with $\mathcal{C}_{3v+l+t-1}(\underline{h}, \underline{t})$ and $\mathcal{C}_{3v+l+t}(\underline{h}, \underline{t})$ respectively for any k and j. One can check that $E'_{k,j} = E_{k,j}$ for $1 \leq k \leq \mu$. Thus $B'_{k,j} = B_{k,j}$ for $1 \leq k \leq \mu$. By definition we have $X'_{k,j} = X_{k,j}$ and $e'_{k,j} = e_{k,j}$ for $1 \leq k < \mu$. Now we consider $X'_{\mu,j}$ and $X_{\mu,j}$. Let $n = 3v+l+t-1$, $d = 2v+l$, $n' = 3v+l+t = n+1$ and $d' = 2v+l = d$. Denote

$$M_0 = \{\,(l,t) \mid l = 0 \text{ or } t \leq 0\,\},$$
$$M_1 = \{\,(l,t) \mid l \geq t \geq 1\,\}.$$

It is clear that $\{\,(l,t) \mid l \geq 0, t \leq l\,\} = M_0 \dot\cup M_1$. Note that

$$\min\{v'_{\mu,j}, n' - d'\} = \begin{cases} v, & \text{if } (l,t) \in M_1 \text{ and } j = \overline{h} \\ v+t, & \text{otherwise} \end{cases},$$

and

$$\min\{v_{\mu,j}, n - d\} = \begin{cases} v, & \text{if } (l,t) \in M_1 \text{ and } j = \overline{h} \\ v+t-1, & \text{otherwise} \end{cases}.$$

It follows that

$$X'_{\mu,j} \setminus X_{\mu,j} = \begin{cases} \varnothing, & \text{if } (l,t) \in M_1 \text{ and } j = \overline{h} \\ \{v+t\}, & \text{otherwise} \end{cases},$$

hence,

$$e'_{\mu,j} - e_{\mu,j} = \begin{cases} 0, \text{ if } (l,t) \in M_1 \text{ and } j = \overline{h} \\ 1, \text{ otherwise} \end{cases}.$$

Then

$$e' - e = \sum_{1 \leq k \leq \mu} \sum_{1 \leq j \leq 2} (e'_{k,j} - e_{k,j}) = \sum_{j=h,\overline{h}} (e'_{\mu,j} - e_{\mu,j}) = \begin{cases} 1, (l,t) \in M_1 \\ 2, (l,t) \in M_0 \end{cases},$$

which leads to the desired result. ∎

Lemma 6. *1. Let $t \leq l - 1$ and $(\underline{h}, \underline{t}) \in HT_{3v+l+t-1,2v+l-1,v}$. Then*

$$|\mathcal{C}_{3v+l+t}(\varphi(\underline{h},\underline{t}))| = \begin{cases} |\mathcal{C}_{3v+l+t-1}(\underline{h},\underline{t})|, & l \geq 1, \varphi = \overline{\xi} \\ |\mathcal{C}_{3v+2l-2}(\underline{h},\underline{t}))|, & l \geq 1, t = l-1, \varphi = \overline{\eta} \\ |\mathcal{C}_{3v+t,2v,v}(\underline{h},\underline{t})|, & l = 1, h_\mu = 2, \varphi = \xi \text{ or } \chi \\ 2\,|\mathcal{C}_{3v+t,2v,v}(\underline{h},\underline{t})|, & l = 1, h_\mu = 1, \varphi = \xi \\ 2\,|\mathcal{C}_{3v,2v,v}(\underline{h},\underline{t})|, & l = 1, t = 0, \varphi = \eta \end{cases}.$$

2. Let $(\underline{h}, \underline{t}) \in HT_{3v+t,2v+t,v+t}$, where $-v \leq t < 0$. Then

$$|\mathcal{C}_{3v+t}(\varrho(\underline{h},\underline{t}))| = |\mathcal{C}_{3v+t}(\underline{h},\underline{t})|.$$

Proof. Here we still exemplify the case that $l \geq 1$ and $\varphi = \overline{\xi}$. Since the proofs of others are similar to the proof of this case, we omit them.

For any $(\underline{h}, \underline{t}) \in HT_{3v+l+t-1,2v+l-1,v}$, let $\overline{\xi}(\underline{h}, \underline{t}) = (\underline{h}', \underline{t}')$. It is known that $(\underline{h}', \underline{t}') \in HT_{3v+l+t,2v+l,v}$. Keep all previous notations for both $(\underline{h}, \underline{t})$ and $(\underline{h}', \underline{t}')$. We claim that $X'_{k,j} = X_{k,j}$ for all (k,j). In fact, based on the relations of the parameters associated to $(\underline{h}', \underline{t}')$ and $(\underline{h}, \underline{t})$ (see comparison in the proof of Lemma 4), wee see

$$B'_{k,j} \begin{cases} \supseteq B_{k,j}, & k+1 = \overline{b} \\ = B_{k,j}, & \text{otherwise} \end{cases}.$$

Now we prove $X'_{k,j} = X_{k,j}$ in the following three cases separately.

1. For the case $k = \bar{b} - 1$. We claim that $x > v_{\bar{b}-1,j}$ if $x \in B'_{k,j} \backslash B_{k,j}$. Hence

$$X'_{\bar{b}-1,j} = B'_{\bar{b}-1,j} \cap [0, v'_{\bar{b}-1,j}] = B'_{\bar{b}-1,j} \cap [0, v_{\bar{b}-1,j}] = B_{\bar{b}-1,j} \cap [0, v_{\bar{b}-1,j}] = X_{\bar{b}-1,j}.$$

In fact, for any $x \in B'_{k,j} \backslash B_{k,j}$, we have

$$(\bar{h}, v + l - 1) = (h_{\bar{b}}, v_{\bar{b}}) \le (j, x) < (h_{\bar{b}}, v'_{\bar{b}}) = (\bar{h}, v + l).$$

If $j = \bar{h}$, then $x \ge v_{\bar{b}} = v_{\bar{b}-1,\bar{h}} + t_{\bar{b}} > v_{\bar{b}-1,\bar{h}}$. If $j = h$, when $l \ge 2$, we have $x \ge v + l - 1 > v \ge v_{\bar{b}-1,h}$. When $l = 1$, we have $h = 1$ and $\bar{h} = 2$. Then $x > v_{\bar{b}} = (v + 1) - 1 = v \ge v_{\bar{b}-1,h}$.

2. For the case $\bar{b} \le k \le \mu$ and $j = \bar{h}$, we have $B'_{k,\bar{h}} = B_{k,\bar{h}}$, $v'_{k,\bar{h}} = v + l$, $v_{k,\bar{h}} = v + l - 1$. So $X_{k,\bar{h}} \subseteq X'_{k,\bar{h}}$. If $x \in B_{k,\bar{h}}$, then $(\bar{h}, x) < (h_{k+1}, v_{k+1}) \le (h, v)$, that is, $x \le v$. Therefore,

$$\begin{aligned} X'_{k,\bar{h}} &= B'_{k,\bar{h}} \cap [0, v'_{k,\bar{h}}] = B_{k,\bar{h}} \cap [0, v + l] \\ &= B_{k,\bar{h}} \cap [0, v] \cap [0, v + l] = B_{k,\bar{h}} \cap [0, v] \\ &\subseteq B_{k,\bar{h}} \cap [0, v_{k,\bar{h}}] = X_{k,\bar{h}} \subseteq X'_{k,\bar{h}}. \end{aligned}$$

Hence $X'_{k,\bar{h}} = X_{k,\bar{h}}$.

3. For the other cases, we have $B'_{k,j} = B_{k,j}$, $v'_{k,j} = v_{k,j}$ and $n' - d' = n - d$. So $X'_{k,j} = X_{k,j}$.

Combine the above cases, and we can get the desired result. ∎

By Lemmas 4 and 6, we have the following result.

Corollary 2. *Denote by $C_{n,d,v}(t_{\bar{b}} \ge 2)$ the subset of $C_{n,d,v}$ made of all possible elements C in $C_{n,d,v}$ with $t_{\bar{b}} \ge 2$, where $t_{\bar{b}}$ is the parameter of $HT(C)$. Similarly, we have the notations $C_{n,d,v}(t_{\bar{b}} \ge 2, \bar{h})$, $C_{n,d,v}(t_{\bar{b}} = 1, \bar{b} = \mu, \bar{h} = 2)$, etc.. Then*

1. *If $l \ge 2$ and $t \le l - 1$, we have*

$$\left| C_{3v+t+l,2v+l,v}(t_{\bar{b}} \ge 2) \right| = \left| C_{3v+t+l-1,2v+l-1,v} \right|,$$
$$\left| C_{3v+2l-1,2v+l,v}(t_{\bar{b}} = 1, \bar{b} = \mu) \right| = \left| C_{3v+2l-2,2v+l-1,v} \right|.$$

2. *If $t \le 0$, we have*

$$\left| C_{3v+t+1,2v+1,v}(t_{\bar{b}} \ge 2, \bar{h} = 2) \right| = \left| C_{3v+t,2v,v} \right|,$$
$$\left| C_{3v+t+1,2v+1,v}(t_{\bar{b}} \ge 2, \bar{h} = 1, \bar{b} = \mu) \right| = 2 \left| C_{3v+t,2v,v}(h_{\mu} = 1) \right|,$$
$$\left| C_{3v+t+1,2v+1,v}(t_{\bar{b}} \ge 2, \bar{h} = 1, \bar{b} < \mu) \right| = \left| C_{3v+t,2v,v}(h_{\mu} = 2) \right|,$$
$$\left| C_{3v+t+1,2v+1,v}(t_{\bar{b}} = 1, \bar{b} = \mu - 1, \bar{h} = 1) \right| = \left| C_{3v+t,2v,v}(h_{\mu} = 2) \right|,$$
$$\left| C_{3v+1,2v+1,v}(t_{\bar{b}} = 1, \bar{b} = \mu, \bar{h} = 2) \right| = \left| C_{3v,2v,v} \right|,$$
$$\left| C_{3v+1,2v+1,v}(t_{\bar{b}} = 1, \bar{b} = \mu, \bar{h} = 1) \right| = 2 \left| C_{3v,2v,v} \right|.$$

3. *If $t < 0$, we have $|\mathcal{C}_{3v+t,2v+t,v+t}| = |\mathcal{C}_{3v+t,2v,v}(t_\mu = -t)|$.*

By Corollaries 1 and 2, we obtain

Corollary 3. *1. If $l \geq 2$, then*

$$|\mathcal{C}_{3v+l+t,2v+l,v}| = \begin{cases} |\mathcal{C}_{3v+l+t-1,2v+l-1,v}|, & t < l-1 \\ 2|\mathcal{C}_{3v+2l-2,2v+l-1,v}|, & t = l-1 \end{cases}.$$

2.

$$|\mathcal{C}_{3v+t+1,2v+1,v}| = \begin{cases} 3|\mathcal{C}_{3v+t,2v,v}|, & t < 0 \\ 6|\mathcal{C}_{3v,2v,v}|, & t = 0 \end{cases}.$$

3. Let $v' = v + t$. Then

$$|\mathcal{C}_{3v+t,2v,v}| = \begin{cases} |\mathcal{C}_{2v,v,0}|, & t = -v \\ 4|\mathcal{C}_{3v+t-1,2v,v}| + |\mathcal{C}_{3v'-2t,2v'-t,v'}|, & -v < t < 0 \end{cases}.$$

By Corollaries 1, 3 and Lemma 5, we obtain

Corollary 4. *1. Let $l \geq 1$. Then*

$$|\mathcal{C}_{3v+l+t,2v+l,v}| = \begin{cases} 2^{2t}|\mathcal{C}_{3v+1,2v+1,v}| = 3 \cdot 2^{2t+3}|\mathcal{C}_{3v-1,2v,v}|, & 0 \leq t \leq l-1 \\ 3|\mathcal{C}_{3v+t,2v,v}|, & t < 0 \end{cases}.$$

2. Let $-v < t < 0$ and $v' = v + t$. Then

$$|\mathcal{C}_{3v+t,2v,v}| = 4|\mathcal{C}_{3v+t-1,2v,v}| + 3 \cdot 2^{-2t+2}|\mathcal{C}_{3v'-1,2v',v'}|.$$

As a consequence of the above Lemmas and Corollaries, we obtain

Lemma 7. *1. Let $l \geq 1$. Then*

$$|\mathcal{C}_{l+t,l,0}| = \begin{cases} 3 \cdot 2^{2t}, & 0 \leq t < l \\ 3 \cdot 2^{2l-1}, & t \geq l \end{cases}. \tag{15}$$

2. Let $v \geq 1$ and $l \geq 0$. Then

$$|\mathcal{C}_{3v+l+t,2v+l,v}| = \begin{cases} \lambda_l \cdot 2^{6v+4t-1}, & t < 0 \\ \lambda_l \cdot 2^{6v+2t-2}, & 0 \leq t < l \\ \lambda_l \cdot 2^{6v+2l-3}, & l \leq t \end{cases} \tag{16}$$

where $\lambda_l = 3$ if $l = 0$; otherwise, $\lambda_l = 9$.

Proof. 1. Note that, when $l = 1$ and $t = 0$, that is, $n = d = 1$, we have the following 3 2-sequences

$$\binom{0}{1}, \binom{1}{1}, \binom{1}{0},$$

the linear complexity of which is equal to 1. Thus, $|\mathcal{C}_{1,1,0}| = 3$. By Lemma 5 and Corollary 4, it gives the immediate result.

2. We first prove that for any $v \geq 1$ and $-v \leq t < 0$,

$$|\mathcal{C}_{3v+t,2v,v}| = 3 \cdot 2^{6v+4t-1}.$$

To do this we will employ the induction theorem for the pair (t, v) according to the natural order defined as before. First, when $v = 1$, it implies that $t = -v = -1$. Thus $|\mathcal{C}_{2,2,1}| = |\mathcal{C}_{2,1,0}| = 6 = 3 \cdot 2^{6 \cdot 1 - 4 \cdot 1 - 1}$. So the conclusion is correct for $(-t, v) = (-1, 1)$. Assume that the conclusion is also correct for all pairs (t', v') such that $v' \geq 1$, $-v' \leq t' < 0$ and $(t', v') < (t, v)$. Then when $t = -v$, by Corollary 3, we have

$$|\mathcal{C}_{2v,2v,v}| = |\mathcal{C}_{2v,v,0}| = 3 \cdot 2^{2v-1} = 3 \cdot 2^{6v-4v-1}.$$

When $-v < t < 0$, by Corollary 4, we have

$$\begin{aligned}
|\mathcal{C}_{3v+t,2v,v}| &= 4 |\mathcal{C}_{3v+t-1,2v,v}| + 3 \cdot 2^{-2t+2} |\mathcal{C}_{3v''-1,2v'',v''}| \\
&= 4 \cdot 3 \cdot 2^{6v-4(-t+1)-1} + 3 \cdot 2^{-2t+2} \cdot 3 \cdot 2^{6(v+t)-4-1} \\
&= 3 \cdot 2^{6v+4t-1},
\end{aligned}$$

where $v'' = v + t$. It shows that the conclusion is correct for (t, v) as well. By the induction theorem, the conclusion follows.

Now let's return to the prove of Item 2. When $l = 0$, if $t < 0$, then

$$|\mathcal{C}_{3v+t,2v,v}| = 3 \cdot 2^{6v+4t-1};$$

if $t \geq 0$, by Corollary 4, then

$$|\mathcal{C}_{3v+t,2v,v}| = 4 |\mathcal{C}_{3v-1,2v,v}| = 3 \cdot 2^{6v-3}.$$

When $l \geq 1$, similarly, by Corollary 4, if $t < 0$, then

$$|\mathcal{C}_{3v+t+l,2v+l,v}| = 3 |\mathcal{C}_{3v+t,2v,v}| = 9 \cdot 2^{6v+4t-1};$$

if $0 \leq t < l$, then

$$|\mathcal{C}_{3v+t+l,2v+l,v}| = 3 \cdot 2^{2t+3} |\mathcal{C}_{3v-1,2v,v}| = 9 \cdot 2^{6v+2t-2};$$

if $t \geq l$, then

$$|\mathcal{C}_{3v+t+l,2v+l,v}| = 2 |\mathcal{C}_{3v+2l-1,2v+l,v}| = 9 \cdot 2^{6v+2l-3}.$$

Combine the above all cases, and we can get the desired conclusion.   ∎

7   The Proof of Main Theorem

Now we come back to the proof of Main Theorem.

Proof. Review Equation (2), and by Lemma 7 we have

$$e(2,n) = \frac{1}{n \cdot 2^{2n}} \sum_{d=1}^{n} \sum_{v=0}^{\lfloor \frac{d}{2} \rfloor} d \ |\mathcal{C}_{n,d,v}|$$

$$= \frac{1}{n \cdot 2^{2n}} \left(\sum_{S_1} + \sum_{S_2} + \sum_{S_3} + \sum_{S_4} + \sum_{S_5} + \sum_{S_6} \right)(2v+l) \ |\mathcal{C}_{n,2v+l,v}|$$

where

$$\begin{aligned}
S_1 &= \{(v,l)|v=0\}, \\
S_2 &= \{(v,l)|v \geq 1, l=0, t<0\}, \\
S_3 &= \{(v,l)|v \geq 1, l=0, t \geq 0\}, \\
S_4 &= \{(v,l)|v \geq 1, l \geq 1, t<0\}, \\
S_5 &= \{(v,l)|v \geq 1, l \geq 1, 0 \leq t < l\}, \\
S_6 &= \{(v,l)|v \geq 1, l \geq 1, t \geq l\}
\end{aligned}$$

and

$$\sum_{S_1} l \ |\mathcal{C}_{n,l,0}| = \sum_{l=1}^{\lfloor \frac{n}{2} \rfloor} 3 \cdot 2^{2l-1} l + \sum_{l=\lfloor \frac{n}{2} \rfloor+1}^{l=n} 3 \cdot 2^{2(n-l)} l,$$

$$\sum_{S_2} (2v+l) \ |\mathcal{C}_{n,2v+l,v}| = \sum_{v=\lfloor \frac{n}{3} \rfloor+1}^{\lfloor \frac{n}{2} \rfloor} 6v \cdot 2^{4n-6v-1}$$

$$\sum_{S_3} (2v+l) \ |\mathcal{C}_{n,2v+l,v}| = \sum_{v=1}^{\lfloor \frac{n}{3} \rfloor} 6v \cdot 2^{6v-3},$$

$$\sum_{S_4} (2v+l) \ |\mathcal{C}_{n,2v+l,v}| = \sum_{l=1}^{n-2} \sum_{v=\lfloor \frac{n-l}{3} \rfloor+1}^{\lfloor \frac{n-l}{2} \rfloor} 9(2v+l)2^{4n-6v-4l-1},$$

$$\sum_{S_5} (2v+l) \ |\mathcal{C}_{n,2v+l,v}| = \sum_{v=1}^{\lfloor \frac{n-1}{3} \rfloor} \sum_{l=\lfloor \frac{n-3v}{2} \rfloor+1}^{n-3v} 9(2v+l)2^{2n-2l-2},$$

$$\sum_{S_6} (2v+l) \ |\mathcal{C}_{n,2v+l,v}| = \sum_{v=1}^{\lfloor \frac{n-2}{3} \rfloor} \sum_{l=1}^{\lfloor \frac{n-3v}{2} \rfloor} 9(2v+l)2^{6v+2l-3}.$$

By straightforward calculation, we can obtain

$$e(2,n) = \frac{2}{3} + \frac{46+3\epsilon_n}{147n} - \left(\frac{3}{14} + \frac{\omega_n}{882n} \right) \cdot 2^{-n} - \left(\frac{1}{7} + \frac{38}{441n} \right) \cdot 2^{-2n},$$

where $0 \leq \epsilon_n \leq 2$, $\epsilon_n = n \pmod 3$, $\omega_n = 200$ if $n \equiv 0 \pmod 2$ and $\omega_n = 151$ if $n \equiv 1 \pmod 2$ ∎

References

1. H. Niederreiter, Some computable complexity measures for binary sequences, in: C. Ding, T. Helleseth, H. Niederreiter(Eds.), Sequences and their applications, Springer, London, 1999, pp.67-78.
2. R.A. Rueppel, Analysis and Design of Stream ciphers, Springer-Verlag, 1986.
3. Z.D.Dai and J.H.Yang, Linear Complexity of Periodically Repeated Random Sequences. Lecture Notes in Computer Science, Advances in EUROCRYPT'91, Spring-Verlag LNCS 547(1991), Editor: D.W.Davies, pp168-175.
4. W. Meidl and H. Niderreiter, The expected value of joint linear complexity of periodic multisequences, Journal of Complexity, 19 (2003), pp61-72.
5. Chaoping Xing, Multi-sequences with Almost Perfect Linear Complexity Profile and Function Fields over Finite Fields, Journal of Complexity 16, pp661-675, 2000.
6. Zongduo Dai, Kunpeng Wang and Dingfeng Ye, m-Continued Fraction Expansions of Multi-Laurent Series, ADVANCES IN MATHEMATICS (CHINA), 2004, Vol.33, No.2, pp246-248.
7. Zongduo Dai, Kunpeng Wang and Dingfeng Ye, Multidimensional Continued Fraction and Rational Approximation, http://arxiv.org/abs/math.NT/0401141.
8. Zongduo Dai, Xiutao Feng and Junhui Yang, Multi-Continued Fraction Algorithm and Generalized B-M Algorithm over F_2, Submitted to 2004 International Conference on Sequences and Their Application (As a full paper of [9]).
9. Zongduo Dai and Xiutao Feng, Multi-Continued Fractions and Generalized B-M Algorithm over F_2, Proceedings (Extended Abstracts) of 2004 International Conference on Sequences and Their Application, pp113-117.

Asymptotic Behavior of Normalized Linear Complexity of Multi-sequences

Zongduo Dai[1], Kyoki Imamura[2], and Junhui Yang[3]

[1] State Key Lab of Information Security, Graduate School of Chinese
Academy of Sciences, Beijing, 100039, China
yangdai@public.bta.net.cn
[2] Department of Computer Science and Electronics, Kyushu Institute
of Technology, Iizuka, Fukuoka 820-8502, Japan
imamura@cse.kyutech.ac.jp
[3] Software Institute, Chinese Academy of Sciences, Beijing, 100080, China

Abstract. Asymptotic behavior of the normalized linear complexity $\frac{L_{\underline{s}}(n)}{n}$ of a multi-sequence $\underline{s}$ is studied in terms of its multidimensional continued fraction expansion, where $L_{\underline{s}}(n)$ is the linear complexity of the length n prefix of $\underline{s}$ and defined to be the length of the shortest multi-tuple linear feedback shift register which generates the length n prefix of $\underline{s}$. A formula for $\limsup_{n \to \infty} \frac{L_{\underline{s}}(n)}{n}$ together with a lower bound, and a formula for $\liminf_{n \to \infty} \frac{L_{\underline{s}}(n)}{n}$ together with an upper bound are given. A necessary and sufficient condition for the existence of $\lim_{n \to \infty} \frac{L_{\underline{s}}(n)}{n}$ is also given.

1 Introduction

For a sequence $s = \{s_t\}_{t \geq 0}$ over a finite field F, the linear complexity of the length n prefix $s^n = (s_0, s_1, \cdots, s_{n-1})$ of s is denoted by $L_s(n)$ and defined to be the length of the shortest linear feedback shift register which generates s^n. Recently the asymptotic behavior of the normalized linear complexity, $\frac{L_s(n)}{n}$, of an ultimately non-periodic infinite binary sequence s has been studied [1] by using continued fractions [5], [6], [7], [8]. We consider only the asymptotic behavior of the normalized linear complexity $\frac{L_s(n)}{n}$ of a non-periodic sequence s, since $\lim_{n \to \infty} \frac{L_s(n)}{n} = 0$ for a periodic sequence s. However its normalized linear complexity $\frac{L_s(n)}{n}$ remains between 0 and 1. Therefore there exists a possibility for the normalized linear complexity to be more useful for discussing randomness of sequences than the linear complexity.

This paper discusses the asymptotic behavior of the normalized linear complexity of an ultimately non-periodic infinite m-tuple multi-sequence (or vector-valued sequence)

$$\underline{s} = \{\underline{s}_t\}_{t \geq 0} = (\{s_{1,t}\}_{t \geq 0}, \cdots, \{s_{m,t}\}_{t \geq 0})^T, \tag{1}$$

T. Helleseth et al. (Eds.): SETA 2004, LNCS 3486, pp. 129–142, 2005.

where $\{s_{i,t}\}_{t\geq 0}$ is an infinite sequence over F, each $\underline{s}_t$ is an m-tuple over F, and the superscript T denotes the transpose.

It is natural to generalize the results on a single sequence case to a multi-dimensional sequence case, since the single sequence case has been satisfactorily solved by using continued fractions [1] and multidimensional continued fraction expansion has been proposed recently [2], [3]. The generalization to the multi-dimensional sequence case makes it possible for us to understand the relation between the single sequence case and multi-dimensional sequence case.

2   Continued Fraction and Linear Complexity

In the case of a single sequence (i.e., $m = 1$) $s = \{s_t\}_{t\geq 0}$, the Laurent series

$$s(z) = \sum_{t\geq 0} s_t z^{-(t+1)}$$

is expanded into a simple continued fraction [5], [6], [7]:

$$C(s) = \cfrac{1}{a_1(z) + \cfrac{1}{a_2(z) + \cfrac{1}{\cdots\cdots + \cfrac{1}{a_k(z) + \cfrac{1}{\cdots\cdots}}}}}$$

$$= [\, 0, a_1(z), a_2(z), \cdots, a_k(z), \cdots \,],$$

where $a_i(z)$ is a polynomial in z over F with degree $t_i = \deg(a_i(z)) \geq 1$. Associated with $C(s)$, we denote

$$d_k = \sum_{1\leq i\leq k} t_i, \quad d_0 = 0, \tag{2}$$

$$n_k = d_k + d_{k-1}, \quad n_0 = 0. \tag{3}$$

The linear complexity of the length n prefix of s can be read out [5], [6], [7], [8] from these parameters, as shown below:

$$L_s(n) = d_k \quad \text{for} \quad n_k \leq n < n_{k+1}, \quad \forall k \geq 0. \tag{4}$$

It has been shown [1] that for an ultimately non-periodic binary sequence s, we have

$$\limsup_{n\to\infty} \frac{L_s(n)}{n} + \liminf_{n\to\infty} \frac{L_s(n)}{n} = 1. \tag{5}$$

If $\{L_s(n)/n\}_{n\geq 1}$ is convergent, then we have [4], [6]

$$\lim_{n\to\infty} \frac{L_s(n)}{n} = \frac{1}{2}. \tag{6}$$

If $m > 1$ and $\underline{s}$ is as in equation (1), then the m-tuple Laurent series

$$\underline{s}(z) = \sum_{t \geq 0} \underline{s}_t z^{-(t+1)}$$

is expanded into a multidimensional continued fraction (m-CF)

$$C(\underline{s}) = \begin{bmatrix} h_1, & h_2, & \cdots, & h_k, & \cdots \\ 0, & \underline{a}_1, & \underline{a}_2, & \cdots & \underline{a}_k, & \cdots \end{bmatrix}, \quad 1 \leq k < \infty,$$

by applying the m-CF algorithm (m-CFA) [2] to $\underline{s}(z)$. Here h_k is an integer between 1 and m, and $\underline{a}_k = (a_{k,1}(z), \cdots, a_{k,m}(z))^T$ is an m-tuple of polynomials over F with $\deg(a_{k,h_k}(z)) \geq 1$.

The outline of the m-CFA [2], [3] for computing $C(\underline{s})$ is as follows. Let $F(z^{-1})$ be a functional field of z over F, i.e., the quotient field of the ring of polynomials in z over F. For an element $\alpha \in F(z^{-1})$ expanded as a formal Laurent series $\alpha = \sum_{t \geq b} a_t z^{-t}$, the discrete valuation of α is denoted by $v(\alpha)$ and defined to be

$$v(\alpha) = \begin{cases} \infty & \text{if} \quad \alpha = 0, \\ b & \text{if} \quad a_b \neq 0. \end{cases} \tag{7}$$

Every element $\alpha \in F((z^{-1}))$ can be written as a sum of two parts as $\alpha = \lfloor \alpha \rfloor + \{\alpha\}$, where its polynomial part $\lfloor \alpha \rfloor$ is defined by $\lfloor \alpha \rfloor = \sum_{t=b \leq 0}^{0} a_t z^{-t}$ if $b = v(\alpha) \leq 0$ and $\lfloor \alpha \rfloor = 0$ if $b = v(\alpha) > 0$, and its remaining part $\{\alpha\}$ is defined by $\{\alpha\} = \alpha - \lfloor \alpha \rfloor$. In this paper we assume that at least one component sequence of (1) is ultimately non-periodic.

Computation of $C(\underline{s})$ is performed as follows [2], which is actually the generalization of the classic continued fraction algorithm from the one-dimensional case to the multi-dimensional case.

We start from the initial values

$$c_{0,1} = \cdots = c_{0,m} = 0; \quad \underline{a}_0 = \underline{0}; \quad \underline{\beta}_0 = (\beta_{0,1}, \cdots, \beta_{0,m})^T,$$

where $\beta_{0,j} = \sum_{t \geq 0} s_{j,t} z^{-(t+1)}$ for $1 \leq j \leq m$.

We repeat the following computation successively for $k \geq 1$.

1. $c_k = \min\{c_{k-1,j} + v(\beta_{k-1,j}) \mid 1 \leq j \leq m\}$
2. $h_k = \min\{j \mid c_{k-1,j} + v(\beta_{k-1,j}) = c_k, \quad 1 \leq j \leq m\}$.
3. $c_{k,j} = c_{k-1,j}$ if $j \neq h_k$ and $c_{k,h_k} = c_k$.
4. Compute $\underline{\rho}_k = (\rho_{k,1}, \cdots, \rho_{k,m})^T$ by $\rho_{k,h_k} = \frac{1}{\beta_{k-1,h_k}}$ and $\rho_{k,j} = \frac{\beta_{k-1,j}}{\beta_{k-1,h_k}}$ if $j \neq h_k$.
5. $\underline{a}_k = (a_{k,1}, \cdots, a_{k,m})^T = \lfloor \underline{\rho}_k \rfloor = (\lfloor \rho_{k,1} \rfloor, \cdots, \lfloor \rho_{k,m} \rfloor)^T$; and
   $\underline{\beta}_k = (\beta_{k,1}, \cdots, \beta_{k,m})^T = \{\underline{\rho}_k\} = \underline{\rho}_k - \underline{a}_k = (\{\rho_{k,1}\}, \cdots, \{\rho_{k,m}\})^T,$
   where $\underline{a}_k = \lfloor \underline{\rho}_k \rfloor$ and $\underline{\beta}_k = \{\underline{\rho}_k\}$ are called the polynomial part and the remaining part of $\underline{\rho}_k$, respectively.

Conversely, given a multidimensional continued fraction C, we can obtain a unique multi-sequence $\underline{s}$ such that C is the multidimensional continued fraction expansion of $\underline{s}$. Define the matrix B_k of order $m+1$ iteratively as below:

$$B_0 = I_{m+1}, \quad B_k = B_{k-1}E_{h_k}A(\underline{a}_k), \quad k \geq 1,$$

where I_{m+1} is the identity matrix of order $m+1$, $E(h_k)$ is a matrix of order $m+1$ which can be obtained by exchanging the h_k-th column and the $(m+1)$-th column of I_{m+1}, and

$$A(\underline{a}_k) = \begin{pmatrix} I_m & \underline{a}_k \\ \mathbf{0} & 1 \end{pmatrix}.$$

Denote by $(\underline{p}_k(z), q_k(z))^T$ the last column vector of the matrix $B_k(z)$, where $\underline{p}_k(z) = (p_{k,1}(z), p_{k,2}(z), \cdots, p_{k,m}(z))^T$ is an m-tuple of polynomials over F and $q_k(z)$ is a polynomial over F. Define

$$\frac{\underline{p}_k(z)}{q_k(z)} = \left(\frac{p_{k,1}(z)}{q_k(z)}, \frac{p_{k,2}(z)}{q_k(z)}, \cdots, \frac{p_{k,m}(z)}{q_k(z)} \right)^T,$$

which is called the k-th approximant of C. It is shown [2, 3] that as k goes into infinity, $\frac{\underline{p}_k(z)}{q_k(z)}$ converges to a multi-sequence $\underline{s} = (s_1, s_2, \cdots, s_m)^T$ in the sense that for any integer n, there exists an integer k_0 such that

$$v\left(s_j - \frac{p_{k,j}(z)}{q_k(z)} \right) > n, \quad 1 \leq j \leq m, \quad \forall k \geq k_0.$$

Associated with $C(\underline{s})$, we denote

$$t_k = \deg(a_{k,h_k}(z))(\geq 1), \tag{8}$$

$$d_k = \sum_{1 \leq i \leq k} t_i, \quad d_0 = 0, \tag{9}$$

$$v_{k,j} = \sum_{1 \leq i \leq k, h_i = j} t_i, \tag{10}$$

$$v_k = v_{k,h_k}, \tag{11}$$

$$n_k = v_k + d_{k-1}. \tag{12}$$

From the definition we have

$$d_k = \sum_{1 \leq j \leq m} v_{k,j} \tag{13}$$

$$d_k \leq n_k < 2d_k, \tag{14}$$

$$\lim_{k \to \infty} n_k = \infty. \tag{15}$$

It is known [2] that

$$n_k \leq n_{k+1} \quad \forall k \geq 1.$$

The linear complexity of the length n prefix $\underline{s}^n$ of an m-tuple sequence $\underline{s}$ is denoted by $L_{\underline{s}}(n)$ and defined to be the length of the shortest m-tuple linear feedback shift register which generates $\underline{s}^n$. As in the one-dimensional case, $L_{\underline{s}}(n)$ can be read out immediately from parameters shown above [3]:

$$L_{\underline{s}}(n) = 0 \quad \text{for} \quad n < n_1,$$
$$L_{\underline{s}}(n) = d_k \quad \text{for} \quad n_k \leq n < n_{k+1}, \quad \forall k \geq 0. \tag{16}$$

3   Main Results

Associated with $C(\underline{s})$, we define parameters $l(k,j)$, $r(k,j)$, J and m' as follows. The integer $l(k,j)$ is the maximum i such that $1 \leq i \leq k$ and $h_i = j$ if such i exists, and $l(k,j) = 0$ otherwise. The integer $r(k,j)$ is the minimum i such that $i > k$ and $h_i = j$ if such i exists, and $r(k,j) = 0$ otherwise. J is the set of integers between 1 and m that appear in the index sequence $\{h_k\}_{1 \leq k < \infty}$ infinitely many times. In other words, $J = \{j \mid 1 \leq j \leq m, \quad r(k,j) \neq 0, \forall k\}$. We let $m' = |J|$.

Lemma 1.

1. $v_{l(k,j)}$ is bounded for $j \notin J$ and $k \geq 1$.

2. $l(k,j) \leq k < r(k,j)$.

3. $\lim_{k \to \infty} l(k,j) = \infty \qquad \forall j \in J$.

4. $h_i \neq j \qquad \text{if } l(k,j) < i < r(k,j),$
   $h_i = j \qquad \text{if } i = l(k,j) \text{ or } i = r(k,j).$

5. $v_{r(k,j)} = v_{l(k,j)} + t_{r(k,j)}.$

6. $n_{r(k,j)} = v_{r(k,j)} + d_{r(k,j)-1} = v_{l(k,j)} + d_{r(k,j)} = n_{l(k,j)} + \sigma_{l(k,j),r(k,j)},$
   where $\sigma_{a,b} = \sum_{a \leq i < b} t_i$ for $1 \leq a < b$.

7. $n_{r(k,h_k)} = n_k + \sigma_k$, where $\sigma_k = \sigma_{k,r(k,h_k)}.$

The main results of this paper are the following three theorems and proved in the next section.

Theorem 1. (Formula for superior limit and inferior limit)

$$\limsup_{n \to \infty} \frac{L_{\underline{s}}(n)}{n} = \limsup_{k \to \infty} \frac{d_k}{n_k} = \frac{1}{1 + \liminf_{k \to \infty} \frac{v_k - t_k}{d_k}},$$

$$\liminf_{n \to \infty} \frac{L_{\underline{s}}(n)}{n} = \liminf_{k \to \infty} \frac{d_{k-1}}{n_k} = \frac{1}{1 + \limsup_{k \to \infty} \frac{v_k}{d_{k-1}}}.$$

If $\{\frac{L_{\underline{s}}(n)}{n}\}_{n\geq 1}$ is convergent, then $\{\frac{d_k}{n_k}\}_{k\geq 1}$ and $\{\frac{d_{k-1}}{n_k}\}_{k\geq 1}$ are convergent, and

$$\lim_{n\to\infty} \frac{L_{\underline{s}}(n)}{n} = \lim_{k\to\infty} \frac{d_k}{n_k} = \lim_{k\to\infty} \frac{d_{k-1}}{n_k},$$

and

$$\lim_{k\to\infty} \frac{t_k}{d_{k-1}} = \lim_{k\to\infty} \frac{t_k}{d_k} = 0.$$

Theorem 2. *(Bounds for superior limit and inferior limit)*
Let $\beta = \limsup_{k\to\infty} \frac{t_k}{d_{k-1}}$. Then

$$\limsup_{n\to\infty} \frac{L_{\underline{s}}(n)}{n} \geq \frac{m'(1+\beta)}{m'(1+\beta)+1} \geq \frac{m'}{m'+1},$$

$$\liminf_{n\to\infty} \frac{L_{\underline{s}}(n)}{n} \leq \frac{m'}{m'+1+\beta} \leq \frac{m'}{m'+1}.$$

As a consequence, if $\{L_{\underline{s}}(n)/n\}_{n\geq 1}$ is convergent, then

$$\lim_{n\to\infty} \frac{L_{\underline{s}}(n)}{n} = \frac{m'}{m'+1}.$$

Theorem 3. *(Necessary and sufficient condition for convergence)*
The sequence $\{L_{\underline{s}}(n)/n\}_{n\geq 1}$ is convergent if and only if

$$\lim_{k\to\infty} \frac{\sigma_k}{d_k} = 0,$$

where $\sigma_k = \sum_{k\leq i\leq r(k,h_k)} t_i$.

From Theorems 2 and 3 we get

Corollary 1. *(Sufficient condition for convergence)*
Assume $\lim_{k\to\infty} \frac{t_k}{d_{k-1}} = 0$. Then

$$\lim_{n\to\infty} \frac{L_{\underline{s}}(n)}{n} = \frac{m}{m+1}$$

if there exists a positive integer B such that $k - l(k,h) < B$ for $\forall k$ and $\forall\, 1 \leq h \leq m$.

Proof of Corollary 1.   From Theorems 2 and 3 it is enough to prove $m' = m$ and $\lim_{k\to\infty} \frac{\sigma_k}{d_k} = 0$.

We have $m' = m$, i.e., every $h(1 \leq h \leq m)$ appears in the index sequence $\{h_k\}_{1\leq k<\infty}$ infinitely many times, since the condition about $l(k,h)$ means that for every $h(1 \leq h \leq m)$ and any $k > B$ there exists at least one $i(k < i \leq k+B)$ such that $h_i = h$.

We claim $\lim_{k\to\infty} \frac{t_{k+i}}{d_k} = 0$ for $\forall i \geq 0$. This can be proved by induction on $i \geq 0$. It is true for $i = 0$, since

$$\frac{t_k}{d_k} = \frac{t_k}{d_{k-1} + t_k} = \frac{\frac{t_k}{d_{k-1}}}{1 + \frac{t_k}{d_{k-1}}}.$$

Assume $\lim_{k\to\infty} \frac{t_{k+i}}{d_k} = 0$, then

$$\lim_{k\to\infty} \frac{t_{k+i+1}}{d_k} = \lim_{k\to\infty} \frac{t_{k+i+1}}{d_{k+1} - t_{k+1}} = \lim_{k\to\infty} \frac{\frac{t_{k+i+1}}{d_{k+1}}}{1 - \frac{t_{k+1}}{d_{k+1}}} = 0.$$

Finally we can prove $\lim_{k\to\infty} \frac{\sigma_k}{d_k} = 0$ by proving $r(k, h_k) \leq k + B$, since $\frac{\sigma_k}{d_k} = \sum_{k \leq i \leq r(k,h_k)} \frac{t_{k+i}}{d_k}$. From the assumption we have $r(k, h_k) - 1 - l(r(k, h_k) - 1, h_k) < B$. From this we have $r(k, h_k) \leq k + B$, since $l(r(k, h_k) - 1, h_k) = k$. □

4   Proof of Theorems 1, 2 and 3

Proof of Theorem 1.   Let

$$n_k \leq n < n_{k+1}. \tag{17}$$

From equation (16), i.e.,

$$\frac{L_{\underline{s}}(n)}{n} = \frac{d_k}{n}$$

we have

$$\frac{L_{\underline{s}}(n_{k+1} - 1)}{n_{k+1} - 1} = \frac{d_k}{n_{k+1} - 1} \leq \frac{L_{\underline{s}}(n)}{n} \leq \frac{d_k}{n_k} = \frac{L_{\underline{s}}(n_k)}{n_k}. \tag{18}$$

We have

$$\limsup_{n\to\infty} \frac{L_{\underline{s}}(n)}{n} = \limsup_{k\to\infty} \frac{d_k}{n_k}, \tag{19}$$

since from equations (17) and (18) we have

$$\limsup_{n\to\infty} \frac{L_{\underline{s}}(n)}{n} \leq \limsup_{k\to\infty} \frac{d_k}{n_k} = \limsup_{k\to\infty} \frac{L_{\underline{s}}(n_k)}{n_k} \leq \limsup_{n\to\infty} \frac{L_{\underline{s}}(n)}{n}.$$

Note that $n_k = d_{k-1} + v_k = d_k + (v_k - t_k)$ from equations (11) and (9). Thus we have

$$\frac{n_k}{d_k} = 1 + \frac{v_k - t_k}{d_k},$$

which leads to

$$\limsup_{k\to\infty} \frac{d_k}{n_k} = \frac{1}{1 + \liminf_{k\to\infty} \frac{v_k - t_k}{d_k}}.$$

Next we have

$$\liminf_{n\to\infty} \frac{L_{\underline{s}}(n)}{n} = \liminf_{k\to\infty} \frac{d_k}{n_{k+1} - 1}, \tag{20}$$

since from equations (17) and (18) we have

$$\liminf_{n\to\infty}\frac{L_{\underline{s}}(n)}{n}\leq\liminf_{k\to\infty}\frac{L_{\underline{s}}(n_{k+1}-1)}{n_{k+1}-1}=\liminf_{k\to\infty}\frac{d_k}{n_{k+1}-1}\leq\liminf_{n\to\infty}\frac{L_{\underline{s}}(n)}{n}.$$

We have

$$\liminf_{k\to\infty}\frac{d_k}{n_{k+1}-1}=\liminf_{k\to\infty}\frac{d_k}{n_{k+1}}, \tag{21}$$

since

$$0\leq\frac{d_k}{n_{k+1}-1}-\frac{d_k}{n_{k+1}}=\frac{d_k}{n_{k+1}(n_{k+1}-1)}\leq\frac{1}{n_{k+1}}$$

and $\lim_{k\to\infty}\frac{1}{n_{k+1}}=0$.

Note that $n_{k+1}=d_k+v_{k+1}$ from equation (12), so we have

$$\frac{n_{k+1}}{d_k}=1+\frac{v_{k+1}}{d_k},$$

which leads to

$$\liminf_{n\to\infty}\frac{L_{\underline{s}}(n)}{n}=\liminf_{k\to\infty}\frac{d_k}{n_{k+1}}=\frac{1}{1+\limsup_{k\to\sup}\frac{v_{k+1}}{d_k}} \tag{22}$$

from equations (20) and (21).

Let us consider the case where $\{\frac{L_{\underline{s}}(n)}{n}\}_{n\geq1}$ is convergent. First we can prove the convergence of $\{\frac{d_k}{n_k}\}_{k\geq1}$ by showing $\limsup_{k\to\infty}\frac{d_k}{n_k}=\liminf_{k\to\infty}\frac{d_k}{n_k}$, since we have

$$\limsup_{k\to\infty}\frac{d_k}{n_k}=\limsup_{n\to\infty}\frac{L_{\underline{s}}(n)}{n}=\liminf_{n\to\infty}\frac{L_{\underline{s}}(n)}{n}=\liminf_{k\to\infty}\frac{d_k}{n_{k+1}}$$

$$\leq\liminf_{k\to\infty}\frac{d_k}{n_k}\leq\limsup_{k\to\infty}\frac{d_k}{n_k}$$

from equations (19) and (22). In a similar way we can prove the convergence of $\{\frac{d_k}{n_{k+1}}\}_{k\geq1}$.

If all of $\{\frac{L_{\underline{s}}(n)}{n}\}_{n\geq1}$, $\{\frac{d_k}{n_k}\}_{k\geq1}$ and $\{\frac{d_k}{n_{k+1}}\}_{k\geq1}$ are convergent, then from (19) and (22) we have

$$\lim_{n\to\infty}\frac{L_{\underline{s}}(n)}{n}=\lim_{k\to\infty}\frac{d_k}{n_k}=\lim_{k\to\infty}\frac{d_{k-1}}{n_k}.$$

We have $\lim_{k\to\infty}\frac{t_k}{d_k}=0$, since using the relation $n_k\leq2d_k$ we have

$$0\leq\lim_{k\to\infty}\frac{t_k}{d_k}\leq2\lim_{k\to\infty}\frac{t_k}{n_k}=2\lim_{k\to\infty}\frac{d_k-d_{k-1}}{n_k}=0.$$

The relation $\lim_{k\to\infty}\frac{t_k}{d_{k-1}}=0$ follows from

$$\frac{t_k}{d_{k-1}}=\frac{t_k}{d_k-t_k}=\frac{\frac{t_k}{d_k}}{1-\frac{t_k}{d_k}}.$$

$\square$

Proof of Theorem 2.   Denote

$$x_k = \frac{v_k - t_k}{d_k}, \qquad y_k = \frac{v_k}{d_{k-1}}$$

and

$$l = \liminf_{k \to \infty} x_k, \qquad u = \limsup_{k \to \infty} y_k$$

From Theorem 1 it is enough to prove

$$l \le \frac{1}{m'(1 + \beta)}, \qquad u \ge \frac{1 + \beta}{m'}. \tag{23}$$

From equations (9), (13) and Lemma 1 we have

$$x_{r(k,j)} = \frac{v_{r(k,j)} - t_{r(k,j)}}{d_{r(k,j)}} = \frac{v_{l(k,j)}}{d_{r(k,j)}} \le \frac{v_{l(k,j)}}{d_{k+1}},$$

and

$$y_{l(k,j)} = \frac{v_{l(k,j)}}{d_{l(k,j)-1}} \ge \frac{v_{l(k,j)}}{d_{k-1}}.$$

Hence

$$\sum_{j \in J} x_{r(k,j)} \le \frac{\sum_{1 \le j \le m} v_{l(k,j)}}{d_{k+1}} = \frac{\sum_{1 \le j \le m} v_{k,j}}{d_{k+1}} = \frac{d_k}{d_{k+1}} = \frac{1}{1 + \frac{t_{k+1}}{d_k}},$$

and

$$\sum_{j \in J} y_{l(k,j)} \ge \frac{\sum_{1 \le j \le m} v_{l(k,j)}}{d_{k-1}} = \frac{\sum_{1 \le j \le m} v_{k,j}}{d_{k-1}} = \frac{d_k}{d_{k-1}} = 1 + \frac{t_k}{d_{k-1}}.$$

Note that

$$\lim_{k \to \infty} r(k,j) = \lim_{k \to \infty} l(k,j) = \infty, \qquad \forall j \in J.$$

We have

$$l = \liminf_{k \to \infty} x_k \le \liminf_{k \to \infty} x_{r(k,j)}, \qquad \forall j \in J,$$

and

$$u = \limsup_{k \to \infty} y_k \ge \limsup_{k \to \infty} y_{l(k,j)}, \qquad \forall j \in J.$$

We can prove equation (23), since we have

$$m'l \le \sum_{j \in J} \liminf_{k \to \infty} x_{r(k,j)} \le \liminf_{k \to \infty} \sum_{j \in J} x_{r(k,j)} \le \liminf_{k \to \infty} \frac{1}{1 + \frac{t_{k+1}}{d_k}}$$

$$= \frac{1}{1 + \limsup_{k \to \infty} \frac{t_{k+1}}{d_k}} = \frac{1}{1 + \beta},$$

and

$$m'u \geq \sum_{j \in J} \limsup_{k \to \infty} y_{l(k,j)} \geq \limsup_{k \to \infty} \sum_{j \in J} y_{l(k,j)} \geq 1 + \limsup_{k \to \infty} \frac{t_k}{d_{k-1}} = 1 + \beta.$$

□

We need the following lemma to prove Theorem 3.

Lemma 2.

$$m'n_{k+1} - \sum_{j \in J} \sigma_{k+1,r(k,j)} \leq (m'+1)d_k - \sum_{j \notin J} v_{l(k,j)} \leq m'n_k + \sum_{j \in J} \sigma_{l(k,j),k}.$$

Proof of Lemma 2.   Define N_k and D_k by

$$N_k = \sum_{j \in J} n_{l(k,j)} \quad \text{and} \quad D_k = (m'+1)n_{k+1} - \sum_{j \notin J} v_{l(k,j)}.$$

We have

$$m'n_{k+1} - \sum_{j \in J} \sigma_{l(k,j),k} - \sum_{j \in J} \sigma_{k+1,r(k,j)} \leq N_k \leq m'n_k, \tag{24}$$

since from Lemma 1 we have

$$N_k = \sum_{j \in J} (n_{r(k,j)} - \sigma_{l(k,j),r(k,j)}) \geq m'n_{k+1} - \sum_{j \in J} \sigma_{l(k,j),r(k,j)}$$

and $N_k \leq \sum_{j \in J} n_k = m'n_k$.
    We have

$$D_k = N_k + \sum_{j \in J} \sigma_{l(k,j),k}, \tag{25}$$

since from equation (13) and Lemma 1 we have

$$N_k + \sum_{j \notin J} v_{l(k,j)} = \sum_{j \in J} (v_{l(k,j)} + d_{l(k,j)-1}) + \sum_{j \notin J} v_{l(k,j)}$$

$$= \sum_{1 \leq j \leq m} v_{l(k,j)} + \sum_{j \in J} d_{l(k,j)-1}$$

$$= d_k + \sum_{j \in J} (d_{l(k,j)-1} + \sigma_{l(k,j),k}) - \sum_{j \in J} \sigma_{l(k,j),k}$$

$$= d_k + \sum_{j \in J} d_k - \sum_{j \in J} \sigma_{l(k,j),k}$$

$$= (m'+1)d_k - \sum_{j \in J} \sigma_{l(k,j),k}.$$

We can prove Lemma 2 from equations (24) and (25).   □

Proof of Theorem 3.

"If part": From Lemma 2 we get

$$\frac{d_k}{n_k} \leq \frac{m'}{m'+1} + \frac{\sum_{j \notin J} v_{l(k,j)}}{(m'+1)n_k} + \frac{\sum_{j \in J} \sigma_{l(k,j),k}}{(m'+1)n_k}. \tag{26}$$

We have

$$\lim_{k \to \infty} \frac{\sigma_{l(k,j)}}{d_{l(k,j)}} = 0, \quad \text{for} \quad j \in J,$$

since $\{\frac{\sigma_{l(k,j)}}{d_{l(k,j)}}\}_{k \geq 1}$ is a subsequence of $\{\frac{\sigma_k}{d_k}\}_{k \geq 1}$.

Note that $\sigma_{l(k,j),k} \leq \sigma_{l(k,j),r(k,j)} = \sigma_{l(k,j)}$ and $d_{l(k,j)} \leq d_k \leq d_k + v_{k-1,h_k} = n_k$, so we have

$$\frac{\sigma_{l(k,j),k}}{(m'+1)n_k} \leq \frac{\frac{\sigma_{l(k,j)}}{d_{l(k,j)}}}{m'+1},$$

and from this we have

$$\lim_{k \to \infty} \frac{\sum_{j \in J} \sigma_{l(k,j),k}}{(m'+1)n_k} = 0. \tag{27}$$

Note that $\sum_{j \notin J} v_{l(k,j)}$ is bounded, so we have

$$\lim_{k \to \infty} \frac{\sum_{j \notin J} v_{l(k,j)}}{(m'+1)n_k} = 0. \tag{28}$$

From equations (26), (27)and (28) we have

$$\limsup_{k \to \infty} \frac{d_k}{n_k} \leq \frac{m'}{m'+1}. \tag{29}$$

From Lemma 2 we also get

$$\frac{m'}{m'+1} \leq \frac{d_k}{n_{k+1}} + \frac{\sum_{j \in J} \sigma_{k+1,r(k,j)}}{(m'+1)n_{k+1}} - \frac{\sum_{j \notin J} v_{l(k,j)}}{(m'+1)n_{k+1}}.$$

We have

$$\frac{m'}{m'+1} \leq \liminf_{k \to \infty} \frac{d_k}{n_{k+1}}, \tag{30}$$

since similarly we have

$$\lim_{k \to \infty} \frac{\sum_{j \in J} \sigma_{k+1,r(k,j)}}{(m'+1)n_{k+1}} = 0 \quad \text{and} \quad \lim_{k \to \infty} \frac{\sum_{j \notin J} v_{l(k,j)}}{(m'+1)n_{k+1}} = 0.$$

From equations (29), (30) and Theorem 1 we have

$$\frac{m'}{m'+1} \leq \liminf_{k \to \infty} \frac{d_k}{n_{k+1}} = \liminf_{k \to \infty} \frac{L_{\underline{s}}(n)}{n} \leq \limsup_{k \to \infty} \frac{L_{\underline{s}}(n)}{n}$$

$$= \limsup_{k \to \infty} \frac{d_k}{n_k} \leq \frac{m'}{m'+1},$$

which leads to

$$\liminf_{k\to\infty} \frac{L_{\underline{s}}(n)}{n} = \limsup_{k\to\infty} \frac{L_{\underline{s}}(n)}{n}.$$

"Only if" part: Denote

$$\Delta_k = \frac{L_{\underline{s}}(n_{r(k,h_k)})}{n_{r(k,h_k)}} - \frac{L_{\underline{s}}(n_k)}{n_k}.$$

By the assumption of convergence of $\frac{L_{\underline{s}}(n)}{n}$ we have

$$\lim_{k\to\infty} \Delta_k = 0.$$

From Theorems 1 and 2 we have

$$\lim_{k\to\infty} \frac{d_{k-1}}{n_k} = \lim_{n\to\infty} \frac{L_{\underline{s}}(n)}{n} = \frac{m'}{m'+1} \quad \text{and} \quad \lim_{k\to\infty} \frac{t_k}{n_k} = 0.$$

Note that

$$\Delta_k = \frac{d_{r(k,h_k)}}{n_{r(k,h_k)}} - \frac{d_k}{n_k} = \frac{d_{k-1} + \sigma_k}{n_k + \sigma_k} - \frac{d_{k-1} + t_k}{n_k}$$

$$= -\frac{t_k}{n_k} + \frac{\sigma_k(n_k - d_{k-1})}{n_k(n_k + \sigma_k)} = -\frac{t_k}{n_k} + \frac{\frac{\sigma_k}{n_k}\left(1 - \frac{d_{k-1}}{n_k}\right)}{1 + \frac{\sigma_k}{n_k}}.$$

From this we get

$$\lim_{k\to\infty} \frac{\frac{\sigma_k}{n_k}}{1 + \frac{\sigma_k}{n_k}} = \lim_{k\to\infty} \frac{\Delta_k + \frac{t_k}{n_k}}{1 - \frac{d_{k-1}}{n_k}} = 0. \tag{31}$$

Let $\alpha = \limsup_{k\to\infty} \frac{\sigma_k}{n_k} \geq 0$. We have

$$\lim_{k\to\infty} \frac{\frac{\sigma_k}{n_k}}{1 + \frac{\sigma_k}{n_k}} = \frac{\alpha}{1 + \alpha},$$

which together with equation (31) leads to $\alpha = 0$. From equation (15) we can prove $\lim_{k\to\infty} \frac{\sigma_k}{d_k} = 0$ if we show $\lim_{k\to\infty} \frac{\sigma_k}{n_k} = 0$, which can be shown by

$$0 \leq \liminf_{k\to\infty} \frac{\sigma_k}{n_k} \leq \limsup_{k\to\infty} \frac{\sigma_k}{n_k} = \alpha = 0.$$

$\square$

5   Conclusion

In this paper the multidimensional continued fraction algorithm [2], [3] is successfully applied to partially extend the results on the normalized linear complexity for a single sequence case [1] to a multi-sequence case.

Recently m-tuple multi-sequences with $\lim_{n\to\infty} \frac{L_{\underline{s}}(n)}{n} = \frac{m}{m+1}$, called multi-sequences with almost perfect linear complexity profile, have been studied [9]. From Theorem 2 the generalization of equation (6) to the m-tuple multi-sequence case shows that $\lim_{n\to\infty} \frac{L_{\underline{s}}(n)}{n}$ takes m different values from $\{\frac{1}{2}, \frac{2}{3}, \cdots, \frac{m}{m+1}\}$. Corollary 1 shows the existence of m-tuple sequences whose normalized linear complexity converges to the value $\frac{m}{m+1}$. The following m-tuple sequence $\underline{s}$ in equation (1) can be shown that its normalized linear complexity converges to the value $\frac{1}{2}$. All the sequences $s_2, \cdots, s_m$ except s_1 are periodic with period $N_2, \cdots, N_m$, respectively and $\lim_{n\to\infty} \frac{L_{s_1}(n)}{n} = \frac{1}{2}$. There exist a pair of polynomials $\{p_{1,n}(z), q_{1,n}(z)\}$ such that they are coprime and $\deg q_{1,n}(z)$ is minimal satisfying

$$v\left(\frac{p_{1,n}(z)}{q_{1,n}(z)} - \sum_{t\geq 0} s_t z^{-t-1}\right) \geq n+1.$$

There also exist $m-1$ polynomial pairs $\{p_2(z), q_2(z)\}, \cdots, \{p_m(z), q_m(z)\}$ such that $p_i(z)$ and $q_i(z)$ are coprime and $\deg q_i(z) = N_i$ $(2 \leq i \leq m)$ satisfying

$$\frac{p_i(z)}{q_i(z)} = \sum_{t\geq 0} s_{i,t} z^{-(t+1)}.$$

We have $\lim_{n\to\infty} \frac{L_{\underline{s}}(n)}{n} = \frac{1}{2}$, since for $n \geq \max\{2N_2, \cdots, 2N_m\}$ we have $\deg q_{1,n}(z) \leq L_{\underline{s}}(n) \leq \deg [q_{1,n}(z)q_2(z)\cdots q_m(z)]$.

Whether some generalization of (5) to the multi-sequence case exists or not is an open problem.

By the relation of multidimensional continued fractions and multi-sequences and the definition of J or m', it is easy to construct many multidimensional continued fractions such that they can take every value among $\frac{1}{2}, \frac{2}{3}, \cdots, \frac{m}{m+1}$, or equivalently, for every $2 \leq k \leq m$, there exist many multi-sequences $\underline{s}$ such that $\{\frac{L_{\underline{s}}(n)}{n}\}_{n\geq 1}$ is convergent and $\lim_{n\to+\infty} \frac{L_{\underline{s}}(n)}{n} = \frac{k}{k+1}$. For example, we consider the multidimensional continued fraction C satisfying the following conditions:

1. $1 \leq h_i \leq k$ and $h_i \equiv i \pmod{k}$ for $\forall i \geq 1$;
2. $\underline{a}_k = (a_{k,1}, a_{k,2}, \cdots, a_{k,m})$, where $a_{k,j} = 0$ if $j \neq h_k$ and $a_{k,h_k} = z$.

Assume that the above multidimensional continued fraction C converges to a multi-sequence $\underline{s}$. Then we have that the normalized linear complexity of $\underline{s}$ is convergent and converges to $\frac{k}{k+1}$.

A fast algorithm for computing the asymptotic value of normalized linear complexity numerically is desirable if we want to apply the results of this paper to practical sequences.

Acknowledgement. The authors wish to thank an anonymous referee for his useful suggestions that improved the presentation of this paper. This work is partly supported by Chinese Natural Science Foundation (Grant No. 60173016) and 973 Foundation (Grant No. 1999035804).

References

1. Z. Dai, S. Jiang, K. Imamura and G. Gong, "Asymptotic behavior of normalized linear complexity of ultimately non-periodic binary sequences", *IEEE Trans. on Information Theory*, vol. 50, pp. 2911–2915, Nov. 2004.
2. Z. Dai, K. Wang and D. Ye, "m-continued fraction expansions of multi-Laurent series", *ADVANCES IN MATHEMATICS (CHINA)*, vol. 33, No.2, pp. 246–248, 2004.
3. Z. Dai, K. Wang and D. Ye, " Multidimensional continued fraction and rational approximation", http://arxiv.org/abs/math.NT/0401141, Jan. 2004.
4. K. Imamura, W. Yoshida and M. Morii, "Two binary sequences and their linear complexities", *Proc. 1988 IEEE Int'l Symp. on Information Theory*, p. 216, Jun. 1988.
5. W.H. Mills, "Continued fractions and linear recurrences", *Math. Comp.*, vol. 29, pp.173–180, Jan. 1975.
6. H. Niederreiter, "The probabilistic theory of linear complexity", *Proc. EURO-CRYPT '88*, LNCS 330, Springer, 1988.
7. L.R. Welch and R.A. Scholtz, "Continued fractions and Berlekamp's algorithm", *IEEE Trans. on Information Theory*, vol. IT-25, pp. 19 -27, Jan. 1979.
8. Z. Dai and K.C. Zeng, "Continued fractions and Berlekamp- Massey algorithm", *Advances in Cryptology - AUSCRYPT '90*, LNCS 453, pp. 24 - 31, Springer-Verlag, 1990.
9. C. Xing, "Multi-sequences with almost perfect linear complexity profile and function field over finite fields", *J. of Complexity*, vol. 16, pp. 661–675, 2000.

A Unified View on Sequence Complexity Measures as Isometries

(Extended Abstract)

Michael Vielhaber[*]

Instituto de Matemáticas, Universidad Austral de Chile,
Casilla 567, Valdivia, Chile
uach@gmx.net

Abstract. We show how to model complexities (linear, 2–adic, tree, etc.), compression schemes (Lempel–Ziv, etc.), and predictors (Markov chains, etc.) in a uniform way as *isometries*.

This isometric setting allows to sort out nonrandom sequences as they violate the bounds of the "Law of the Iterated Logarithm".

We also consider the computational complexity of calculating the isometric models, as well as how to deal with finite sequences.

1 Introduction

Let $A = \{0, 1, \ldots, |A| - 1\} \subset \mathbb{N}_0$ be an alphabet. Given a finite prefix s of an infinite word a, $A^* \ni s \lhd a \in A^\omega$, we may <u>predict</u> what symbol comes next in a. If we are right (more often than wrong) we may <u>compress</u> the sequence since its <u>complexity</u> is low.

In Sections 3–6 we model several complexity measures (see [13]) as isometries. In Sections 7–9 we state bounds for the number of nonzeroes and lengths of zero runs as Lévy classes which permits us to recognize nonrandom sequences. The final sections cover finite sequences and computational complexity.

2 Rankings and Isometries

We model prediction, online (no preview) compression, and (prefix) complexity by a ranking $C: A^* \to S_A \cong S_{|A|}$ (symmetric group), where for each string $w \in A^*$ the permutation $C(w)$ on the alphabet A orders the symbols in decreasing probability of occurence after w.

Let $\overline{C} : A^* \to \mathbb{R}$ be a complexity measure. Then $\overline{C}$ induces a ranking $C: A^* \to S_A$ as follows: Given $w \in A^*$, we compute $\overline{C}(w0), \overline{C}(w1), \ldots, \overline{C}(w(|A| - 1))$ and sort these $|A|$ numbers as $\overline{C}(wa_0) \leq \overline{C}(wa_1) \leq \ldots \leq \overline{C}(wa_{|A|-1})$ (if a value occurs twice, the further ordering can be *e.g.* according to the alphabet A).

[*] Supported by Project FONDECYT 2001, No. 1010533 of CONICYT, Chile. Partially supported by DID–UACH.

T. Helleseth et al. (Eds.): SETA 2004, LNCS 3486, pp. 143–153, 2005.

We thus obtain the permutation $C(w) \in S_A$ of A as $C(w) : a_i \mapsto i$. Hence, $C(w)^{-1}(0) = a_0$ is the symbol predicted after w by $\overline{C}$, a_0 has highest probability to occur, wa_0 has lowest complexity.

Definition 1. *We define a "degree" on A^ω as $|a| = -k$, when $a_i = 0$ for $1 \leq i < k$ and $a_k \neq 0$, and $|0^\omega| = -\infty$. We define the "difference" $a - b$ componentwise, then $|a - b| = -k$ if and only if $a_i = b_i, i < k$, and $a_k \neq b_k$.*

An isometry is a function $\mathbf{C}: A^\omega \to A^\omega$ that preserves distance that is for $a, b \in A^\omega$ with $a_i = b_i, 1 \leq i < k$ and $a_k \neq b_k$, also $\mathbf{C}(a)_i = \mathbf{C}(b)_i, 1 \leq i < k$ and $\mathbf{C}(a)_k \neq \mathbf{C}(b)_k$, or shorter $|\mathbf{C}(a) - \mathbf{C}(b)| = |a - b|$.

Every ranking C induces an isometry $\mathbf{C}: A^\omega \to A^\omega$ on the set of infinite sequences:

$$\mathbf{C}(a_1 a_2 \ldots) = (C(a_1 \ldots a_{n-1})(a_n))_{n=1}^\infty$$

$$= \Big(C(\varepsilon)(a_1), C(a_1)(a_2), C(a_1 a_2)(a_3), \ldots \ldots, C(a_1 \ldots a_{n-1})(a_n), \ldots \Big)$$

We have $\mathbf{C}(a)_n = 0$, if and only if a_n has been predicted by C after $(a_1 \ldots a_{n-1})$ as having highest probability.

We will not be interested in the complexities as such but only in the behaviour of the induced isometries. As $\mathbf{C}$ is an isometry, the result $\mathbf{C}(a)$ of a randomly drawn (*i.i.d.*) sequence $a \in A^\omega$ will behave like a Bernoulli process. This allows, *regardless* of the development of the actual complexity, to apply the sharp bounds known for Bernoulli experiments to all complexity models.

In particular, we have a *uniform* treatment of the *borderline* between random and nonrandom behaviour (which is just the most interesting case: deterministically produced – hence nonrandom – sequences that should look like random ones).

In general the bordercase decisions for arbitrary complexities are difficult, but the Law of the Iterated Logarithm will give an easy-to-use criterion.

3   Linear and Jump Complexity [15, 22, 23]

For $A \equiv \mathbb{F}_q, a \in \mathbb{F}_q^\omega$, let $G(a) = \sum_{i=1}^\infty a_i x^{-i} = \cfrac{1}{A_1(x) + \cfrac{1}{A_2(x) + \cfrac{1}{A_3(x) + \ldots}}}$ be the generating function and its continued fraction expansion with nonconstant polynomials $A_i(x) \in \mathbb{F}_q[x] \backslash \mathbb{F}_q$. The sequence (A_i) is finite iff $G(a)$ is rational, that is a is ultimately periodic. Let $\pi: \mathbb{F}_q[x] \backslash \mathbb{F}_q \ni \sum_{i=0}^d r_i x^i \mapsto 0^{d-1} r_d r_{d-1} \ldots r_0 \in \mathbb{F}_q^{2d}$ and $\pi_D: \mathbb{F}_q[x] \backslash \mathbb{F}_q \ni \sum_{i=0}^d r_i x^i \mapsto 0^{d-1} r_d \in \mathbb{F}_q^d$ be the encoding of a nonconstant polynomial and its degree part, resp., by symbols from $\mathbb{F}_q$.

Let $\mathbf{K}(0^\omega) := 0^\omega$, and otherwise map a to the concatenation of the encodings

$$\mathbb{F}_q^\omega \ni a \mapsto G(a) \equiv \left\{ \begin{array}{l} (A_i)_{i=1}^n \mapsto (\pi(A_i))_{i=1}^n | 0^\omega \\ (A_i)_{i=1}^\infty \mapsto (\pi(A_i))_{i=1}^\infty \end{array} \right\} =: \mathbf{K}(a) \in \mathbb{F}_q^\omega$$

(where the upper concatenation refers to the rational case, the lower one to irrational $G(a)$), also let $\mathbf{K_D}(a)$ be the concatenation of the $\pi_D(A_i)$.

Proposition 1.

(i) $\mathbf{K}$ is an isometry on $\mathbb{F}_q^\omega$.

(ii) $\mathbf{K_D}$ is equidistributed in the sense of $|\{a \in A^{2n} \mid \mathbf{K_D}(a)_{i,i=1...n} = b\}| = 2^n$ for all $b \in A^n$.

Another isometry is $a \mapsto \mathbf{K_{BMA}}(a)$, where $\mathbf{K_{BMA}}(a)$ is just the discrepancy sequence of the original Berlekamp–Massey–Algorithm (and on the other hand, $\mathbf{K}$ can be obtained as the discrepancy sequence of a modified BMA) [4] [15].

The usual "LFSR" *linear complexity profile* (l.c.p.) of a sequence $a \in \mathbb{F}_q^\omega$ then is given by $L(a, n) = \sum_{i=1}^{k} deg(A_i)$, where A_k is the last partial denominator whose leading coefficient is encoded in $\mathbf{K}(a)_{i,i=1...n}$. $(L(a, n))$ is monotonously increasing and jumps, $L(a, i) > L(a, i-1)$, where $\mathbf{K}(a)_i$ encodes a leading coefficient.

The usual definition in linear feedback shift register (LFSR) theory is equivalent:

Theorem 1. [22]
$L(a, n)$ denotes the length of a shortest LFSR that produces $a_1 \ldots a_n$.

For later use we define the *linear complexity deviation*

$$m(a, n) = 2L(a, n) - n \in \mathbb{Z}.$$

m oscillates around zero and shows the deviation from a perfect l.c.p.

The *jump complexity* is the number of jumps in the l.c.p.,

$$J_\mathbf{K}(a, n) := |\; \{i \mid i \le n, \; L(a, i-1) < L(a, i)\}\;| = |\{i \mid i \le \frac{n}{2}, \; \mathbf{K_D}(a)_i \neq 0\}| + \delta$$

for some $\delta \in \{0, 1\}$.

Proposition 2. The largest jump in the l.c.p. $(= \max_{i \le k} deg(A_i))$ up to n is
$1 + \max\limits_{i \le n/2} \{l \mid \mathbf{K_D}(a)_{i-l} = \ldots = \mathbf{K_D}(a)_{i-1} = 0, \mathbf{K_D}(a)_i \neq 0, \}$.

4   2–Adic Complexity [9, 10]

For $A \equiv \mathbb{F}_2$, let $a \in \mathbb{F}_2^\omega$, $z = \sum_{i=1}^{\infty} a_i 2^{i-1} \in \mathbb{Z}_2$ be a dyadic integer. For $n \in \mathbb{N}$, let $\mathcal{L}(z, n) = \{(p, q) \in \mathbb{Z}^2 \mid p \equiv q \cdot z \bmod 2^n\}$ be the lattice of approximations of z up to a_n. Let $(c_n, d_n) \in \mathcal{L}(z, n) \backslash \{(0, 0)\}$ be a minimal approximation in the sense of $\max\{|c_n|, |d_n|\} \le \max\{|p|, |q|\}$ for all $(p, q) \in \mathcal{L}(z, n) \backslash \{(0, 0)\}$. Setting the initial value $(c_0, d_0) := (0, 1)$, we define a function $\mathbf{A} \colon \mathbb{F}_2^\omega \to \mathbb{F}_2^\omega$ by

$$\mathbf{A}(a)_i = \begin{cases} 0, & \text{if } (c_i, d_i) = (c_{i-1}, d_{i-1}) \\ 1, & \text{if } (c_i, d_i) \neq (c_{i-1}, d_{i-1}) \end{cases}$$

Theorem 2. [22] $\mathbf{A}$ is an isometry on $\mathbb{F}_2^\omega$.

We further define the *2–adic jump complexity* $J_\mathbf{A}$ as counting the number of changes in the 2–adic complexity profil: $J_\mathbf{A}(a)(n) := \sum_{i=1}^{n} \mathbf{A}(a)_i$, which should behave like $J_\mathbf{A}(n) \approx \frac{n}{2}$. In analogy to m, we also define the *2–adic jump complexity deviation* $m_\mathbf{A}(n) := 2 \cdot J_\mathbf{A}(n) - n \in \mathbb{Z}$.

Remark. Whereas the original 2–adic complexity $\phi(a, n) = \log_2 \max(|c_n|, |d_n|)$ is not usually integral, $J_\mathbf{A}$ and $m_\mathbf{A}$ get us again in the world of coin tossing.

5   Tree Complexity [14]

Tree complexity (a similar concept is automaticity [21]) determines the number of distinct subtrees arranging $s \in A^*$ in a "heap". Let $s = a_1 \ldots a_n \in A^*$ and put $a_i = \#$ for $i > n$. The subtree patterns of $a = s\#^\omega$ are $P''(s) = \{(a_k, a_{2k}, a_{2k+1}, \ldots, a_{2^i k}, \ldots, a_{2^i k + 2^i - 1}, \ldots) \mid 1 \leq k \leq n\} \subset (A \cup \{\#\})^\omega$.

Heap:

Let $P'(s) = \{v \in A^* \mid v\#^\omega \in P''(s)\}$, and $P(s) = \{v \in P'(s) \mid \nexists v' \neq v \in P'(s) : v \lhd v'\}$ (prefix free). Then $\overline{T}(s) := |P(s)|$ is the tree complexity of s.

Two examples: Let $|A| = 10$ and $s_1 = 123456789$, then $P'(s_1) = \{123456789, 24589, 367, 489, 5, 6, 7, 8, 9\} = P(s_1)$ and thus $\overline{T}(s_1) = 9$.

For $|A| = 2$ and $s_2 = 101000110$, $P'(s_2) = \{101000110, 00010, 101, 010, 0, 1\}$ and $P(s_2) = \{101000110, 00010, 010\}$, whence $\overline{T}(s_2) = 3$.

We define a ranking T and an isometry $\mathbf{T}$ on A^ω as follows: For all $s \in A^*$ and $\alpha \in A$, compute the set $P(s\alpha)$. Define $T(s) \in S_A$ by $T(s)(\alpha) < T(s)(\beta)$ if and only if $\overline{T}(s\alpha) < \overline{T}(s\beta)$, or $\overline{T}(s\alpha) = \overline{T}(s\beta)$ and $\alpha < \beta$. $\mathbf{T}$ is defined in terms of T as in section 2 for arbitrary $\mathbf{C}$.

Let α_s be the predicted symbol that is $\mathbf{T}(s)(\alpha_s) = 0$. Then $\overline{T}(s\alpha_s) = \overline{T}(s)$.

6   Prediction and Compression

When working over the binary alphabet $A = \{0, 1\}$, we can turn a predictor function $\overline{P} : A^* \rightarrow A$, $w \mapsto \overline{P}(w)$ where $\overline{P}(w)$ is the symbol predicted after w, into a ranking $P : A^* \rightarrow S_A$, $P(w)(\overline{P}(w)) = 0$, the symbol predicted after w receives rank 0, and $P(w)(1 - \overline{P}(w)) = 1$, the other symbol then has rank 1.

Since $\overline{P}$ singles out only *one* symbol, for $|A| > 2$ it would have no unique induced ranking. From P we get an isometry $\mathbf{P}$ as before,

$$\mathbf{P}(a_1, a_2, a_3, \ldots) = \Big(P(\varepsilon)(a_1), P(a_1)(a_2), P(a_1 a_2)(a_3), \ldots, P(a_1 \ldots a_{n-1})(a_n), \ldots \Big).$$

A compression scheme (e.g. Lempel–Ziv LZ) can be modelled as a function $\overline{L} : A^* \rightarrow \mathbb{R}$, $w \mapsto \overline{L}(w) = |\mathrm{LZ}(w)|$, the length of the compressed output.

This function also induces a ranking: For every prefix w, we sort A according to increasing complexity, $\overline{L}(wa_0) \leq \overline{L}(wa_1) \leq \overline{L}(wa_2) \leq \ldots$, and then set $L(w)(a_i) = i$. We then obtain the isometry $\mathbf{L}$ as usual.

7   Lévy Classes

We have seen how to model compression, prediction, and complexity as rankings and then as isometries. In the sequel we will recall the Law of the Iterated Logarithm and its consequences for isometries.

In order to describe the "typical" behaviour of a function $f(a, n)$, for $a \in A^\omega, n \in \mathbb{N}$, we introduce Lévy classes (Upper and Lower Class, resp.) of real sequences:

(i) $UUC(f) = \{x \in \mathbb{R}^\mathbb{N} \mid \forall_\mu\, a, \exists n_0 \in \mathbb{N}, \forall n > n_0 : f(a, n) < x_n\}$

(ii) $ULC(f) = \{x \in \mathbb{R}^\mathbb{N} \mid \forall_\mu\, a, \forall n_0 \in \mathbb{N}, \exists n > n_0 : f(a, n) \geq x_n\}$

(iii) $LUC(f) = \{x \in \mathbb{R}^\mathbb{N} \mid \forall_\mu\, a, \forall n_0 \in \mathbb{N}, \exists n > n_0 : f(a, n) \leq x_n\}$

(iv) $LLC(f) = \{x \in \mathbb{R}^\mathbb{N} \mid \forall_\mu\, a, \exists n_0 \in \mathbb{N}, \forall n > n_0 : f(a, n) > x_n\}$

Thus for all choices $x^{(1)} \in LLC(f)$, $x^{(2)} \in LUC(f)$, $x^{(3)} \in ULC(f)$, $x^{(4)} \in UUC(f)$ and for almost all sequences $a \in A^\infty$, we have $x_n^{(1)} < f(a)(n) < x_n^{(4)}$ asymptotically, but μ–almost all sequences will make f oscillate so much as to repeatedly leave the interval $(x^{(2)}, x^{(3)})$ of unavoidable oscillation.

8   Lévy Classes for Occurrence Counts

In this and the following section we give known results as compiled by Révész. Most theorems are given only for the practically most important case $|A| = 2$.

We partition A into the two classes A_0 (successes) and $A_1 = A \backslash A_0$ (failures), e.g. $A_0 = \{0\}$ for correct prediction, and set $p_i = |A_i|/|A|$.

We then count the successes up to n as

$$S(a, n) = |\{i \leq n \mid a_i \in A_0\}|.$$

The Law of the Iterated Logarithm now is stated as:

Theorem 3. *Law of the Iterated Logarithm* ([7], Section VIII.5.)

(i) $\overline{\lim}_{n \to \infty} (S(a, n) - n \cdot p_0) / \sqrt{n \cdot p_0 \cdot p_1 \cdot 2 \cdot \log\log n} = +1$ μ − a.e.

(ii) $\underline{\lim}_{n \to \infty} (S(a, n) - n \cdot p_0) / \sqrt{n \cdot p_0 \cdot p_1 \cdot 2 \cdot \log\log n} = -1$ μ − a.e.

Slightly sharper bounds for the case $|A| = 2, A_i = \{i\}$ use the difference between successes and failures

$$D(a, n) = |\{i \leq n \mid a_i \in A_0\}| - |\{i \leq n \mid a_i \in A_1\}|.$$

Theorem 4. *Law of the Iterated Logarithm for tossing a fair coin* ([5],[8],[11], [19–5.2])

$$f(t) \in UUC(D(a, t)/\sqrt{t}) \iff \sum_{n=1}^\infty \frac{f(n)}{n} \cdot e^{-\frac{f(n)^2}{2}} < \infty$$
$$f(t) \in ULC(D(a, t)/\sqrt{t}) \iff \sum_{n=1}^\infty \frac{f(n)}{n} \cdot e^{-\frac{f(n)^2}{2}} = \infty$$
$$f(t) \in LUC(D(a, t)/\sqrt{t}) \iff -f(t) \in ULC(D(a, t)/\sqrt{t})$$
$$f(t) \in LLC(D(a, t)/\sqrt{t}) \iff -f(t) \in UUC(D(a, t)/\sqrt{t})$$

Some example functions bounding $D(a, t)/\sqrt{t}$ show that we can not avoid oscillations on the order of the "iterated logarithm" $\log\log(t)$ times $\sqrt{t}$.

Example 1. For all $\varepsilon > 0$ we have:

$$
\begin{aligned}
(2 \cdot \log \log(t) + (3 + \varepsilon) \cdot \log \log \log t)^{1/2} &\in UUC(D(a,t)/\sqrt{t}) \\
(2 \cdot \log \log(t) + \qquad\quad \log \log \log t)^{1/2} &\in ULC(D(a,t)/\sqrt{t}) \\
-(2 \cdot \log \log(t) + \qquad\quad \log \log \log t)^{1/2} &\in LUC(D(a,t)/\sqrt{t}) \\
-(2 \cdot \log \log(t) + (3 + \varepsilon) \cdot \log \log \log t)^{1/2} &\in LLC(D(a,t)/\sqrt{t})
\end{aligned}
$$

As a modified version of Theorem 3 we obtain in the setting of isometries:

Theorem 5. *The Law of the Iterated Logarithm for Isometries*
For an isometry $\mathbf{C}$ *on* A^{ω}, *for any partition* $A = A_0 \cup A_1$ *of* A, *let* $b = \mathbf{C}(a)$
and $D(b, n)$ *as before. Then*

$$
\left.
\begin{aligned}
&(i) \quad \overline{\lim}_{n \to \infty} \\
&(ii) \quad \underline{\lim}_{n \to \infty}
\end{aligned}
\right\}
\quad
\frac{D(b,n)|A| - n \cdot (|A_0| - |A_1|)}{\sqrt{8|A_0||A_1| \cdot n \cdot \log \log n}}
\quad =
\begin{cases}
+1 \\
-1
\end{cases}
\mu - \text{a.e.}
$$

Remark 1. The preceeding theorem sorts out two classes of sequences: Those too bad (rational ones under $\mathbf{K}$ give an excess of zeroes, perfect l.c.p.'s typically give an excess of ones), but also those *too good* to be true: e.g., if all partial denominators had degree 2, $\mathbf{K}(a)$ would be of the form $\mathbf{K}(a) = (01**)^{\infty}$, which for random $*$ leads to equidistribution, but the fluctuations depend only on *half* of the symbols and thus

$$
\overline{\lim}_{n \to \infty} D(b,n)|A| - n \cdot (|A_0| - |A_1|) \big/ \sqrt{8|A_0||A_1| \cdot n/2 \cdot \log \log n/2} = 1 \; \mu - \text{a.e.}
$$

and hence

$$
\overline{\lim}_{n \to \infty} D(b,n)|A| - n \cdot (|A_0| - |A_1|) \big/ \sqrt{8|A_0||A_1| \cdot n \cdot \log \log n} = \frac{1}{\sqrt{2}} \; \mu - \text{a.e.},
$$

similarly for $\underline{\lim}$. So these sequences can be identified as nonrandom, as their fluctuations are too *small*. A more trivial example for "too good" a behaviour is the sequence $(01)^{\omega}$ with $\overline{\lim} \ldots = \underline{\lim} \ldots = 0$.

For the jump complexity $J_{\mathbf{K}}$, only half of the symbols of $\mathbf{K}$ are considered in $\mathbf{K_D}$ and thus

Corollary 1.

$$
(i) \; \overline{\lim}_{n \to \infty} \left(J_{\mathbf{K}}(a,n) - \frac{n}{2} \cdot \frac{q-1}{q} \right) \big/ \sqrt{\frac{q-1}{q^2} n \cdot \log \log \frac{n}{2}} = +1 \; \mu - \text{a.e.}
$$

$$
(ii) \; \underline{\lim}_{n \to \infty} \left(J_{\mathbf{K}}(a,n) - \frac{n}{2} \cdot \frac{q-1}{q} \right) \big/ \sqrt{\frac{q-1}{q^2} n \cdot \log \log \frac{n}{2}} = -1 \; \mu - \text{a.e.}
$$

We also have

Corollary 2. *Law of the Iterated Logarithm for the 2–adic Jump Complexity Deviation*

$$
(i) \; \overline{\lim}_{n \to \infty} m_{\mathbf{A}}(n) \big/ \left(\sqrt{2n \cdot \log \log n} \right) = +1 \; \mu - \text{a.e.}
$$

$$
(ii) \; \underline{\lim}_{n \to \infty} m_{\mathbf{A}}(n) \big/ \left(\sqrt{2n \cdot \log \log n} \right) = -1 \; \mu - \text{a.e.}
$$

9   Maximum Run Lengths for $|A| = 2$

Let the length of the largest sequence of zeroes in $a_1, \ldots, a_n$ be $Z(a, n)$. This maximum run length behaves roughly like $\log n$, more precisely:

Theorem 6. *Lévy classes for $Z(a, n)$ for $|A| = 2$ [6, 20]*

$$\sum_{n=1}^{\infty} 2^{-f(n)} \begin{cases} < \infty \\ = \infty \end{cases} \iff f(n) \in \begin{cases} UUC(Z(a, n)), \ e.g. \ f(n) = \log_2(n)(1 + \varepsilon) \\ ULC(Z(a, n)), \ e.g. \ f(n) = \log_2(n) \end{cases}$$

$$f(n) = \lfloor \log_2(n) - \log_2 \log_2 \log_2(n) + \log_2 \log_2(e) - 1 + \varepsilon \rfloor \in LUC(Z(a, n)), \forall \varepsilon > 0$$

$$f(n) = \lfloor \log_2(n) - \log_2 \log_2 \log_2(n) + \log_2 \log_2(e) - 2 - \varepsilon \rfloor \in LLC(Z(a, n)), \forall \varepsilon > 0$$

Corollary 3. [22]
(i) The sequences with d–perfect linear complexity profile comprise a set of measure $\mu = 0$.
(ii) μ–almost all sequences have a good linear complexity profile

The largest jump in the l.c.p. up to n has height $1 + Z(b, n/2)$ by Proposition 2.

Corollary 4. *For μ–almost all $a \in A^\omega$ and all n, the largest jump $1 + Z(b, n/2)$ must be of the order $(\log_2 n)(1 + o(1))$.*

Theorem 7. *(i) For the lengths $Z_2(n), Z_3(n), \ldots$ of the second, third, $\ldots$ largest run of zeroes, Deheuvels [3] has found the following functions (where $\log_2^{(j)} :=$ $\log_2(\log_2^{(j-1)})$ with $\log_2(x) = 0$ for $x < 1$, $\underline{\log_2}(x) = \log_2(x)$ for $x \geq 1$).*
For all $k \in \mathbb{N}, r \geq 2, \varepsilon > 0$ we have:

$$f(n) = \log_2(n) + \frac{1}{k} \cdot (\log_2^{(2)}(n) + \ldots + \log_2^{(r-1)}(n) + (1 + \varepsilon) \log_2^{(r)}(n))$$
$$\in UUC(Z_k(n))$$

$$f(n) = \log_2(n) + \frac{1}{k} \cdot (\log_2^{(2)}(n) + \ldots + \log_2^{(r-1)}(n) + \log_2^{(r)}(n)) \in ULC(Z_k(n))$$

$$f(n) = \lfloor \log_2(n) - \log_2 \log_2 \log_2(n) + \log_2 \log_2(e) + \varepsilon \rfloor \in LUC(Z_k(n))$$

$$f(n) = \lfloor \log_2(n) - \log_2 \log_2 \log_2(n) + \log_2 \log_2(e) - 2 - \varepsilon \rfloor \in LLC(Z_k(n))$$

(ii) Hence for the second, etc., largest degree $d_k(t)$ in the continued fraction expansion, we obtain:

$$f(t) = 1 + \log_2(\frac{t}{2}) + \frac{1}{k} \cdot (\log_2^{(2)}(\frac{t}{2}) + \ldots + \log_2^{(r-1)}(\frac{t}{2}) + (1 + \varepsilon) \log_2^{(r)}(\frac{t}{2}))$$
$$\in UUC(d_k(t))$$

$$f(t) = 1 + \log_2(\frac{t}{2}) + \frac{1}{k} \cdot (\log_2^{(2)}(\frac{t}{2}) + \ldots + \log_2^{(r-1)}(\frac{t}{2}) + \log_2^{(r)}(\frac{t}{2})) \in ULC(d_k(t))$$

$$f(t) = 1 + \lfloor \log_2(\frac{t}{2}) - \log_2 \log_2 \log_2(\frac{t}{2}) + \log_2 \log_2(e) + \varepsilon \rfloor \in LUC(d_k(t))$$

$$f(t) = 1 + \lfloor \log_2(\frac{t}{2}) - \log_2 \log_2 \log_2(\frac{t}{2}) + \log_2 \log_2(e) - 2 - \varepsilon \rfloor \in LLC(d_k(t))$$

10   The Finite Case

The Law of the Iterated Logarithm *in rigor* does not make any statement at all about finite sequences as we only deal with asymptotic behaviour.

"Finite Non/Random Sequences" live in the twilight zone described by Laplace's *"Sixth Principle"*, and the often–cited John v. Neumann *"Anyone who considers arithmetic methods of producing random digits is, of course, in a state of sin"*.

What we can do anyway in view of the Law of the Iterated Logarithm is to *declare* as nonrandom every finite sequence (and its corresponding cylinder set of continuations, hence we will throw away a set of strictly positive measure μ instead of μ-almost nothing) that does not fulfill the inequality

$$|S(a, n) - n \cdot p_0| \leq \sqrt{2 \cdot n \cdot \log \log n} + \Delta$$

for some $n \geq 3$ (for $n \leq 2$, $\sqrt{\log \log}$ is undefined), where $\Delta \in \mathbb{R}^+$ is some parameter chosen by us.

In the case of $A = \{0, 1\}$, for different Δ and sequence lengths n, we then throw away the following fractions μ of all sequences:

Δ	0.748809	0.74881	1	2	3	4	5
$n \mid \mu$							
32	0.25131	0.00508	0.00281	0.00031	0.000035	0.000004	0.0000003
64	0.25206	0.00626	0.00373	0.00064	0.000139	0.000032	0.000007
128	0.25271	0.00720	0.00450	0.00102	0.000315	0.000108	0.000038
256	0.25319	0.00785	0.00510	0.00139	0.000534	0.000233	0.000108
512	0.25355	0.00834	0.00556	0.00171	0.000756	0.000385	0.000210
1024	0.25382	0.00871	0.00590	0.00199	0.000966	0.000546	0.000332

We should obviously choose a Δ above $0.74881 = 3/2 - \sqrt{2 \cdot 3 \cdot \log \log 3}$.

11   Structured Sequences

The most structured sequences according to a ranking C, or an isometry $\mathbf{C}$, are those with finite support $\mathbf{C}(a) \in \varphi_0 := \{b \in A^\omega \mid \exists n_0, \forall i > n_0 : b_i = 0\}$ that is which are completely recognized by the model underlying C after n_0 symbols.

We have the following description of $\mathbf{C}^{-1}(\varphi_0)$ for some $\mathbf{C}$:

K,

A,

L – rationals,

K ∘ **K** – quadratic algebraics (Lagrange [12], those a with rational $b = \mathbf{K}(a)$),

T – algebraics of any order (Christol *et al.* [2]), and

χ (Turing–Kolmogorov–Chaitin) – every sequence with finite description.

Rational, algebraic refers to the induced generating function over some $\mathbb{F}(x)$.

Hence (apart from the noncomputable χ) **T** is the most "efficient" measure in terms of the size of $\mathbf{C}^{-1}(\varphi_0)$.

Also, the combination $\mathbf{T} \circ \mathbf{K}$ may be very effective: $\mathbf{K}$ *rapidly* identifies rationals and then $\mathbf{T}$ furthermore sorts out all a with algebraic $\mathbf{K}(a)$.

12   Computational Complexity

The computational (*e.g.* bit) complexity $\| \bullet \|$ of calculating an isometry $\mathbf{C}$ up to n input symbols, $\|\mathbf{C}\|(n)$, is upperbounded by

$$\|\mathbf{C}\|(n) = O(|A| \cdot \|\overline{C}\|(n)).$$

It is sufficient to compute the values $\overline{C}(a_1 \ldots a_n \alpha)$ for all prefixes $s_n = a_1 \ldots a_n$ and every element $\alpha \in A$, and then to sort the $\overline{C}$–complexities in constant time in order to obtain the ranking C at s_n.

Hence, investing a constant factor $|A|$ allows us for *any* compression, complexity etc. to apply the precise laws for Bernoulli trials to the resulting isometry.

For instance $\|\mathbf{K}\|(n)$ is $O(n^2)$, see [1, 15, 16].

13   Shifted Sequences

We may want to evaluate the isometry $\mathbf{C}$ for all shifted sequences $\mathbf{C}(a_1, \ldots, a_n)$, $\mathbf{C}(a_2, \ldots, a_n), \ldots$ up to $\mathbf{C}(a_n)$ (for instance, to consider different starting points for the linear complexity profile, see [18]).

In this case we can use the *Shift Commutator*

$$[\mathbf{C}^{-1}, \sigma] = \mathbf{C} \circ \sigma^{-1} \circ \mathbf{C}^{-1} \circ \sigma,$$

again an isometry [1]. Let some $w = \mathbf{C}(v)$ be given, *e.g.* $\varepsilon = \mathbf{C}(\varepsilon)$, then $\mathbf{C}(av) = [\mathbf{C}^{-1}, \sigma](aw)$ for $a \in \mathbb{F}_q$, since

$$aw \xrightarrow{\sigma} w \xrightarrow{\mathbf{C}^{-1}} v \xrightarrow{\sigma_a^{-1}} av \xrightarrow{\mathbf{C}} \mathbf{C}(av) = [\mathbf{C}^{-1}, \sigma](aw)$$

Iterating this procedure, we obtain the following upper bound for the combined bit complexity:

$$\sum_{i=1}^{n} \|\mathbf{C}(a_i \ldots a_n)\|(n - i + 1) \le n \cdot \|[\mathbf{C}^{-1}, \sigma]\|(n)$$

In the case of $\mathbf{C} = \mathbf{K}$, the shift commutator $[\mathbf{K}^{-1}, \sigma]$ can be computed by a transducer with finite state space and an up-down-counter in amortized linear time $\|[\mathbf{K}^{-1}, \sigma]\|(n) \le 7.5n$ (for a detailed description see [15] for the case $\mathbb{F}_2$ and [16] for general finite fields).

Hence we can compute $\mathbf{K}$ for all shifts simultaneously in amortized quadratic time (the same as *one* run of the Berlekamp–Massey–Algorithm) by repeated application of $[\mathbf{K}^{-1}, \sigma]$ via the above formula.

We thereby obtain with no additional cost *all* continued fractions of *all* shifted sequences $(a_1, \ldots, a_n)$, $(a_2, \ldots, a_n) \ldots (a_{n-1}, a_n)$ as well.

14   Conclusion

We can use many different ways to assess the randomness of a sequence $a \in A^\omega$ like various complexity measures, compression schemes, and predictors. When interfacing them by isometrization, we then have a uniform decision procedure to accept / reject the resulting sequence by the Law of the Iterated Logarithm.

Much less is known about the actual *e.g.* 2–adic or tree complexities ϕ and $\overline{T}$ and its admissible fluctuations for large $n \to \infty$. The additional time complexity for isometrization is bounded by a factor $|A|$.

A sequence $a \in A^\omega$ must pass at least the tests by the $\mathbf{K}, \mathbf{K} \circ \mathbf{K}, \mathbf{A}, \mathbf{L}, \mathbf{T}$, and $\mathbf{T} \circ \mathbf{K}$ isometries, staying within the bounds of the respective Laws of the Iterated Logarithm, to be considered random or pseudo–random.

For finite sequences $a \in A^n$, some $n \in \mathbb{N}$, in particular, we uniformly sort out the *same* fraction (measure) of "nonrandom" ones, for every complexity measure.

Acknowledgements

I would like to thank Harald Niederreiter, Matthew Parker, Zongduo Dai, and The Anonymous Referees for their comments (or questions) improving the paper, and Servet Martínez as well as Alejandro Maass for their hospitality at Núcleo Milenio "Información y Aleatoriedad", DIM, Universidad de Chile, Santiago.

References

1. M. del P. Canales Chacón, M. Vielhaber, "Structural and Computational Complexity of Isometries and their Shift Commutators", *Electronic Colloquium on Computational Complexity*, ECCC TR04–057, 2004.
2. G. Christol, T. Kamae, M. Mendès–France, G. Rauzy, "Suites algébriques, automates et substitutions", *Bull. Soc. Math. France* **108**, 401 – 419, 1980.
3. P. Deheuvels, "On the Erdős–Rényi theorem for random fields and sequences and its relationship with the theory of runs and spacings", *Z. Wahrschein. verw. Gebiete* **70**, 91 – 115, 1985.
4. J. L. Dornstetter, "On the equivalence between Berlekamp's and Euclid's algorithm", *IEEE Trans. Inform. Th.* **33(3)**, 428 – 431, 1987.
5. P. Erdős, "On the law of the iterated logarithm", *Ann. Math.* **43**, 419 – 436, 1942.
6. P. Erdős, P. Révész, "On the length of the longest head–run", in: *Colloq. Math. Soc. J. Bolyai*, vol. 16, Topics in Information Theory, 219 – 228, Keszthely, 1975.
7. W. Feller, *An Introduction to Probability Theory and Its Applications I*, 3rd Ed., John Wiley & Sons, New York, 1968.
8. W. Feller, "The general form of the so–called law of the iterated logarithm", *Trans. Amer. Math. Soc.* **51**, 373 – 402, 1943.
9. A. Klapper, M. Goresky, "2–adic shift registers", in: *Fast Software Encryption*, Cambridge Security Workshop Proc., Springer, 1994.
10. A. Klapper, M. Goresky, "Feedback Shift Registers, Combiners with Memory, and 2–adic Span", *J. Cryptology* **10(2)**, 111–147, 1997.

11. A. Kolmogoroff, "Über das Gesetz des iterierten Logarithmus", *Math. Ann.* **101**, 126 – 135, 1929.
12. J. L. Lagrange, Additions au mémoire sur la résolution des équations numériques, *Mém. Berl.* **24**, 1770.
13. H. Niederreiter, "Some computable complexity measures for binary sequences", in: *Sequences and their Applications* SETA 98 (C. Ding, T. Helleseth, H. Niederreiter, eds.), Springer, 1999.
14. H. Niederreiter, M. Vielhaber, "Tree complexity and a doubly exponential gap between structured and random sequences", *J. Complexity* 12, **3**, 187 – 198, 1996.
15. H. Niederreiter, M. Vielhaber, "Simultaneous shifted continued fraction expansions in quadratic time", *AAECC* **9, (2)**, 125 – 138, 1998.
16. H. Niederreiter, M. Vielhaber, "An algorithm for shifted continued fraction expansions in parallel linear time", *Theoretical Computer Science* **226**, 93-104, 1999.
17. O. Perron, *Die Lehre von den Kettenbrüchen I*, Teubner, Stuttgart, 1954/1977.
18. F. Piper, "Stream ciphers", *Elektrotechnik und Maschinenbau* **104**, 564 – 568, 1987.
19. P. Révész, *Random walk in random and non–random environments*, World Scientific, Singapore, 1990.
20. P. Révész, "Strong theorems on coin tossing", *Proc. Int. Cong. of Mathematicians*, 749 – 754, Helsinki, 1978.
21. J. Shallit, Y. Breitbart, "Automaticity I: Properties of a Measure of Descriptional Complexity", *J. Comput. Syst. Sci.* **53(1)**, 10–25, 1996.
22. M. Vielhaber, "Continued Fraction Expansion as Isometry: The Law of the Iterated Logarithm for Linear, Jump, and 2–Adic Complexity", Submitted.
23. M. Wang, "Linear Complexity Profiles and Continued Fractions", in: *Advances in cryptology – EUROCRYPT* '89, (J. J. Quisquater, J. Vandewalle, eds.), LNCS 434, 571 – 585, Springer, Berlin, 1990.

One-Error Linear Complexity over F_p of Sidelnikov Sequences*

Yu-Chang Eun[1],[**], Hong-Yeop Song[2], and Gohar M. Kyureghyan[3]

[1] SAMSUNG ELECTRONICS CO., LTD., Dong Suwon P.O.BOX-105, 416
Maetan-3Dong, Paldal-Gu, Suwon-City, Gyeonggi-Do, Korea, 442-600
yc.eun@samsung.com
[2] Center for Information Technology of Yonsei University, Coding and Information
Theory Lab, Department of Electrical and Electronics Engineering,
Yonsei University, 134 Shinchon-dong Seodaemun-gu, Seoul, Korea, 120-749
hy.song@coding.yonsei.ac.kr
[3] Otto-von-Guericke University, Magdeburg, Faculty of Mathematics,
Postfach 4120, Magdeburg, Germany 39016
gohar.kyureghyan@mathematik.uni-magdeburg.de

Abstract. Let p be an odd prime and m be a positive integer. In this paper, we prove that the one-error linear complexity over F_p of Sidelnikov sequences of length $p^m - 1$ is $(\frac{p+1}{2})^m - 1$, which is much less than its (zero-error) linear complexity.

1 Introduction

Let p be an odd prime and m be a positive integer. Let F_{p^m} be the finite field with p^m elements, and α be a primitive element of F_{p^m}. The Sidelnikov sequence $S = \{s(t) : t = 0, 1, 2, ..., p^m - 2\}$ of period $p^m - 1$ is defined as [1]

$$s(t) = \begin{cases} 1 & \text{if } \alpha^t + 1 \in \mathrm{N} \\ 0 & \text{otherwise} \end{cases} \tag{1}$$

where $\mathrm{N} = \{\alpha^{2t+1} : t = 0, 1, ..., \frac{p^m-1}{2} - 1\}$ is the set of quadratic nonresidues over F_{p^m}. In [1], it was shown that S has the optimal autocorrelation and balance property. Sidelnikov sequences were rediscovered by Lempel *et al* [2], and Sarwate pointed out that the sequences described by Lempel *et al* were in fact the same as the ones by Sidelnikov [3]. Sidelnikov sequences are a special case of the construction by No *et al* [4].

* This work was supported by Korea Research Foundation Grant (KRF-2003-041-D00417).
** He was with Dept. of Electrical and Electronics Engineering, Yonsei University, while he was doing this research.

T. Helleseth et al. (Eds.): SETA 2004, LNCS 3486, pp. 154–165, 2005.

Helleseth and Yang [5] originated the study of the linear complexity of Sidelnikov sequences over F_2. They found also a representation of the sequences using the indicator function $I(\cdot)$ and the quadratic character $\chi(\cdot)$ as

$$s(t) = \frac{1}{2}\left(1 - I(\alpha^t + 1) - \chi(\alpha^t + 1)\right),\qquad(2)$$

where $I(x) = 1$ if $x = 0$ and $I(x) = 0$ otherwise, and $\chi(x)$ denotes the quadratic character of $x \in F_{p^m}$ defined by

$$\chi(x) = \begin{cases} +1, & \text{if } x \text{ is a quadratic residue} \\ 0, & \text{if } x = 0 \\ -1, & \text{if } x \text{ is a quadratic nonresidue.} \end{cases}$$

Kyureghyan and Pott [6] have extended the calculation of the linear complexity of the sequences over F_2 following the results in [5]. However, the determination of the linear complexity of S over F_2 turns out to be difficult since the characteristic of the field, which is 2, divides the length of the sequence [6].

Observing that it is more natural to consider the linear complexity over F_p since the sequences are constructed over F_p, Helleseth et al [7] derived the linear complexity over F_p (not over F_2) of the sequence S of length $p^m - 1$ as well as its trace representation for $p = 3, 5$, and 7, and finally, Helleseth et al [8] finished the calculation of the linear complexity over F_p of the sequence of length $p^m - 1$ for all odd prime p.

According to the results in both [7] and [8], the linear complexity over F_p is roughly the same as the period, and the sequences can be thought of having an "excellent" linear complexity. We noted that the linear complexity of the sequences obtained by deleting the term $I(\alpha^t + 1)$ in (2) is much smaller than the one of the original sequence. For example, the sequence of length $3^3 - 1 = 26$

$$1\,1\,1\,1\,0\,1\,1\,0\,1\,1\,0\,0\,0\,0\,0\,1\,0\,0\,1\,1\,1\,0\,0\,0\,1\,0$$

has linear complexity 23 over F_3. But the sequence obtained by deleting the term $I(\alpha^t + 1)$ in (2) is

$$1\,1\,1\,1\,0\,1\,1\,0\,1\,1\,0\,0\,0\,2\,0\,1\,0\,0\,1\,1\,1\,0\,0\,0\,1\,0$$

which has linear complexity 7 over F_3. We conjectured that this phenomenon may persist in all cases of Sidelnikov sequences, and this paper is the result of this investigation. In this paper we show that the value $(\frac{p+1}{2})^m - 1$, first appeared in [7] in the middle of the calculations, is indeed the one-error linear complexity over F_p of the sequence of period $p^m - 1$ for all odd prime p and all positive integers $m \geq 1$.

We give some notation and basic techniques for the calculation of the linear complexity of the sequences over F_p in Section 2. In Section 3, we prove that the "upper bound" on the one-error linear complexity of Sidelnikov sequences over F_p of period $p^m - 1$ is $(\frac{p+1}{2})^m - 1$, by constructing explicitly a one-error sequence. Note that this is already surprising enough since the true value of the one-error linear complexity is at most this number. In Section 4, we prove that the equality holds in the upper bound.

2   Preliminaries

Let p be an odd prime and $m \geq 1$. Denote the linear complexity over F_p of Sidelnikov sequence S defined in (1) or (2) by $L(S)$. Let $Z = \{z(t) \ : \ t = 0, 1, 2, ..., p^m - 2\}$ be a sequence of length $p^m - 1$ over F_p. Then the k-error linear complexity [9][10] of Sidelnikov sequence of length $p^m - 1$ over F_p is defined as

$$L_k(S) = \min_{0 \leq \mathrm{WH}(Z) \leq k} L(S + Z) \tag{3}$$

where $\mathrm{WH}(Z)$ denotes the Hamming weight of Z, i.e., the number of components of Z that are non-zero. Assume $k = 1$ in (3) and

$$z^{(\tau, \lambda)}(t) = \frac{\lambda}{2} I(\alpha^{t-\tau} + 1), \quad 0 \leq \tau < p^m - 1, \quad \lambda \in F_p.$$

Then, any sequence over F_p of length $p^m - 1$ with Hamming weight ≤ 1 can be represented by the sequence $Z^{(\tau, \lambda)} = \{z^{(\tau, \lambda)}(t)|t = 0, 1, ..., p^m - 2\}$ for some $0 \leq \tau < p^m - 1$ and $\lambda \in F_p$.

Let $S_Z^{(\tau, \lambda)} = \{s_z^{(\tau, \lambda)}(t) \ : \ t = 0, 1, 2, ..., p^m - 2\}$ be defined as

$$
\begin{aligned}
s_z^{(\tau, \lambda)}(t) &\triangleq s(t) + z^{(\tau, \lambda)}(t) \\
&= \frac{1}{2} \left(1 - I(\alpha^t + 1) - \chi(\alpha^t + 1)\right) + \frac{\lambda}{2} I(\alpha^{t-\tau} + 1).
\end{aligned}
\tag{4}
$$

Then the one-error linear complexity of S can be represented as

$$L_1(S) = \min_{\substack{\lambda \in F_p \\ 0 \leq \tau \leq p^m - 2}} L(S_Z^{(\tau, \lambda)}). \tag{5}$$

To compute the linear complexity in general, we use the Fourier transform in the finite field F_{p^m} defined for a p-ary sequence $Y = \{y(t)\}$ of period $n = p^m - 1$ by

$$A_i = \frac{1}{n} \sum_{t=0}^{n-1} y(t) \alpha^{-it}$$

where α is a primitive element of F_{p^m} and $A_i \in F_{p^m}$ [11][12]. The inverse Fourier transform is similarly represented as

$$y(t) = \sum_{t=0}^{n-1} A_i \alpha^{it}. \tag{6}$$

Then the linear complexity of Y is defined as [11][12]

$$L(Y) = |\{ \ i \ | \ A_i \neq 0, \ 0 \leq i \leq n - 1 \ \}|.$$

3   Main Results

The Fourier transform of the Sidelnikov sequences is given in [7].

Lemma 1. [7] *Let the p-adic expansion of an integer i, where $0 \leq i \leq p^m - 2$, be given by*

$$i = \sum_{a=0}^{m-1} i_a p^a$$

where $0 \leq i_a \leq p - 1$. Then the Fourier coefficient $A_{-i} \in F_{p^m}$ of the Sidelnikov sequence defined in (2) of period $p^m - 1$ is given by

$$A_{-i} = \frac{(-1)^i}{p-2} \left(-1 + (-1)^{-\frac{p^m-1}{2}} \prod_{a=0}^{m-1} \binom{i_a}{\frac{p-1}{2}} \right). \tag{7}$$

Then it is straightforward, that the Fourier coefficients of the one-error allowed Sidelnikov sequences are given as follows.

Lemma 2. *The Fourier coefficient $A_{-i}(\tau, \lambda)$ of the one-error allowed Sidelnikov sequence $S_Z^{(\tau,\lambda)}$ defined in (4) is given by*

$$A_{-i}(\tau, \lambda) = \frac{(-1)^i}{p-2} \left(-1 + \lambda \alpha^{\tau i} + (-1)^{-\frac{p^m-1}{2}} \prod_{a=0}^{m-1} \binom{i_a}{\frac{p-1}{2}} \right) \in F_{p^m} \tag{8}$$

where i_a is defined in Lemma 1.

Consider the case $\alpha^\tau = 1$ (or $\tau = 0$) and $\lambda = 1$. In this case we have

$$s_z^{(0,1)}(t) = \frac{1}{2}(1 - \chi(\alpha^t + 1)),$$

and

$$L\left(S_Z^{(0,1)}\right) = |\{ i \ : \ A_{-i}(0,1) \neq 0, \ 0 \leq i < p^m - 1 \}|$$

$$= |I_{\mathrm{nz}}| = \left(\frac{p+1}{2}\right)^m - 1 \tag{9}$$

where

$$I_{\mathrm{nz}} \triangleq \left\{ i \ : \ \prod_{a=0}^{m-1} \binom{i_a}{\frac{p-1}{2}} \neq 0, \ 0 \leq i < p^m - 1 \right\}. \tag{10}$$

Note that I_{nz} contains all the i's in the range $i = 0, 1, 2, ..., p^m - 2$ that satisfy $\frac{p-1}{2} \leq i_a \leq p - 1$ for all a.

Table 1. Comparison of L_0 and L_1 when $p = 3$

m	L_0	L_1	$n = 3^m - 1$	L_0/n (%)	L_1/n (%)
2	7	3	8	87.5	37.5
3	23	7	26	88.5	26.9
4	73	15	80	91.3	18.8
5	227	31	242	93.8	12.8
6	697	63	728	95.7	8.7
7	2123	127	2186	97.1	5.8
8	6433	255	6560	98.1	3.9

Table 2. Comparison of L_0 and L_1 when $p = 5$

m	L_0	L_1	$n = 5^m - 1$	L_0/n (%)	L_1/n (%)
2	21	8	24	87.5	33.3
3	117	26	124	94.4	21.0
4	608	80	624	97.4	12.8
5	3083	244	3124	98.7	7.8
6	15501	728	15624	99.2	4.7
7	77717	2186	78124	99.5	2.8
8	389248	6560	390624	99.6	1.7

Alternatively, without specifically calculating $A_{-i}(0,1)$ for all i, we have

$$
\begin{aligned}
s_z^{(0,1)}(t) &= \frac{1}{2}\left(1 - \chi(\alpha^t + 1)\right) = \frac{1}{2}\left(1 - (\alpha^t + 1)^{\frac{p^m-1}{2}}\right) \\
&= \frac{1}{2}\left(1 - (\alpha^t + 1)^{\sum_{k=0}^{m-1}(\frac{p-1}{2})p^k}\right) \\
&= \frac{1}{2}\left(1 - \prod_{k=0}^{m-1}(\alpha^t + 1)^{(\frac{p-1}{2})p^k}\right) \\
&= \frac{1}{2}\left(1 - \prod_{k=0}^{m-1}(a_0 + a_1\alpha^t + \cdots + a_{\frac{p-1}{2}}\alpha^{\frac{p-1}{2}t})^{p^k}\right).
\end{aligned}
\tag{11}
$$

where $a_i = \binom{\frac{p-1}{2}}{i}$. Since the characteristic is p and $a_i \not\equiv 0 \pmod{p}$ we obtain the same linear complexity as (9) by just counting all the sum-terms when (11) is represented as (6). This construction provides an upper bound on the one-error linear complexity of the Sidelnikov sequences.

Theorem 1. *Let S be the Sidelnikov sequence of period $p^m - 1$ for some odd prime p and a positive integer m. Then for the one-error linear complexity $L_1(S)$ of S it holds*

$$
L_1(S) \leq \left(\frac{p+1}{2}\right)^m - 1.
$$

Even though the above bound was not explicitly mentioned in [7], we would like to add that it was first calculated there in the middle of the calculations. It is very surprising to have such an upper bound for $L_1(S)$. In fact there is an equality in Theorem 1, which may not be very unexpected.

Theorem 2 (main). *Let p be an odd prime and $m \geq 1$. Let S be the Sidelnikov sequence of period $p^m - 1$. Then the one-error linear complexity of S is*

$$L_1(S) = \left(\frac{p+1}{2} \right)^m - 1.$$

Tables I and II show some numerical data for $p = 3, 5$ and $1 < m \leq 8$. Observe that for $p = 5$ and $m = 8$, the one-error linear complexity becomes less than 2% of the period.

4   Proof of Main Theorem

Note first that it is enough to show that, for all τ and λ,

$$L(S_Z^{(\tau,\lambda)}) \geq \left(\frac{p+1}{2} \right)^m - 1,$$

where $S_Z^{(\tau,\lambda)}$ is given in (4). For this, we will denote α^τ by β, and take care of all possible cases of β and λ as follows:

1. CASE $\beta \notin F_p$ and $\lambda \neq 0$.
2. CASE $\beta \in F_p$.
   (a) case $\lambda = 0$.
   (b) case $\lambda \neq 0$. This case is further divided into the following:
      i. subcase $\beta = 1$.
      ii. subcase $\beta \neq 1$. This subcase is treated by several different methods according to the values of m as follows:
         A. for $m \geq 3$.
         B. for $m = 2$, or all even values of $m \geq 2$.
         C. for $m = 1$.

4.1   CASE $\beta \notin F_p$ and $\lambda \neq 0$

Note that if $\beta^i \notin F_p$, then we have $A_{-i}(\tau, \lambda) \neq 0$. Therefore,

$$L(S_Z^{(\tau,\lambda)}) \geq \left| \{\, i \, : \, \beta^i \notin F_p, \ 0 \leq i < p^m - 1 \,\} \right| \triangleq N.$$

If we let d be the least positive integer such that $\beta^d \in F_p$, then $d \geq 2$, and hence,

$$N = (p^m - 1)\left(1 - \frac{1}{d} \right) \geq \frac{p^m - 1}{2} \geq \left(\frac{p+1}{2} \right)^m - 1.$$

4.2   CASE $\beta \in F_p$

We will use

$$L(S_Z^{(\tau,\lambda)}) = n - |C| = p^m - 1 - |C|,\tag{12}$$

where

$$C \triangleq \{ i \ : \ A_{-i}(\tau,\lambda) = 0, \ 0 \leq i < p^m - 1 \}$$

and where $A_{-i}(\tau,\lambda)$ is given in Lemma 2. Observe that

$$C = \left\{ i \ : \ \prod_{a=0}^{m-1} \binom{i_a}{\frac{p-1}{2}} = (-1)^{\frac{p^m-1}{2}}(1 - \lambda\beta^i), \ 0 \leq i < p^m - 1 \right\}.\tag{13}$$

Recall that, from earlier notation,

$$I_{nz} = \left\{ i \ : \ \prod_{a=0}^{m-1} \binom{i_a}{\frac{p-1}{2}} \neq 0, \ 0 \leq i < p^m - 1 \right\} \quad \text{and} \quad |I_{nz}| = \left(\frac{p+1}{2}\right)^m - 1.$$

We will also consider its complement as follows:

$$I_{nz}^C \triangleq \{0, 1, ..., p^m - 2\} \backslash I_{nz} \quad \text{and hence} \quad |I_{nz}^C| = p^m - \left(\frac{p+1}{2}\right)^m.$$

Then, it is not difficult to show that

$$|I_{nz}| \leq |I_{nz}^C|.$$

Therefore, it is sufficient to prove that either $|C| \leq |I_{nz}|$ or $|C| \leq |I_{nz}^C|$, since for both cases we have $|C| \leq |I_{nz}^C|$, and therefore,

$$L(S_Z^{(\tau,\lambda)}) = p^m - 1 - |C| \geq p^m - 1 - |I_{nz}^C| = |I_{nz}| = \left(\frac{p+1}{2}\right)^m - 1.$$

4.2.(a) case $\lambda = 0$.

For $\lambda = 0$, we have

$$C = \left\{ i \ : \ \prod_{a=0}^{m-1} \binom{i_a}{\frac{p-1}{2}} = \pm 1, \ 0 \leq i < p^m - 1 \right\},$$

which implies $|C| \leq |I_{nz}|$. We will assume that $\lambda \neq 0$ in the remaining of the proof.

4.2.(b) case $\lambda \neq 0$.

subcase $\beta = 1$.

   If $\lambda = 1$, then $1 - \lambda\beta^i = 1 - \lambda = 0$, and hence, $|C| = |I_{nz}^C|$. If $\lambda \in F_p \backslash \{0,1\}$, then $1 - \lambda\beta^i = 1 - \lambda \neq 0$, and hence, $|C| \leq |I_{nz}|$.

subcase $\beta \neq 1$.

   Note that in this case we have an initial estimation of the size of C from (13) as follows:

$$|C| \leq |\{ i \ : \ \beta^i = \lambda^{-1} \} \cap I_{nz}^C| + |\{ i \ : \ \beta^i \neq \lambda^{-1} \} \cap I_{nz}|.\tag{14}$$

Let $e > 1$ be the order of β over F_p, and hence, note that $e | (p-1)$. If there does not exist an integer u satisfying $\lambda^{-1} = \beta^u$ and $0 \leq u < e$, then

$$|C| \leq \left| \{ i \; : \; \beta^i \neq \lambda^{-1} \} \cap I_{\mathrm{nz}} \right| \leq |I_{\mathrm{nz}}| \, .$$

If such u exists, then (14) becomes,

$$|C| \leq \left| \left\{ i : \sum_{a=0}^{m-1} i_a \equiv u(\mathrm{mod}\ e) \right\} \cap I_{\mathrm{nz}}^{\mathrm{C}} \right| + \left| \left\{ i : \sum_{a=0}^{m-1} i_a \not\equiv u(\mathrm{mod}\ e) \right\} \cap I_{\mathrm{nz}} \right|,$$

$$(15)$$

since

$$i = \sum_{a=0}^{m-1} i_a p^a \equiv \sum_{a=0}^{m-1} i_a \quad (\mathrm{mod}\ e).$$

We need the following observation:

Lemma 3. *Let A be a set of k consecutive integers and e be a divisor of k, then*

$$\left| \left\{ (x_0, \ldots, x_{m-1}) \in A^m : \sum_{j=0}^{m-1} x_j \equiv u \quad (\mathrm{mod}\ e) \right\} \right| = k^{m-1} \frac{k}{e},$$

for any $0 \leq u \leq e - 1$. If e is not a divisor of k, then the above cardinality is $\geq k^{m-1} \lfloor \frac{k}{e} \rfloor$ and $\leq k^{m-1} \lceil \frac{k}{e} \rceil$.

Proof. If we take any $m - 1$ elements $x_0, x_1, \ldots, x_{m-2}$ from A, there are still k/e choices for x_{m-1}. ∎

Now, we try to estimate both terms on the RHS of the inequality (15) as follows. The first term is bounded as follows:

$$\left| \left\{ i : \sum_{a=0}^{m-1} i_a \equiv u \quad (\mathrm{mod}\ e) \text{ and there is } i_a \text{ with } 0 \leq i_a < \frac{p-1}{2} \right\} \right|$$

$$= \left| \left\{ i : \sum_{a=0}^{m-1} i_a \equiv u \quad (\mathrm{mod}\ e), 0 \leq i_a \leq p-1 \right\} \right|$$

$$- \left| \left\{ i : \sum_{a=0}^{m-1} i_a \equiv u \quad (\mathrm{mod}\ e), \frac{p-1}{2} \leq i_a \leq p-1 \right\} \right|$$

$$\leq p^{m-1} \left\lceil \frac{p}{e} \right\rceil - \left(\frac{p+1}{2} \right)^{m-1} \left\lfloor \frac{p+1}{2e} \right\rfloor,$$

where the last inequality follows from Lemma 3. The second term on the RHS of the inequality (15) is bounded as follows:

$$\left| \left\{ i : \sum_{a=0}^{m-1} i_a \not\equiv u \pmod{e} \text{ with } \frac{p-1}{2} \le i_a \le p-1 \text{ for all } i_a \right\} \right|$$

$$= |I_{\text{nz}}| - \left| \left\{ i : \sum_{a=0}^{m-1} i_a \equiv u \pmod{e} \text{ with } \frac{p-1}{2} \le i_a \le p-1 \text{ for all } i_a \right\} \right|$$

$$\le \left(\frac{p+1}{2} \right)^m - \left(\frac{p+1}{2} \right)^{m-1} \left\lfloor \frac{p+1}{2e} \right\rfloor.$$

Therefore, the inequality (15) becomes

$$|C| \le p^{m-1} \left\lceil \frac{p}{e} \right\rceil + \left(\frac{p+1}{2} \right)^m - 2 \left(\frac{p+1}{2} \right)^{m-1} \left\lfloor \frac{p+1}{2e} \right\rfloor \qquad (16)$$

$$\le p^{m-1} \left(\frac{p-1}{e} + 1 \right) + \left(\frac{p+1}{2} \right)^m - \left(\frac{p+1}{2} \right)^{m-1} \left(\frac{p-1}{e} - 1 \right). \qquad (17)$$

Observe, that for $p = 3$ (and thus $e = 2$) (16) directly implies that

$$|C| \le 3^m - 2^m = |I_{\text{nz}}^C|, \quad \text{for all} \quad m \ge 3.$$

Now, it is not difficult to show, if $p \ge 5$ and $m \ge 3$, then (17) does not exceed $p^m - \left(\frac{p+1}{2} \right)^m$. For this, we need to show that

$$\left(\frac{p+1}{2} \right)^{m-1} \left(2\frac{p+1}{2} - \frac{p-1}{e} + 1 \right) \le p^{m-1} \left(p - \frac{p-1}{e} - 1 \right)$$

which is the same as

$$\left(\frac{p+1}{2p} \right)^{m-1} \le \frac{p - \frac{p-1}{e} - 1}{p - \frac{p-1}{e} + 2}.$$

Note that, for $m \ge 3$ and $p \ge 5$, we have

$$\left(\frac{p+1}{2p} \right)^{m-1} \le \left(\frac{p+1}{2p} \right)^2 \le \left(\frac{3}{5} \right)^2 = \frac{9}{25},$$

and therefore it is enough to prove

$$\frac{p - \frac{p-1}{e} - 1}{p - \frac{p-1}{e} + 2} \ge \frac{6}{25}.$$

The last inequality holds, since

$$e \ge 2 > \frac{p-1}{p-2} > \frac{19p - 19}{19p - 37}$$

for $p \ge 5$.

The case $m = 2$ can be covered by direct calculations, using (15). Or, we may consider the following, which, in fact, works for all $p \geq 3$ and even values of $m \geq 2$. Let

$$H \triangleq \left\{ i \ : \ 0 \leq i_a \leq \frac{p-1}{2}, \ 0 \leq i < p^m - 1, \ i \neq \frac{p^m - 1}{2} \right\}. \tag{18}$$

Then

$$\left| \left\{ i : \sum_{a=0}^{m-1} i_a \not\equiv u \pmod{e} \right\} \cap I_{\text{nz}} \right| = \left| \left\{ i : \sum_{a=0}^{m-1} i_a \not\equiv u \pmod{e} \right\} \cap H \right| \tag{19}$$

since

$$I_{\text{nz}} = \left\{ i \ : \ \frac{p-1}{2} \leq i_a \leq p-1, \ 0 \leq i < p^m - 1 \right\}$$

and

$$\sum_{a=0}^{m-1} i_a = \sum_{a=0}^{m-1} \left(i_a - \frac{p-1}{2} \right) + m \frac{p-1}{2} \equiv \sum_{a=0}^{m-1} \left(i_a - \frac{p-1}{2} \right) \pmod{e}.$$

Since $H \subset I_{\text{nz}}^C$, the second term of (15) is upper bounded by

$$\left| \left\{ i \ : \ \sum_{a=0}^{m-1} i_a \not\equiv u \pmod{e} \right\} \cap I_{\text{nz}}^C \right|.$$

Therefore,

$$|C| \leq |I_{\text{nz}}^C|.$$

The proof will be complete if we show the following, for the case $m = 1$.

Lemma 4. *Let p be an odd prime and $\lambda \neq 0, \beta \in F_p$, and $\beta \neq 1$. Then,*

$$|C| = \left| \left\{ i : 0 \leq i \leq p-2, \ \binom{i}{\frac{p-1}{2}} \equiv (-1)^{\frac{p-1}{2}}(1 - \lambda \beta^i) \pmod{p} \right\} \right| \leq \frac{p-1}{2}.$$

Proof. Let $e > 1$ be the order of β. If there is no u with $1 - \lambda \beta^u = 0$, then obviously, by setting $(-1)^{\frac{p-1}{2}}(1 - \lambda \beta^i) = d(i) \pmod{p}$,

$$|C| = \left| \left\{ i : \frac{p-1}{2} \leq i \leq p-2, \ \binom{i}{\frac{p-1}{2}} = d(i) \not\equiv 0 \pmod{p} \right\} \right| \leq \frac{p-1}{2}.$$

Suppose, there is $0 \leq u < e$ with $1 - \lambda \beta^u = 0$, implying $1 - \lambda \beta^w = 0$ for any $w \equiv u \pmod{e}$, $0 \leq w \leq p-2$. Then

$$\begin{aligned} |C| = &\left| \left\{ i : 0 \leq i < \frac{p-1}{2}, \binom{i}{\frac{p-1}{2}} \equiv d(i) \equiv 0 \pmod{p} \right\} \right| \\ &+ \left| \left\{ i : \frac{p-1}{2} \leq i \leq p-2, \binom{i}{\frac{p-1}{2}} \equiv d(i) \not\equiv 0 \pmod{p} \right\} \right|. \end{aligned} \tag{20}$$

Since it is obvious $\left(\frac{i}{\frac{p-1}{2}}\right) \neq d(i)$ for $i = \frac{p-1}{2}$, this case can be excluded from the second term of (20). Then the second term is equal to

$$\left|\left\{i: \frac{p-1}{2} < i \leq p-2\right\}\right| - \left|\left\{i: \frac{p-1}{2} < i \leq p-2, \; i \equiv u \pmod{e}\right\}\right|$$
$$= \frac{p-1}{2} - 1 - \left\lfloor \frac{p-1}{2e} - \frac{1}{e} \right\rfloor.$$

This yields

$$|C| \leq \left\lceil \frac{p-1}{2e} \right\rceil + \frac{p-1}{2} - 1 - \left\lfloor \frac{p-1}{2e} - \frac{1}{e} \right\rfloor. \tag{21}$$

If $2e|p-1$, RHS of (21) is obviously equal to $\frac{p-1}{2}$. If not, it is enough to prove

$$\left\lfloor \frac{p-1}{2e} - \frac{1}{e} \right\rfloor = \left\lfloor \frac{p-1}{2e} \right\rfloor.$$

Let $p-1 \equiv k \pmod{2e}$. Since k is even and ≥ 2, we get

$$\left\lfloor \frac{p-1}{2e} \right\rfloor = \frac{p-1}{2e} - \frac{k}{2e} \leq \frac{p-1}{2e} - \frac{1}{e}.$$

Together with

$$\left\lfloor \frac{p-1}{2e} - \frac{1}{e} \right\rfloor \leq \left\lfloor \frac{p-1}{2e} \right\rfloor,$$

we can complete the proof.

References

1. V. M. Sidelnikov, "Some k-valued pseudo-random and nearly equidistant codes," *Probl. Pered. Inform.*, vol. 5, no. 1, pp. 16–22, 1969.
2. A. Lempel, M. Cohn, and W. L. Eastman, "A class of balanced binary sequences with optimal autocorrelation properties," *IEEE Trans. Inform. Theory*, vol. 23, no. 1, pp. 38–42, Jan. 1977.
3. D. V. Sarwate, "Comments on 'A Class of Balanced Binary Sequences with Optimal Autocorrelation Properties'," *IEEE Trans. Inform. Theory*, vol. 24, no. 1, pp. 128–129, Jan. 1978.
4. J.-S. No, H. Chung, H.-Y. Song, K. Yang, J.-D. Lee and T. Helleseth, "New Construction for Binary Sequences of period $p^m - 1$ with Optimal Autocorrelation Using $(z + 1)^d + az^d + b$," *IEEE Trans. Inform. Theory*, vol. 47, no. 4, pp. 1638–1644, May 2001.
5. T. Helleseth and K. Yang, "On binary sequences of period $p^m - 1$ with optimal autocorrelation," in *Proc. 2001 Conf. Sequences and Their Applications* (SETA '01), Bergen, Norway, May 13-17 2001, pp. 29–30.
6. G. M. Kyureghyan and A. Pott, "On the linear complexity of the Sidelnikov-Lempel-Cohn-Eastman sequences," *Design, Codes and Cryptography.*, vol. 29, pp. 149–164, 2003.
7. T. Helleseth, S.-H. Kim, and J.-S. No, "Linear complexity over F_p and trace representation of Lempel-Cohn-Eastman sequences," *IEEE Trans. Inform. Theory*, vol. 49, no. 6, pp. 1548–1522, June 2003.

8. T. Helleseth, M. Maas, J.E. Mathiassen and T. Segers, "Linear complexity over F_p of Sidelnikov sequences," *IEEE Trans. Inform. Theory*, vol. 50, no. 10, pp. 2468-2472, Oct. 2004.

9. M. Stamp and C. F. Martin, "An algorithm for the k-linear complexity of binary sequences with period 2^n," *IEEE Trans. Inform. Theory*, vol. 39, no. 4, pp. 1398–1401, July. 1993.

10. T. W. Cusick, C. Ding, and A. Renvall, *Stream Ciphers and Number Theory*. North-Holland Mathematical Library 55. Amsterdam: North-Holland/Elsevier, 1998.

11. R. E. Blahut, "Transform techniques for error control codes," *IBM J. Res. Develop.*, vol. 23, pp. 299–315, 1979.

12. R. E. Blahut, *Theory and Practice of Error Control Codes*. New York: Addison-Wesley, 1983.

On the Generalized Lauder-Paterson Algorithm and Profiles of the k-Error Linear Complexity for Exponent Periodic Sequences

Takayasu Kaida

Department of Information and Electronic Engineering,
Yatsushiro National College of Technology,
Yatsushiro, Kumamoto 866-8501, Japan
kaida@m.ieice.org, kaida@as.yatsushiro-nct.ac.jp
http://y-page.yatsushiro-nct.ac.jp/u/kaida/index.html

Abstract. The Lauder-Paterson algorithm gives the profile of the k-error linear complexity for a binary sequence with period 2^n. In this paper a generalization of the Lauder-Paterson algorithm into a sequence over $GF(p^m)$ with period p^n, where p is a prime and m, n are positive integers, is proposed. We discuss memory and computation complexities of proposed algorithm. Moreover numerical examples of profiles for balanced binary and ternary exponent periodic sequences, and proposed algorithm for a sequence over $GF(3)$ with period $9(= 3^2)$ are given.

Keywords: exponent periodic sequence, Games-Chan algorithm, k-error linear complexity, Lauder-Paterson algorithm, pseudo-random sequence, Stamp-Martin algorithm.

1   Introduction

In 1993 M.Stamp and C.Martin proposed the k-error linear complexity (k-LC) for periodic sequences as one of measurements for randomness [10]. The k-LC is a generalization of the linear complexity (LC) in order to guard from instability properties of the LC [11, 12, 1]. At the same time a fast algorithm (Stamp-Martin algorithm) of the k-LC for a binary sequence with period 2^n is shown [10]. Although the sphere complexity, as similar as the k-LC, was proposed earlier than the k-LC [2], we use the k-LC in sense of a natural extension of the LC. We generalized the Stamp-Martin algorithm into two algorithms for a sequence over $GF(p^m)$ with period p^n, where p is a prime and m, n are positive integers [4, 5]. One of them should be called the generalized Stamp-Martin algorithm because this algorithm becomes the Stamp-Martin algorithm in case of binary sequences [5]. Another one has the same function and does not use concepts named "shift" and "offset" [4]. The procedure of "shift" changes the cost matrix to fit the input sequence at that step by the cyclic shift for each columns of the cost matrix. After this, all elements of the value at the first row are same value and the minimum through all elements of that shifted cost matrix. Therefore we can set

T. Helleseth et al. (Eds.): SETA 2004, LNCS 3486, pp. 166–178, 2005.

all-zero at the first row by subtracting that value from all elements of the cost matrix, called the procedure of "offset". For binary sequences, "shift" and "offset" are very effective because these work for changing the cost matrix into the cost vector, and decreasing input value k. However in non-binary case, there is not so much benefits only dropping one row of the cost matrix. In calculations of the k-LC for non-binary exponent periodic sequences, the algorithm without "shift" and "offset" is simpler than the generalized Stamp-Martin algorithm with "shift" and "offset". It is important for applications that a pseudorandom sequence has good profile of the k-LC, which is the decrease points of the k-LC against increase of k [2, 8, 6]. Unfortunately, the Stamp-Martin algorithm and two generalized Stamp-Martin algorithms answer the k-LC against only one fixed k and one fixed binary sequence with period 2^n or one fixed sequence over $GF(p^m)$ with period p^n, respectively. Recently A.Lauder and K.Paterson proposed a fast algorithm (Lauder-Paterson algorithm) computing the profile of the k-LC, i.e., the k-LC for all $k \geq 0$, for a fixed binary sequence with period 2^n [9].

In this paper the LC and its fast algorithm such as the generalized Games-Chan algorithm, and the k-LC and the generalized k-LC algorithm are recalled in Section 2 and Section 3, respectively. For preliminaries of a generalization of the Lauder-Paterson algorithm, we describe the profiles of the k-LC and the Lauder-Paterson algorithm in Section 4. The main theorem and proposed generalized Lauder-Paterson algorithm for a sequence over $GF(p^m)$ with period p^n are given in Section 5. Because of complications in the algorithm with "shift" and "offset", we propose the generalized Lauder-Paterson algorithm without "shift" and "offset" although the original Lauder-Paterson algorithm uses "shift" and "offset". In Section 6 some numerical examples for profiles for balanced binary and ternary exponent periodic sequences, and proposed algorithm for a sequence over $GF(3)$ with period $9(= 3^2)$ are given. Finally conclusion and future works are shown in Section 7.

2 Linear Complexity and Generalized Games-Chan Algorithm

We define the linear complexity of a sequence and recall the generalized Games-Chan algorithm [3, 2] computing the LC for a sequence over $GF(p^m)$ with period p^n.

We consider an infinite sequence $S = (s_0, s_1, \cdots)$ over a finite field K through this paper.

Definition 1. *The linear complexity (LC) of S is defined as*

$$L(S) = \min\{\deg f(x) | f(x) \in G(S)\},$$

where the set $G(S)$ consists of the generator polynomial,

$$f(x) = f_L x^L + f_{L-1} x^{L-1} + \cdots + f_1 x + 1 \in K[x],$$

of S such that

$$s_{L+i} + f_1 s_{L+i-1} + \cdots + f_{L-1} s_{i+1} + f_L s_i = 0 \tag{1}$$

for all integer $i \geq 0$. □

Definition 2. *If there exists an integer N such that $s_i = s_{N+i}$ for all $i \geq 0$ then N is defined the period of S.*   □

In this paper we call the period of S only the minimum of N satisfying above condition the period of S.

We denote one period (or subsequence) with length N of an infinite sequence S by $S^{(N)}$, i.e., $S^{(N)} = (s_0, s_1, \cdots, s_{N-1})$, and an infinite sequence repeating a finite sequence F by F^∞. Hence we can rewrite an infinite sequence S with period N by $S = (S^{(N)})^\infty$.

The LC can be also defined for a finite sequence with length N by satisfying (1) for $0 \leq i \leq N - L$ instead of all $i \geq 0$. However we only consider the LC of infinite sequences in this paper then let F be a finite sequence, we simply denote $L(F)$ and $L_k(F)$ instead of $L(F^\infty)$ and $L_k(F^\infty)$, respectively. ($L_k(F)$ will be defined in next section.)

Definition 3. *Let $K = GF(p^m)$ with a prime p and a positive integer m. If a sequence S over K has the period $N = p^n$ with a positive integer n then S is called an exponent periodic sequence.*   □

For exponent periodic sequences the generalized Games-Chan algorithm is known as one of fast algorithms computing the LC.

Definition 4. *For an exponent periodic sequence S over $K = GF(p^m)$ with period $N = p^n = pM$, we define one period of S by*

$$S^{(N)} = (s(0)^{(M)}, s(1)^{(M)}, \cdots, s(p-1)^{(M)}),$$

i.e., $s(j)^{(M)} = (s_{jM}, s_{jM+1}, \cdots, s_{(j+1)M-1})$ for $0 \leq j < p$ and a vector $b^{(M,u)}$ with length M over K is defined by

$$b^{(M,u)} = F_u(s(0)^{(M)}, s(1)^{(M)}, \cdots, s(p-1)^{(M)}) \tag{2}$$

for $0 \leq u < p$, where

$$F_u(s) = \sum_{j=0}^{p-u-1} \binom{p-j-1}{u} s_j \tag{3}$$

of $s = (s_0, s_1, \cdots, s_{p-1}) \in K^p$ is applied componentwise and $\binom{p-j-1}{u}$ means the binomial coefficients of $p - j - 1$ and u.   □

We recall the generalized Games-Chan algorithm shown in Fig.1. It is obvious that the generalized Games-Chan algorithm is induced by Definition 4. The final L of the generalized Games-Chan algorithm indicates the LC of an infinite sequence S with its one period $S^{(N)}$.

```
input: S^(N) = (s_0, s_1, ···, s_{N-1}), N = p^n
M = p^{n-1}, L = 0, s^(N) = S^(N),
for j = n - 1 down to 0
    b^(M,u) for 0 ≤ u ≤ p - 2 from s^(pM) by (2)
    if b^(M,0) ≠ 0 then case 1
    if b^(M,u) = 0 for 0 ≤ u ≤ w - 2, b^(M,w-1) ≠ 0 then case w
    if b^(M,u) = 0 for 0 ≤ u ≤ p - 2 then case p
    if case w then s^(M) = b^(M,w-1) from (2) and L = L + (p - w)M
    if M ≠ 1 then M = M/p
if s_0^(1) ≠ 0 then L = L + 1
```

Fig. 1. Generalized Games-Chan algorithm

3 k-Error Linear Complexity and Generalized k-LC Algorithm

In this section the k-LC and the generalized k-LC algorithm (see Fig.2) is also recalled in order to derive a generalization of the Lauder-Paterson algorithm.

Definition 5. *The k-error linear complexity (k-LC) of a periodic sequence S over K with period N is defined as*

$$L_k(S) = \min\{LC(S + E)|W(E^{(N)}) \le k\},$$

where a periodic sequence E over K has period N or the divisor of N, $W(E^{(N)})$ is the Hamming weight of $E^{(N)} = (e_0, e_1, ···, e_{N-1})$ and a sequence $S + E = (s_0 + e_0, s_1 + e_1, ···)$ over K. □

We have $L_{k-1}(S) \ge L_k(S)$ for $1 \le k \le W = W(S^{(N)})$, $L_0(S) = L(S)$ and $L_k(S) = 0$ for $W \le k \le N$ from Definition 5 obviously.

If a sequence S is an exponent periodic sequence then we can apply the generalized k-LC algorithm to S.

Definition 6. *Let a sequence S be an exponent periodic sequence over $K = GF(p^m)$ with period $N = p^n$ and a $q \times N$ matrix $\Sigma = [\sigma(h, i)]$ for $1 \le h \le q$, $0 \le i < N$ over the integers, where $q = p^m$, and Σ is called a cost matrix of S. Moreover let α be a primitive element over K. We need that h-th row of Σ ($1 \le h \le q$) corresponds to an element α_h in K, then we set $\alpha_1 = 0$ and $\alpha_h = \alpha^{h-2}$ for $2 \le h \le q$.* □

When the initial cost matrix $\Sigma = [\sigma(h, i)]$ is defined as

$$\sigma(h, i) = \begin{cases} 0 \text{ if } h = 1, \\ 1 \text{ if } h \neq 1 \end{cases} \tag{4}$$

for $1 \le h \le q, 0 \le i < N$, an element $\sigma(h, i)$ of the cost matrix Σ indicates the number of changing element at the original sequence with length N for substituting $s_i^{(M)}$ into $s_i^{(M)} + \alpha_h$ at that depth with length M and keeping the final LC by a previous depth.

From the generalized Games-Chan algorithm we need to set ranges of the LC. At the step M, meaning its input sequence with length pM, the value $T^{(M,u)}$, defined in next definition, means the minimum changing number of the LC range increasing LC value from $(p-w-1)M$ up to $(p-w)M$, and the set $D_i^{(M,u)}$ consists of all error pattern collecting error values at position $i, M+i, \ldots, (p-1)M+i$ at its input sequence with length pM satisfying above condition.

Definition 7. *Let a sequence S be an exponent periodic sequence over $K = GF(p^m)$ with period $N = p^n$ and a $q \times N$ matrix $\Sigma^{(N)}$, and let $M = N/p = p^{n-1}$. For $0 \le u < p - 1$*

$$T^{(M,u)} = \sum_{i=0}^{M-1} B_i^{(M,u)} \tag{5}$$

is calculated from S and $\Sigma^{(N)} = [\sigma(h,i)^{(N)}]$ by

$$B_i^{(M,u)} = \min \left\{ \sum_{j=0}^{p-1} \sigma(h_j, jM+i) \,\middle|\, e \in D_i^{(M,u)} \right\}$$

for $0 \le i < M$, where $e = (\alpha_{h_0}, \alpha_{h_1}, \cdots, \alpha_{h_{p-1}}) \in K^p$ and

$$D_i^{(M,u)} = \{e | F_j(e) + b_i^{(M,j)} = 0 \ (0 \le j \le u)\}$$

from (2) and (3).   □

Next calculations of the cost matrix $\Sigma_w^{(M)}$ for next step by case w, its input sequence with length M, in the generalized k-LC algorithm are defined as follows:

Definition 8. *Let a sequence S be an exponent periodic sequence over $K = GF(p^m)$ with period $N = p^n$ and a $q \times N$ matrix $\Sigma^{(N)}$, and let $M = N/p = p^{n-1}$. Then $\Sigma_w^{(M)} = [\sigma(h,i)_w^{(M)}]$ is calculated from S and $\Sigma^{(N)}$ by*

$$\sigma(h,i)_w^{(M)} = \min \left\{ \sum_{j=0}^{p-1} \sigma(h_j, jM+i)^{(N)} \,\middle|\, e \in \hat{D}(h,i)_w^{(M)} \right\}, \tag{6}$$

where $e = (\alpha_{h_0}, \alpha_{h_1}, \cdots, \alpha_{h_{p-1}}) \in K^p$ and

$$\hat{D}(h,i)_1^{(M)} = \{e \in K^p | F_0(e) - \alpha_h = 0\},$$
$$\hat{D}(h,i)_w^{(M)} = \left\{e \in K^p \,\middle|\, \begin{array}{l} F_j(e) + b_i^{(M,j)} = 0 \ (0 \le j < w-1), \\ F_{w-1}(e) - \alpha_h = 0 \end{array}\right\}$$

for $2 \le w \le p$.   □

These calculations propagate information about the number of change at the original input sequence with length N from step M to step M/p.

After above preparations, we show the generalized k-LC algorithm of the k-LC without shift and offset in Fig.2. The final value L of this algorithm is the k-LC with a fixed k of its input exponent periodic sequence S. This algorithm and the Lauder-Paterson algorithm, shown in next section, are used for proposed generalization of the lauder-Paterson algorithm.

---

input: $k, S^N = (s_0, s_1, \cdots, s_{N-1}), N = p^n$

$M = p^{n-1}, L = 0, s^{(N)} = S^N,$

$\Sigma^{(N)} = [\sigma(h, i)^{(N)}], \sigma(h, i)^{(N)} = \begin{cases} 0, \text{ if } h = 1, \\ 1, \text{ if } h \neq 1, \end{cases}$

for $j = n - 1$ **down to** 0

    $T^{(M,u)}$ for $u = 0, \cdots, p - 2$ from (5)

    if $k < T^{(M,0)}$ **then** case 1

    if $T^{(M,w-2)} \leq k < T^{(M,w-1)}$ **then** case w

    if $T^{(M,p-2)} \leq k$ **then** case p

    if case w **then** $s^{(M)} = b^{(M,w-1)}$ from (2) and $L = L + (p - w)M$

    Set $\Sigma^{(M)} = \Sigma_w^{(M)}$ from $\Sigma^{(pM)}$ by (6)

    if $M \neq 1$ **then** $M = M/p$

if $\sigma(h, 0) > k$ such that $\alpha_h - s_1^{(0)} = 0$ **then** $L = L + 1$

Fig. 2. Generalized k-LC algorithm

4   Profile of k-LC and Lauder-Paterson Algorithm

In this section we recall the profile of the k-LC and the Lauder-Paterson algorithm. The Lauder-Paterson algorithm, which gives the profile of the k-LC for a binary exponent sequence with period 2^n, is shown in Fig.3.

Definition 9. *Let a triple $\hat{S} = (S, \sigma, N)$ with a binary sequence S with length N, a vector σ over the integers[1] with length N and $N = 2^n$ be a cost binary sequence. We define $B(\hat{S}) = (B(S), B(\sigma), N/2)$ with length $N/2$ by*

$$B(S)_i = s_i + s_{i+(N/2)}, \quad B(\sigma)_i = \min\{\sigma_i, \sigma_{i+(N/2)}\}$$

for $0 \leq i < N/2$. And $L(\hat{S}) = (L(S), L(\sigma), N/2)$ with length $N/2$ is defined by

$$L(S)_i = \begin{cases} s_i \text{ if } s_i = s_{i+(N/2)} \text{ or } \sigma_i > \sigma_{i+(N/2)}, \\ s_{i+(N/2)} \text{ otherwise,} \end{cases}$$
$$L(\sigma)_i = \begin{cases} \sigma_i + \sigma_{i+(N/2)} \text{ if } s_i = s_{i+(N/2)}, \\ |\sigma_i - \sigma_{i+(N/2)}| \text{ otherwise} \end{cases} \quad (7)$$

for $0 \leq i < N/2$, where $|a|$ is the absolute value of a. □

---

[1] The vector σ is originally defined over real numbers [9]. However σ is enough to over integers in this paper.

---

input: $\text{LP}(\hat{S}, t, r, c)$

---
if $\ell > 1$ then
    $T = \sum_{B(S)_i = 1} B(\sigma)_i$ for $i = 0$ to $\ell/2 - 1$
    if $T > 0$ then $\text{LP}(B(\hat{S}), t, \min\{r, t + T - 1\}, c + (\ell/2))$
    if $t + T \leq r$ then $\text{LP}(L(\hat{S}), t + T, r, c)$
else   /* $\ell = 1$ */
    if $s_0 = 0$ then output (t, c)
    if $s_0 = 1$ and $\sigma_0 > 0$ then output $(t, c + 1)$
    if $s_0 = 1$ and $t + \sigma_0 \leq r$ then output $(t + \sigma_0, c)$

Fig. 3. Lauder-Paterson algorithm

The Lauder-Paterson algorithm is a recurrent algorithm (see Fig.3) and its final output from the initial input $\text{LP}(\hat{S} = (S^{(N)}, \sigma = (1, 1, \cdots, 1), N), 0, N, 0)$ is equal to the extended decrease set $EDS(S)$, defined by

$$EDS(S) = \{(0, L(S))\} \cup \{(k, L_k(S)) \mid L_k(S) < L_{k-1}(S), 1 \leq k \leq W(S^N)\}$$

for an exponent periodic sequence S with period $N = 2^n$. From the definition of the k-LC, $EDS(S)$ shows complete profile of the k-LC for a cost binary sequence $\hat{S}$ with period 2^n.

5   Generalization of Lauder-Paterson Algorithm

In this section we propose a generalized Lauder-Paterson algorithm (see Fig.4) which is not used the concepts of shift and offset as same as the generalized k-LC algorithm proposed in [4], although the Lauder-Paterson algorithm is using them as same as the Stamp-Martin algorithm [10] and its second generalization [5].

We can construct proposed algorithm as same as the Lauder-Paterson algorithm which is a recurrent algorithm. we need to consider p branches in each depth and some conditions are decided by their cost matrix in similar to the generalized Games-Chan algorithm. Moreover from the generalized k-LC algorithm and Definition 5 (the definition of the k-LC), for instance, if $r = T^{(M,0)}$ or $T^{(M,w-1)} = T^{(M,w)}$ or $T^{(M,p-2)} = t$ in case of $p = 3$ then there is no decrease point in the corresponding range decided the k-LC. Next main theorem is derived from above discussion.

Theorem 1. Let $(S^{(N)}, \Sigma, N, 0, 0, N + 1)$ be an input of the generalized Lauder-Paterson algorithm shown in Fig.4, where $\Sigma = [\sigma(h, i)]$ is defined from (4). The final output of the algorithm indicates the extended decrease set $EDS(S)$ of an exponent periodic sequence S with period N.

(Sketch of Proof): At first calculations of borders $T^{(N,u)}$ by Definition 7 is correct from the correctness of the generalized k-LC algorithm. Moreover it is obvious that p branches are needed at the generalized Lauder-Paterson algorithm from definition of the k-LC, the Lauder-Paterson algorithm and the generalized k-LC algorithm. Because the lower border and the upper border to keep that condition

---

```
input: GLP(S, Σ, N, c, r, t)
```

$M = N/p,\ S^{(N)} = S,\ \Sigma^{(N)} = \Sigma$
$T^{(M,u)}$ for $u = 0$ to $p - 2$ by (5)
if $N > 1$ then
   if $r < T^{(M,0)} = t'$ then GLP($b^{(M,0)}, \Sigma_1^{(M)}, M, c + (p-1)M, r, t'$)
   for $w = 1$ to $p - 2$
   if $r' = T^{(M,w-1)} < T^{(M,w)} = t'$ then GLP($b^{(M,w)}, \Sigma_{w+1}^{(M)}, M, c + (p-w-1)M, r', t'$)
   if $r' = T^{(M,p-2)} < t$ then GLP($b^{(M,p-1)}, \Sigma_p^{(M)}, M, c, r', t$)
else   /* $N = 1$ */
   $\alpha_h = -s_0^{(1)}$
   $m = \min\{\sigma(\ell, 0) | 1 \le \ell \le q, \ell \ne h\}$
   if $m < \sigma(h, 0)$ then output($m, c + 1$)
   if $\sigma(h, 0) < t$ then output($\sigma(h, 0), c$)

Fig. 4. Generalized Lauder-Paterson algorithm

as similar as the Lauder-Paterson algorithm, it is induced the correctness of proposed algorithm from the generalized k-LC algorithm and the Lauder-Paterson algorithm. □

Next we analyze memory and computation complexities about the generalized Lauder-Paterson algorithm.

Firstly we consider memory complexity in the single step of the algorithm. Four values M, c, r, t, each elements of $\Sigma^{(M)}$ and $p - 1$ times $T^{(M,u)}$ are integers less than or equal to N, and the number of them is $p^m M + p + 3$. The elements of sequence $S^{(M)}$ has M elements over $GF(p^m)$. If we can use p-state memory, we need $n(p^m M + p + 3) + Mm$ memories from $N = p^n$ in the worst case of the single step. Because the algorithm runs from $M = p^{n-1}$ to 1 (n steps), we need $n^2(p + 3) + p^n(np^m + m)$ memories in the worst case of the whole algorithm.

Secondly we consider computation complexity in the single step of the algorithm. Mp^{mp+1} times addition operations are needed for one $T^{(M,u)}$ calculation. Since u runs $p - 1$ times, we need $(p - 1)Mp^{mp+1}$ additions over $GF(p^m)$ from Definition 7. Moreover we need $p^m Mp^{mp+1} = Mp^{m(p+1)+1}$ additions for $\Sigma_w^{(M)}$ and Mp^2 times additions for $b^{M,w}$. Consequently about $Mp^{m(p+1)+2}$ additions are needed in the single step of the algorithm. If we decide one extended decrease point, the algorithm runs from $M = p^{n-1}$ to $M = 1$ (n steps). Hence we need about $p^n p^{m(p+1)+2} = p^{n+m(p+1)+2}$ additions for one extended decrease point. Since the number of the extended decrease set is N in the worst case, the computation complexity of the algorithm is about $Np^{n+m(p+1)+2} = p^{2n+m(p+1)+2}$ additions.

6   Numerical Examples

In this section we consider balanced exponent periodic sequences [7] defined as follows.

Definition 10. *If an exponent periodic sequence S over $K = GF(p^m)$ has period p^n with $n \ge m$ and same distributions for all elements in K, i.e., the number of an* *element within one period is p^{n-m} for all elements in K, then a sequence S is called a balanced exponent periodic sequence (BEPS).*   □

Especially, if a binary exponent periodic sequence S is balanced sequence we call S a balanced binary exponent periodic sequence (BBEPS).

6.1   Binary Exponent Periodic Sequences with Period 16

In this section a numerical example of BEPS with period 16 using the LPA [9] is given in order to study about distributions on the profile of the k-LC. We show all profiles of them in Table.1, where # is the number of BBEPS with that profile except the periodic isomorphism, and selected lines in Fig.5. In except the periodic isomorphism [7], the number of all BBEPS is 800 and the number of BBEPS with condition of Seq.5 at Table.5 is 16.

Table 1. k-LC of BBEPS ($N = 16$)

k	0	2	4	6	#	k	0	2	4	6	#
Seq.1	15	10	5	2	128		14	3	3	3	8
	15	10	3	2	64	Seq.3	13	13	3	3	64
	15	9	9	2	128		13	13	2	2	32
	15	6	3	2	32	Seq.4	12	9	9	5	32
	15	5	5	2	32		12	7	2	2	8
	15	3	3	2	8		12	6	3	3	4
Seq.2	15	2	2	2	8		12	5	5	5	8
	14	11	5	3	64		11	11	5	5	16
	14	11	2	2	32		11	11	2	2	8
	14	9	9	3	64		10	10	5	5	8
	14	7	2	2	16		10	10	3	3	4
	14	5	5	3	16	Seq.5	9	9	9	9	16
									Total		800

6.2   Exponent Periodic Sequences over $GF(3)$ with Period 9

In order to study about distributions on the profile of the k-LC, we show numerical examples to apply the generalized Lauder-Paterson (LP) algorithm to a balanced exponent periodic sequence (BEPS) S over $GF(3)$ with period $9(= 3^2)$. Note that $[0\ 3\ 3]^t$ shows a matrix $\begin{bmatrix} 0 \\ 3 \\ 3 \end{bmatrix}$ in next example 1.

Example 1: [Example of Generalized LP Algorithm]

$$S^{(9)} = (220211010), \quad \Sigma^{(9)} = \begin{bmatrix} 000000000 \\ 111111111 \\ 111111111 \end{bmatrix}$$

[Depth 1]:

$$b^{(3,0)} = (111), \quad b^{(3,1)} = (021), \quad b^{(3,2)} = (220),$$

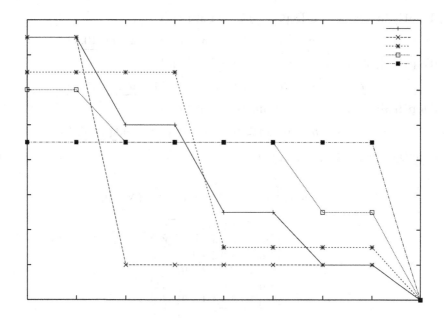

Fig. 5. Profile of k-LC for BBEPS ($N = 16$)

$$c = 0, \quad r = 0, \quad T^{(3,0)} = 3, \quad T^{(3,1)} = 3, \quad t = 10$$

$$\text{Case 1}: \text{GLP}(\boldsymbol{b}^{(3,0)} = (111), \begin{bmatrix} 000 \\ 111 \\ 111 \end{bmatrix}, 3, 6, 0, 3),$$

$$\text{Case 3}: \text{GLP}(\boldsymbol{b}^{(3,2)} = (220), \begin{bmatrix} 121 \\ 232 \\ 313 \end{bmatrix}, 3, 0, 3, 10)$$

[Depth 2]: of Case 1 at Depth 1

$$\boldsymbol{b}^{(1,0)} = (0), \quad \boldsymbol{b}^{(1,1)} = (0), \quad \boldsymbol{b}^{(1,2)} = (1),$$

$$c = 6, \quad r = 0, \quad T^{(1,0)} = 0, \quad T^{(1,1)} = 0, \quad t = 3$$

$$\text{Case 3}: \text{GLP}(\boldsymbol{b}^{(1,2)} = (1), [0\ 3\ 3]^t, 1, 6, 0, 3)$$

[Depth 2]: of Case 3 at depth 1

$$\boldsymbol{b}^{(1,0)} = (1), \quad \boldsymbol{b}^{(1,1)} = (0), \quad \boldsymbol{b}^{(1,2)} = (2),$$

$$c = 0, \quad r = 3, \quad T^{(1,0)} = 3, \quad T^{(1,1)} = 6, \quad t = 10$$

$$\text{Case 2}: \text{GLP}(\boldsymbol{b}^{(1,1)} = (0), [6\ 6\ 3]^t, 1, 1, 3, 6),$$

$$\text{Case 3}: \text{GLP}(\boldsymbol{b}^{(1,2)} = (2), [6\ 6\ 6]^t, 1, 0, 6, 10)$$

[Depth 3]: of Case 3 at Dep.2 and Case 1 at Dep.1

$$c = 6, \quad m = 0, \quad \sigma(3,0) = 3, \quad t = 3, \quad \underline{\text{output}(0,7)}$$

[Depth 3]: of Case 2 at Dep.2 and Case 3 at Dep.1

$$c = 1, \quad m = 3, \quad \sigma(1,0) = 6, \quad t = 6, \quad \underline{\text{output}(3,2)}$$

[Depth 3]: of Case 3 at Dep.2 and Case 3 at Dep.1

$$c = 0, \quad m = 6, \quad \sigma(2,0) = 6, \quad t = 10, \quad \underline{\text{output}(6,0)}$$

$$EDS(S^{(9)}) = \{(0,7), (3,2), (6,0)\} \qquad \qquad \Box$$

Table 2. k-LC of BEPS over $GF(3)$ ($N = 9$)

k	0 1 2 3 4 5 6	#
Seq.1	8 8 4 4 2 2 0	972
Seq.2	8 8 2 2 2 2 0	162
Seq.3	7 7 7 2 2 2 0	342
Seq.4	6 6 4 4 4 4 0	162
Seq.5	4 4 4 4 4 4 0	54
	Total	1674

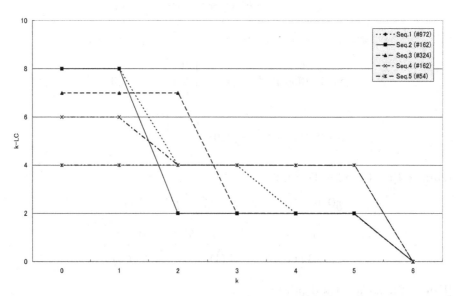

Fig. 6. Profile of k-LC for BEPS over $GF(3)$ ($N = 9$)

Example 2: [Profiles of the k-LC for BEPS over $GF(3)$ with Period 9]
A numerical example of BEPS over $GF(3)$ with $N = 9$ using the LPA [9]. We show all profiles of them in Table.2, where # is the number with that profile of BEPS's except the periodic isomorphism [13, 7], and all lines in Fig.6. $\qquad \Box$

7   Conclusion

In this paper we proposed the generalized Lauder-Paterson algorithm computing the profile of the k-LC for an exponent periodic sequence over a finite field. In order to derive proposed algorithm we recalled the generalized Games-Chan algorithm of the LC and the generalized k-LC algorithm for an exponent periodic sequence. The analysis of memory and computation complexities of the generalized Lauder-Paterson algorithm is given. Numerical examples of proposed algorithm for a BEPS over $GF(3)$ with period $9(= 3^2)$ is given to confirm the algorithm and all profiles and the number of them for BEPS over $GF(3)$ with period 9.

This proposed algorithm should be called a generalized k-LC spectrum [9] algorithm because this algorithm does not use the concepts of shift and offset. We may be able to rewrite proposed algorithm into an algorithm using the concepts of shift and offset.

Future works are fast algorithms of the LC and the k-LC for sequences with arbitrary period for fast algorithms for the k-LC. Moreover a generalization of the Lauder-Paterson algorithm using the concepts of shift and offset, remaining profiles of the k-LC for BEPS's and investigations of the k-LC and their profiles for non-binary sequences and non-exponent periodic sequences are also future works.

Acknowledgements

The author would like to thank Dr. Kenny Paterson for helpful discussion and bringing my attention to the paper [9] during the Second International Conference on Sequences and their Applications (SETA'01) in May, 2001 at Bergen, Norway. The author would like to thank Prof. Kyoki Imamura for encouragement and discussion to solve some problems on this area.

References

1. Z.Dai, and K.Imamura, "Linear Complexity for One-Symbol Substitution of a Periodic Sequence over $GF(q)$", *IEEE Trans. on Information Theory*, vol.44, pp.1328-1331, May, 1998.
2. C.Ding, G.Xiao, and W.Shan, *The Stability Theory of Stream Ciphers*, Lecture Notes in Computer Science, vol.561, Springer-Verlag, 1991.
3. R.Games, and A.Chan, "A Fast Algorithm for Determining the Complexity of a Binary Sequence with Period 2^n", *IEEE Trans. on Information Theory*, vol.IT-29, pp.144-146, Jan., 1983.
4. T.Kaida, S.Uehara, and K.Imamura, "An Algorithm for the k-Error Linear Complexity of Sequences over $GF(p^m)$ with Period p^n, p a Prime", *Information and Computation*, Vol.151, pp.134-147, Academic Press, May, 1999.
5. T.Kaida, S.Uehara, and K.Imamura, "A New Algorithm for the k-Error Linear Complexity of Sequences over $GF(p^m)$ with Period p^n", *Sequences and their Applications - Proceedings of SETA'98*, pp.284-296, Springer-Verlag, 1999.
6. T.Kaida, S.Uehara, and K.Imamura, "On the Profile of the k-Error Linear Complexity and the Zero Sum Property for Sequences over $GF(p^m)$ with Period p^n", *Sequences and their Applications - Proceedings of SETA'01*, pp.218-227, Springer-Verlag, 2001.

7. T.Kaida, "A Typical Profile of the k-Error Linear Complexity for Balanced Binary Sequences with Period 2^n", submitted to *IEICE Trans. Fundamentals*, Special Section of Cryptography and Information Security, Jan., 2005.

8. K.Kurosawa, F.Sato, T.Sakata, and W.Kishinoto, "A Relationship between Linear Complexity and k-Error Linear Complexity", *IEEE Trans. on Information Theory*, vol.46, pp.694-698, Mar., 2000.

9. A.G.B.Lauder, K.G.Paterson, "Computing the Error Linear Complexity Spectrum of a Binary Sequence of Period 2^n", *IEEE Trans. Inf. Theory*, vol.49, pp.273-280, Jan., 2003.

10. M.Stamp, and C.Martin, "An Algorithm for the k-Error Linear Complexity of Binary Sequences with Period 2^n", *IEEE Trans. on Information Theory*, vol.39, pp.1398-1401, July, 1993.

11. S.Uehara, and K.Imamura, "Linear Complexity of Periodic Sequences Obtained from $GF(q)$ Sequences with Period $q^n - 1$ by One-Symbol Insertion", *IEICE Trans. on Fundamentals*, vol.E79-A, pp.1739-1740, Oct., 1996.

12. S.Uehara, and K.Imamura, "Linear Complexity of Periodic Sequences Obtained from $GF(p)$ Sequences with Period $p^n - 1$ by One-Symbol Deletion", *IEICE Trans. on Fundamentals*, vol.E80-A, pp.1164-1166, June, 1997.

13. S.Uehara, K.Imamura, T.Kaida, "Value distribution of linear complexity for q-ary periodic sequences with period p^n, p a prime", *IEICE Trans. Fundamentals*, vol.E80-A, pp.920-921, May, 1997.

On the Computation of the Linear Complexity and the k-Error Linear Complexity of Binary Sequences with Period a Power of Two

Ana Sălăgean

Department of Computer Science,
Loughborough University, UK
A.M.Salagean@lboro.ac.uk

Abstract. The linear Games-Chan algorithm for computing the linear complexity $c(s)$ of a binary sequence s of period $\ell = 2^n$ requires the knowledge of the full sequence, while the quadratic Berlekamp-Massey algorithm only requires knowledge of $2c(s)$ terms. We show that we can modify the Games-Chan algorithm so that it computes the complexity in linear time knowing only $2c(s)$ terms. The algorithms of Stamp-Martin and Lauder-Paterson can also be modified, without loss of efficiency, to compute analogues of the k-error linear complexity and of the error linear complexity spectrum for finite binary sequences viewed as initial segments of infinite sequences with period a power of two. Lauder and Paterson apply their algorithm to decoding binary repeated-root cyclic codes of length $\ell = 2^n$ in $\mathcal{O}(\ell(\log_2 \ell)^2)$ time. We improve on their result, developing a decoding algorithm with $\mathcal{O}(\ell)$ bit complexity.

1 Introduction and Notation

We denote by $\mathcal{S}$ the set of all (infinite) linearly recurrent sequences over $\mathbb{F}_2$. Let $s \in \mathcal{S}$, $s = s_0, s_1, s_2 \ldots$. We will say that a polynomial $f \in \mathbb{F}_2[x]$, $f = x^m + a_{m-1}x^{m-1} + \cdots + a_1 x + a_0$ is an *annihilator polynomial* for s if s satisfies the linear recurrence given by the coefficients of f i.e. $s_{i+m} + a_{m-1}s_{i+m-1} + \cdots + a_1 s_{i+1} + a_0 s_i = 0$ for $i = 0, 1, 2, \ldots$. The monic annihilator polynomial of minimal degree is unique and is called the *characteristic polynomial* of s. We will denote it by $\sigma(s)$. The characteristic polynomial generates the ideal of all annihilator polynomials. The linear complexity of s is the degree of the characteristic polynomial and will be denoted by $c(s)$.

Denote by P_N the set of sequences in $\mathcal{S}$ having (not necessarily minimal) period N. If $s \in P_N$ then $\sigma(s) | x^N - 1$. We will denote by $\mathcal{T}$ the set of binary sequences with period any power of two, i.e. $\mathcal{T} = \cup_{i=0}^{\infty} P_{2^i}$. For any $s \in \mathcal{T}$, the linear complexity of s equals c if and only if $\sigma(s) = (x-1)^c$, i.e.knowing the linear complexity is tantamount to knowing the characteristic polynomial.

The linear complexity of a finite sequence $z = (z_0, z_1, \ldots, z_{t-1}) \in \mathbb{F}_2^t$ viewed as an initial segment of an infinite sequence in a set $A \subseteq \mathcal{S}$, denoted $c(z, A)$, is

T. Helleseth et al. (Eds.): SETA 2004, LNCS 3486, pp. 179–184, 2005.

defined as the minimum linear complexity of all sequences in A which have z as an initial segment i.e. $c(z, A) = \min\{c(s)|s \in A, s_i = z_i \text{ for } i = 0, \ldots, t - 1\}$.

The Berlekamp-Massey algorithm, [1, 8], computes the linear complexity of finite sequences, i.e. it computes $c(z, \mathcal{S})$ for any finite sequence $z \in \mathbb{F}_2^t$. The complexity of the algorithm is quadratic in the length t of the finite sequence. It is well known that if $s \in A$ and $t \geq 2c(s)$ then $c((s_0, \ldots, s_{t-1}), A) = c(s)$. So one can think of the Berlekamp-Massey algorithm as computing the linear complexity $c(s)$ of an infinite sequence s knowing only the first $2c(s)$ terms of the sequence. For infinite sequences with period a power of two, Games-Chan, [4], developed a linear algorithm which computes the linear complexity of the sequence. The whole sequence needs to be known.

The k-error linear complexity of a periodic sequence was defined in [11] and is closely related to previously defined notions of sphere complexity [3] and weight complexity [2]. The k-error linear complexity of a sequence $s \in P_N$, as a sequence of period N, denoted by $c_{k,N}(s)$, is defined as the minimum complexity that s can have after modifying k bits of a period i.e.

$$c_{k,N}(s) = \min\{c(s + e)|e \in P_N, \text{wt}((e_0, e_1, \ldots, e_{N-1})) \leq k\}.$$

The definition can be extended to a costed sequence with the cost given by a vector $cost \in \mathbb{R}^N$ as

$$c_{k,N}(s, cost) = \min\{c(s + e)|e \in P_N, \sum_{0 \leq i < N, e_i \neq 0} cost[i] \leq k\}.$$

The Stamp-Martin algorithm, [11], computes the k-error linear complexity for any binary (costed) sequence with period a power of two, i.e. for any $s \in P_{2^n}$ it computes $c_{k,2^n}(s)$ or $c_{k,2^n}(s, cost)$ respectively. As in the Games-Chan algorithm, the whole sequence needs to be known and the time (bit operations) and space complexity of the algorithm is linear in the period $\ell = 2^n$ of the sequence.

The error linear complexity spectrum of a sequence $s \in P_N$ is defined as the set of pairs $\{(k, c_{k,N}(s))|0 \leq k \leq \text{wt}((s_0, \ldots, s_{N-1})\}$. It can be computed for The Lauder-Paterson algorithm, [7], computes the error linear complexity spectrum of any binary sequence with period $\ell = 2^n$. The bit complexity of the algorithm is $\mathcal{O}(\ell(\log \ell)^2)$.

The k-error linear complexity of a finite sequence $z = (z_0, z_1, \ldots, z_{t-1}) \in \mathbb{F}_2^t$ viewed as an initial segment of a sequence in $A \subseteq \mathcal{S}$, denoted $c_k(z, A)$, will be defined as

$$c_k(z, A) = \min\{c(z + e, A)|e \in \mathbb{F}_2^t, \text{wt}(e) \leq k\}.$$

Note that while for an infinite sequence s, $c(s)$ can be computed given a finite segment of at least $2c(s)$ terms, we cannot expect to be able to compute $c_{k,N}(s)$ given less than the whole sequence, as we cannot know how many of the errors in an error pattern that minimises linear complexity will fall outside our known portion of the sequence.

2   Computing the Linear Complexity and k-Error Linear Complexity for Finite Sequences

In this section our goal is to develop an algorithm which computes the linear complexity and the k-error linear complexity of a finite sequence viewed as an initial segment of a binary sequence with period a power of two (we do not need to know which power though). The following two theorems allow us to do so. The proofs are elementary; full details will appear in [10].

Theorem 1. *Let $t > 0$ and $z = (z_0, \ldots, z_{t-1}) \in \mathbb{F}_2^t$. Define $u = \lceil \log_2 t \rceil$ and define the infinite sequence s' of period 2^u as follows: $s'_i = z_i$ for $i = 0, 1, \ldots, t-1$ and $s'_i = z_{i-2^{u-1}}$ for $i = t, t+1, \ldots, 2^u - 1$. Then*

(i) *If $c(z, \mathcal{T}) \leq \frac{t}{2}$ then $c(z, \mathcal{T}) = c(s')$.*

(ii) *If $c(z, \mathcal{T}) > \frac{t}{2}$ then $c(s') > \frac{t}{2}$.*

Theorem 2. *Let $t > 0$ and $z = (z_0, \ldots, z_{t-1}) \in \mathbb{F}_2^t$. Define $u = \lceil \log_2 t \rceil$ and define the infinite costed sequence s' of period 2^u as follows: $s'_i = z_i$ and $cost[i] = 1$ for $i = 0, 1, \ldots, t-1$ and s'_i have arbitrary values and $cost[i] = 0$ for $i = t, t+1, \ldots, 2^u - 1$. Then $c_k(z, \mathcal{T}) = c_{k,2^u}(s', cost)$ for all $k = 0, 1, \ldots, \text{wt}(z)$. In particular, $c(z, \mathcal{T}) = c_{0,2^u}(s', cost)$.*

Hence by setting up (in linear time) an infinite costed sequence s' of period $2^{\lceil \log_2 t \rceil}$ as in Theorem 2 and then applying the Stamp-Martin algorithm to compute $c_{k,2^{\lceil \log_2 t \rceil}}(s', cost)$ we obtain in fact $c_k(z, \mathcal{T})$ and in particular, for $k = 0$ we obtain $c(z, \mathcal{T})$. The resulting algorithm obviously runs in $\mathcal{O}(t)$ time and is thus a more efficient alternative to the Berlekamp-Massey algorithm for the particular class of binary sequences with period a power of two.

However, the cryptographic applications of this results are limited. Namely, assume it is known that sequences with period a power of two are used. An opponent intercepts a finite segment of t terms of a sequence and wants to recover the whole sequence (to break a stream cipher for example). They could simply assume $(x - 1)^t$ is an annihilating polynomial and, as long as $t \geq c(s)$ (rather than $t \geq 2c(s)$) they would recover the correct sequence.

The Games-Chan and Stamp-Martin algorithms have been extended to sequences over $\mathbb{F}_{p^m}$ with period $\ell = p^n$, where p is a prime, [6, 3]. Theorem 1 does not hold for $p > 2$ but Theorem 2 does. Hence we can use it to compute the complexity and k-error linear complexity of finite sequences over $\mathbb{F}_{p^m}$, viewed as initial segments of infinite sequences with period $\ell = p^n$.

3   Decoding Repeated-Root Cyclic Codes

Repeated-root binary codes with length a power of two have been introduced in [9]. It is shown, *loc. cit.*, that these codes are subcodes of Reed-Muller codes, and it is proposed that they be decoded by majority logic, just like the Reed-Muller codes. An alternative decoding algorithm with bit complexity $\mathcal{O}(\ell(\log \ell)^2)$, where $\ell = 2^n$ is the length of the code, is proposed in [7]. In this section we develop a linear, $\mathcal{O}(\ell)$, decoding algorithm for these codes.

A binary repeated-root cyclic code of length 2^n can be described as being the set of all binary sequences of period 2^n having complexity at most c, for a given parameter c. Decoding a received r amounts to finding e of minimal weight such that $c(r + e) \leq c$.

An explicit algorithm for computing an error pattern of minimal weight which brings the linear complexity of the sequence below a given value c is given below. It is similar to the Stamp-Martin algorithm. We also need to compute the error pattern (rather than just the number of errors). We used a technique similar to the so-called L-pullup and B-pullup of [7], in a more compact and efficient form. Alternatively the method of [5] could be similarly adapted.

Algorithm 3. (Computing a minimum weight sequence e such that $c(s+e) \leq c$)

Input: n, c positive integers and $s = (s_0, s_1, \ldots, s_{2^n-1})$ a sequence of period
         2^n given by its first 2^n terms
Output: e a sequence of period 2^n given by its first 2^n terms and an integer
         $k = \mathrm{wt}(e)$ such that e is of minimal weight such that $c(s + e) \leq c$.

begin
$a \leftarrow s$; $\ell \leftarrow 2^n$; $c' \leftarrow 0$; $k' \leftarrow 0$,
for $i = 0$ **to** $\ell - 1$ **do** $cost[i] \leftarrow 1$ **endfor**
for $j = 0$ **to** $n - 1$ **do**
    $flag[j] \leftarrow 0$
    for $i = 0$ **to** $2^{n-j} - 1$ **do** $error[j][i] \leftarrow 0$ **endfor**
endfor
for $j = 0$ **to** $n - 1$ **do**
    $\ell \leftarrow \ell/2$ % now $\ell = 2^{n-j-1}$
    $L = a_0 a_1 \ldots a_{\ell-1}$; $R = a_\ell a_{\ell+1} \ldots a_{2\ell-1}$;
    $b \leftarrow L + R$; $T \leftarrow \sum_{i=0}^{\ell-1} b_i \min(cost[i], cost[i + \ell])$
    if $T = 0$ **or** $c' + \ell \geq c$ **then**
        $k' \leftarrow k' + T$; $flag[j] \leftarrow 1$
        for $i = 0$ **to** $\ell - 1$ **do**
            if $b_i = 1$ **then**
                if $cost[i] \leq cost[i + \ell]$
                then $a_i \leftarrow R_i$; $cost[i] \leftarrow cost[i + \ell] - cost[i]$; $error[j][i] \leftarrow 1$
                else $a_i \leftarrow L_i$; $cost[i] \leftarrow cost[i] - cost[i + \ell]$; $error[j][i + \ell] \leftarrow 1$
                endif
            else $a_i \leftarrow L_i$; $cost[i] \leftarrow cost[i] + cost[i + \ell]$
            endif
        endfor
    else
        $c' \leftarrow c' + \ell$
        for $i = 0$ **to** $\ell - 1$ **do**
            $a_i \leftarrow b_i$;

```
            if cost[i] ≤ cost[i + ℓ] then error[j][i] ← 1
            else cost[i] ← cost[i + ℓ]; error[j][i + ℓ] ← 1
            endif
        endfor
    endif
endfor
if a₀ = 1 and c' + 1 > c then k' ← k' + cost[0]; e ← 1
else e ← 0
endif
k ← k'
for j = n − 1 downto 0 do
    e ← duplicate(e)
    if flag[j] = 1 then e ← e XOR error[j]
    else e ← e AND error[j]
    endif
endfor
return(e, k)
end
```

The function *duplicate* simply duplicates a binary string, i.e. concatenates it with a copy of itself. The XOR and AND operators are bitwise operators between binary strings of equal lengths.

The correctness of the algorithm can be proved in a similar manner as the correctness of the Stamp-Martin algorithm and of the L-pullup and B-pullup constructions (see [11, 7]). We also proved that the time and space bit complexities of Algorithm 3 and of the Stamp-Martin algorithm are linear in the period $\ell = 2^n$ of the sequence. Full details of the proofs will appear in [10].

We expect that the algorithms of [6, 3] for $p > 2$ could be modified along the lines of Algorithm 3 to decode repeated-root codes over $\mathbb{F}_{p^m}$ with length p^n.

Acknowledgements. We would like to thank the anonymous referees for pointing out an error in the interpretation of one of the theorems and for suggesting directions for further research.

References

1. E.R. Berlekamp. *Algebraic Coding Theory*. McGraw Hill, 1968.
2. C. Ding. Lower bounds on the weight complexities of cascaded binary sequences. In J. Seberry and J. Pieprzyk, editors, *Advances in Cryptology – AUSCRYPT'90*, volume 453 of *LNCS*, pages 39–43. Springer Verlag, 1991.
3. C. Ding, G. Xiao, and W. Shan. *The stability Theory of Stream Ciphers*, volume 561 of *LNCS*. Springer Verlag, 1991.
4. R.A. Games and A.H. Chan. A fast algorithm for determining the complexity of a binary sequence with period 2^n. *IEEE Trans. Inform Theory*, 29:144–146, 1983.

5. T. Kaida, S. Uehara, and K. Imamura. Computation of the k-error linear complexity of binary sequences with period 2^n. In R.C.H. Yap, editor, *Concurrency and Parallelism, Programming, Networking*, volume 1179 of *LNCS*, pages 182–191. Springer Verlag, 1996.

6. T. Kaida, S. Uehara, and K. Imamura. An algorithm for the k-error linear complexity of sequences over $GF(p^m)$ with period p^n, p a prime. *Inform. Comput.*, 151:134–147, 1999.

7. A.G.B. Lauder and K.G. Paterson. Computing the error linear complexity spectrum of a binary sequence of period 2^n. *IEEE Trans. Inform Theory*, 49:273–280, 2003.

8. J.L. Massey. Shift register synthesis and BCH decoding. *IEEE Trans on Inform Theory*, 15:122–127, 1969.

9. J.L. Massey, D.J. Costello, and J. Justesen. Polynomial weights and code constructions. *IEEE Trans on Inform Theory*, 19:101–110, 1973.

10. A.M. Sălăgean. On the computation of the linear complexity and the k-error linear complexity of binary sequences with period a power of two. *IEEE Trans on Inform Theory*, 51:1145–1150, 2005.

11. M. Stamp and C.F. Martin. An algorithm for the k-error linear complexity of binary sequences of period 2^n. *IEEE Trans. Inform Theory*, 39:1398–1401, 1993.

On the 2-Adic Complexity and the k-Error 2-Adic Complexity of Periodic Binary Sequences*

Honggang Hu[1,2] and Dengguo Feng[1]

[1] State Key Laboratory of Information Security (Graduate School of Chinese Academy of Sciences), Beijing, 100049, China
[2] Institute of Electronics, Chinese Academy of Sciences, Beijing, 100080, China
hghu@ustc.edu, feng@is.iscas.ac.cn

Abstract. In this paper, we point out a significant difference between the linear complexity and the 2-adic complexity of periodic binary sequences. The concept of the symmetric 2-adic complexity of periodic binary sequences is presented based on this observation. We determine the expected value of the 2-adic complexity and derive a lower bound on the expected value of the symmetric 2-adic complexity of periodic binary sequences. Because the 2-adic complexity of periodic binary sequences is unstable, we present the concepts of the k-error 2-adic complexity and the k-error symmetric 2-adic complexity, and lower bounds on them are also derived.

1   Introduction

Klapper and Goresky introduced a new feedback architecture for shift register generation of pseudorandom binary sequences called feedback with carry shift register (FCSR) [7]. See also their other work on this FCSRs [2, 3, 4, 8]. An FCSR is determined by r coefficients $q_1, q_2, ..., q_r$, where $q_i \in \{0, 1\}$, $i = 1, 2, ..., r$, and an initial memory integer m_{r-1}. If the contents of the register at any given time are $(a_{n-1}, a_{n-2}, ..., a_{n-r+1}, a_{n-r})$ and the memory integer is m_{n-1}, then the operation of the shift register is defined as follows:

1. Form the integer sum $\sigma_n = \sum_{k=1}^{r} q_k a_{n-k} + m_{n-1}$.
2. Shift the contents one step to the right, outputting the rightmost bit a_{n-r}.
3. Place $a_n = \sigma_n \pmod 2$ into the leftmost cell of the shift register.
4. Replace the memory integer m_{n-1} with $m_n = (\sigma - a_n)/2$.

The integer $q = -1 + q_1 2 + q_2 2^2 + ... + q_r 2^r$ is called the connection integer of the FCSR.

---

* This work was supported by National Key Foundation Research 973 Project (No. G1999035802), National Nature Science Foundation Project (No. 60273027) and National Distinguished Youth Science Foundation Project (No. 60025205) of China.

T. Helleseth et al. (Eds.): SETA 2004, LNCS 3486, pp. 185–196, 2005.

Sequences generated by FCSR share many important properties with sequences genetated by LFSR (see [8]). Klapper and Goresky discussed some basic properties of FCSR sequences [2, 3, 4, 7, 8], such as their periods, rational expressions, exponential representations, rational approximation algorithms and their randomness. Considering FCSR's particular structure, Klapper and Goresky also introduced the concept of the 2-adic complexity, which is the arithmetic analog of the linear complexity. The 2-adic complexity is very useful in the study of the security of pseudorandom sequences for cryptographic applications. For example, by means of the rational approximation algorithm (an analogue of the Berlekamp-Massey algorithm, see [8]), the summation cipher proposed by Rueppel [14] can be easily attacked. Meidl presented an FCSR analog of (extended) Games-Chan algorithm [11], which efficiently yields an upper bound for the 2-adic complexity of a periodic binary sequence with period p^n.

Any infinite binary sequence $S = s_0, s_1, s_2, \ldots$ can be identified with the element $\alpha = \sum_{i=0}^{\infty} s_i 2^i$ in the ring Z_2 of 2-adic numbers. For a comprehensive survey of p-adic numbers please refer to [9]. The sequence S is eventually periodic if and only if the 2-adic number α is rational, i.e., there exist integers p, q such that $\alpha = -p/q \in Z_2$. In particular, if S is strictly periodic with minimal period T, then

$$\alpha = \sum_{i=0}^{\infty} s_i 2^i = -\frac{\sum_{i=0}^{T-1} s_i 2^i}{2^T - 1}.$$

Let us write $\alpha = -p/q$ as a fraction reduced to lowest terms with q positive and $0 \leq p \leq q$. Then $T = ord_q(2)$, where $ord_q(2)$ is the minimal integer t such that $2^t \equiv 1 \pmod{q}$. According to [8], q is the connection integer of the smallest FCSR, i.e., the FCSR with minimal number r of coefficients q_i, which can generate the binary sequence S. Notice that we have $r = \lfloor \log_2(q+1) \rfloor$.

The following two definitions were given by Klapper and Goresky in [8].

Definition 1. *If $S = s_0, s_1, s_2, \ldots$ is a periodic binary sequence, and $-p/q$ is the fraction in lowest terms whose 2-adic expansion agrees with the sequence S, then the 2-adic complexity $\Phi(S)$ of S is the real number $log_2(max(p,q))$.*

Remark 1. If S is the all-0 sequence, we put $\Phi(S) = 0$.

Remark 2. If S is strictly periodic with minimal period $T \geq 2$, then $\Phi(S) = log_2 q$.

Definition 2. *An l-sequence is a periodic sequence which is obtained from an FCSR with connection integer q for which 2 is a primitive root.*

Remark 3. In this case, q must be of the form $q = r^e$ with r an odd prime and $e \geq 1$. When $e \geq 3$, 2 is primitive modulo r^e if and only if 2 is primitive modulo r^2 [5].

l-sequences are the arithmetic analogs of m-sequences. Many of their properties are analogous to those of m-sequences. They are very important in the study of FCSR sequences. Some lower bounds on the linear complexity of l-sequences were given in [15]. Qi and Xu studied the partial period distribution of l-sequences by means of exponential sum [13].

The linear complexity $L(S)$ of a periodic binary sequence S is the least order of a linear recurrence relation satisfied by S. By the Berlekamp-Massey algorithm only a knowledge of $2L(S)$ bits is needed to reproduce the binary sequence S [10]. Similarly, by the rational approximation algorithm, only a knowledge of $\lceil 2\Phi(S)\rceil + 2$ bits is needed (see [8], Theorem 10.2). Therefore any binary sequence with low 2-adic complexity is insecure for cryptographic applications.

Meidl and Niederreiter determined the expected value of the linear complexity of T-periodic sequences explicitly by the generalized discrete Fourier transform [12] and confirmed Rueppel's conjecture [14]. Motivated by their work, we study the expected value of the 2-adic complexity and the symmetric 2-adic complexity of periodic binary sequences.

Like the linear complexity, the 2-adic complexity of periodic sequence is unstable: a small perturbation within one period may cause an extreme increase or decrease. To measure the instability, we propose the concept of k-error 2-adic complexity and k-error symmetric 2-adic complexity in this paper. There has been a lot of work on the k-error linear complexity. However there has been little study carried out on the k-error 2-adic complexity and the k-error symmetric 2-adic complexity.

In the following sections we only consider strictly periodic sequences, and we just call them periodic sequences for simplicity. The organization of this paper is as follows. In Section 2 we point out a significant difference between linear complexity and 2-adic complexity and give some examples to specify this. Based on this observation, the concept of the symmetric 2-adic complexity of periodic binary sequences is presented. In Section 3, we study the expected value of the 2-adic complexity and the symmetric 2-adic complexity of periodic binary sequences. In Section 4, we study the k-error 2-adic complexity and the k-error symmetric 2-adic complexity of periodic binary sequences. Finally, Section 5 concludes this paper.

2   The 2-Adic Complexity of Periodic Binary Sequences

Let $S = s_0, s_1, s_2, \ldots$ be a periodic binary sequence with period T. Since S is completely determined by its first T terms, we can describe S by the notation $S = (s_0, s_1, \ldots, s_{T-1})^\infty$. We define $S^T(x)$ to be the polynomial $S^T(x) = s_0 + s_1 x + \ldots + s_{T-1} x^{T-1}$. The minimal polynomial $f(x)$ of S is given by

$$f(x) = \frac{x^T - 1}{gcd(x^T - 1, S^T(x))} = \frac{x^T + 1}{gcd(x^T + 1, S^T(x))}.$$

Let $\widehat{S}$ be the inverse sequence of S, i.e., $\widehat{S} = (s_{T-1}, s_{T-2}, \ldots, s_1, s_0)^\infty$, then the minimal polynomial $\widehat{f}(x)$ of $\widehat{S}$ is given by

$$\widehat{f}(x) = \frac{x^T - 1}{gcd(x^T - 1, \widehat{S}^T(x))} = \frac{x^T + 1}{gcd(x^T + 1, \widehat{S}^T(x))},$$

where

$$\widehat{S}^T(x) = s_{T-1} + s_{T-2}x + \ldots + s_1 x^{T-2} + s_0 x^{T-1} = x^{T-1} S^T(\frac{1}{x}).$$

Let $g(x) = gcd(x^T + 1, S^T(x))$, $\widehat{g}(x) = gcd(x^T + 1, \widehat{S}^T(x))$. Then one can check that

$$\widehat{g}(x) = x^{deg(g(x))} g(\frac{1}{x}).$$

Hence $deg(f(x)) = deg(\widehat{f}(x))$ and $L(S) = L(\widehat{S})$. However when we consider the case of the 2-adic complexity, this is not valid. I.e., $\Phi(S)$ may be different from $\Phi(\widehat{S})$. For example, $S = (101110010001)^\infty$ is an l-sequence with minimal period $T = 12$ and its rational expression is

$$-\frac{2^0 + 2^2 + 2^3 + 2^4 + 2^7 + 2^{11}}{2^{12} - 1} = -\frac{7}{13}.$$

On the other hand, the rational expression of $\widehat{S}$ is

$$-\frac{2^0 + 2^4 + 2^7 + 2^8 + 2^9 + 2^{11}}{2^{12} - 1} = -\frac{47}{65}.$$

Therefore $\Phi(S) \neq \Phi(\widehat{S})$. Moreover, because 2 is not primitive modulo 65, $\widehat{S}$ is not an l-sequence.

The difference between the linear complexity and the 2-adic complexity results from the difference between polynomials and integers. Suppose the polynomial $h(x)$ over F_2 is an irreducible polynomial. Then

$$\widehat{h}(x) = x^{deg(h(x))} h(\frac{1}{x})$$

is also an irreducible polynomial. However, if $h(2)$ is a prime number, $\widehat{h}(2)$ may not be a prime number. (Here, $h(x)$ and $\widehat{h}(x)$ are thought of as polynomials over the ring Z of integers.) For example, $h(x) = 1 + x + x^4$ is irreducible, and $\widehat{h}(x) = 1 + x^3 + x^4$ is also irreducible. On the other hand, $h(2) = 19$ is prime, but $\widehat{h}(2) = 25$ is composite.

If we can reproduce the sequence $\widehat{S}$ by the rational approximation algorithm, we can also get the sequence S. According to [8], only a knowledge of $\lceil 2\Phi(\widehat{S}) \rceil + 2$ bits is needed to find the minimal FCSR which can generate the sequence $\widehat{S}$. Therefore $min(\lceil 2\Phi(S) \rceil + 2, \lceil 2\Phi(\widehat{S}) \rceil + 2)$ bits are sufficient for reproducing the sequence S. Thus the following concept of symmetric 2-adic complexity is a better measure for assessing the strength of a periodic sequence against the rational approximation attack.

Definition 3. *Let S be a periodic binary sequence with minimal period $T \geq 2$ and rational expression $-p/q$, where $0 < p < q$ and $gcd(p,q)=1$. Let $\widehat{S}$ be its inverse sequence with rational expression $-p'/q'$, where $0 < p' < q'$ and $gcd(p',q')=1$. Then the symmetric 2-adic complexity $\overline{\Phi}(S)$ of S is the real number $min(log_2 q, log_2 q')$.*

Because l-sequences are a significant kind of FCSR sequences, we will discuss their symmetric 2-adic complexity below.

Proposition 1. *Let S be an l-sequence generated by a FCSR with prime connection integer q. Then $\overline{\Phi}(S) = \log_2 q$.*

Proof. The minimal period of S is $T = q - 1$. Suppose the rational expression of $\widehat{S}$ is $-p'/q'$, where $0 < p' < q'$ and $\gcd(p', q')=1$. Then $T = ord_{q'}(2)$. So $q - 1 = ord_{q'}(2) \leq q' - 1$. Thus $q \leq q'$ and $\overline{\Phi}(S) = \log_2 q$. This completes the proof.

3   The Expected Value of the 2-Adic Complexity of Periodic Binary Sequences

In this section, the underlying stochastic model is that each binary sequence with period T has the same probability 2^{-T}.

Let $S = s_0, s_1, s_2, ...$ be a periodic binary sequence with period $T \geq 2$. Suppose that $2^T - 1 = p_1^{e_1} p_2^{e_2} ... p_h^{e_h}$, where p_i are prime numbers with $p_1 < p_2 < ... < p_h, e_i \geq 1, i = 1, 2, ..., h$. If S is not the all-0 sequence or the all-1 sequence, then we have

$$\alpha = -\frac{S^T(2)}{2^T - 1} = -\frac{a}{p_{i_1}^{f_{i_1}} p_{i_2}^{f_{i_2}} ... p_{i_t}^{f_{i_t}}}, \tag{1}$$

where $1 \leq t \leq h, 1 \leq i_1 < i_2 < ... < i_t \leq h, 1 \leq f_{i_j} \leq e_{i_j}, 1 \leq a < \prod_{j=1}^{t} p_{i_j}^{f_{i_j}}$, and $\gcd(a, \prod_{j=1}^{t} p_{i_j}^{f_{i_j}}) = 1$. In (1), there are $\prod_{j=1}^{t} \phi(p_{i_j}^{f_{i_j}})$ integers a such that $1 \leq a < \prod_{j=1}^{t} p_{i_j}^{f_{i_j}}$ and $\gcd(a, \prod_{j=1}^{t} p_{i_j}^{f_{i_j}}) = 1$, where ϕ is the Euler function. Thus, there are $\prod_{j=1}^{t} \phi(p_{i_j}^{f_{i_j}})$ binary sequences S with period T such that the 2-adic complexity of S is $\Phi(S) = \sum_{j=1}^{t} f_{i_j} \log_2 p_{i_j}$. Moreover, for any h integers $f_1, f_2, ..., f_h$ with $0 \leq f_i \leq e_i, i = 1, 2, ..., h$, and at least one of them nonzero, one can easily check that there are $\prod_{i=1}^{h} \phi(p^{f_i})$ binary sequences S with period T such that $\Phi(S) = \sum_{i=1}^{h} f_i \log_2 p_i$.

Let E_T denote the expected value of the 2-adic complexity of periodic binary sequences with period T. The identity below is useful when computing E_T.

Lemma 1. *For any real number $x \neq 0, 1$,*

$$\sum_{n=1}^{e} nx^{n-1} = \frac{1 - (e+1)x^e + ex^{e+1}}{(x-1)^2}.$$

Theorem 1. *Suppose that $T \geq 2$ and $2^T - 1 = p_1^{e_1} p_2^{e_2} ... p_h^{e_h}$, where p_i are prime numbers with $p_1 < p_2 < ... < p_h, e_i \geq 1, i = 1, 2, ..., h$. Then the expected value E_T of the 2-adic complexity of binary sequences with period T is given by*

$$E_T = (1 - \frac{1}{2^T}) \log_2(2^T - 1) - (1 - \frac{1}{2^T}) \sum_{i=1}^{h} \frac{(1 - p_i^{-e_i}) \log_2 p_i}{p_i - 1}.$$

Proof. If S is the all-0 sequence or the all-1 sequence, $\Phi(S) = 0$. So we only need to consider the other cases of S. From the discussion above, we have

$$E_T = \frac{1}{2^T} \sum_{f_1=0}^{e_1} \sum_{f_2=0}^{e_2} \cdots \sum_{f_h=0}^{e_h} \phi(p_1^{f_1})\phi(p_2^{f_2})\ldots\phi(p_h^{f_h})(f_1 \log_2 p_1 + \ldots + f_h \log_2 p_h)$$

$$= \frac{1}{2^T} \sum_{i=1}^{h} \sum_{f_i=0}^{e_i} \phi(p_i^{f_i})f_i \log_2 p_i \prod_{j=1,j\neq i}^{h} \sum_{f_j=0}^{e_j} \phi(p_j^{f_j})$$

$$= \frac{1}{2^T} \sum_{i=1}^{h} \sum_{f_i=0}^{e_i} \phi(p_i^{f_i})f_i \log_2 p_i \prod_{j=1,j\neq i}^{h} p_j^{e_j}$$

$$= \frac{1}{2^T} \sum_{i=1}^{h} (p_i - 1) \sum_{f_i=1}^{e_i} f_i p_i^{f_i-1} \log_2 p_i \prod_{j=1,j\neq i}^{h} p_j^{e_j}. \tag{2}$$

By Lemma 1 and (2),

$$E_T = \frac{2^T - 1}{2^T} \sum_{i=1}^{h} \frac{e_i p_i^{e_i+1} - (e_i + 1)p_i^{e_i} + 1}{(p_i - 1)p_i^{e_i}} \log_2 p_i$$

$$= (1 - \frac{1}{2^T}) \sum_{i=1}^{h} [e_i \log_2 p_i - \frac{(1 - p_i^{-e_i}) \log_2 p_i}{p_i - 1}]$$

$$= (1 - \frac{1}{2^T}) \log_2(2^T - 1) - (1 - \frac{1}{2^T}) \sum_{i=1}^{h} \frac{(1 - p_i^{-e_i}) \log_2 p_i}{p_i - 1}.$$

This completes the proof.

Now we study the expected value $\overline{E}_T$ of the symmetric 2-adic complexity of periodic binary sequences with period T. Although we are unable to find the exact value of $\overline{E}_T$, we can derive a nontrivial lower bound on $\overline{E}_T$.

There are $\phi(2^T - 1)$ binary sequences S with period T such that $gcd(S^T(2), 2^T - 1) = 1$. Among them, there are at least $\phi(2^T - 1) - [2^T - 1 - \phi(2^T - 1)] = 2\phi(2^T - 1) - (2^T - 1)$ sequences S such that $gcd(\widehat{S}^T(2), 2^T - 1) = 1$. Hence we get the following lemma.

Lemma 2. *There are at least $2\phi(2^T - 1) - (2^T - 1)$ binary sequences S with period T such that the symmetric 2-adic complexity of S is $\overline{\Phi}(S) = \log_2(2^T - 1)$.*

Theorem 2. *The expected value $\overline{E}_T$ of the symmetric 2-adic complexity of binary sequences with period T satisfies*

$$\overline{E}_T \geq [\frac{\phi(2^T - 1)}{2^{T-1}} - 1 + \frac{1}{2^T}] \log_2(2^T - 1).$$

Proof. By Lemma 2,

$$\overline{E}_T = \frac{1}{2^T} \sum_S \overline{\Phi}(S)$$

$$\geq \frac{2\phi(2^T - 1) - (2^T - 1)}{2^T} \log_2(2^T - 1)$$

$$= [\frac{\phi(2^T - 1)}{2^{T-1}} - 1 + \frac{1}{2^T}] \log_2(2^T - 1).$$

The corollary below is a consequence of Theorem 2.

Corollary 1. *Suppose that $T \geq 2$ and $2^T - 1 = p_1^{e_1} p_2^{e_2} ... p_h^{e_h}$, where p_i are prime numbers with $p_1 < p_2 < ... < p_h$, $e_i \geq 1, i = 1, 2, ..., h$. We have*

$$\overline{E}_T > [\frac{2^T - 1}{2^{T-1}}(1 - \frac{1}{p_1})^h - 1 + \frac{1}{2^T}] \log_2(2^T - 1).$$

Proof. By Theorem 2,

$$\overline{E}_T \geq [\frac{\phi(2^T - 1)}{2^{T-1}} - 1 + \frac{1}{2^T}] \log_2(2^T - 1)$$

$$= [\frac{2^T - 1}{2^{T-1}}(1 - \frac{1}{p_1})(1 - \frac{1}{p_2})...(1 - \frac{1}{p_h}) - 1 + \frac{1}{2^T}] \log_2(2^T - 1)$$

$$> [\frac{2^T - 1}{2^{T-1}}(1 - \frac{1}{p_1})^h - 1 + \frac{1}{2^T}] \log_2(2^T - 1).$$

Remark 4. If T is large enough, we have

$$[\frac{2^T - 1}{2^{T-1}}(1 - \frac{1}{p_1})^h - 1 + \frac{1}{2^T}] \log_2(2^T - 1) \approx [2(1 - \frac{1}{p_1})^h - 1] \log_2(2^T - 1).$$

Moreover, if p_1 is also large enough,

$$[2(1 - \frac{1}{p_1})^h - 1] \log_2(2^T - 1) \approx [2(1 - \frac{h}{p_1}) - 1] \log_2(2^T - 1)$$

$$= (1 - \frac{2h}{p_1}) \log_2(2^T - 1).$$

If $2h/p_1 \ll 1$, then $(1 - 2h/p_1) \log_2(2^T - 1)$ is close to $\log_2(2^T - 1)$. Hence the lower bound is nontrivial.

Remark 5. The bound in Theorem 2 may be very weak. For example, if $3, 5$ and 7 all divide $2^T - 1$, then the bound is negative, hence vacuous.

We give one example below.
Let $T = 37$. Then $2^T - 1 = 223 \times 616318177$.

By Theorem 1,

$$E_{37} > (1 - \frac{1}{2^{37}}) \log_2(2^{37} - 1) - (1 - \frac{1}{2^{37}})(\frac{\log_2 223}{222} + \frac{\log_2 616318177}{616318176})$$

$$\approx \log_2(2^{37} - 1) - (\frac{\log_2 223}{222} + \frac{\log_2 616318177}{616318176})$$

$$\approx 36.9649.$$

By Corollary 1,

$$\overline{E}_{37} > [\frac{2^{37} - 1}{2^{36}}(1 - \frac{1}{223})^2 - 1 + \frac{1}{2^{37}}] \log_2(2^{37} - 1)$$

$$\approx [2(1 - \frac{1}{223})^2 - 1] \log_2(2^{37} - 1)$$

$$\approx (1 - \frac{2}{223}) \log_2(2^{37} - 1)$$

$$= \frac{221}{223} \log_2(2^{37} - 1)$$

$$\approx 36.6682.$$

4   The k-Error 2-Adic Complexity of Periodic Binary Sequences

It is well known that the linear complexity of a periodic sequence is unstable under small perturbations [1]. This is also true for the case of the 2-adic complexity. For example, let $S = (1, 0, 0, ..., 0)^\infty$ or $(0, 1, 1, ..., 1)^\infty$ with period T. Then $\Phi(S) = \log_2(2^T - 1)$. However, after changing 1 bit within every period, $\Phi(S)$ become 0. Hence it is interesting to investigate the properties of the k-error 2-adic complexity and the k-error symmetric 2-adic complexity of periodic binary sequences. In the following, we will give the formal definitions of these concepts.

Definition 4. Let $S = (s_0, s_1, s_2, ..., s_{T-1})^\infty$ be a periodic binary sequence with period T and k be an integer with $0 \leq k \leq T$. Then the k-error 2-adic complexity $\Phi_{T,k}(S)$ (or the k-error symmetric 2-adic complexity $\overline{\Phi}_{T,k}(S)$) of S is $\min_P \Phi(P)$ (or $\min_P \overline{\Phi}(P)$), where the minimum is extended over all periodic binary sequences $P = (t_0, t_1, t_2, ..., t_{T-1})^\infty$ with period T for which the Hamming distance of the vectors $(s_0, s_1, s_2, ..., s_{T-1})$ and $(t_0, t_1, t_2, ..., t_{T-1})$ is at most k.

We have the simple proposition below.

Proposition 2. Let $S = (s_0, s_1, s_2, ..., s_{T-1})^\infty$ be a periodic binary sequence with period $T \geq 2$ and let $2^T - 1$ be a prime number. Let $W_H(S)$ denotes the Hamming weight of the vector $(s_0, s_1, s_2, ..., s_{T-1})$. Then we have

$$\Phi_{T,k}(S) = \begin{cases} \log_2(2^T - 1), & \text{if } 0 \leq k < min(W_H(S), T - W_H(S)); \\ 0, & \text{if } min(W_H(S), T - W_H(S)) \leq k \leq T. \end{cases}$$

and

$$\overline{\Phi}_{T,k}(S) = \begin{cases} \log_2(2^T - 1), & if \ \ 0 \le k < min(W_H(S), T - W_H(S)); \\ 0, & if \ \ min(W_H(S), T - W_H(S)) \le k \le T. \end{cases}$$

Lemma 3. *Let p, q be two positive integers, where $0 < p < q$. Let h be a nonzero integer and $(ph)_{mod \, q}/q = p'/q'$, where the notation $(ph)_{mod \, q}$ means the reduced residue of ph modulo q, and $0 < p' < q'$. Then*

$$\frac{q'}{gcd(p', q')} \le \frac{q}{gcd(p, q)}. \tag{3}$$

The equality holds in (5) if and only if

$$gcd(h, \frac{q}{gcd(p, q)}) = 1.$$

Proof. Put $d = gcd(p, q)$. Then $p = dp_1, q = dq_1$, and $gcd(p_1, q_1) = 1$. We have

$$\frac{(ph)_{mod \, q}}{q} = \frac{(dp_1 h)_{mod \, (dq_1)}}{dq_1} = \frac{(p_1 h)_{mod \, (q_1)}}{q_1}.$$

Thus

$$\frac{q'}{gcd(p', q')} \le q_1 = \frac{q}{gcd(p, q)}.$$

The equality holds if and only if

$$gcd(h, \frac{q}{gcd(p, q)}) = gcd(h, q_1) = 1.$$

This completes the proof.

The following two theorems are the main results of this section.

Theorem 3. *Let $S = (s_0, s_1, s_2, ..., s_{T-1})^\infty$ be a periodic binary sequence with period $T \ge 2$. Suppose that the rational expression of S is $-p/q$, where $0 < p < q$, and $gcd(p, q) = 1$. Then the k-error 2-adic complexity $\Phi_{T,k}(S)$ of S satisfies*

$$\log_2(2^T - 1) \ge \Phi_{T,k}(S) > \log_2(2^T - 1) - \frac{(k-1)T}{k} - 1 - \Phi(S). \tag{4}$$

Proof. Suppose that $P = (t_0, t_1, t_2, ..., t_{T-1})^\infty$ is obtained from S by changing k bits within every period, $1 \le k \le T$, and the k positions where bits are changed are $j_1, j_2, ..., j_k, 0 \le j_1 < j_2 < ... < j_k \le T - 1$. I.e., $t_i = s_i, i \ne j_1, j_2, ..., j_k$, $t_i = s_i + 1$ otherwise.

The corresponding 2-adic number associated with S is $-S^T(2)/(2^T - 1)$, where $S^T(x) = s_0 + s_1 x + ... + s_{T-1} x^{T-1}$. The corresponding 2-adic number associated with P is

$$-\frac{S^T(2) + \sum_{i=1}^{k} (-1)^{s_{j_i}} 2^{j_i}}{2^T - 1}.$$

Because $-S^T(2)/(2^T - 1) = -p/q$, we have $S^T(2) = p(2^T - 1)/q$. Hence

$$-\frac{S^T(2) + \sum_{i=1}^{k}(-1)^{s_{j_i}}2^{j_i}}{2^T - 1} = -\frac{\frac{p(2^T-1)}{q} + \sum_{i=1}^{k}(-1)^{s_{j_i}}2^{j_i}}{2^T - 1}.$$

Suppose that

$$-\frac{p_1}{q_1} = -\frac{\frac{p(2^T-1)}{q} + \sum_{i=1}^{k}(-1)^{s_{j_i}}2^{j_i}}{2^T - 1},$$

where $0 < p_1 < q_1$, and $gcd(p_1, q_1) = 1$. By Lemma 3, we have

$$q_1 \geq \frac{2^T - 1}{gcd(2^T - 1, q|\sum_{i=1}^{k}(-1)^{s_{j_i}}2^{j_i}|)} \geq \frac{2^T - 1}{q|\sum_{i=1}^{k}(-1)^{s_{j_i}}2^{j_i}|}.$$

It is reasonable to put $j_1 = 0$ and $T - j_k = max\ (T - j_k, j_k - j_{k-1}, ..., j_3 - j_2, j_2 - j_1)$. So $T - j_k \geq T/k$. It follows that $j_k \leq (k-1)T/k$ and

$$\log_2 |\sum_{i=1}^{k}(-1)^{s_{j_i}}2^{j_i}| < \frac{(k-1)T}{k} + 1.$$

Hence we get

$$\log_2 q_1 > \log_2(2^T - 1) - \frac{(k-1)T}{k} - 1 - \log_2 q.$$

This completes the proof.

Theorem 4. Let $S = (s_0, s_1, s_2, ..., s_{T-1})^{\infty}$ be a periodic binary sequence with period $T \geq 2$. $\widehat{S}$ is the inverse sequence of S. Then the k-error symmetric 2-adic complexity $\overline{\Phi}_{T,k}(S)$ of S satisfies

$$\log_2(2^T - 1) \geq \overline{\Phi}_{T,k}(S) > \log_2(2^T - 1) - \frac{(k-1)T}{k} - 1 - max(\Phi(S), \Phi(\widehat{S})).$$

In particular, if $S = \widehat{S}$, then

$$\log_2(2^T - 1) \geq \overline{\Phi}_{T,k}(S) > \log_2(2^T - 1) - \frac{(k-1)T}{k} - 1 - \Phi(S).$$

Proof. By (6), we have

$$\Phi_{T,k}(S) > \log_2(2^T - 1) - \frac{(k-1)T}{k} - 1 - \Phi(S),$$

and

$$\Phi_{T,k}(\widehat{S}) > \log_2(2^T - 1) - \frac{(k-1)T}{k} - 1 - \Phi(\widehat{S}).$$

Hence

$$\overline{\Phi}_{T,k}(S) = min(\Phi_{T,k}(S), \Phi_{T,k}(\widehat{S}))$$
$$> \log_2(2^T - 1) - \frac{(k-1)T}{k} - 1 - max(\Phi(S), \Phi(\widehat{S})).$$

If $S = \widehat{S}$, then $\max(\Phi(S), \Phi(\widehat{S})) = \Phi(S)$. Hence

$$\overline{\Phi}_{T,k}(S) > \log_2(2^T - 1) - \frac{(k-1)T}{k} - 1 - \Phi(S).$$

This completes the proof.

Remark 6. If $k \geq min(W_H(S), T - W_H(S))$, then $\Phi_{T,k}(S) = 0$, and $\overline{\Phi}_{T,k}(S) = 0$.

As was shown by Jiang, Dai, and Imamura [6], although the bounds given in the above two theorems are not always tight because we do not use the information about the change values, they can tell us how low the 2-adic complexity and the symmetric 2-adic complexity become as k increases.

Acknowledgment

The authors wish to express their deep gratitude to the referee for many helpful comments and suggestions.

References

1. C. Ding, G. Xiao, and W. Shan, The Stability Theory of Stream Ciphers (LNCS), Berlin, Germany: Springer-Verlag, 1991, vol 561
2. M. Goresky, A. Klapper, Feedback registers based on ramified extensions of the 2-adic numbers, Advances in Cryptology-Eurocrypt'94, LNCS, vol, 950, Springer-Verlag, Berlin, 1995, pp. 215-222
3. M. Goresky, A. Klapper, Large periods nearly de Bruijn FCSR sequences, Advances in Cryptology-Eurocrypt'95, LNCS, vol. 921, Springer-Verlag, Berlin, 1995, pp. 263-273
4. M. Goresky, A. Klapper, Cryptanalysis based on 2-adic rational approximation, Advances in Cryptology-Crypt'95, LNCS, vol. 963, Springer-Verlag, Berlin, 1995, pp. 262-273
5. K. Ireland and M. Rosen, A Classical Introduction to Modern Number Theory (second edition), in GTM. New York: Springer Verlag, 1990, vol. 84
6. S. Jiang, Z. Dai and K. Imamura, Linear complexity of a sequence obtained from a periodic sequence by either substituting, inserting ,or deleting k symbols within one period, IEEE Trans. Inform. Theory, vol. 46, pp. 1174-1177, May 2000
7. A. Klapper, M. Goresky, 2-adic shift registers, in: R. Anderson (Ed.), Fast Software Encryption, Lecture Notes in Computer Science, Vol. 809, Springer, New York, 1994, pp. 174-178.
8. A. Klapper, M. Goresky, Feedback shift registers, 2-adic span, and combiners with memory, J. Cryptology, vol. 10, pp. 111-147, 1997.
9. N. Koblitz, p-Adic Numbers, p-Adic Analysis, and Zeta Functions, Graduate Texts in Mathematics, Vol. 58, Springer, New York, 1984.
10. James L. Massey, Shift-register synthesis and BCH decoding, IEEE Trans. Info. Theory, vol. IT-15, pp. 122-127, January 1969.
11. Wilfried Meidl, Extended Games-Chan algorithm for the 2-adic complexity of FCSR-sequences,Theoretical Computer Science, Volume 290, 2003, Pages 2045-2051

12. W. Meidl, H. Niederreiter, On the expected value of the linear complexity and the k-error linear complexity of periodic sequences, IEEE Trans. Inform. Theory 48(2002) 2817-2825.

13. Wenfeng Qi, Hong Xu, Partial period distribution of FCSR sequences ,IEEE Trans. Inform. Theory, vol. 49, March 2003, Pages 761-765

14. R. A. Rueppel, Analysis and Design of Stream Cipher. Berlin, Germany: Springer-Verlag, 1986

15. Changho Seo, Sangjin Lee, Yeoulouk Sung, Keunhee Han, Sangchoon Kim, A lower bound on the linear span of an FCSR, IEEE Trans. Info. Theory, vol. IT-46, pp. 691-693, March 2000.

Almost-Perfect and Odd-Perfect Ternary Sequences

Evgeny I. Krengel⋆

Kedah Electronics Engineering, Zelenograd, Korpus 445,
Moscow 124498, Russia
legnerk@gagarinclub.ru

Abstract. Ternary $(-1, 0, +1)$ almost-perfect and odd-perfect autocorrelation sequences are applied in many communication, radar and sonar systems, where signals with good periodic autocorrelation are required. New families of almost-perfect ternary (APT) sequences of length $N = (p^n - 1)/r$, where r is an integer, and odd-perfect ternary (OPT) sequences of length $N/2$, derived from the decomposition of m-sequences of length $p^n - 1$ over $GF(p)$, $n = km$, with p being an odd prime, into an array with $T = (p^n - 1)/(p^m - 1)$ rows and $p^m - 1$ columns, are presented. In particular, new APT sequences of length $4(p^n - 1)/(p^m - 1)$, $(p^m + 1) = 2 \bmod 4$, and OPT sequences of length $2(p^n - 1)/(p^m - 1)$ with peak factor close to 1 when p becomes large, are constructed. New perfect 4-phase and 8-phase sequences with some zeroes can be derived from these OPT sequences. The obtained APT sequences of length $4(p^m + 1)$, $p > 3$ and m – any even positive integer, possess length uniqueness in comparison with known almost-perfect binary sequences.

1 Introduction

Binary $(-1, +1)$ and ternary $(-1, 0, +1)$ – sequences with low autocorrelations are widely used in spread spectrum communication, radar and sonar systems [1]. In many applications it is important to have sequences with perfect, or almost-perfect periodic, or perfect odd-periodic autocorrelation functions [1–3]. A sequence is called a perfect [1], if all its out-of-phase autocorrelation coefficients are 0, almost-perfect if all its out-of-phase autocorrelation coefficients except one are 0 [4], and odd-perfect if all its out-of-phase odd-periodic autocorrelation coefficients are 0 [5].

A large family of almost-perfect binary (APB) sequences of length $2(p^m + 1)$, with p an odd prime, $m = 1, 2, 3, \ldots$ has been obtained and studied by Ipatov [6, Theorem 3.12], Pott and Bradley [4] and Wolfmann [7]. APT sequences of length $2(q^k - 1)/(q - 1), q = p^m$ with $(2k, q - 1) = 2$ have been found by

---

⋆ The partial results of the paper were presented at 6th International Conference on Digital Signal Processing and its Applications, Moscow, Russia, March 31–April 2 2004.

T. Helleseth et al. (Eds.): SETA 2004, LNCS 3486, pp. 197–207, 2005.

Langevin [8]. Later Schotten and Lüke have constructed APT sequences of length $2(q^k - 1)/(q - 1)$ for all $k > 1$ [9]. OPT sequences of length $(q^k - 1)/(q - 1)$ with $(q^{k-1} - 1)/(q - 1)$ zero elements and in particular OPT sequences of length $q + 1$ with one zero element (odd-perfect almost-binary correlation sequence) have been constructed in [9,5].

In the present work, we use some properties of shift sequences of m-sequences over $GF(p)$ of length $p^n - 1$, where p is an odd prime [10,11] to construct new APT and OPT sequence sets, and in particular APT sequences of length $4(p^m + 1)$, $(p^m + 1) = 2 \bmod 4$, and OPT sequence of length $2(p^m + 1)$ with two zero elements. Also, we show that there are many new APT and OPT sequences with lengths different from those of known APB and OPT sequences.

This paper is organized as follows: Section 2 describes different constructions of APT sequences based on m-sequences over $GF(p)$ of length $p^n - 1$. Section 3 shows how by using APT sequences of length N, OPT sequences of length $N/2$ can be produced and also how by using these OPT sequences, perfect polyphase sequences of length $N/2$ can be produced. In Section 4, we demonstrate a length uniqueness for large number of the new APT sequences, i.e. that their lengths do not coincide with lengths of any known APB sequence and give examples of such sequences. In Section 5, some configuration properties of APT sequences are considered. All APT sequence constructions from Sections 2–3 are illustrated by appropriate examples.

2   Construction of APT Sequences

Notations:

- $Tr_m^n(x) = \sum\limits_{i=0}^{n/m-1} x^{p^{im}}$, the trace function of an element $x \in GF(p^n)$ to $GF(p^m)$;

- $\lfloor y \rfloor$, the maximum $(v|v \le y, v - \text{an integer})$;

- $ind_\beta z$, the index (logarithm) function z to base β ;

- $\theta(l) = \sum\limits_{i=0}^{N-1} a_i a_{i+l}^*$, the periodic autocorrelation function (ACF) of a sequence $a = \{a_i\}$ with period N;

- $\hat{\theta}(l) = \sum\limits_{i=0}^{N-l-1} a_i a_{i+l}^* + \sum\limits_{i=N-l}^{N-1} a_i \bar{a}_{i+l-N}^*$, the odd-periodic ACF of a sequence $a = \{a_i\}$;

- $Nmax(a_i^2) \Big/ \sum\limits_{i=0}^{N-1} a_i^2$ the peak factor of a sequence a with period N.

By analogy with balanced m-sequences [1], a ternary sequence with elements in $\{-1, 0, 1\}$ will be called balanced if $\sum\limits_{i=0}^{N-1} a_i = 0$.

Theorem 1. *Let $n = mk$, $m \ge 1$, $k > 1$ positive integers and p odd prime, $(p^m + 1) = 2 \bmod 4$. Let α be a primitive element of $GF(p^n)$, β be a primitive element of $GF(p^m)$ and let $T = (p^n - 1)/(p^m - 1)$. Then a sequence given by*

$$w_i = \psi(Tr_m^n(\alpha^i)), \quad i = 0, 1, \ldots, 4T - 1 \qquad (1)$$

and

$$\psi(z) = \begin{cases} (-1)^{\lfloor((ind_\beta z) \bmod 4)/2\rfloor}, & \text{if } z \neq 0 \\ 0, & \text{if } z = 0 \end{cases} \tag{2}$$

is a balanced APT sequence of length $N = 4(p^n - 1)/(p^m - 1)$ *with* $4(p^{n-m} - 1)/(p^m - 1)$ *zero elements and autocorrelation peak* $4p^{n-m}$.

Proof. Let $\mathbf{b}$ be an m-sequence over $GF(p)$ of length $p^n - 1$ with elements $b_i = Tr_1^n(\alpha^i)$, $0 \leq i < p^n - 1$. Fold $\mathbf{b}$ into a decomposition array $\mathbf{B}$ by columns [10] with $T = (p^n - 1)/(p^m - 1)$ rows and $p^m - 1$ columns. Its rows are either null rows or cyclic shifts of an m-sequence of length $p^m - 1$ over $GF(p)$. For the case $GF(p^n)$ a shift sequence is defined by

$$\mathbf{e} = \{e_i\} = \begin{cases} \infty, & \text{if } Tr_m^n(a^i) = 0 \\ ind_\beta(Tr_m^n(a^i)), & \text{if } Tr_m^n(a^i) \neq 0 \end{cases}, \tag{3}$$

where $0 \leq i < p^n - 1$ [11]. The first T of its elements give all these cyclic shifts relative to a reference short m-sequence of length $p^m - 1$ or point (element ∞) to the null rows. According to m-sequence properties, the number of non-zero rows is p^{n-m}. From (3) we have $e_{k+jT} = (e_k + j) \bmod p^m - 1$ for $k = 0, 1, 2, \ldots, T-1$, $j = 0, 1, \ldots, T - 3$. It can be proved that for any $i \in Z(p^n - 1)$, each element of $Z(p^m - 1)$ appears as a differences $(e_{i+k} - e_k) \bmod (p^m - 1)$, $k \in Z(T)$ exactly p^{n-2m} times [11]. Let us form a binary short sequence $\mathbf{c}$ of length $p^m - 1$ with elements $c_j = (-1)^{\lfloor((e_0+j) \bmod 4)/2\rfloor}$, $j \in Z(p^m - 1)$ and replace all non-zero rows of the array $\mathbf{B}$ by this binary short sequence with the same shift sequence $\mathbf{e}$. As a result, we get a new array $\mathbf{W}$ which consists of repeated columns of period 4. Then, according to the shift sequence property, a ternary sequence of length $p^n - 1$ associated with the array $\mathbf{W}$ has zero ACF for all shifts $i \neq jT$, $j \in Z(p^m - 1)$. Since the short sequence $\mathbf{c}$ has zero ACF for odd shifts, the autocorrelation of the ternary sequence for shifts $i = (2t + 1)T$ is also zero. Let $\mathbf{w}$ be a ternary sequence associated with the first four columns of the array $\mathbf{W}$. Then ACF of the sequence $\mathbf{w}$ is zero for all shifts $0 \leq i < 4T$ except $i = 0$ and $i = 2T$ in which $\theta(0) = -\theta(2T) = 4p^{n-m}$. Balance property of the sequence $\mathbf{w}$ follows from its construction. □

Corollary 1. *For* $n = 2m$, *APT sequences (1) have length* $4(p^m + 1)$, *autocorrelation peak* $4p^m$ *and 4 zero elements.*

Corollary 2. *Taking* $w_{T/2} = w_{5T/2} = \pm 1$ *or* $w_{3T/2} = w_{7T/2} = \pm 1$, *we get an APT sequence with two zero elements and* $\theta(0) = 4p^m + 2$ *and* $\theta(2T) = -4p^m + 2$.

From Theorem 1, we have that the number of nonzero elements of APT sequence (1) is $4p^{n-m}$. Then the peak factor of the sequences (1) is equal to $(p^n - 1)/(p^n - p^{n-m})$, i.e. close to 1 as p becomes large.

Example 1. Let $p = 17$, $n = 2$, $m = 1$ and $x^2 + x + 3$ be a primitive polynomial over $GF(17)$. Then the shift sequence of an m-sequence of length 288 over

$GF(17)$ associated with this polynomial is 1, 11, 0, 13, 14, 4, 6, 9, 2, ∞, 11, 3, 1, 0, 11, 11, 15, 11. After substituting the short sequence 1, 1, -1, -1 of length 4 by Theorem 1 we get a balanced APT sequence of length 72: -1 1 1 -1 -1 -1 1 -1 -1 -1 0 1 1 -1 1 1 1 1 1 -1 1 -1 -1 1 -1 1 -1 1 0 1 1 -1 -1 1 1 1 1 1 -1 -1 1 1 1 -1 1 1 1 0 -1 -1 1 1 -1 -1 -1 -1 1 1 -1 1 1 -1 1 1 -1 0 -1 -1 1 1 1 -1 -1 -1 -1 with peak 68.

By the Corollary 1 we suppose in (1) $w_9 = w_{45} = 1$. Then a sequence -1 1 1 -1 -1 1 -1 -1 -1 1 1 1 1 -1 1 1 1 1 1 1 -1 1 -1 -1 1 -1 1 -1 1 0 1 1 -1 -1 1 1 1 1 1 -1 -1 1 1 1 -1 1 1 1 1 1 -1 -1 1 -1 -1 -1 -1 -1 1 1 -1 1 1 -1 1 1 -1 0 -1 -1 1 1 1 -1 -1 -1 -1 -1 is also an APT sequence but with two zero elements and $\theta(0) = 70$ and $\theta(36) = -66$.

Example 2. Let $p = 73$, $n = 2$, $m = 1$ and $x^2 + x + 11$ be a primitive polynomial over $GF(73)$. The shift sequence for two-dimensional array 74×72 of an m-sequence of length $73^2 - 1 = 5328$ over $GF(73)$ based on the polynomial is: 65, 21, 6, 25, 20, 28, 61, 68, 70, 19, 61, 6, 9, 1, 47, 59, 10, 61, 62, 34, 38, 19, 9, 43, 3, 61, 49, 67, 9, 14, 39, 63, 29, 55, 26, 36, 1, ∞, 38, 2, 65, 23, 70, 33, 10, 58, 54, 41, 24, 37, 52, 21, 60, 71, 19, 16, 45, 45, 67, 45, 34, 61, 70, 68, 52, 11, 63, 62, 56, 24, 17, 23, 5, 21. According to Theorem 1 we get a balanced APT sequence of length 296: -1 -1 -1 -1 1 1 -1 1 -1 1 -1 -1 -1 -1 1 1 -1 -1 -1 -1 -1 1 -1 1 1 1 -1 -1 1 1 -1 -1 1 1 -1 1 1 -1 1 -1 0 -1 -1 -1 1 1 -1 -1 -1 -1 -1 -1 1 1 -1 1 1 -1 1 1 1 1 1 -1 -1 1 1 -1 -1 -1 -1 -1 -1 1 1 1 1 -1 1 1 -1 1 1 -1 -1 -1 -1 1 1 -1 -1 -1 -1 -1 1 1 1 -1 1 1 -1 -1 1 1 1 1 -1 1 1 1 1 1 -1 1 1 1 -1 1 -1 -1 1 1 -1 1 1 1 1 -1 1 1 1 -1 -1 0 1 1 -1 1 1 1 -1 1 1 1 1 -1 -1 -1 -1 -1 -1 1 1 1 -1 -1 -1 1 1 -1 1 -1 -1 1 1 1 1 1 -1 -1 1 1 -1 1 1 -1 1 1 1 1 1 -1 -1 1 1 1 1 1 1 -1 1 1 -1 -1 -1 1 1 1 -1 1 1 1 -1 -1 1 1 -1 1 1 -1 1 0 1 1 1 -1 1 1 1 1 1 1 1 -1 1 1 -1 1 1 -1 -1 -1 -1 -1 1 1 1 -1 1 1 1 1 1 1 -1 -1 -1 -1 -1 1 1 -1 -1 1 1 -1 -1 1 1 -1 1 1 1 1 1 -1 1 1 1 1 1 1 -1 -1 1 1 -1 1 1 -1 -1 -1 -1 1 1 -1 1 1 -1 -1 -1 -1 -1 1 1 1 1 1 -1 -1 1 1 1 1 -1 -1 -1 1 1 1 1 -1 -1 -1 1 1 1 1 -1 1 1 1 with four zero elements 1 with peak values $\theta(0) = 292$ and $\theta(148) = -292$.

Supposing $w_{111} = w_{259} = \pm 1$ by Corollary 1 we have the following APT sequences: -1 -1 -1 -1 1 1 -1 1 -1 1 -1 -1 -1 -1 1 1 -1 -1 -1 -1 -1 1 1 -1 1 1 -1 1 -1 -1 1 1 -1 1 1 -1 1 -1 0 -1 -1 -1 1 -1 -1 -1 -1 -1 -1 1 1 -1 1 1 -1 1 1 1 1 1 -1 -1 1 1 -1 1 -1 -1 -1 -1 1 1 1 1 1 -1 -1 1 1 -1 1 -1 -1 -1 -1 1 -1 -1 -1 -1 -1 1 1 1 -1 1 1 -1 1 1 -1 1 1 1 1 -1 1 1 1 1 -1 1 1 1 1 1 -1 1 1 1 -1 -1 1 1 -1 1 1 1 -1 -1 1 1 1 1 -1 -1 1 1 1 1 -1 -1 1 1 1 1 -1 1 1 1 -1 1 1 1 1 -1 1 -1 -1 1 1 1 -1 -1 -1 1 1 -1 -1 1 1 1 1 1 -1 -1 1 1 -1 1 1 -1 1 1 -1 1 1 1 1 1 1 -1 -1 1 1 1 1 1 1 -1 1 1 -1 1 -1 -1 1 1 -1 1 1 -1 1 -1 1 1 0 1 1 1 -1 1 1 1 1 1 1 1 -1 1 1 -1 1 1 -1 -1 1 1 -1 -1 -1 -1 1 1 1 1 1 -1 1 -1 -1 1 1 -1 1 1 1 1 1 1 -1 -1 1 1 -1 1 1 1 -1 -1 -1 -1 1 1 -1 1 1 -1 -1 1 1 1 1 1 -1 1 1 1 1 1 1 -1 -1 1 1 1 1 -1 1 1 and -1 -1 -1 -1 1 1 1 -1 1 -1 1 1 -1 -1 -1 -1 1 1 1 -1 -1 -1 -1 1 -1 -1 1 1 -1 1 1 1 -1 1 1 -1 1 -1 1 1 -1 1 0 -1 -1 1 1 -1 1 -1 -1 -1 -1 -1 -1 1 1 -1 1 1 -1 1 1 1 -1 -1 1 1 -1 -1 -1 -1 -1 1 1 1 1 1 -1 1 1 1 -1 1 1 -1 -1 1 1 -1 -1 1 1 -1 1 1 -1 1 -1 -1 -1 1 1 1 -1 1 1 1 1 -1 1 1 1 -1 -1 1 1 -1 -1 1 1 -1 1 1 1 1 -1 -1 -1 1 1 -1 1 1 1 -1 -1 -1 -1 1 1 1 1 -1 -1 1 1 1 1 1 1 -1 1 1 1 -1 -1 -1 1 1 -1 -1 -1 1 1 -1 -1 1 1 -1 1 1 1 1 -1 -1 -1 -1 -1 1 1 1 1 -1 -1 1 1 1 1 1 -1 -1 1 1 1 1 1 1 -1 1 1 -1 1 -1 -1 -1 1 1 -1 1 1 -1 1 1 -1 1 1 0 1 1 1 -1 1 1 1 1 1 1 1 -1 -1 1 -1 1 1 1 1 1 1 1 1 -1 -1 1 1 -1 -1 -1 -1 1 1 1 -1 1 1 1 1 1 -1 -1 -1 -1 1 -1 -1 1 1 -1 1 1 1 1 1 -1 1 1 1 1 1 1 -1 -1 1 1 -1 -1 1 1 -1 -1 -1 -1 1 1 -1 -1 1 1 1 -1 -1 1 1 -1 -1 1 1 -1 1 1 -1 -1 -1 -1 1 1 -1 1 1 1 -1 -1 -1 1 1 -1 -1 -1 -1 -1 1 1 1 1 -1 -1 1 1 -1 -1 -1 -1 1 1 -1 1 1 1

1 1 1 1 -1 -1 1 1 1 -1 1 -1 1 -1 1 1 -1 -1 -1 1 1 1 -1 1 1 with two zeroes and
$\theta(0) = 294$ and $\theta(148) = -290$.

Example 3. Let $p = 13$, $n = 3$, $m = 1$ and $x^3 + x^2 + 2$ be a primitive polynomial
over $GF(13)$. The shift sequence of the m-sequence of length 2196 over $GF(13)$
in this case is 0, ∞, ∞, 1, 7, 1, 11, 4,0, ∞, 5, 2, 8, 4, 7, 11, 0, 9, 11, 3, 3, 8,
5, 8, 6,∞, 9, 5, 11, 2, 9, 11, 6, 1, 4, 6, 3, 7, ∞, 4, 11, 5, ∞, 0, 1, 7, 8, 9, 0,
5, 4, 9, 2, 4, 5, 1, 8, 4, 0, 5, ∞, 1, 0, 6, 3, 11, 8, 5, 5, 0, 1, 0, 10, 5, 2, 7, 10,
9, 0, 3, 3, 11, 10, 11, 9, 1, 4, 5, 10, 10, 7, 2, 7, 5, 1, 1, 0, 4, 0, 10, 10, 0, 3, 0,
10, 11, 3, 0, ∞, 4, 9, 3, 7, 9, 9, 0, 8, 0, 10, 1, ∞, 11, 1, 7, 6, 3, 2, 1, 3, ∞, 2,
6, 0, 2, 1, ∞, 3, 6, 0, 7, ∞, 1, 1, 7, 7, 7, 5, 7, 5, 3, 2, 9, 9, 10, ∞, 10, 8, 2, 7,
11, 6, 10, 2, 1, 9, 10, 5, 4, 4, 8, 1, 8, 6, 3, 4, 6, 2, 4, 6, 11, 0, 0, ∞. Then by
Theorem 1 we have the following APT sequence of length 732 with 56 zero
elements and peak 676 : 1 0 0 1 -1 1 -1 1 1 0 1 -1 1 1 -1 -1 1 1 -1 -1 -1 1 1 1 -1 0
1 1 -1 -1 1 -1 -1 1 1 -1 -1 -1 0 1 -1 1 0 1 1 -1 1 1 1 1 1 1 -1 1 1 1 1 1 1 0 1 1 -1
-1 -1 1 1 1 1 1 1 -1 1 -1 -1 -1 1 1 -1 -1 -1 -1 -1 1 1 1 1 -1 -1 -1 -1 -1 1 1 1 1 1 1 -1
-1 1 -1 1 -1 -1 -1 1 1 0 1 1 -1 -1 1 1 1 1 1 -1 1 1 0 -1 1 -1 -1 -1 -1 1 -1 -1 0 -1 -1 1 -1 1 1 0
-1 -1 1 -1 0 1 1 -1 -1 -1 1 1 -1 1 -1 -1 1 1 1 -1 0 -1 1 -1 -1 -1 -1 -1 -1 1 1 1 -1 1 1 1 1 1 1
1 -1 -1 1 -1 -1 1 -1 -1 1 1 1 0 1 0 0 -1 1 -1 1 1 1 0 -1 -1 1 1 1 1 1 -1 1 1 1 1 -1 1 1 -1
0 -1 -1 1 1 -1 -1 1 1 -1 -1 1 1 -1 1 1 1 0 1 1 -1 0 1 -1 1 1 1 -1 1 1 -1 1 1 -1 1 1 -1 -1 1 1 1 1 -1 0
-1 1 1 -1 1 1 1 1 -1 -1 1 1 -1 1 1 -1 -1 -1 -1 1 1 -1 -1 1 1 1 1 1 -1 1 1 -1 -1 1 1 -1 -1 -1 -1 1 1 -1 1 1 -1 -1 -1
-1 1 1 1 1 -1 -1 1 1 1 1 -1 1 1 1 1 0 1 -1 1 1 1 -1 -1 1 1 1 1 -1 -1 0 1 -1 1 1 -1 1 1 -1 -1 1 1 0 -1
-1 1 1 -1 -1 0 1 -1 1 1 1 0 -1 -1 1 1 1 1 -1 1 1 -1 1 1 -1 -1 -1 -1 -1 0 -1 1 -1 1 1 1 -1 -1 -1 -1 -1 -1
-1 -1 1 1 1 1 -1 1 1 -1 1 1 1 -1 -1 1 1 -1 1 1 1 1 0 -1 0 0 -1 1 -1 1 1 -1 -1 1 0 -1 1 1 -1 -1 1 1 1 -1 -1 -1
1 1 1 -1 -1 -1 1 1 0 -1 1 1 1 1 -1 1 1 1 1 -1 -1 1 1 1 1 0 -1 1 1 -1 0 -1 -1 1 1 -1 -1 -1 -1 -1 -1 -1 1 1
-1 -1 -1 -1 -1 -1 -1 0 -1 -1 1 1 1 1 -1 -1 -1 -1 -1 -1 1 1 -1 1 1 1 1 -1 -1 1 1 1 1 1 1 -1 -1 -1
-1 1 1 1 1 1 1 -1 -1 -1 -1 -1 -1 1 1 1 -1 1 1 -1 1 1 1 1 -1 0 -1 -1 1 1 1 -1 -1 -1 -1 -1 1 1 -1 0 1
-1 1 1 1 1 1 -1 1 1 0 1 1 -1 1 1 -1 0 1 1 -1 1 1 0 -1 -1 1 1 1 1 -1 1 1 -1 1 1 1 -1 -1 1 1 0 1 -1 1 1 1 1
1 1 1 -1 -1 1 1 -1 -1 -1 -1 -1 -1 -1 1 1 1 -1 1 1 1 -1 1 1 1 -1 -1 1 0 -1 0 0 1 -1 1 1 -1 -1 -1 1 0 1 1
-1 -1 -1 -1 -1 1 1 -1 -1 -1 -1 1 1 -1 1 1 0 1 1 -1 1 1 1 -1 1 1 1 -1 1 1 -1 -1 -1 0 -1 -1 1 1 0 -1 1 -1
-1 1 1 -1 1 1 1 -1 1 1 1 -1 -1 -1 1 1 0 1 -1 1 1 -1 -1 -1 1 1 1 -1 1 1 -1 1 1 1 1 -1 1 1 1 -1 -1 -1 -1 -1
-1 1 1 -1 1 1 1 -1 1 1 1 1 -1 1 1 -1 1 1 1 1 -1 -1 -1 1 1 1 -1 -1 -1 1 1 -1 -1 -1 0 -1 1 1 -1 -1 1 1 1 -1
-1 -1 1 1 1 0 -1 1 1 -1 1 1 -1 1 1 1 -1 1 0 1 1 -1 1 1 1 0 -1 1 1 -1 -1 1 0 1 1 -1 -1 -1 1 1 -1 1 1 -1 1 1 1
1 1 0 1 -1 1 1 -1 -1 1 1 1 1 1 1 1 1 -1 -1 -1 1 1 -1 1 1 -1 -1 1 1 1 -1 1 1 -1 1 -1 -1 -1 1 0.

In Section 5 it is demonstrated that all the sequences from the Examples 1-3 are
"unique".

In [8] some APT sequences of length $(p^{mk} - 1)/(p^m - 1)$ with $(2k, p^m - 1) = 2$
have been constructed by using the multiplicative characters over $GF(p^m)$.
Schotten and Lüke [9] derived larger set of APT sequences of length
$2(p^{mk} - 1)/(p^m - 1)$ which include Langevin APT sequences [8]. In a more
explicit form, their result can be expressed by the following theorem.

Theorem 2. *Let $n = mk$, $m \geq 1$, $k > 1$ positive integers and p odd prime. Let
α be a primitive element of $GF(p^n)$, β be a primitive element of $GF(p^m)$ and
let $T = (p^n - 1)/(p^m - 1)$. Then a sequence given by*

$$\bar{w}_i = \varphi(Tr_m^n(\alpha^i)) , \quad i = 0, 1, \ldots, 2T - 1 \tag{4}$$

and

$$\varphi(z) = \begin{cases} (-1)^{(ind_\beta z) \bmod 2}, & \text{if } z \neq 0 \\ 0, & \text{if } z = 0 \end{cases} \tag{5}$$

is a balanced APT sequence of length $N = 2(p^n - 1)/(p^m - 1)$ *with* $2(p^{n-m} - 1)/(p^m - 1)$ *zero elements and autocorrelation peak* $2p^{n-m}$.

Proof. The proof is similar to Theorem 1, the only difference being that here, we use the sequence: $(-1)^j$, $j = 0, 1, \ldots, p^m - 2$ as the short sequence.   □

Corollary 3. *For a case* $n = 2m$ *APT sequences (4) have length* $2(p^m + 1)$, *autocorrelation peak* $2p^m$ *and 2 zero elements.*

Corollary 4. *Let* $n = 2m$. *Then a sequence given by*

$$\acute{w}_i = \varphi(Tr_m^n(\alpha^i)) , \quad i = 0, 1, \ldots, 2p^m + 1 \tag{6}$$

and

$$\varphi(z) = \begin{cases} (-1)^{(ind_\beta z) \bmod 2}, & \text{if } z \neq 0 \\ \pm 1, & \text{if } z = 0 \end{cases} \tag{7}$$

is an APB sequence of length $N = 2(p^m + 1)$.

First this result was obtained in [12]. Besides, it was shown there that APB sequences (6) and APB sequences [4,6,7] are the same.

Generalization of Theorem 1 and Theorem 2 is given by the following theorem.

Theorem 3. *Let* $p > 2$ *be a prime and let* **b** *be an m-sequence of length* $p^n - 1$ *over* $GF(p)$, *where* $n = mk$, $m \geq 1$, $k > 1$ *folded into Baumert decomposition array with* $T = (p^n - 1)/(p^m - 1)$ *rows and* $p^m - 1$ *columns. Let* $\mathbf{z} = \{z_i\}$ *be a balanced almost-perfect binary or ternary sequence of length* $h = (p^m - 1)/r$, *where* r *is an integer, with autocorrelation peak* P *and let* $\mathbf{z}^+$ *be a sequence of length* $p^m - 1$, *consisting of consecutive periods of the sequence* $\mathbf{z}$. *Let* $\mathbf{y} = \{y_i\}$, $0 \leq i < p^n - 1$ *be a ternary sequence obtained by substitution of non-zero rows with the sequence* $\mathbf{z}^+$ *with the same shift sequence. Then sequence* $\mathbf{w} = \{w_i\}$ *of length* $(p^n - 1)/r$ *with elements* $w_i = y_i$, $0 \leq i < (p^n - 1)/r$ *is a balanced almost-perfect ternary sequence with autocorrelation peak* $P * p^{(k-1)m}$.

Example 4. For the case of Example 1 with $p = 17$, $n = 2$, $m = 2$, $k = 1$ and the polynomial $x^2 + x + 3$ take $r = 1$. Then by Theorem 3 length of a sequence **w** is 288. Let a short balanced APT sequence of length 16 be 0 1 -1 -1 -1 1 -1 -1 0 -1 1 1 1 -1 1 1. As a result, we get a balanced APT sequence of length 288: 1 1 0 -1 1 -1 -1 -1 -1 0 1 -1 1 0 1 1 1 1 -1 1 1 1 1 1 -1 1 -1 0 1 -1 -1 1 1 1 0 1 -1 -1 1 -1 1 0 -1 0 1 -1 0 -1 1 -1 -1 -1 -1 1 -1 -1 1 -1 0 1 -1 -1 1 1 1 0 1 -1 -1 -1 1 1 -1 1 1 1 -1 1 -1 0 1 -1 -1 0 1 -1 1 1 -1 1 1 -1 1 -1 0 1 -1 -1 -1 1 1 -1 0 0 0 -1 1 0 0 -1 0 -1 1 -1 -1 -1 1 1 1 1 0 0 1 -1 -1 -1 1 1 1 1 0 -1 -1 -1 1 1 -1 0 -1 0 -1 1 0 -1 -1 -1 -1 -1 -1

-1 0 1 -1 1 1 1 1 0 -1 1 -1 0 -1 -1 -1 -1 1 -1 -1 -1 -1 -1 1 -1 1 0 -1 1 1 -1 -1 -1 0 -1
1 1 1 -1 0 1 0 -1 1 0 1 -1 1 1 1 1 -1 1 1 1 -1 1 1 0 -1 1 1 -1 -1 0 -1 1 1 1 -1 -1 1 -1 -1
-1 1 -1 1 0 -1 1 1 0 -1 1 -1 1 -1 -1 1 1 -1 1 0 -1 1 1 1 -1 -1 1 0 0 0 1 -1 0 0 1 0 1 -1
1 1 1 -1 -1 -1 0 0 -1 1 1 1 -1 -1 -1 -1 0 1 1 1 -1 -1 1 1 0 1 0 1 -1 0 1 1 1 1 1 with 50
zeroes and peak 238.

Example 5. Now consider the case of Example 2 with $p = 73$, $n = 2$, $m = 1$, $k = 2$ and the polynomial $x^2 + x + 11$. Then, for $r = 9$ take a short balanced APT sequence of length 8: 0 1 -1 -1 0 -1 1 1. Applying Theorem 3 we get a balanced APT sequence of length 592: 1 -1 1 1 0 0 -1 0 1 -1 -1 1 1 1 1 -1 -1 -1 1 1
-1 1 -1 1 -1 -1 -1 1 -1 1 1 1 1 -1 1 -1 0 1 0 1 -1 1 1 1 1 -1 -1 1 1 0 -1 0 -1 0 1 -1 0
-1 -1 -1 -1 -1 -1 1 0 0 -1 1 1 0 0 1 1 -1 -1 -1 1 1 -1 -1 -1 1 -1 1 0 1 1 -1 -1 0 0 -1 1
1 -1 1 1 0 -1 0 0 1 -1 0 -1 1 0 0 1 0 -1 -1 -1 0 1 -1 -1 0 1 -1 -1 -1 1 -1 1 1 1 -1 1 -1 0
0 1 1 1 0 1 -1 1 1 1 -1 -1 0 0 1 1 1 -1 0 1 1 -1 1 1 0 -1 1 1 1 1 0 -1 1 1 0 -1 -1 1 -1 0 1
0 0 0 -1 -1 -1 -1 1 1 -1 -1 -1 0 1 1 1 1 0 1 -1 0 0 0 -1 1 0 -1 0 0 0 -1 -1 1 1 1 1 1 -1
-1 1 1 1 -1 1 0 1 0 1 1 -1 1 1 0 -1 -1 -1 1 1 1 1 0 0 1 0 1 1 0 1 1 1 0 1 0 0 -1 1 -1 0 1 -1
1 1 0 1 1 0 0 1 0 1 -1 -1 0 -1 -1 1 0 0 1 -1 0 -1 1 1 0 -1 -1 1 0 -1 0 1 0 1 -1 1 -1 1 0 0
1 0 -1 0 1 1 1 1 -1 1 -1 -1 0 -1 0 0 -1 1 -1 -1 0 0 1 0 -1 1 1 -1 -1 -1 -1 1 1 1 1 -1 1
-1 1 -1 1 1 1 1 -1 1 -1 -1 -1 -1 -1 1 1 -1 1 0 -1 0 -1 1 -1 -1 -1 -1 1 1 1 -1 -1 0 1 0 1 0 -1 1 0
1 1 1 1 1 1 -1 0 0 1 -1 -1 0 0 -1 1 1 1 1 -1 -1 1 1 1 1 -1 1 -1 0 -1 -1 1 1 1 0 0 1 -1 -1
1 -1 0 1 0 0 -1 1 0 1 -1 0 0 -1 0 1 1 1 0 -1 1 1 1 0 -1 1 1 1 1 -1 1 1 -1 -1 1 1 -1 1 0 0 -1 -1
-1 0 -1 1 -1 -1 1 1 0 0 -1 -1 -1 1 1 0 -1 -1 1 1 -1 0 1 -1 -1 -1 -1 0 1 -1 0 1 1 -1 1 1 0 -1 0
0 0 1 1 1 1 -1 1 1 1 1 0 -1 -1 -1 -1 0 -1 1 0 0 0 1 -1 0 1 0 0 0 1 1 -1 -1 -1 -1 -1 1 1 1
-1 -1 1 1 -1 0 -1 0 -1 -1 1 1 -1 0 1 1 1 -1 -1 -1 1 0 0 -1 0 -1 -1 0 -1 -1 -1 0 -1 0 0 1 -1 1 1
0 -1 1 -1 -1 0 -1 -1 0 0 -1 0 -1 1 1 0 1 1 -1 0 0 -1 1 0 1 -1 0 1 1 -1 0 1 0 -1 0 -1 0 -1 1
-1 1 1 0 0 -1 0 1 0 -1 -1 -1 -1 1 -1 1 1 1 0 1 0 0 with 154 zeroes and the peak 438.

3 Construction of OPT Sequences

From Theorem 1–3, it follows that all the constructed APT sequences have an inverse-repeat property, i.e. $w_i = -w_{i+N/2}$, $0 \le i < N/2$. Using this fact and the decomposition properties of m-sequences we can prove the following theorem.

Theorem 4. *Let* $\mathbf{w} = \{w_i\}$, $i=0,1,\ldots,N-1$ *be an APT sequence, constructed by Theorem 1 or 2 or 3 . Then a ternary sequence*

$$\hat{\mathbf{w}} = \{w_i\}, i = 0, 1, \ldots, N/2 - 1 \qquad (8)$$

is an OPT sequence of length $N/2$.

It is obvious that Theorem 4 includes both, the well-known OPT sequences of length $(q^k - 1)/(q - 1)$ [5,9] and new OPT sequences of length $(q^k - 1)/r$, where $r < (q - 1)$. A new family of OPT sequences is given below.

Corollary 5. *Let* $n = 2m$ *and* $\mathbf{w} = \{w_i\}$, $i = 0, 1, .., 4p^m + 3$ *be an APT sequence (1). Then a sequence*

$$\hat{\mathbf{w}} = \{w_i\}, i = 0, 1, \ldots, 2p^m + 1 \qquad (9)$$

is an OPT sequence of length $2(p^m + 1)$ *with two zero elements.*

Example 6. From Example 1 by Theorem 4 we get the following OPT sequence of length 36: -111-1-11-1-1-1011-111111-11-1-11-11-11011-1-11111. Computer calculations verify this.

As was shown in [13], if the even-odd transformation (EOT) is applied to an odd-perfect sequence, then, as a result, a perfect polyphase sequence will be produced. EOT with an integer parameter t for any sequence $\mathbf{s} = \{s_j\}$ of length N is given by [13]

$$s_j\langle t\rangle = s_j \exp^{i\pi j(2t+1)/N}, \quad j = 0, 1, \ldots, N-1, \ i = \sqrt{-1} \tag{10}$$

Consider some applications of EOT to OPT sequences of length $N = 2(p^n - 1)/(p^m - 1)$ derived from APT sequences (1). Let $N/2 = 2t + 1$, i.e. $N/2 = 1 \bmod 2$. It is obvious that $k = n/m$ must be odd. Hence, we obtain new perfect 4-phase sequences with $2(p^{n-m} - 1)/(p^m - 1)$ zeroes. In the case when $N/4 = 2t + 1$, (it is possible that $k/2$ is odd) we can produce new perfect 8-phase sequences. The case $k/2 = 1$, $n = 2m$ is of special interest. In this case, we have perfect 8-phase sequences with only two zeroes.

Example 7. Take the OPT sequence of length 36 with $n = 2m$ from the previous example. From (10) we have the following perfect 8-phase sequence $\hat{\mathbf{w}}\langle 4\rangle = \{\hat{w}_j \exp^{i2\pi j/8}\}$, $\{\hat{w}_j\} =$-1 1 1 -1 -1 1 -1 -1 -1 0 1 1 -1 1 1 1 1 1 -1 1 -1 -1 1 -1 1 -1 1 0 1 1 -1 -1 1 1 1 1, $j = 0, 1, \ldots, 35$. It is more convenient to present this sequence in a form $\hat{w}\langle 4\rangle = \{\exp^{i2\pi u_j/8}\}$, where $u = \{u_j\}$ is a sequence of length 36 with elements from $Z(8)$: 4 1 2 7 0 5 2 3 4 - 2 3 0 5 6 7 0 1 6 3 0 1 6 3 0 5 2 - 4 5 2 3 0 1 2 3 and $\hat{w}_j\langle 4\rangle = 0$ for $u_j =$-.

Example 8. Consider the APT sequence of length 732 (Example 3). From it we have an OPT sequence of length 366: 1 0 0 1 -1 1 -1 1 1 0 1 -1 1 1 -1 -1 1 1 -1 -1 -1 1 1 1 1 -1 0 1 1 -1 -1 1 1 -1 -1 1 1 -1 -1 -1 0 1 -1 1 0 1 1 -1 1 1 1 1 1 1 -1 1 1 1 1 1 1 1 0 1 1 -1 -1 -1 1 1 1 1 1 1 -1 1 -1 -1 -1 1 1 -1 -1 -1 -1 -1 1 1 1 1 -1 -1 -1 -1 1 1 1 1 1 1 -1 -1 1 1 -1 1 -1 -1 -1 1 0 1 1 -1 -1 1 1 1 1 1 1 -1 1 0 -1 1 -1 -1 -1 -1 1 1 -1 0 -1 -1 1 -1 1 0 -1 -1 1 -1 0 -1 -1 1 -1 0 1 1 -1 -1 -1 1 1 -1 1 -1 -1 1 1 -1 0 -1 1 -1 -1 -1 -1 -1 -1 1 1 -1 1 1 1 1 1 1 -1 -1 1 -1 -1 1 -1 -1 1 1 1 0 1 0 0 -1 1 -1 1 1 1 0 -1 -1 1 1 1 1 -1 1 1 1 1 -1 1 1 -1 0 -1 -1 1 -1 -1 1 1 -1 -1 1 -1 1 1 0 1 1 1 -1 -1 1 1 -1 1 -1 -1 1 1 -1 -1 1 1 1 1 1 1 -1 0 -1 1 1 -1 1 1 -1 1 -1 -1 1 1 -1 -1 1 1 -1 -1 1 1 -1 -1 -1 1 1 -1 -1 1 1 1 1 1 -1 1 1 1 1 -1 -1 1 1 1 1 -1 1 1 1 1 0 1 -1 1 1 1 -1 -1 1 1 1 1 -1 -1 1 0 1 -1 1 -1 1 -1 -1 1 1 0 -1 -1 1 1 -1 -1 1 0 1 -1 1 1 0 -1 -1 1 1 1 1 -1 1 -1 -1 1 -1 -1 -1 -1 -1 -1 1 1 1 1 -1 1 1 -1 1 1 1 -1 -1 1 1 -1 1 1 1 0.

Taking $2t + 1 = 183$, in this case we obtain a perfect 4-phase sequence $\hat{w}\langle 91\rangle = \{\exp^{i2\pi u_j/4}\}$, where $u = \{u_j\}$ is a sequence of length 366 with elements from $Z(4)$: 0 - - 3 2 1 0 3 0 - 2 1 0 1 0 1 0 1 0 1 2 1 2 3 2 - 2 3 2 3 2 1 2 1 2 1 2 3 - 3 2 1 - 3 0 3 2 3 0 1 2 3 2 1 2 3 0 1 2 3 - 1 2 1 2 3 2 3 0 1 2 3 2 1 0 1 2 1 2 1 2 3 0 1 0 1 2 3 2 3 0 1 2 1 2 3 0 1 2 1 2 1 0 3 2 3 0 3 - 1 2 1 2 1 2 3 0 1 0 3 - 3 2 1 2 3 0 3 2 - 0 1 0 3 2 - 2 3 2 1 - 1 2 1 2 3 2 1 0 3 0 3 0 3 - 1 0 3 0 1 2 3 0 3 0 3 2 3 0 1 2 3 2 3 2 1 2 1 0 1 0 1 - 3 - - 0 3 2 1 2 3 - 3 0 3 0 1 2 3 2 1 2 3 0 3 2 1 - 3 0 3 2 3 2 1 2 1 0 3 0 - 2 3 2 - 2 1 0 1 0 3 2 1 0 1 0 3 0 3 0 1 0 - 2 1 0 3 0 1.

0 1 0 3 2 1 2 3 2 1 2 1 2 3 0 3 2 1 2 1 0 1 2 1 0 3 2 3 0 3 0 1 0 1 0 1 2 1 0 1 2 - 0
3 2 3 2 3 2 3 0 3 0 - 0 3 2 1 0 3 0 3 - 3 0 3 2 3 - 3 2 1 2 - 2 3 2 3 0 3 2 1 0 3 0 1
2 - 0 3 2 1 2 1 2 3 0 1 2 3 2 3 0 3 2 1 0 1 0 1 0 3 2 3 0 -.

4   Length Uniqueness

It is interesting to compare lengths of the new APT sequences with ones of the known APB sequences of length $2(p^m + 1)$ [4,6,7]. It was found that for all APT sequences (1) of length $N < 100$ there is only one of length $N = 72$ for which APB sequences do not exit. However, there are a lot of APT sequence lengths that don't coincide with lengths of APB sequences. To prove this, take $m = 0$ mod 2. It is easy to see that in this case the sequences (1) exist for all primes $p \geq 3$.

Theorem 5. *Let $p > 3$ be a prime. Then the lengths of APT sequences (1) with $m=0$ mod 2 and APB sequences [4,6,7] don't coincide.*

Proof. Let us suppose that there are such primes $p_1 > 3, p_2 > 3$ and positive integers w, u for which $4(p_1^{2w} + 1) = 2(p_2^u + 1)$. Then $2p_1^{2w} + 1 = p_2^u$. But this is impossible, as $2p_1^{2w} + 1$ is divisible by 3 for $p_1 > 3$.   □

Note that in the case $p_2 = 3$ and $u \leq 69$, equation $2p_1^{2w} + 1 = 3^u$ is true only for $u = 5$, $p_1 = 11$ and $w = 1$.

Now we will show that APT sequences (1) with parameters $m = 1$, $k = 2$, $p = 10t + 7 = 1$ mod 4 and length $N = 4(p + 1) = 40t + 32$ are also unique. To demonstrate it we will prove that $N/2 - 1$ is not a power of a prime number. Since $N/2 - 1 = 20t + 15$ is divisible by 5, suppose that $20t + 15 = 5^w$ for an integer $w > 3$. Then, the last two digits of $N/2 - 1$ must be 25. But it is impossible, as

Table 1. Unique APT sequences of length $N = 4(p^m + 1) < 2000$

p	m	N	p	m	N	p	m	N
17	1	72	181	1	728	349	1	1400
5	2	104	193	1	776	353	1	1416
37	1	152	197	1	792	19	2	1448
7	2	200	229	1	920	373	1	1496
61	1	248	241	1	968	389	1	1560
73	1	296	257	1	1032	397	1	1592
97	1	392	269	1	1080	401	1	1608
101	1	408	277	1	1112	409	1	1640
109	1	440	17	2	1160	421	1	1688
137	1	552	313	1	1256	433	1	1736
149	1	600	317	1	1272	449	1	1800
157	1	632	337	1	1352	457	1	1832
13	2	680	7	3	1376	461	1	1848

Table 2. Some unique APT sequences of length $N = 4(p^{mk} - 1)/(p^m - 1)$ with odd k

p	m	k	N	Z	p	m	k	N	Z
13	1	3	732	56	61	1	3	15132	248
5	1	5	3124	624	73	1	3	21612	296
37	1	3	5628	152	81	1	3	26572	328
41	1	3	6892	164	3	2	5	29524	3280
7	2	3	9804	200	89	1	3	32044	360
53	1	3	11452	216	101	1	3	41212	408

$(N/2 - 1) - 5 = 20t + 10$ is not divisible by 20, that leads to a contradiction. Hence, APT sequences (1) with parameters $m = 1$, $k = 2$, $p = 10t + 7 = 1 \bmod 4$ are unique.

All unique APT sequences of length $N = 4(p^m + 1) < 2000$ are presented in Table 1. Some other unique APT sequences of length $N = 4(p^n - 1)/(p^m - 1)$ with odd k are given in Table 2 (Z is the number of zeroes).

Further, as OPT sequences of length $p^m + 1$ [5] can be derived from APT sequences of length $2(p^m + 1)$ [9] then the OPT sequences (9) with $m = 0 \bmod 2$ are also unique like the above APT sequences (1).

5   Some Configuration Properties of APT Sequences

APT sequences (1) are invariant to the following transformations:

- shifts : $\{w_{i+k}\}$, $0 \leq k < N$;
- inverse: $\{\bar{w}_i\}$;
- reverse: $w_{N-1}, w_{N-2}, \ldots, w_1, w_0$;
- "even" inverse: $\bar{w}_0, w_1, \bar{w}_2, \ldots, \bar{w}_{N-2}, w_{N-1}$;
- "odd" inverse: $w_0, \bar{w}_1, w_2, \ldots, w_{N-2}, \bar{w}_{N-1}$.

Indeed, the first three cases directly follow from properties of the periodic correlation function. In the two last cases note that by Theorem 3 "even" or "odd" inverse of APT sequences are equivalent to a transformation of the shift sequence $e' = \{e'_j\}$, where $e'_j = e_j \bmod h$, $h = (p^m - 1)/r$, $0 \leq j < T$ into a sequence $e^1 = \{e^1_j\} = e'_0 + h/2, e'_1, e'_2 + h/2, \ldots, e'_{T-2} + h/2, e'_{T-1}$ and a shift sequence $e^2 = \{e^2_j\} = e'_0, e'_1 + h/2, e'_2, e'_3 + h/2, \ldots, e'_{T-2}, e'_{T-1} + h/2$ respectively. It is easy to see that sequences e^1 and e^2 possess the shift sequence property, i.e. for any $i \in Z(p^n - 1)$ each element of $Z(h)$ appears in differences $e^1_{i+k} - e^1_k \bmod h$ and $e^2_{i+k} - e^2_k \bmod h$, $k \in Z(T)$ exactly rp^{n-2m} times. Therefore, applying "even" and "odd" inverse transformations we also get APT sequences.

6   Conclusions

The constructions described in this paper allow to get both known APT sequences [8,9], OPT sequences [5,9] and new ones. New families of APT sequences of length $4(p^n - 1)/(p^m - 1)$, $(p^m + 1) = 2 \bmod 4$, and OPT sequences of length $2(p^n - 1)/(p^m - 1)$ have the peak factor close to 1, when p becomes large. For the case $n = 2m$ new APT sequences of length $4(p^m + 1)$ and OPT sequences of length $2(p^m + 1)$ with two zeros are obtained. Subsets of these APT and OPT sequences for all even m are unique in the sense that there are no APB sequences [4,6,7] or OPT sequences [5] having the same length. Using Mow even-odd transformation we can produce new perfect polyphase sequences with some zeroes from the OPT sequences. The constructed sequences can be used in wireless broadband mobile communication systems, for synchronization and channel estimation as well as in radar and sonar systems for ranging.

Acknowledgment

The author thanks the referees for their careful and helpful comments.

References

1. Fan P. and Darnell M.: "Sequence Design for Communications Applications", Research Studies Press Ltd., London, 1996.
2. Luke H.D.: "Binary Alexis sequences with perfect correlation", IEEE Transactions on Communications, vol. 49, No. 6, pp. 966–968, June 2001.
3. Krämer G.: "Application of Lüke–Schotten codes to Radar", Proceedings German Radar Symposium GRS 2000, Berlin, Germany, October 11–12 2000.
4. Pott A. and Bradley S.: "Existance and nonexistence of almost-perfect autocorrelation sequences", IEEE Transaction on Information Theory, vol. IT-41, No. 1, pp. 301–304, 1995.
5. Lüke H.D. and Schotten H.D.: "Odd-perfect almost binary correlation sequences", IEEE Trans. Aerosp. Electron. Syst., vol. 31, pp. 495–498, Jan. 1995.
6. Ipatov V.P.: "Periodic discrete signals with optimal correlation properties", Moscow, "Radio i svyaz", ISBN-5-256-00986-9, 1992.
7. Wolfmann J.: "Almost perfect autocorrelation sequenc es", IEEE Transaction on Information Theory, vol. IT-38, No. 4, pp. 1412–1418, 1992.
8. Langevin P.: "Some sequences with good autocorrelation properties", in Finite Fields, vol.168, pp. 175–185, 1994.
9. Schotten H.D. and Lüke H.D.: "New perfect and w-cyclic-perfect sequences", in Proc. 1996 IEEE International Symp. on Information Theory, pp. 82–85, 1996.
10. Baumert L.D.: "Cyclic difference sets", Berlin, Springer–Verlag, 1971.
11. R.A Games.: "Crosscorrelation of m-sequences and GMW sequences with the same primitive polynomial", Discrete Applied Mathematics, 12, pp. 139–146, 1985.
12. Luke H.D.: "Almost-perfect polyphase sequences with small phase alphabet", IEEE Transaction on Information Theory, vol. IT-43, No. 1, pp. 361–363, January 1997.
13. Mow W.H.: "Even-odd transformation with application to multi-user CW radars", 1996 IEEE 4th International Symposium on Spread Spectrum Techniques and Applications Proceedings, Mainz, Germany, pp. 191–193, September 22-25 1996.

Cross-Correlation Properties
of Perfect Binary Sequences

Doreen Hertel

Institut for Algebra and Geometry, Otto-von-Guericke University Magdeburg,
Postfach 4120, 39016 Magdeburg, Germany
doreen.hertel@mathematik.uni-magdeburg.de

Abstract. Binary sequences with good autocorrelation properties are
widely used in cryptography. If the autocorrelation properties are optimum, then the sequences are called perfect. In the last few years, new
constructions for perfect sequences have been found. In this paper we
investigate the cross-correlation properties between perfect sequences.
We give a lower bound for the maximum cross-correlation coefficient between arbitrary perfect sequences. We conjecture that this bound is not
best possible. Furthermore, we determine perfect sequences with prov-
able good correlation properties.

1   Introduction

Sequences $a = (a_i)_{i \geq 0}$ are called **periodic** with period n if $a_i = a_{i+n}$ for all
i and **binary** if $a_i \in \{0, 1\}$ for all i. The **autocorrelation** (AC) of a binary
sequence a with period n is defined by

$$c_t(a) := \sum_{i=0}^{n-1} (-1)^{a_i + a_{i+t}}. \tag{1}$$

We define the **shift** $a^{[t]} = (a_i^{[t]})_{i \geq 0}$ of the sequence a by $a_i^{[t]} := a_{i+t}$. The autocorrelation function of a sequence a is a measure for how much the given sequence
differs from its shifts $a^{[t]}$, $1 \leq t \leq n - 1$. Since the sequences are n-periodic, we
may compute the indices modulo n.

  In this paper we call a sequence with n odd and $c_t(a) = -1$ for $1 \leq t \leq n-1$
perfect, see [12] for more background on perfect sequences. It is well known
and easy to see, that $c_t(a) \equiv n \bmod 4$, since we consider only sequences with
$n \equiv 3 \bmod 4$. A sequence a is called **balanced** if the number of ones and zeros
in each period is $\frac{n}{2}$ if n is even or $\frac{n \pm 1}{2}$ if n is odd. For each perfect sequence a
we have $\sum_{i=0}^{n-1} (-1)^{a_i} = \pm 1$, because

$$\left(\sum_{i=0}^{n-1} (-1)^{a_i} \right)^2 = \sum_{i=0}^{n-1} \sum_{t=0}^{n-1} (-1)^{a_i + a_{i+t}} = \sum_{t=0}^{n-1} (c_t(a)) = (-1)(n-1) + n = 1,$$

since $c_0(a) = n$. Therefore perfect sequences are always balanced. Balanced (or
almost balanced) sequences with autocorrelation -1 (or low autocorrelation) are

T. Helleseth et al. (Eds.): SETA 2004, LNCS 3486, pp. 208–219, 2005.
© Springer-Verlag Berlin Heidelberg 2005

widely used in communications and cryptography. We call two sequences a and b **shift distinct**, if a is not equal to $b^{[t]}$ for all $t = 0, ..., n-1$.

We define the d-**decimation** $a^{(d)} = (a_i^{(d)})_{i \geq 0}$ of an n-periodic sequence a by $a_i^{(d)} := a_{id}$ and the **binary complement** $\bar{a} = (\bar{a}_i)_{i \geq 0}$ by $\bar{a}_i := a_i + 1 \bmod 2$. Note, if a is perfect, then all shifts $a^{[t]}$, all decimations $a^{(d)}$ with $\gcd(d, n) = 1$ and all their binary complements are perfect, again. We call two sequences a and b **equivalent**, if a can transformed into b by cyclic shift, decimation and/or taking the binary complement. We are interested in constructions, which produce inequivalent perfect sequences. The most known examples for perfect sequences have period $n = 2^m - 1$ and are constructed using finite fields of characteristic 2.

The finite field with 2^m elements is denoted by $\mathbb{F}_{2^m}$ and its multiplicative group by $\mathbb{F}_{2^m}^*$. Let α be a primitive element of $\mathbb{F}_{2^m}$, then any binary sequence of period $n = 2^m - 1$ describes a function $f : \mathbb{F}_{2^m}^* \to \mathbb{F}_2$ such that $f(\alpha^i) = a_i$. Conversely any function $f : \mathbb{F}_{2^m} \to \mathbb{F}_2$ describes a binary sequence, where the value $f(0)$ is irrelevant.

In the following we fix a primitive element α and identify sequences with the corresponding functions. If the sequence is balanced, resp. perfect, then we call the corresponding function balanced, resp. perfect. We choose $f(0) \in \{0, 1\}$ such that $\sum_{x \in \mathbb{F}_{2^m}} (-1)^{f(x)} = 0$, which is always possible if f is balanced.

The autocorrelation of a Boolean function f is defined by

$$c_y(f) := \sum_{x \in \mathbb{F}_{2^m}} (-1)^{f(x)+f(yx)}. \tag{2}$$

For $y = \alpha^t$ we have $c_{\alpha^t}(f) = c_t(a) + 1$. Thus, a function f is perfect iff

$$\sum_{x \in \mathbb{F}_{2^m}} (-1)^{f(x)+f(ax)} = \begin{cases} 0 & \text{if } a \neq 1 \\ 2^m & \text{if } a = 1. \end{cases} \tag{3}$$

The *trace* function is the linear mapping $tr : \mathbb{F}_{2^m} \to \mathbb{F}_2$ defined by $tr(x) = \sum_{i=0}^{m-1} x^{2^i}$. It is well known that the mappings tr_β, $\beta \in \mathbb{F}_{2^m}$, defined by $tr_\beta(x) = tr(\beta x)$ are linear, again, and all 2^m linear mappings $\mathbb{F}_{2^m} \to \mathbb{F}_2$ can be represented like this. The sequences $a = (a_i)_{i \geq 0}$ with $a_i = tr_\beta(\alpha^i)$ and $\beta \in \mathbb{F}_{2^m}^*$ are called m-**sequences**. They are perfect, see [11].

Other classes of perfect sequences are known. We refer the reader to the chapter on difference sets in [4] since perfect sequences correspond to a certain classes of cyclic difference sets, see also [12]. Using finite fields there are four more constructions known. Let us briefly recall these known series of perfect sequences: One important class are the Gordon-Mills-Welch-sequences (GMW-sequences) [10]. Maschietti [13] construct perfect sequences by taking the characteristic functions f of a set $D \subset \mathbb{F}_{2^m}^*$ (i.e. $f(x) = 1$ iff $x \in D$), where D is defined by $D = \{\varphi(x) | x \in \mathbb{F}_{2^m}\}$ for a certain mapping φ. Maschietti used the function $\varphi(x) = x + x^d$. If $\gcd(2^m - 1, d) = 1$ and φ is a 2-1-mapping, then the construction gives a perfect sequence. No, Chung and Yun [14] conjectured, that $\varphi(x) = x^d + (x+1)^d$, where $d = 2^{2k} - 2^k + 1$ is the so called Kasami exponent, yields a perfect sequence if $m \equiv 3k \pm 1$. The conjecture is proved in [6] and [7]. Dillon and Dobbertin showed in [7], that the function

$\varphi(x) = x^d + (x+1)^d + 1$ with $d = 2^{2k} - 2^k + 1$ also yields a perfect sequence, if $\gcd(k, m) = 1$. We call these sequences the Dillon-Dobbertin (DD) sequences.

We look at the cross-correlation between arbitrary perfect sequences. Similarly to the autocorrelation, we define the **cross-correlation** (CC) between two functions $f, g : \mathbb{F}_{2^m} \to \mathbb{F}_2$ by

$$c_y(f, g) := \sum_{x \in \mathbb{F}_{2^m}} (-1)^{f(x) + g(yx)} \tag{4}$$

for all $y \in \mathbb{F}_{2^m}$. It is also possible to express the cross-correlation similar to (1). The set of all CC-coefficient $Sp(f, g) := \{c_y(f, g) | y \in \mathbb{F}_{2^m}^*\}$ is called the **CC-spectrum** of f and g. Trivially, we have $c_0(f, g) = (-1)^{f(0) + g(0)} c_0(f)$.

It would be interesting to know the CC of all perfect sequences, but this seems to be illusive. So far, the following is known: The CC of m-sequences with their decimations was first 1968 been examined and today they are the most examined and best-known perfect sequences, see [11] for instance. The CC between an m-sequence and a GMW-sequence has been investigate in [16] and [17]. Antweiler [3] shows, that the calculation of the CC between GMW sequences can be reduced to the calculation of the CC between m-sequences and their decimations. The calculation of the CC between No-Chung-Yun sequences (Dillon-Dobbertin sequences) and certain m-sequences is contained in [7].

In chapter 2 we study the CC of arbitrary perfect sequences. We tried to generalize or to disprove known results for m-sequences to the whole class of perfect sequences. In chapter 3 we consider the CC of a special family of perfect sequences.

2   Cross-Correlation of Perfect Sequences

In the following we restrict ourselves to the description of sequences via functions $\mathbb{F}_{2^m} \to \mathbb{F}_2$. We need the following Lemma, which is a straightforward generalization of the Parseval formula.

Lemma 1. Let $f, h, g : \mathbb{F}_{2^m} \to \mathbb{F}_2$ be functions and g be perfect, then

$$\sum_{x \in \mathbb{F}_{2^m}} (-1)^{f(x)} (-1)^{h(x)} = \frac{1}{2^m} \sum_{y \in \mathbb{F}_{2^m}} c_y(f, g) \, c_y(h, g). \tag{5}$$

We call (5) the **generalized Parseval formula**.

Proof. For $f, g, h, k : \mathbb{F}_{2^m} \to \mathbb{F}_2$ and $z \in \mathbb{F}_{2^m}^*$ we have [2]

$$\sum_{x \in \mathbb{F}_{2^m}} c_x(f, g) c_{yx}(h, k) = \sum_{x \in \mathbb{F}_{2^m}} c_x(f, h) c_{yx}(g, k).$$

For the case $g = k$ we get $\sum_{x \in \mathbb{F}_{2^m}} c_x(f, g) c_{yx}(h, g) = \sum_{x \in \mathbb{F}_{2^m}} c_x(f, h) c_{yx}(g)$. If g is perfect, then $c_1(g) = 2^m$ and $c_z(g) = 0$ for all $z \neq 1$. Hence

$$2^m c_{y^{-1}}(f, h) = \sum_{x \in \mathbb{F}_{2^m}} c_x(f, g) c_{yx}(h, g),$$

and for $y = 1$ we get (5). $\square$

In the next proposition we list some basic (known) properties of the CC-function between perfect functions.

Proposition 1. *Let* $f, g : \mathbb{F}_{2^m} \to \mathbb{F}_2$ *be functions,* $\gcd(d, 2^m - 1) = 1$ *and* $y \in \mathbb{F}_{2^m}^*$.

1. *We have* $c_y(f, g^{(d)}) = c_{y^d}(f^{(1/d)}, g)$.
2. *If* f *and* g *are balanced, then* $c_y(f, g) \equiv 0 \bmod 4$.
3. *If* g *is perfect, then for all* $i \in \{0, ..., m - 1\}$ *exists* $y_i \in \mathbb{F}_{2^m}^*$ *such that* $c_y(f, g^{(2^i d)}) = c_{y_i}(f, g^{(d)})$.
4. *If* f *and* g *are perfect, then for all* $i \in \{0, ..., m - 1\}$ *exists* $y_i \in \mathbb{F}_{2^m}^*$ *such that* $c_{y^{2^i}}(f, g) = c_{y_i}(f, g)$.
5. *If* g *is perfect, then* $\sum\limits_{y \in \mathbb{F}_{2^m}} c_y(f, g) = 2^m (-1)^{f(0)+g(0)}$.
6. *If* g *is perfect, then* $\sum\limits_{y \in \mathbb{F}_{2^m}} (c_y(f, g))^2 = 2^{2m}$.
7. *If* f *and* g *are perfect, then* $\sum\limits_{y \in \mathbb{F}_{2^m}} c_y(f, g) c_{ay}(f, g) = \begin{cases} 0 & \text{if } a \in \mathbb{F}_{2^m} \backslash \{1\} \\ 2^{2m} & \text{if } a = 1 \end{cases}$.

We say, $s \in \{2, ..., n-1\}$ is a **multiplier** of an n-periodic sequences a, if there exists t with $a_{si} = a_{i+t}$ for all $i \geq 0$. The properties (3) and (4) holds, since 2 is always a multiplier of a perfect sequence of period $2^m - 1$, see [4]. The properties (4), (5) and (6) can be easily proved by the generalized Parseval formula.

Let $d_y(f, g) := |\{x \in \mathbb{F}_{2^m} | f(x) \neq g(yx)\}|$. This measures the distance between $f(x)$ and the translate $g(yx)$ of $g(x)$. In this paper we are interested in perfect Boolean functions f and g, whose number $d_y(f, g)$ and $d_y(f, \bar{g})$, where $\bar{g}$ is the complement of g (i.e. $\bar{g}(x) = g(x) + 1$) is as large as possible for all $y \in \mathbb{F}_{2^m}^*$. It is easy to see that $d_y(f, g) = 2^m - d_y(f, \bar{g})$ holds for $y \in \mathbb{F}_{2^m}^*$.

For two Boolean functions $f, g : \mathbb{F}_{2^m} \to \mathbb{F}_2$, we have $c_y(f, g) = 2^m - 2d_y(f, g)$ and $c_y(f, \bar{g}) = 2^m - 2d_y(f, \bar{g}) = -(2^m - 2d_y(f, g))$ for $y \in \mathbb{F}_{2^m}^*$. Hence we try to find functions f and g such that

$$\mathcal{M}(f, g) := \max_{y \in \mathbb{F}_{2^m}^*} \left| \sum_{x \in \mathbb{F}_{2^m}} (-1)^{f(x)+g(yx)} \right|$$

is as small as possible. The maximal CC-coefficient (in absolut value) is a measure for how much g may be used to approximate f. We are interested in lower bounds for the maximum CC-coefficient between two perfect functions.

The next Theorem gives a lower bound for the maximal CC-coefficient (in absolut value) of two Boolean functions, if one of the functions is perfect.

Theorem 1. *Let* $f, g : \mathbb{F}_{2^m} \to \mathbb{F}_2$ *be functions and* g *be perfect, then*

$$\max_{y \in \mathbb{F}_{2^m}} |c_y(f, g)| \geq 2^{m/2} \tag{6}$$

and

$$\mathcal{M}(f, g) \geq \sqrt{\frac{2^{2m} - |c_0(f)|^2}{2^m - 1}}. \tag{7}$$

Proof. We have $\sum_{y\in\mathbb{F}_{2^m}}(c_y(f,g))^2 = 2^{2m}$ by property (6) of Proposition 1. In the sum on the left side we have 2^m non-negative terms, which explains (6).

We also have $\sum_{y\in\mathbb{F}_{2^m}^*}(c_y(f,g))^2 = 2^{2m} - |c_0(f)|^2$. In the sum on the left side we have $2^m - 1$ non-negative terms, which explains (7).   □

Corollary 1. *Let $f, g : \mathbb{F}_{2^m} \to \mathbb{F}_2$ be functions and g be perfect. If $|c_0(f)| < 2^{m/2}$ then*

$$\mathcal{M}(f,g) > 2^{m/2}. \tag{8}$$

In particular, if f is balanced or perfect we have $\mathcal{M}(f,g) > 2^{m/2}$.

If g is linear, then Theorem 1 as well as Lemma 1 are not new. More precisely, let $F : \mathbb{F}_{2^m} \to \mathbb{C}$ be a function, the **Walsh transform** $\mathcal{W}(F)$ of F is the mapping $\mathbb{F}_{2^m} \to \mathbb{C}$ defined by $\mathcal{W}(F)(y) = \sum_{x\in\mathbb{F}_{2^m}} F(x)(-1)^{tr(yx)}$, see [11] for more information. If $F(x) = (-1)^{f(x)}$, we simply write $\mathcal{W}(f)$ instead of $\mathcal{W}(F)$. The **linearity** is the maximum Walsh coefficient (in absolute value) of f. If g is the *trace* function, then

$$\mathcal{W}(f)(y) = c_y(f,g).$$

Thus, the CC of a function f with the *trace* function is related to the linearity of the function f and (5) is the usual Parseval formula. The maximum Walsh coefficient of a function $f : \mathbb{F}_{2^m} \to \mathbb{F}_2$ is $\max_{y\in\mathbb{F}_{2^m}} |\mathcal{W}(f)(y)| \geq 2^{\frac{m}{2}}$, where equality occurs iff f is bent, see [11]. Since bent-functions are not balanced, we have $\max_{y\in\mathbb{F}_{2^m}^*} |\mathcal{W}(f)(y)| > 2^{\frac{m}{2}}$ for balanced functions f, compare with Corollary 1.

If the sequences corresponding to f and g are both m-sequences, then it is well-known that $\mathcal{M}(f,g) \geq 2^{\frac{m+1}{2}}$, and if equality occurs (of course, only for m odd), then $c_y(f,g) \in \{0, \pm 2^{\frac{m+1}{2}}\}$ for all $y \in \mathbb{F}_{2^m}$, see [5]. This property is not true for arbitrary perfect functions. Antweiler shows in [3], that there exists GWM-functions f and g with $\mathcal{M}(f,g) = 2^{(m+1)/2}$ and their CC-spectrum is not 3-valued.

The bound (8) seems to be bad if f and g are both perfect. We have no examples of perfect functions f and g with $\mathcal{M}(f,g) < 2^{(m+1)/2}$, hence we would like to ask the following question:

Question 1. Let f and g be two perfect functions. Is it true, that

$$\mathcal{M}(f,g) \geq 2^{\frac{m+1}{2}} \ ? \tag{9}$$

It is known, that the CC-spectrum of two perfect functions f and g, where the corresponding sequences a and b (and $\bar{a}$ and b) are shift distinct, contains at least three different values. Helleseth proved this for the corresponding functions of m-sequences in [1], but the proof is also true for arbitrary perfect functions. For the proof we only need the properties (5) and (6) of Proposition 1.

It is interesting to look for 3-valued CC-spectra where $\mathcal{M}(f,g)$ is small. In view of Question 1, we consider functions with $\mathcal{M}(f,g) = 2^{(m+1)/2}$.

The following proposition is well-known if g is linear.

Proposition 2. *Let $f, g : \mathbb{F}_{2^m} \to \mathbb{F}_2$ be functions, let g be perfect and f be balanced. If the CC-spectrum between f and g takes only the three values $\pm c$ and 0, then $c = 2^{\frac{m+k}{2}}$ with $k \in \mathbb{N}_0$ and the multiplicities are:*

cross-correlation value	multiplicity
0	$2^m - 2^{m-k} - 1$
$+2^{\frac{m+k}{2}}$	$2^{m-1-k} + 2^{\frac{m-k}{2}-1}(-1)^{f(0)+g(0)}$
$-2^{\frac{m+k}{2}}$	$2^{m-1-k} - 2^{\frac{m-k}{2}-1}(-1)^{f(0)+g(0)}.$

Proof. Note, since f is balanced, we have $c_0(f, g) = 0$. Let x denotes the number of CC-coefficient $\pm c$. By the generalized Parseval formula we get $2^{2m} = \sum_{z \in \mathbb{F}_{2^m}} (c_z(f, g))^2 = c^2 x$. This shows, that c^2 has to divide 2^{2m} and therefore c is a power of 2.

By Theorem 1 we have $c \geq 2^{m/2}$. We define $z_1 := |\{z \in \mathbb{F}_{2^m}^* | c_z(f, g) = 0\}|$, $z_2 := |\{z \in \mathbb{F}_{2^m}^* | c_z(f, g) = 2^{\frac{m+k}{2}}\}|$ and $z_3 := |\{z \in \mathbb{F}_{2^m}^* | c_z(f, g) = -2^{\frac{m+k}{2}}\}|$. Obviously we have

$$z_1 + z_2 + z_3 = 2^m - 1. \tag{10}$$

On the one hand we get $\sum_{z \in \mathbb{F}_{2^m}} c_z(f, g) = (-1)^{f(0)+g(0)} 2^m$ by property (5) of Proposition 1, since g is perfect. On the other hand we have $\sum_{z \in \mathbb{F}_{2^m}} c_z(f, g) = (z_2 - z_3) \cdot 2^{\frac{m+k}{2}}$. Consequently we get

$$z_2 - z_3 = (-1)^{f(0)+g(0)} 2^{\frac{m-k}{2}}. \tag{11}$$

By property (6) of Proposition 1 we get $\sum_{z \in \mathbb{F}_{2^m}} (c_z(f, g))^2 = 2^{2m}$ and, on the other hand $\sum_{z \in \mathbb{F}_{2^m}} (c_z(f, g))^2 = (z_2 + z_3) \cdot 2^{m+k}$. Therefore we have

$$z_2 + z_3 = 2^{m-k}. \tag{12}$$

The equations (10), (11) and (12) show the multiplicities.   □

If we have such a 3-valued CC-spectrum between two perfect Boolean functions, then the sequence $b = (b_i)_{i \geq 0}$ defined by

$$b_i := \frac{c_{\alpha^i}(f, g)}{2^{(m+k)/2}} \tag{13}$$

is a perfect ternary sequence, see [18]. Thus, it is also interesting to look for 3-valued CC-spectra between arbitrary perfect functions in order to get new perfect ternary sequences.

3   Good Cross-Correlation Spectra

In this chapter we look at the Dillon-Dobbertin (DD) functions f and g with $\mathcal{M}(f, g) = 2^{(m+1)/2}$. With Question 1 in chapter 2 it is possible, that this case is optimum. So, in the following we restrict ourselves to the case m odd.

We recall the definition of the DD-functions, see Chapter 1. Let $d := 2^{2k} - 2^k + 1$ and $\gcd(k, m) = 1$, the DD-function b_k is defined by

$$b_k(x) := \begin{cases} 0 \text{ if } x \in B_k \\ 1 \text{ if } x \notin B_k \end{cases}$$

where $B_k := \{x^d + (x+1)^d + 1 | x \in \mathbb{F}_{2^m}\}$. These functions are perfect [7]. Note, that b_1 is the *trace* function. It is shown in [7], that $B_2^{(5)} = \{\varphi(x) | x \in \mathbb{F}_{2^m}\}$ where $\varphi(x) = x + x^6$ is a Maschietti function.

Let $\gcd(s, 2^m - 1) = 1$. For $m \le 17$ odd we list all the functions b_k and $b_l^{(s)}$ with $M(b_k, b_l^{(s)}) = 2^{\frac{m+1}{2}}$, which were found by computer:

m	k	l	s	
5	1	2	1	$\star_2$
5	1	2	5	$\star_1$
5	1	2	7	$\star_3$
7	1	2	1	$\star_2$
7	1	2	5	$\star_1$
7	1	2	43	
7	1	3	1	$\star_2$
7	1	3	9	$\star_1$
7	1	3	15	
7	1	3	27	
7	1	3	43	
7	2	3	1	$\star_2$
7	2	3	27	
7	3	3	19	$\circ_1$

m	k	l	s	
9	1	2	5	$\star_1$
9	1	4	17	$\star_1$
11	1	2	5	$\star_1$
11	1	3	1	$\star_2$
11	1	3	9	$\star_1$
11	1	4	1	$\star_2$
11	1	4	17	$\star_1$
11	1	5	33	$\star_1$
11	2	3	1	$\star_2$
11	2	5	1	$\star_2$
11	4	5	1	$\star_2$
13	1	2	5	$\star_1$
13	1	3	1	$\star_2$
13	1	3	9	$\star_1$

m	k	l	s	
13	1	4	1	$\star_2$
13	1	4	17	$\star_1$
13	1	5	33	$\star_1$
13	1	6	65	$\star_1$
13	2	5	1	$\star_2$
13	2	6	1	$\star_2$
13	3	4	1	$\star_2$
13	5	6	1	$\star_2$
15	1	2	5	$\star_1$
15	1	4	17	$\star_1$
15	1	7	129	$\star_1$
17	1	2	5	$\star_1$
17	1	3	1	$\star_2$
17	1	3	9	$\star_1$

m	k	l	s	
17	1	4	17	$\star_1$
17	1	5	33	$\star_1$
17	1	6	1	$\star_2$
17	1	6	65	$\star_1$
17	1	7	129	$\star_1$
17	1	8	257	$\star_1$
17	2	5	1	$\star_2$
17	2	6	1	$\star_2$
17	3	8	1	$\star_2$
17	4	5	1	$\star_2$
17	4	7	1	$\star_2$
17	7	8	1	$\star_2$

We do not consider the case $k = l = 1$, because it is the CC between m-sequences and these sequences are already analysed, see [11] for instance. It is interesting, that in all cases listed in the table, except the case $\circ_1$, we have a 3-valued CC-spectrum with the values $\pm 2^{\frac{m+1}{2}}$ and 0 and the multiplicities from Proposition 2. In the case $\circ_1$ we have the CC-spectrum $Sp(b_3, b_3^{(19)}) = \{-16(21), -8(28), 0(21), 8(28), 16(29)\}$, where the numbers in the brackets are the multiplicities.

Each value in the table represents a whole class of values, which also have the same CC-spectrum: We only list the values k and l such that $1 \le k \le l \le \frac{m-1}{2}$, because $b_k^{(s)} = b_{m-k}^{(s)}$ and for each s with $\gcd(s, 2^m - 1) = 1$ we have $c_y(b_k, b_l^{(s)}) = c_{y'}(b_l, b_k^{(1/s)})$ with $y' = y^{-1/s}$. For s we only consider the smallest value in the set $\{2^i s | i = 0, ..., m - 1\}$, since property (3) of Proposition 1 holds.

We can explain the values in the table indicated by a star. We do not only show, that the maximal CC-coefficient is $2^{(m+1)/2}$. We prove, that the CC-spectrum contains only the three values $\pm 2^{(m+1)/2}$ and 0. The proof contains a method related to the concept of Hadamard-equivalence introduced in [6] and

[7]. Therefore we define the function $m^{(d)} : \mathbb{F}_{2^m} \to \mathbb{F}_2$ by $m^{(d)}(x) = tr(x^d)$ for all $x \in \mathbb{F}_{2^m}$. We show, that for some special k, l and s the CC of the DD-functions b_k and $b_l^{(s)}$ is equivalent to the Walsh transform of $m^{(d)}$, where $d = 2^k + 1$ or $d = 2^{2k} - 2^k + 1$ with $\gcd(k, m) = 1$. The Walsh spectrum of such functions $m^{(d)}$ has only the three values $\pm 2^{\frac{m+1}{2}}$ and 0, see [11] for instance. Note, the function $m^{(d)}$ is the corresponding function of the d-decimation of an m-sequence.

The following highly nontrival result is the major step in [7] to prove that b_k is a perfect function.

Result 1. (Dillon and Dobbertin [7]) *Let m be odd and $\gcd(k, m) = 1$, then*

$$\mathcal{W}(b_k^{(2^k+1)})(y) = \mathcal{W}(m^{(3)})(y^{(2^k+1)/3}) \quad \text{for all } y \in \mathbb{F}_{2^m}.$$

Result 1 explains all entries in the table indicated by $\star_1$, since $c_y(b_1, b_k^{(s)}) = \mathcal{W}(b_k^{(s)})(y^{-1})$. In particular, we have

$$c_y(b_1, b_k^{(2^k+1)}) \in \{\pm 2^{(m+1)/2}, 0\}.$$

Theorem 2. *Let m be odd and $\gcd(k, m) = \gcd(l, m) = 1$, then*

$$c_y(b_k, b_l) = \mathcal{W}(m^{((2^k+1)/(2^l+1))})(y^{-1/(2^k+1)}) \quad \text{for all } y \in \mathbb{F}_{2^m}.$$

A function $\varphi : \mathbb{F}_{2^m} \to \mathbb{F}_{2^m}$ is called **maximal nonlinear**, if the condition $\max_{a,b \in \mathbb{F}_{2^m}, b \neq 0} |\sum_{x \in \mathbb{F}_{2^m}} (-1)^{tr(ax+b\varphi(x))}| = 2^{(m+1)/2}$ holds.

Corollary 2. *Let m be odd and $\gcd(k, m) = \gcd(l, m) = 1$, we have that the maximal CC-coeffient of b_k and b_l is $2^{(m+1)/2}$ if and only if the function $x \mapsto x^{(2^k+1)/(2^l+1)}$ is maximal nonlinear.*

The following corollary explains all entries in the table indicated by $\star_2$.

Corollary 3. *Let $m \equiv \pm 1 \bmod 3$ and $\gcd(l, m) = 1$, then*

$$c_y(b_{3l}, b_l) \in \{\pm 2^{(m+1)/2}, 0\} \text{ for all } y \in \mathbb{F}_{2^m}.$$

Before we present the proof of Theorem 2 we write each CC-coefficient between two DD-functions in terms of CC-coefficients between m-sequences and their decimations. We have

$$c_a(b_k, b_l^{(s)}) = \sum_{x \in \mathbb{F}_{2^m}} (-1)^{b_k(x)+b_l^{(s)}(ax)} = \sum_{x \in \mathbb{F}_{2^m}} (-1)^{b_k(x)+b_l((ax)^s)}.$$

We apply the generalized Parseval formula with $g = m^{(1/(2^k+1))}$ in the first step and $g = m^{(s/(2^l+1))}$ in the second step and get

$$
2^{2m} c_a(b_k, b_l^{(s)}) \overset{(5)}{=} 2^m \sum_{x \in \mathbb{F}_{2^m}} \left(\sum_{y \in \mathbb{F}_{2^m}} (-1)^{b_k(y)+tr(xy^{1/(2^k+1)})} \right)
$$
$$
\cdot \left(\sum_{z \in \mathbb{F}_{2^m}} (-1)^{b_l(a^s z^s)+tr(xz^{1/(2^k+1)})} \right)
$$
$$
\overset{(5)}{=} \sum_{x,z \in \mathbb{F}_{2^m}} \left(\sum_{y \in \mathbb{F}_{2^m}} (-1)^{b_k(y)+tr(xy^{1/(2^k+1)})} \right)
$$
$$
\cdot \left(\sum_{w \in \mathbb{F}_{2^m}} (-1)^{b_l(a^s w^s)+tr(zw^{s/(2^l+1)})} \right)
$$
$$
\cdot \left(\sum_{v \in \mathbb{F}_{2^m}} (-1)^{tr(xv^{1/(2^k+1)}+zv^{s/(2^l+1)})} \right)
$$
$$
= \sum_{x,z \in \mathbb{F}_{2^m}} \left(\sum_{y \in \mathbb{F}_{2^m}} (-1)^{b_k(y)+tr(xy^{1/(2^k+1)})} \right)
$$
$$
\cdot \left(\sum_{w \in \mathbb{F}_{2^m}} (-1)^{b_l(w)+tr(za^{-s/(2^l+1)}w^{1/(2^l+1)})} \right)
$$
$$
\cdot \left(\sum_{v \in \mathbb{F}_{2^m}} (-1)^{tr(v^t+xz^{-1/t}v)} \right),
$$

where $t := s(2^k+1)/(2^l+1)$. We use the Result 1 twice and get

$$
2^{2m} c_a(b_k, b_l^{(s)}) = \sum_{x,z \in \mathbb{F}_{2^m}} \left(\sum_{y \in \mathbb{F}_{2^m}} (-1)^{tr(y^3+x^{(2^k+1)/3}y)} \right)
$$
$$
\cdot \left(\sum_{w \in \mathbb{F}_{2^m}} (-1)^{tr(w^3+z^{(2^l+1)/3}a^{-s/3}w)} \right)
$$
$$
\cdot \left(\sum_{v \in \mathbb{F}_{2^m}} (-1)^{tr(v^t+xz^{-1/t}v)} \right)
$$
$$
= \sum_{x,z \in \mathbb{F}_{2^m}} \mathcal{W}(m^{(3)})(x^{(2^k+1)/3}) \cdot \mathcal{W}(m^{(3)})(z^{(2^l+1)/3}a^{-s/3})
$$
$$
\cdot \mathcal{W}(m^{(t)})(xz^{-1/t}) \tag{14}
$$

This shows, that the calculation of the CC between two DD-functions is reduced to the calculation of the CC between m-sequences and their decimations. Now we prove Theorem 2.

Proof. We define $c_h := \mathcal{W}(m^{((2^k+1)/(2^l+1))})(h)$ and transform

$$2^{2m} c_a(b_k, b_l)$$

$$= \sum_{x,z \in \mathbb{F}_{2^m}} \left(\sum_{y \in \mathbb{F}_{2^m}} (-1)^{tr(y^3 + x^{(2^k+1)/3}y)} \right) \left(\sum_{w \in \mathbb{F}_{2^m}} (-1)^{tr(w^3 + z^{(2^l+1)/3}a^{-1/3}w)} \right)$$

$$\cdot \underbrace{\left(\sum_{v \in \mathbb{F}_{2^m}} (-1)^{tr(v^{(2^k+1)/(2^l+1)} + xz^{-(2^l+1)/(2^k+1)}v)} \right)}_{= c_h \text{ with } x = hz^{(2^l+1)/(2^k+1)}}$$

$$= \sum_{h,z \in \mathbb{F}_{2^m}} \left(\sum_{y \in \mathbb{F}_{2^m}} (-1)^{tr(y^3 + h^{(2^k+1)/3}z^{(2^l+1)/3}y)} \right)$$

$$\cdot \left(\sum_{w \in \mathbb{F}_{2^m}} (-1)^{tr(w^3 + z^{(2^l+1)/3}a^{-1/3}w)} \right) c_h$$

$$= \sum_{h,w \in \mathbb{F}_{2^m}} \left(\sum_{y \in \mathbb{F}_{2^m}} (-1)^{tr(y^3 + w^3)} \right) \underbrace{\left(\sum_{z \in \mathbb{F}_{2^m}} (-1)^{tr(z(h^{(2^k+1)/3}y + a^{-1/3}w))} \right) c_h}_{= \begin{cases} 2^m & \text{if } w = a^{1/3}yh^{(2^k+1)/3} \\ 0 & \text{otherwise} \end{cases}}$$

$$= \sum_{h \in \mathbb{F}_{2^m}} \left(\sum_{y \in \mathbb{F}_{2^m}} (-1)^{tr(y^3 + ay^3 h^{(2^k+1)})} \right) c_h 2^m$$

$$= \sum_{h \in \mathbb{F}_{2^m}} \underbrace{\left(\sum_{y \in \mathbb{F}_{2^m}} (-1)^{tr(y(1 + ah^{(2^k+1)}))} \right)}_{= \begin{cases} 2^m & \text{if } h = a^{-1/(2^k+1)} \\ 0 & \text{otherwise} \end{cases}} c_h 2^m$$

$$= 2^{2m} c_{a^{-1/(2^k+1)}}.$$

We obtain $c_a(b_k, b_l) = \mathcal{W}(m^{((2^k+1)/(2^l+1))})(a^{-1/(2^k+1)})$. $\qquad\square$

We can also explain $\star_3$. It is known, that b_2 is a quadratic residue sequence of period 31, and $b_2 = b_2^{(7)}$ since 7 is a quadratic residue modulo 31. Thus, for $m = 5$ we have $Sp(b_1, b_2^{(7)}) = Sp(b_1, b_2)$, which is 3-valued with $\pm 2^{(m+1)/2}$ and 0 by Result 1.

4   Open Problems

We list some intersting questions about the CC between DD-functions, which follow from the table above and from (3).

Question 2. Do we have more examples of DD-functions $b_k, b_l^{(s)}$ with $\mathcal{M}(b_k, b_l^{(s)}) = 2^{(m+1)/2}$ and the CC-spectrum is not 3-valued, except the case $\circ_1$ in the table.

The corresponding ternary sequences of the CC of the DD-functions, which are indicated by a star in the table, are not new. These ternary sequences are equivalent to the ternary sequences obtained from the CC between m-sequences and their decimations. For $m = 7$ there are two more inequivalent ternary sequences (the open cases), which are not equivalent to a ternary sequence corresponding to the CC between an m-sequence and one of its decimations.

Question 3. Let $m > 5$. Do we have examples of DD-function b_k with s is a multiplier of b_k and $s \notin \{1, 2, ..., 2^{m-1}\}$.

Question 4. Let b_k and $b_l^{(s)}$ be two DD-functions, then the CC-coefficients between b_k and $b_l^{(s)}$ take only the three values $\pm 2^{(m+1)/2}$ and 0 only if $x \mapsto x^{s(2^k+1)/(2^l+1)}$ is a maximal nonlinear function.

It is easy to see, that in Question 4 the converse is not true: Let $m = 11, k = 2, l = 3$ and $s = 9$, then is $x^{s(2^k+1)/(2^l+1)} = x^5$ maximal nonlinear, but the CC-spectrum $Sp(b_2, b_3^{(9)})$ contains more then three values.

In the case $s = 1$ the answer to Question 4 is yes and we even have "if and only if", see Corollary 2.

Acknowledgments

The author wish to thank Prof. D. V. Sarwate for helpful comments.

References

1. Helleseth, T.: Some results about the cross-correlation function between two maximal linear sequences. Discrete Mathematics, Vol. 16. (1976) 209–232
2. Pursley, M. B., Sarwate, D. V.: Performance Evaluation for Phase-code Spread-spectrum Muiltiple-access Communication. II: Code Sequence Analysis. IEEE Transactions Communication, Vol. 25 (1977) 800-803
3. Antweiler, M.: Cross-correlation of p-ary GMW Sequences. IEEE Transactions on Information Theory, Vol. 40. (1994) 1253–1261
4. Beth, T., Jungnickel, D., Lenz, H.: Design Theory. Encyclopedia of Mathematics and its Applications. Cambridge University Press, Vol. 1, 2nd ed., Cambridge (1999)
5. Chabach, F., Vaudenay, S.: Links between differtial and linear cryptanalysis. In: Advances in Cryptology - Eurocrypt'94, A. D. Santis, ed., Vol. 950 of Lecture in Computer Science, New York, Springer-Verlag (1995) 356–365
6. Dillon, J.F.: Multiplicative Difference Sets via Additive Charakters. Designs, Codes and Cryptography, Vol. 17 (1999) 225–235
7. Dillon, J.F., Dobbertin, H.: New Cyclic Difference Sets with Singer Parameters. preprint (1999)
8. Dobbertin, H.: Construction of bent functions and balanced boolean functions with high nonlinearity. Fast software encrytion, Lecture Notes in Computer Science 1008. (1995) 61–74
9. Games, R.A.: Crosscorrelation of m-Sequences and GMW-sequences with the same primitive polynomial. Discrete Applied Mathematics, Vol. 12 (1985) 139–146

10. Gordon, B., Mills, W.H., Welch, L.R.: Some new difference sets. Canadian Journal of Mathematics, Vol. 14 (1962) 614–625

11. Helleseth, T., Kumar, P.V.: Sequences with low correlation. In: Handbook of Coding Theory, Vol. 1,2, North-Holland, Amsterdam (1998) 1065–1138

12. Jungnickel, D., Pott, A.: Perfect and almost perfect sequences. Discrete Applied Mathematics, Vol. 95 (1999) 331–359

13. Maschietti, A.: Difference Sets and Hyperovals. Designs, Codes and Cryptography, Vol. 14 (1998) 89–98

14. No, J.S., Chung, H., Yun, M.S.: Binary Pseudorandom Sequences of Period $2^m - 1$ with Ideal Autocorrelation Generated by the Polynomial $z^d + (z + 1)^d$. IEEE Transactions on Information Theory, Vol. 44 (1998) 1278–1282

15. Patterson, N. J., Wiedemann, D. H.: The covering radius of the $(2^{**}(15),16)$ Reed-Muller code is at least 16276. IEEE Transactions on Information Theory, Vol. 29 (1983) 354–356

16. Games, R. A.: Crosscorrelation of m-sequences and GMW-sequences with the same Primitive Polynomial. Discrete Applied Mathematics, Vol. 12 (1985) 139–146

17. Chan, A. H., Goresky, M., Klapper, A.: Correlation Functions of Geometric Sequences. In: Advances in Cryptology - Eurocrypt'90, Spinger-Verlag, (1990) 214–221

18. Shedd, D. A., Sarwate, D. V.: Construction of Sequences with Good Correlation Properties. IEEE Transactions on Information Theory, Vol. it-25, No. 1 (1979) 94–97

19. Golomb, S.: Shift Register Sequences. Oakland, CA: Holden-Day, 1967. Revised edition: Laguna Hills, CA: Aegean Park Press.

New Sets of Binary and Ternary Sequences with Low Correlation

Evgeny I. Krengel[1], Andrew Z. Tirkel[2], and Tom E. Hall[2]

[1] Kedah Electronics Engineering, Zelenograd, Moscow 124498, Russia
legnerk@gagarinclub.ru
[2] School of Mathematical Sciences, Monash University,
PO Box 28M, Victoria 3800, Australia
atirkel@bigpond.net.au

Abstract. New binary and ternary sequences with low correlation and simple implementation are presented. The sequences are unfolded from arrays, whose columns are cyclic shifts of a short sequence or constant columns and whose shift sequence (sequence of column shifts) has the distinct difference property. It is known that a binary m-sequence/GMW sequence of length $2^{2m} - 1$ can be folded row-by-row into an array of $2^m - 1$ rows of length $2^m + 1$. We use this to construct new arrays which have at most one column matching for any two dimensional cyclic shift and therefore have low off-peak autocorrelation. The columns of the array can be multiplied by binary orthogonal sequences of commensurate length to produce a set of arrays with low cross-correlation. These arrays are unfolded to produce sequence sets with identical low correlation.

Outline

Sequences with low correlation and large linear complexity are widely used in spread spectrum communication systems [1,2]. Here, we present new binary sequences, with low even periodic cross-correlation, low off-peak autocorrelation and simple implementation. This paper generalizes a construction of sequence sets introduced by Gong [7], called interleaved sequences. Such sequence sets are obtained by writing an m/GMW sequence of length $p^{km} - 1$ row-by-row into an array, where the columns are cyclic shifts of a single m/GMW sequence of length $p^m - 1$ or constants. Games [4] introduced the shift sequence to describe the sequence of cyclic shifts of the columns. The shift sequence has some remarkable properties, which are crucial to our construction. Where the number of m/GMW columns is commensurate with the length of a pseudonoise sequence, or an integer multiple of periods of a pseudonoise sequence, these columns can be multiplied by cyclic shifts of that sequence. The resultant arrays can be unfolded to produce sequences with identical correlation. The off-peak autocorrelation and cross-correlation of such arrays and sequences is low, because the number of phase-matched columns is constrained by a property of the shift sequence. Our paper is organized in two parts. In Sections 1-4, we find suitable multiplication

T. Helleseth et al. (Eds.): SETA 2004, LNCS 3486, pp. 220–235, 2005.

sequences, commensurate with m/GMW arrays, as described above. We restrict our results to binary and ternary sequences. In Sections 5-10 we **synthesize** shift sequences with desirable properties and construct arrays with commensurate pseudonoise column sequences and multiplication sequences. We then unfold the arrays into long sequences with good auto and cross-correlation. We find an abundance of such long sequences and restrict ourselves to purely binary cases.

1   Sequence Construction

Here, we present new binary and ternary sequences of length $2^n - 1$, with low even periodic cross-correlation, low off-peak autocorrelation and simple implementation. We fold a long m or GMW sequence of length $2^n - 1$, with even n, into an array and multiply the columns of the array by commensurate pseudonoise sequences. When the parent sequence is an m-sequence, the implementation is simple, but linear complexity is low. GMW parent sequences require more complex implementation, but the linear complexity of the resulting sequence is much larger.

1.1   Binary Sequence

Let n be even, $T = 2^{n/2} + 1$, and let $q = T/3$ be a prime of type $4t + 3$. Let $a = \{a_i\}$, $i = 0, 1, 2, \ldots, 2^n - 2$, $a_i \in \{-1, 1\}$ be a binary m-sequence (or GMW sequence) of length $N = 2^n - 1$. Let $b = \{b_i\}$, $i = 0, 1, 2, \ldots, q-1$, $b_i \in \{-1, 1\}$, be an infinite periodic binary Hall or Legendre sequence with period q [3]. Let b_j be the sequence b shifted by j units. We form the sequence $c_j = b_j a$, with entries $c_{ji} = b_{j+i} a_i$, $i = 0, 1, 2, \ldots, 2^n - 2$. We denote the set $\{c_j: j = 0, 1, 2, \ldots, q-1\}$ simply by $\{c_j\}$ and we adjoin a to this set, and denote the final set by $\{c_j, a\}$. The number of different sequences in the set is $M = q + 1 = (2^{n/2} + 1)/3 + 1$.

Example 1. Let $n = 6$ and $a = $ *1 1 1 1 1-1-1-1-1 1 1 1-1 1 1 1-1 1-1 1-1 1 1-1-1* *1-1 1 1 1 1 1-1 1 1 1 1-1 1-1-1 1-1-1-1-1-1-1 1-1 1-1-1-1 1 1 1 1-1-1 1 1 1-1-1-1-1 1-1-1* and $b = $ *11-1*. Sequence a is an m-sequence of length 63 and b is a Legendre sequence of length 3. Then $c_0 = b_0 a = $ *1 1-1 1 1 1-1-1 1 1 1 1 1 1 1 1-1-1-1* *1-1-1-1-1 1-1 1 1 1 1 1-1-1 1 1-1 1 1-1-1 1-1-1-1-1 1 1-1-1-1 1 1 1-1-1-1-1* *1 1-1 1 ; $c_1 = b_1 a = $ *1-1 1 1 1-1-1-1 1-1 1-1-1 1-1-1 1 1 1-1-1 1-1 1 1-1-1 1 1-1-1* *1-1 1-1-1-1-1-1-1 1-1-1 1 1-1-1-1-1-1 1 1 1-1-1-1-1 1-1 1-1 1 1-1*; $c_2 = b_2 a = $ *-1* *1 1-1 1-1 1-1-1-1 1-1-1 1-1-1-1 1 1 1 1 1 1-1 1 1 1 1 1-1 1-1-1-1 1 1 1 1-1 1 1-1-1* *1-1 1 1 1-1 1-1 1-1 1-1 1 1 1 1 1-1-1-1-1-1.*

1.2   Ternary Sequence Construction

Let n be even, $T = 2^{n/2} + 1$ be a prime. New ternary sequences result from the multiplication of a long binary m/GMW sequence $a = \{a_i\}$ with $a_i \in \{-1, 1\}$ and length $2^n - 1$, with $2^{n/2} - 1$ repeats of a short ternary Legendre sequence $d = \{d_i\}$ with $d_i \in \{0, -1, 1\}$ and length $t = 2^{n/2} + 1$. We apply all shifts of this short sequence and form the ternary sequence $t_j = d_j a$, with entries $t_{ji} = d_{j+i} a_i$, $i = 0, 1, 2, \ldots, 2^n - 2$, $j = 0, 1, 2, \ldots, 2^{n/2}$. Thus, we obtain a new set

$\{t_j, \boldsymbol{a}\}$ consisting of $2^{n/2} + 1$ ternary sequences and the reference long sequence, i.e. $M = 2^{n/2} + 2$ distinct sequences. When $2^{n/2} + 1$ is prime, it is called a Fermat prime [2]. Currently five Fermat primes are known for $n = 2, 4, 8, 16, 32$.

Example 2. Let $n = 4$. $\boldsymbol{a} =$ *-1 1 1 1 -1 1 1 -1 -1 1 1 -1 -1 -1 -1* is an m-sequence of length 15. $\boldsymbol{d} =$ *0 1-1 -1 1* is a ternary Legendre sequence of length 5. Then

$$t_0 = d_0 \boldsymbol{a} = 01 - 1 - 1 - 1011110111 - 1$$
$$t_1 = d_1 \boldsymbol{a} = -1 - 1 - 1101 - 11 - 10 - 1 - 11 - 10$$
$$t_2 = d_2 \boldsymbol{a} = 1 - 110 - 1 - 1 - 1 - 1011 - 1 - 10 - 1$$
$$t_0 = d_0 \boldsymbol{a} = 11011 - 110 - 1 - 1110 - 11$$
$$t_4 = d_4 \boldsymbol{a} = -101 - 1110 - 11 - 1 - 10 - 111$$

2   Correlation Properties

2.1   Binary Sequence

Now fold the binary sequence c_j of Subsection 1.1, row-by-row into a two-dimensional array $\boldsymbol{C}_j$ of size $(2^{n/2} + 1)(2^{n/2} - 1)$ in accordance with Games' representation [4]. When the sequence c_j is shifted by any u $(= kT + l)$ places, each new column is a cyclic shift of a column of $\boldsymbol{C}_j$ [4]. We use Games' representation also to form an array $\boldsymbol{A}$ from a sequence $\boldsymbol{a}$, of the same size as $\boldsymbol{C}_j$. From [4] the array $\boldsymbol{A}$ contains one column with entries being 1, and the remaining $2^{n/2}$ columns are cyclic shifts of a short m-sequence of length $2^{n/2} - 1$. As shown in [4], for any cyclic shift of sequence $\boldsymbol{a}$, the shifted array agrees with the unshifted array $\boldsymbol{A}$ in exactly one column.

The autocorrelation function (AC) of sequence $c_j = b_j \boldsymbol{a}$ is:

$$AC_j(u) = \sum_{i=0}^{T-1} \sum_{s=0}^{T-3} b_{j+i} b_{j+l+i} a_{i+sT} a_{i+l+(s+k)T}, \tag{1}$$

where $u = kT + l$, $0 \le k < T\text{-}2$, $0 \le l < T$.

Similarly, the cross-correlation function (CC) of sequences $c_{j_1} = b_{j_1} \boldsymbol{a}$ and $c_{j_2} = b_{j_2} \boldsymbol{a}$ can be expressed by

$$CC_{j_1 j_2}(u) = \sum_{i=0}^{T-1} \sum_{s=0}^{T-3} b_{j_1+i} b_{j_2+l+i} a_{i+sT} a_{i+l+(s+k)T}. \tag{2}$$

Theorem 1. *Autocorrelation of the new binary sequences has 4 levels, with the following values:*
$$-2^{n/2} + 3; \quad -1; \quad 2^{n/2} + 3; \quad 2^n - 1.$$

Proof. Consider two cases: **(1)** $l = 0 \bmod T$ and **(2)** $l \ne 0 \bmod T$.
Case 1. If $k = 0$ then $AC(0) = 2^n - 1$. If $k \ne 0$, only the single column of all ones or all minus ones is matched and contributes $T - 2$ to the AC. All other columns are mismatched and hence $AC(u) = T - 2 - (T - 1) = -1$ for all non-zero $u = kT$.

Case 2. In this case, only one m-sequence column can be matched in phase, but we have same or opposite polarity, depending on the sign of $b_{j+i}b_{j+i+l}$. The total number of matched and mismatched pairs b_{j+i} and b_{j+i+l} is respectively $(T-3)/2$ and $(T+3)/2$, since b is a Hall or Legendre sequence over $\{-1, +1\}$. Denote by i' the column of C_j matching column $(i'+l)$ of the shifted C_j.

Let $b_{j+i'}b_{j+i'+l} = 1$. The contribution to the autocorrelation number by the matching pair of columns is $2^{n/2} - 1$. The term $b_{j+i}b_{j+i+l}$ is 1 for $(T-3)/2 - 1$ columns other than column i', and is -1 for $(T+3)/2$ columns. Thus

$$AC(u) = (2^{n/2} - 1) - ((T-3)/2 - 1) + (T+3)/2 = 2^{n/2} + 3.$$

Let $b_{j+i'}b_{j+i'+l} = -1$. Now the matching pair of columns contributes $-(2^{n/2}-1)$. The term $b_{j+i}b_{j+i+l}$ is -1 for $(T+3)/2-1$ values of $i \neq i'$, and is $+1$ for $(T-3)/2$ values of i. Thus

$$AC(u) = -(2^{n/2} - 1) + ((T+3)/2 - 1) - (T-3)/2 = -2^{n/2} + 3.$$

$\square$

A calculation of CC similar to that above for autocorrelation yields the following four values of CC: $-3(2^{n/2} - 1)$; $-2^{n/2} + 3$; -1; $2^{n/2} + 3$. The analysis shows that the CC peak of two sequences from the same set appears at zero shift: this property can be used in Quasi-Synchronous CDMA systems, where zero phase shifts between sequences can be avoided.

2.2 Ternary Sequence Correlation

The autocorrelation (AC) of sequence $t_j = d_j a$ is:

$$AC_j(u) = \sum_{i=0}^{T-1}\sum_{s=0}^{T-3} d_{j+i}d_{j+l+i}a_{i+sT}a_{i+l+(s+k)T}, \tag{3}$$

where $u = kT+l$, $0 \leq k < T\text{-}2$, $0 \leq l < T$.

Similarly, the cross-correlation (CC) of sequences $t_{j_1} = d_{j_1}a$ and $t_{j_2} = d_{j_2}a$ can be expressed by

$$CC_{j_1 j_2}(u) = \sum_{i=0}^{T-1}\sum_{s=0}^{T-3} d_{j_1+i}d_{j_2+l+i}a_{i+sT}a_{i+l+(s+k)T}. \tag{4}$$

Theorem 2. *Autocorrelation of the new ternary sequences has 4 or 5 levels, with the following values:*

1. *The acf of a sequence having only one constant column with all null elements is 4-level, with values:* $-2^{n/2}$; $-(2^{n/2} - 1)$; $2^{n/2} + 1$; $2^n - 2^{n/2}$
2. *The acf of the other $2^{n/2}$ sequences having two constant columns (the first column with all null elements and the second column with all non-null elements) is 5-level, with values:* $-(2^{n/2} - 1)$; 0; 1; $2^{n/2} + 1$; $2^n - 2^{n/2}$.

Proof. Fold the sequence t_j row-by-row into a two-dimensional array $\boldsymbol{T}_j$ of size $(2^{n/2} + 1)(2^{n/2} - 1)$ in accordance with Games' representation [4]. When the sequence t_j is shifted by any u $(= kT + l)$ places, each new column is a cyclic shift of a column of $\boldsymbol{T}_j$ [4].

We use Games' representation [4] also to form an array $\boldsymbol{A}$, say, from the m-sequence $\boldsymbol{a}$, of the same size as $\boldsymbol{T}_j$, again by placing the consecutive entries of $\boldsymbol{a}$ along row 0, then along row 1, $\dots$, and finally along row $T - 3 = 2^{n/2} - 2$.

From Games [4], the array $\boldsymbol{A}$ contains one column with entries being 1, and the remaining $2^{n/2}$ columns are cyclic shifts of a short m-sequence of length $2^{n/2} - 1$. Games [4] showed that, for any cyclic shift of sequence $\boldsymbol{a}$, the shifted array agrees with the unshifted array $\boldsymbol{A}$ only in exactly one column.

Consider the sequence t_j in "square" array form. The columns of the array are short m-sequences or constants. The autocorrelation of the new sequence is the same as that of the array, and can be obtained by summing the autocorrelations of the columns.

The sequences t_j can be divided into two classes:

Case 1. The null of the multiplying Legendre sequence falls upon the constant column of the m-array. The resultant new sequence has only one constant column, consisting of all zeroes.

Case 2. The null of the multiplying Legendre sequence does not fall on the constant column of the m-array. The array versions of these sequences have two constant columns: one being all zeroes, and the other being all +1 or −1, depending on the value of the Legendre sequence for that column.

The long sequence t_j is shifted snake-like through the array. Shifts which are integer multiples of the row width appear to shift the long sequence vertically. For zero shift, the autocorrelation is just the number of non-zero entries in the array i.e. $2^n - 1 - (2^{n/2} - 1) = 2^n - 2^{n/2}$.

In **Case 1.** for zero vertical shift, and any non-zero horizontal shift, the total number of non-null overlying column pairs is $T - 1$. None of these are phase-matched. Therefore each column correlation contributes -1. Hence $AC(u)=-(T-1)= -2^{n/2}$. For nonzero vertical shifts and any horizontal shifts we have two cases:

1. phase-matched columns with coefficient pair $b_{j+i'} b_{j+i'+l} = -1$. Then among the remaining columns, we have $(T - 5)/2$ matched coefficient pairs and $(T - 5)/2$ mismatched coefficient pairs and $AC(u) = -(2^{n/2} - 1)$. (Two columns contribute null correlations.)
2. phase-matched columns with coefficient pair $b_{j+i'} b_{j+i'+l} = 1$. Then we have $(T - 3)/2$ mismatched coefficient pairs and $(T - 5)/2 - 1$ matched coefficient pairs and thus:

$$AC(u) = T - 2 + (T - 3)/2 - ((T - 5)/2 - 1) = T = 2^{n/2} + 1.$$

In **Case 2.** we get the following off-peak autocorrelation values: $-(2^{n/2} - 1)$; 0; 1; $2^{n/2} + 1$; $2^n - 2^{n/2}$. □

Then, a similar argument shows that the cross-correlation of a sequence having only one constant column (with all null elements) and all other sequences is

4-level with values: $-(2^{n/2}-1)$; 0; 1; $2^{n/2}+1$. The cross-correlation of sequences having two constant columns is 5-level, with values: $-2^{n/2}$; $-(2^{n/2}$; -1; 0; 1; $2^{n/2}+1$.

Analysis shows that cross-correlation of the reference long m-sequence and any of the new sequences has 3 values: $-2^{n/2}; 0; 2^{n/2}$. As a result, the sequence set $\{t_j\}$ has 4-level auto and 5-level cross-correlation, with an upper bound of $2^{n/2}+1$.

3   Linear Complexity

The linear complexity of a sequence is the length of the shortest linear feedback shift register that generates this sequence. First, consider the construction based on a long m-sequence. From $q = (2^{n/2}+1)/3$ being prime, and $n/2$ being odd, it follows that $q = 8r+3$. As shown in [5,6] the linear complexity of Legendre and Hall sequences of length $q = 8r+3$ is q and they have the same feedback polynomial $g(x) = x^q - 1$. The feedback polynomial $f(x)$ of any m-sequence is a primitive polynomial of degree n and all its roots are primitive elements of $GF(2^n)$. Hence $gcd(f(x),g(x))=1$. However, from [7], the feedback polynomial of sequence $c_j = a+b_j$ is $f(x)g(x)$. Consequently, the linear complexity of the sequence c_j is $L(c_j) = L(a)+L(b)$, where $L(a)$ and $L(b)$ are the respective linear complexities of sequences a and b respectively.

For the construction based on a long GMW sequence, any GMW sequence can be expressed as a sum of decimated m-sequences of the same length. So, as in the previous case, the roots of the feedback polynomial of GMW sequence are primitive elements of $GF(2^n)$. Thus, in this case, $L(c_j) = L(a)+L(b)$ as well.

Example 3. Let n $= 14$. Consider the two following cases.

1. Let a be an m-sequence of length $2^{14} - 1$ and let b be a Legendre/Hall sequence of length 43 with $L(a) = 14$ and $L(b) = 43$. Then $L(c_j) = 14+43 = 57$.
2. Let a be a GMW sequence of length 16383 with $L(a)= 1232$. Then $L(c_j) = 1232 + 43 = 1275$.

4   Number of Distinct Sequence Sets

The number of sequences within a set $\{t_j, a\}$ *is* $M = T+1 = 2^{n/2}+2$. The total number of long sequences is: $W = (|GMW_n|+1)\frac{\varphi(2^n-1)}{n}$. Here, $|GMW_n|$ is the number of different classes of GMW sequences [8], based on the number of distinct column substitutions in the array decomposition of the parent m-sequence of length $2^n - 1$ [8]; $\varphi(2^n-1)/n$ is the number of distinct m-sequences of length $2^n - 1$, resulting in distinct arrays and φ is the Euler totient function. Let U be the number of distinct short sequences of length $l = (2^n+1)/3$. U takes on the values 1 (for $n = 3$), 2 and 8. The total number of distinct new sequence sets is $P_S = W \times U$. So for $n = 14$, with $\varphi(2^{14}-1)/14 = 756$, $U = 8$ (two

Table 1. Table 1 Parameters of Binary Sequences

n	M	U	Θ_A	Θ_C
6	4	1	11	21
10	12	2	35	93
14	44	8	131	381
22	684	2	2051	8141
26	2732	8	8195	24573
34	43692	2	131075	393213
38	174764	2	524291	1572861
46	2796204	2	8388611	25165821

Table 2. Parameters of Ternary Sequences

n	M	$|GMW|$	P_S	Θ_A	Θ_C
4	6	0	4	5	5
8	18	1	64	17	17
16	258	63	262144	257	257
32	65538	135167	18141941858304	65537	65537

Legendre and six Hall sequences) and $|GMW_n| = 79$ [8], we have $P_S = 483840$. The number of distinct short ternary Legendre sequences of length $2^{n/2} + 1$ is 2. Hence, the total number of distinct ternary sequence sets is $P_S = 2W$. So for $n = 8, \varphi(2^n - 1)/n = 16$ and $|GMW_n| = 1$ [8], so we have $P_S = 64$. The parameters of new ternary sequence sets for all currently available n are shown in Table 2. P_S values were calculated using results in [8,9,10]. Θ_A and Θ_C are maximum off-peak autocorrelation and maximum cross-correlations respectively.

5   Our Sequences in Array Format

Our construction relies on the existence of arrays with a special property. Such arrays are composed of columns which are cyclic shifts of a shorter sequence, or a constant column. Therefore, they are described by the column sequence and the sequence of cyclic column shifts, henceforth called the shift sequence. A constant column is represented in the shift sequence by a blank. For an array of T columns of length v, the shift sequence for columns $j = 0, ..., T - 1$ is denoted as f_j, which is evaluated modulo v. Desirable correlation values ensue when the shift sequence has the distinct difference property. Consider differences in the shift sequence for separation k, i.e. $f_{j+k} - f_j$. If these differences are all distinct modulo v, then the array overlaid on a shift of itself (for $k \neq 0 \bmod T$) has zero or one column matching in its cyclic shift. For $k = 0 \bmod T$ the constant column matches (when it exists) or no columns match. For pseudonoise columns, the off-peak autocorrelation of the arrays is $-T$ or $v + 1 - T$. We exploit this property to construct new arrays, with good auto and cross-correlation. This is done by multiplying the columns of the array by any set of orthogonal or near orthogonal sequences, such as the shifts of a binary or ternary pseudonoise sequence of length T. This does not alter the number of shift matched columns, merely influences their polarity and hence the polarity of the off-peak autocorrelation and cross-correlation. The resulting arrays can be unfolded to yield sequence sets.

6   Sequence Folding/Unfolding

6.1   Row-by-Row Folding

The row by row folding/unfolding requires array shifts to be constrained to shifts of this type, i.e. snake-line shifts modulo vT. Array dimensions are not restricted.

Example 4. Consider $p = 7, n = 2$. An m-sequence of length 48 can be written as an 6×8 array, where the columns are m-sequences of length 6. The shift sequence is: $f_j = 2, 3, 2, 0, \infty, 4, 1, 3$. γ is a primitive element of $GF(7)$.

$$
\begin{array}{cccccccc}
\gamma^5 & \gamma^4 & \gamma^5 & \gamma^3 & 0 & \gamma^6 & \gamma^1 & \gamma^4 \\
\gamma^1 & \gamma^5 & \gamma^1 & \gamma^2 & 0 & \gamma^4 & \gamma^3 & \gamma^5 \\
\gamma^3 & \gamma^1 & \gamma^3 & \gamma^6 & 0 & \gamma^5 & \gamma^2 & \gamma^1 \\
\gamma^2 & \gamma^3 & \gamma^2 & \gamma^4 & 0 & \gamma^1 & \gamma^6 & \gamma^3 \\
\gamma^6 & \gamma^2 & \gamma^6 & \gamma^5 & 0 & \gamma^3 & \gamma^4 & \gamma^2 \\
\gamma^4 & \gamma^6 & \gamma^4 & \gamma^1 & 0 & \gamma^2 & \gamma^5 & \gamma^6
\end{array}
$$

6×8 **Row-by-Row Array**

6.2   Diagonal Folding

For a $T \times v$ array $A = (a_{ij}$ with $gcd(T, v) = 1$, there is a diagonal containing each entry exactly once, defined as the sequence $a_{0,0}, a_{1,1}a_{2,2}..., a_{i,i}, ..., a_{T-1,v-1}$, where the first index i of $a_{i,i}$ is calculated modulo T, the second modulo v. More general diagonals are of the form $a_{0,0}, a_{k,l}a_{2k,2l}...$, where $gcd(k, T = gcd(l, v) = 1$ [14] Diagonal folding/unfolding allows the array to be shifted in a standard, two-dimensional manner (modulo v and modulo T respectively). This kind of folding/unfolding is restricted to arrays with co-prime dimensions $gcd(v, T) = 1$, where single pass diagonals through the matrix exist. The number of sequences equals the number of diagonals. An array with $gcd(v, T) > 1$ can be unfolded row by row, but then, the number of shift matched columns may be doubled [11], so this method is not used.

Example 5. Consider a long binary m-sequence of length 63 i.e. $(2^6 - 1)$, written row-by-row as an array of 7 rows ($v = 2^3 - 1 = 7$) each of length 9 (i.e. $T = \frac{2^6-1}{2^3-1} = 9$). This array is shown below:

```
0 1 1 1 1 1 1 0 1
0 1 1 1 0 0 0 1 1
0 0 1 1 1 0 1 1 0
0 0 0 0 1 1 1 1 0
0 1 0 0 1 0 1 0 1
0 0 1 1 0 1 0 0 0
0 1 0 0 0 1 0 1 1
```

Binary 7×9 Row-by-Row Array

The shift sequence is: $f_j = \infty, 4, 5, 5, 0, 3, 0, 6, 4$. Unlike in [12], here, we do not use the extended shift sequence, but just the non-repeating part. The reference shift for the column sequence is (arbitrarily) chosen as : 1,0,1,1,1,0,0. This is just the matrix representation of the sequences introduced in Subsection 1.1 in Example 1. Since $gcd(9,7) = 1$, it is also possible to write the long sequence down a (1:1) diagonal. This is shown below.

```
0 0 0 1 0 0 1 0 0
0 0 1 0 1 1 0 1 0
0 1 0 0 1 1 0 0 1
0 0 1 1 1 1 1 1 0
0 1 1 0 0 0 0 1 1
0 1 1 1 0 0 1 1 1
0 1 0 1 1 1 1 0 1
```

Binary 7×9 "Diagonal" Array

The (symmetric) shift sequence is: $f_j = \infty, 2, 1, 3, 6, 6, 3, 1, 2$. The relationship between the two shift sequences is analyzed in [12]. One long m-sequence can generate two different arrays. Arrays unfolded by an incompatible method result in long sequence correlation deterioration. This should be avoided. The columns of the above array can be multiplied by 3 periods of binary m-sequence of length 3: 0,1,1. There are 3 cyclic shifts of that sequence, so there are 3 such multiplied arrays, as shown below. Here, we map (0,1) to (+1,-1).

b_j	0	1	1	0	1	1	0	1	1

b_j	1	1	0	1	1	0	1	1	0

b_j	1	0	1	1	0	1	1	0	1

```
0 1 1 1 1 1 1 1 1      1 1 0 0 1 0 0 1 0      1 0 1 0 0 1 0 0 1
0 1 0 0 0 0 0 0 1      1 1 1 1 0 1 1 0 0      1 0 0 1 1 0 1 1 1
0 0 1 0 0 0 0 1 0      1 0 0 1 0 1 1 1 1      1 1 1 1 1 0 1 0 0
0 1 0 1 0 0 1 0 1      1 1 1 0 0 1 0 0 0      1 0 0 0 1 0 0 1 1
0 0 0 0 1 1 0 0 0      1 0 1 1 1 0 1 0 1      1 1 0 1 0 1 1 1 0
0 0 0 1 1 1 1 0 0      1 0 1 0 1 0 0 0 1      1 1 0 0 0 1 0 1 0
0 0 1 1 0 0 1 1 0      1 0 0 0 0 1 0 1 1      1 1 1 0 1 0 0 0 0
```

3 Arrays (7×9) with their Multiplication Sequences b_j

The 4 arrays can be unfolded (diagonally) to produce 4 sequences of length 63, with good auto and cross-correlation.

7 Shift Sequences with DDP

Shift sequences with distinct difference property (DDP) are known to arise from the following arrays:

7.1 An m/GMW-Sequence of Length $p^{2m} - 1$

This is embedded row-by-row into an array of $p^m + 1$ columns each of length $p^m - 1$. The leading column is constant (all 1's) whilst the remainder are all shifts of a short pseudonoise sequence. The case where $p = 2$ is analyzed in Part I of this paper. For $p > 2$, the columns in the parent array are non-binary. However, sometimes their length is commensurate with binary sequences, so a column substitution can be performed, with no effect on the correlation values. This occurs for $p = 5$, where the columns are of length 4 and therefore can be substituted by the only perfect binary sequence in existence. The columns of the m-array can be lengthened, as described later.

Example 6. An m-sequence of length 24 can be written as a 4×6 array ($p = 5, n = 2$), where the shift sequence is : $121 - 02$. This is shown in the parent array, shown below (left). The columns can be substituted by cyclic shifts of the perfect binary sequence $0, 1, 1, 1$ as shown in the array on the right. The columns of this (right) array can be multiplied by two repeats of the binary m-sequence $0, 1, 1$. There are 2 other shifts of such a multiplication sequence, so that this method yields 4 binary arrays with good auto and cross-correlation. The arrays can be unfolded row-by row (consistent with array folding!) to produce 4 binary sequences of length 24, with good auto and cross-correlation. γ is a primitive element of $GF(5)$.

$$
\begin{array}{cccccc}
\gamma^1 & \gamma^3 & \gamma^1 & 1 & \gamma^2 & \gamma^3 \\
\gamma^2 & \gamma^1 & \gamma^2 & 1 & \gamma^4 & \gamma^1 \\
\gamma^4 & \gamma^2 & \gamma^4 & 1 & \gamma^3 & \gamma^2 \\
\gamma^3 & \gamma^4 & \gamma^3 & 1 & \gamma^1 & \gamma^4
\end{array}
\qquad
\begin{array}{cccccc}
1 & 1 & 1 & 0 & 0 & 1 \\
0 & 1 & 0 & 0 & 1 & 1 \\
1 & 0 & 1 & 0 & 1 & 0 \\
1 & 1 & 1 & 0 & 1 & 1
\end{array}
$$

(4×6) Parent Array (left) Binary Column Substitution Array (right)

7.2 A Quadratic Shift Array of p (an Odd Prime) Columns of Length p

Here $f_j = aj^2$ where a is any non-zero integer in Z_p. The number of matching columns between an array and a two-dimensional matrix shift of itself (k, l) is determined by the solutions of:

$$ f_{j+k} - f_j \equiv l \quad or \quad (j + k)^2 - j^2 \equiv l. \tag{5} $$

$$ \text{Hence} \quad 2kj + k^2 - l \equiv 0. \tag{6} $$

This equation has a single solution for all $k \neq 0$. For $k = l = 0$ all columns match, whilst for $k = 0, l \neq 0$, no columns match. Hence the shift sequence has DDP. If the columns are pseudonoise, with peak correlation p and off-peak of -1, the autocorrelation values are p^2 for full match, $-p$ for purely vertical shifts and $p-(p-1) = +1$ for all other shifts. Such arrays are studied in detail in [13]. This array can be unfolded row by row, but this results in the undesirable potential doubling of the number of shift matched columns. Later, we show, how to modify these arrays, by lengthening their columns, without sacrificing DDP.

Example 7. Consider $p = 7, a = 1$ and an m-sequence column of length 7. The shift sequence is $f_j = 0, 1, 4, 2, 2, 4, 1$ and the matrix looks as below:

$$
\begin{array}{ccccccc}
1 & 0 & 1 & 0 & 0 & 1 & 0 \\
0 & 1 & 1 & 0 & 0 & 1 & 1 \\
1 & 0 & 0 & 1 & 1 & 0 & 0 \\
1 & 1 & 0 & 0 & 0 & 0 & 1 \\
1 & 1 & 1 & 1 & 1 & 1 & 1 \\
0 & 1 & 0 & 1 & 1 & 0 & 1 \\
0 & 0 & 1 & 1 & 1 & 1 & 0 \\
\end{array}
$$

(7×7) **Quadratic Shift Array**

7.3   An Exponential Shift Array of $p - 1$ Columns

Consider columns of length p, with p an odd prime: $f_j = g^j$, where g is a primitive root of Z_p. Since $gcd(p - 1, p) = 1$, such arrays possess a single pass diagonal and can therefore be unfolded into sequences without penalty. Actually, this is the optimum array for our construction, since it involves the shortest columns commensurate with DDP, with the array being suitable for direct unfolding. Additionally, there are many suitable binary pseudonoise columns available for column substitution. It is unnecessary to lengthen the columns in this case.

Example 8. Consider $p = 7$ and a primitive root $g = 3$. An exponential shift sequence is $f_j = 3, 2, 6, 4, 5, 1$. Using the same column sequence as above, the matrix looks as shown below.

$$
\begin{array}{cccccc}
1 & 0 & 0 & 1 & 1 & 0 \\
0 & 0 & 1 & 1 & 1 & 1 \\
0 & 1 & 1 & 0 & 1 & 0 \\
1 & 0 & 1 & 0 & 0 & 1 \\
0 & 1 & 0 & 1 & 0 & 1 \\
1 & 1 & 0 & 0 & 1 & 1 \\
1 & 1 & 1 & 1 & 0 & 0 \\
\end{array}
$$

(6×7) **Exponential Shift Array**

There are also two types of logarithmic shift sequence with DDP (index function and Zech logarithm). Both result in a column of zeros, so they can only be used to construct ternary arrays.

8   Matrix Modifications Which Preserve DDP

The three types of arrays with DDP can be augmented by modifications of the above, which preserve DDP. We have examined the following modifications:

8.1   Column Deletion (Matrix Puncture)

A shift sequence with DDP has an upper bound on the number of shift matched columns of 1. For a single column deletion, each pair of shift matched columns corresponds to a solution for j of:

$$f_{j+k} - f_j \equiv l, \text{where } f \text{ is evaluated mod } v, \text{ whilst } j \text{ is evaluated mod } T \quad (7)$$

Puncturing the matrix reduces the number of columns from T to $T-1$. Therefore, there are two possibilities for matching columns:

$$f_{j+k} - f_j \equiv l, j + k < T \text{ (no wraparound)} \quad (8)$$

$$f_{j+k+T-1} - f_j \equiv l, j + k > T \text{ (with wraparound)} \quad (9)$$

Therefore, an upper bound on the number of shift matched columns rises to 2. Any further deletions cause further deterioration.

8.2   Column Insertion

Column insertions double the upper bound due to the change of T, as in puncturing, and may include additional matches due to the inserted columns.

8.3   Column Shortening

Consider a matrix whose columns are shortened from v to $u > v/2$. There are four possible cases of columns matching, when the equation is reduced from modulo v to modulo u.

$$f_{j+k} - f_j \equiv l - u \quad \text{for} \quad v > l > u, \quad f_{j+k} - f_j \equiv l \quad \text{for} \quad u > l > 0. \quad (10)$$

$$f_{j+k} - f_j \equiv l + u \quad \text{for} \quad 0 > l > -u, \quad f_{j+k} - f_j \equiv l + 2u \quad \text{for} -v > l > -u. \quad (11)$$

Therefore, an upper bound on the number of shift matched columns rises to 4. Shortening to less than $v/2$ makes matters even worse.

8.4   Column Lengthening

Lengthening columns from v to s results in:

$$f_{j+k} - f_j \equiv l \quad \text{for} \quad l > 0 \quad or \quad f_{j+k} - f_j \equiv l + s \quad \text{for} \quad 0 > l. \quad (12)$$

Clearly, if $s > 2v$, the smallest value of $l + s$ is greater than v, whilst the largest value of l is less than v and thus the two sets of solutions are disjoint and hence DDP is preserved. Unfortunately, doubling the column length doubles its contribution to the correlation and therefore, this result is no better than column deletion. However, it is possible for DDP to be preserved for some specific values of $v < s < 2v$. Clearly, it is desirable to have s as close to v as possible. In addition, if the shift sequence is derived from the quadratic or exponential parent matrix, it is required that $gcd(s, T) = 1$, so that a single pass diagonal exists and enables the unfolding of the new matrix into sequences. A method of obtaining low values of s begins with the examination of differences between entries in the shift sequence. The shift sequence has its desirable property when evaluated modulo v. Its entries can be positive or negative numbers modulo v. Now consider what happens if the columns were lengthened to infinity. DDP would remain unaffected. Evaluate the greatest difference between the entries and call it Δ_{max}. Clearly, the largest negative difference is $-\Delta_{max}$. Therefore, if the shift sequence is re-expressed modulo $s = 2\Delta_{max}$, DDP must be preserved. The objective of the construction is to minimize Δ_{max}. The authors have attempted using a systematic approach, such as a greedy algorithm to compute the smallest Δ_{max} for different starting values of v and T for the parent array, and found this approach unreliable. However, for small arrays, the results are easy to deduce by inspection, and this is what we did to obtain the arrays, whose properties are listed in Tables 3 and 4.

Example 9. Consider lengthening the columns of the 7×7 quadratic shift array. The shift sequence mod 7 is $f_j = 0, 1, 4, 2, 2, 4, 1$. $\Delta_{max} = 4$, so this shift sequence has DDP for all lengths greater than $s = 2\Delta_{max} + 1 = 9$. There exists a binary Legendre sequence of length 11, $(1, 1, 0, 1, 1, 1, 0, 0, 0, 1, 0)$, so it can be used to construct the array below (left). Another 7 arrays are produced by multiplying the columns by cyclic shifts of the m-sequence of length $7 : 1, 0, 1, 1, 1, 0, 0$. A typical array with multiplied columns is shown on the right (with multiplication sequence above).

b_j	1	0	1	1	1	0	0

```
1 0 0 1 1 0 0        0 0 1 0 0 0 0
1 1 0 0 0 0 1        0 1 1 1 1 0 1
0 1 1 1 1 1 1        1 1 0 0 0 1 1
1 0 0 1 1 0 0        0 0 1 0 0 0 0
1 1 1 0 0 1 1        0 1 0 1 1 1 1
1 1 1 1 1 1 1        0 1 0 0 0 1 1
0 1 0 1 1 0 1        1 1 1 0 0 0 1
0 0 1 1 1 1 0        1 0 0 0 0 1 0
0 0 1 0 0 1 0        1 0 0 1 1 1 0
1 0 1 0 0 1 0        0 0 0 1 1 1 0
0 1 0 0 0 0 1        1 1 1 1 1 0 1
```

7×7 **Arrays Lengthened to** 11×7**: multiplication sequence** b_j **(left)**

There are 8 such arrays. They can be unfolded diagonally into into 8 sequences of length 77, with auto and cross correlation values of: +77 (full match for autocorrelation), +13 (one column match, with multiplication sequence agreement), +1

(0 column match for non-zero horizontal shifts), -7 (0 column match for purely vertical shifts), -11, (one column match, multiplication sequence disagreement).

9   Array Construction

Tables 3 and 4 describe the smallest arrays constructed by the above methods. a denotes column sequence, b_j denotes multiplication sequence. These sequences are labeled: m - m-sequence, L - Legendre, H - Hall, PB - Perfect Binary. The types of shift sequences are: e - exponential, m - m-array, l - lengthened quadratic. For symmetric shift sequences only half is shown.

Table 3. Shift Sequences with DDP and Array Details

T	v	a	b_j	Shift Sequence
3	4	PB	m	0,1,1 (l)
4	7	m	PB	0,1,3,2 (ad hoc)
6	4	PB	2m(3)	1,2,1,∞,0,2 (m)
6	7	m	2 m(3)	3,2,6,4,5,1 (e)
9	7	m	3 m(3)	∞,1,2,0,4,4,0,2,1 (m)(d)
7	11	L	m	0,1,4,2,2,4,1 (l)
11	15	m	L	0,1,4,13,5,3,3,5,13,4,1 (l)
12	15	m	3PB	0,5,3,2,3,0,∞ ,6,0,0,2,5 (m)
19	31	m/L	L	0,1,4,9,28,6,29,11,7,5 ,5,7,11,29,6,28,9,4,1 (l)
22	23	L	2L(11)	Exponential
23	35	L	T P	0,1,4,9,16,2,13,3,18, 12,8,6,6,8,12,18 ,3,13,2,16,9,4,1 (l)
33	31	m	3L(11)	m-array (TBC)
46	47	L	2 L(23)	Exponential
58	59	L	2L(29)	Exponential
70	71	L	2TP(35)	Exponential
118	119	L	2L (59)	Exponential
127	127	m/L	m/L	Cubic Unfold by rows
129	127	m	3L/H(43)	m-array

10   Large Array

The 127×127 array in Tables 3 and 4 is a special case, where the commensurate folding/unfolding rule is violated. 127 is a prime, so it supports m-sequences, binary Legendre sequences, Hall sequences and Baumert-Fredricksen sequences as column and multiplication sequences. Additionally, polynomial shift sequences

Table 4. Correlation Properties of Unfolded Sequence Sets

Length	N	Auto	Cross
12	4	+4,0,-4	+4,0,-4
28	5	+8,0,-4,-8	+8,0,-4,-8
24	4	+4,0,-4	+4,0,-4,-8
42	4	+10,+2,-6	+10,+2,-6,-14
63	4	+11, -1,-5	+11,-1,-5,-21
77	8	+13,+1,-7,-11	+13,+1,-7,-11
99	4	+15,+3 ,-9	+15,-9,-33
165	12	+17,+1,-11,-15	+17,+1,-11,-15
180	5	+16,+4,0,-12,-16	+16,0,-12,-16
589	20	+33,+1,-19,-31	+33,+1,-19,-31
506	12	+26,+2,-22	+26,+2,-22,-46
805	24	+37,+1,-23,-35	+37,+1,-23,-35
1023	12	+35,-1,-29	+35,-1,-29,-93
2162	24	+50,+2,-46	+50,+2,-46,-94
3244	30	+62,+2,-58	+62,+2,-58,-108
4970	36	+74,+2,-70	+74,+2,-70,-142
14042	60	+122,+2,-118	+122,+2,-118,-238
16129	2064384	<3.1%	<3.1%
16383	44	+131,+1,-125	+131,+1,-125,-381

can be used to construct these arrays. Consider a matrix A generated by a polynomial shift sequence f_j (Sections 2-5), say

$$f_j \equiv a_m j^m + a_{m-1} j^{m-1} + \ldots + a_1 j + a_0. \qquad (13)$$

The array has column j matching column $j + k$ in a (k, l) shift B of itself if:

$$a_m(j + k)^m + a_{m-1}(j + k)^{m-1} + \ldots + l \equiv a_m(j)^m + a_{m-1}(j)^{m-1} + \ldots \qquad (14)$$

Collecting terms and rearranging [15]: $a_m k(j)^{m-1} + \ldots \equiv 0$. There are at most $m - 1$ values of j which can satisfy the above equation, so this is an upper bound on the number of shift matched columns. If the matrix is read out row-by-row, this bound is doubled [11]. For a cubic shift sequence, this is equivalent to 4 columns matching in shift, as a worst case. These columns can contribute at most 4×127 to the correlation, whose peak autocorrelation is 127×127. The remaining columns contribute only -1 or $+1$ depending on the multiplication sequence. Therefore all off-peak autocorrelation and non-zero shift cross-correlation values are constrained to below roughly 0.031 (normalized). For zero shift in cross-correlation, all columns are shift matched, but the multiplication sequences are not, so the cross-correlation is equivalent to 1 column: -127. Therefore, all off-peak correlations are bounded above by about 0.031 (normalized). The total number of matrices can be computed to be $N = (127^2 - 1) \times 128 = 2064384$. There are $127^2 - 1$ shift distinct matrices, 127 shifts of the multiplication sequence and one non-multiplied entry. This sequence set is large! The off-peak autocorrelation/cross-correlation and the number of sequences in the set can be traded off, by choosing appropriate degree polynomials.

11   Conclusions

The "interleaved" binary sequence sets (Gong [7]) can be generalized to a variety of constructions, based on the existence of shift sequences with DDP, of length T, expressed modulo v, where binary pseudonoise sequences of length T and v exist.

Acknowledgments

The authors thank the (anonymous) referees for their helpful suggestions.

References

1. Fan P and Darnell M. Sequence Design for Communications Applications. - Research Studies Press Ltd., London, 1996.
2. Schroeder M.R. 'Number theory in science and communication', 3rd Edition Springer-Verlag 1996.
3. Baumert L.D. : 'Cyclic difference sets', - Berlin, Springer-Verlag, 1971.
4. Games R.A. : 'Crosscorrelation of M-Sequences and GMW Sequences With the Same Primitive Polynomial', - Discrete Applied Mathematics, 1985, 12, 139-146.
5. Ding C., Helleseth T.: 'On the linear complexity of Legendre sequences',IEEE IT, 1998, 44, 1276-1278.
6. Kim J-H and Song H-Y. : 'On the linear complexity of Hall's sextic Residue sequences', IEEE - IT, No.5, 2001, 47, 2094-2096.
7. Gong G. : 'New Designs for Signal Sets with Low Cross-correlation, Balance Property and Large Linear Span: GF(p) Case', IEEE–IT, 11, 2002, 48, pp. 2847-2867.
8. Krengel E.I.:'O chisle psewdoslyhchainih posledowatelnostey Gordona,Milza,Welcha', Tehnika sredstv svyazi, Seriya Tehnika radioswyazi, Wipusk 1979, 3, 17-30 (in Russian)
9. No J.S., Golomb S., Gong G., Lee H.K.,Gaal P.: 'Binary pseudorandom sequences of period 2^n-1 with ideal autocorrelation', IEEE-IT,44, 1998, 814-817.
10. Meshkovskii K.A. and Krengel E.I. : 'About GMW sequence generation', Radiotehnika, No.6, 1998 (in Russian).
11. Scholtz R.A., Kumar P.V. and Corrada Bravo C.J. "Signal Design for Ultra-Wideband Radio," Sequences and Their Applications (SETA '01), May, 2001.
12. Tirkel A.Z. and Hall T.E. New Quasi-Perfect and Perfect Sequences of Roots of Unity and Zero. (SETA'04)
13. Hall T.E., Osborne C.F. and Tirkel A.Z., "Families of matrices with good auto and cross-correlation", Ars Combinatoria, 61(2001), 187-196.
14. Tirkel A.Z., Hall T.E., and Osborne C.F., "A New Class of Spreading Sequences", Fifth International Symposium on Spread Spectrum Techniques and Applications (ISSSTA;98) Sun City, South Africa, September 2-4, 1998, Vol 1, p. 46-50.
15. Tirkel A.Z. and Hall T.E., "A unique watermark for every image", IEEE Multimedia, October to December, 2001, 30-37.

Improved p-ary Codes and Sequence Families from Galois Rings

San Ling[1] and Ferruh Özbudak[2]

[1] Department of Mathematics, National University of Singapore,
2 Science Drive 2, Singapore 117543, Republic of Singapore
`matlings@nus.edu.sg`*
[2] Department of Mathematics and Institute of Applied Mathematics,
Middle East Technical University, İnönü Bulvarı, 06531, Ankara, Turkey
`ozbudak@math.metu.edu.tr`†

Abstract. In this paper, a recent bound on some Weil-type exponential sums over Galois rings is used in the construction of codes and sequences. The bound on these type of exponential sums provides a lower bound for the minimum distance of a family of codes over $\mathbb{F}_p$, mostly nonlinear, of length p^{m+1} and size $p^2 \cdot p^{m\left(D - \lfloor \frac{D}{p^2} \rfloor\right)}$, where $1 \leq D \leq p^{m/2}$. Several families of pairwise cyclically distinct p-ary sequences of period $p(p^m - 1)$ of low correlation are also constructed. They compare favorably with certain known p-ary sequences of period $p^m - 1$. Even in the case $p = 2$, one of these families is slightly larger than the family $Q(D)$ of [H-K, Section 8.8], while they share the same period and the same bound for the maximum non-trivial correlation.

1 Introduction

Bounds on exponential sums over finite fields, such as the Weil-Carlitz-Uchiyama bound, have been found to be useful in applications such as coding theory and sequence designs. The analog of the Weil-Carlitz-Uchiyama bound for Galois rings was presented by [K-H-C]. An improved bound for a related Weil-type exponential sum over Galois rings of characteristic 4, which is also sometimes called the trace of exponential sums, was obtained in [H-K-M-S] and was used in [S-K-H] to construct a family of binary codes with the same length and size as the Delsarte-Goethals codes, but whose minimum distance is significantly bigger. The shortening of these codes also leads to efficient binary sequences.

* The research of this author is partially supported by NUS-ARF research grant R-146-000-029-112 and DSTA research grant POD0411403.
† The research of this author is partially supported by the Turkish Academy of Sciences in the framework of Young Scientists Award Programme (F.Ö./TÜBA-GEBIP/2003-13).

T. Helleseth et al. (Eds.): SETA 2004, LNCS 3486, pp. 236–242, 2005.
© Springer-Verlag Berlin Heidelberg 2005

Recently, an analog of the bound of [H-K-M-S] was obtained for Galois rings of characteristic p^2, for all primes p [L-O]. In this paper, we explore some applications of this bound to the construction of codes and sequences.

We fix the following conventions throughout the paper: p is a prime number; $m \geq 2$ is an integer; $\mathbb{F}_p$ and $\mathbb{F}_{p^m}$ are finite fields of cardinality p and p^m; $\mathrm{GR}(p^2, m)$ is a Galois ring of characteristic p^2 with cardinality p^{2m}; $\mathbb{Z}_{p^2}$ is the ring of integers modulo p^2; $\mathrm{Tr}_m : \mathrm{GR}(p^2, m) \rightarrow \mathbb{Z}_{p^2}$ is the trace map from $\mathrm{GR}(p^2, m)$ onto $\mathbb{Z}_{p^2}$; Γ_m is the Teichmüller set in $\mathrm{GR}(p^2, m)$; β is a primitive $(p^m - 1)$-th root of unity in $\mathrm{GR}(p^2, m)$; $\rho : \mathrm{GR}(p^2, m) \rightarrow \mathrm{GR}(p^2, m)/p\mathrm{GR}(p^2, m)$ $\cong \mathbb{F}_{p^m}$ is reduction modulo p map in $\mathrm{GR}(p^2, m)$. We extend ρ to the polynomial ring mapping $\rho : \mathrm{GR}(p^2, m)[x] \rightarrow \mathbb{F}_{p^m}[x]$ by its action on the coefficients. Let Frob be the Frobenius operator on $\mathrm{GR}(p^2, m)$ (cf. [K-H-C], [L-O]). Frob is extended to $\mathrm{GR}(p^2, m)[x]$ naturally. A polynomial $f(x) \in \mathrm{GR}(p^2, m)[x]$ is called *non-degenerate* if it cannot be written in the form $f(x) = \mathrm{Frob}(g(x)) - g(x) + u \bmod p^2$, where $g(x) \in \mathrm{GR}(p^2, m)[x]$ and $u \in \mathrm{GR}(p^2, m)$.

2 $\mathbb{Z}_{p^2}$-Linear Codes

Definition 1. *For a finite $\mathbb{Z}_{p^2}$-module $S \subseteq \mathrm{GR}(p^2, m)[x]$, let*

$$S_0 = \{a(x) \in \Gamma_m[x] : \text{there exists } b(x) \in \Gamma_m[x] \text{ such that } a(x) + pb(x) \in S\},$$

and

$$S_1 = \{b(x) \in \Gamma_m[x] : \text{there exists } a(x) \in \Gamma_m[x] \text{ such that } a(x) + pb(x) \in S\}.$$

For a prime number p, the weight function w_p on $\mathbb{N}$ is defined as the sum of digits of the representation of $u \in \mathbb{N}$ in base p. For every $f(x) = a(x) + pb(x) \in \mathrm{GR}(p^2, m)[x]$, where $a(x), b(x) \in \Gamma_m[x]$ are uniquely determined, we recall that the *weighted degree* D_f of $f(x)$ is

$$D_f = \max\{p \deg(a(x)), \deg(b(x))\}.$$

For a positive integer D, let $I(D)$ be the set of positive integers

$$I(D) = \{i : i \not\equiv 0 \mod p \text{ and } 0 \leq i \leq D\}$$

and let $S(D) \subseteq \mathrm{GR}(p^2, m)[x]$ be the finite $\mathbb{Z}_{p^2}$-module

$$S(D) = \{f(x) \in \mathrm{GR}(p^2, m)[x] : f(x) = \sum_{i \in I(D)} f_i x^i \text{ and } D_f \leq D\}.$$

Let $f(x) = a(x) + pb(x)$ be a non-degenerate polynomial with $a(x), b(x) \in \Gamma_m[x]$. We recall some definitions which depend on $f(x)$. Let $I_f, J_f \subseteq \mathbb{N}$ be subsets defined as

$$a(x) = \sum_{i \in I_f} a_i x^i \text{ and } b(x) = \sum_{j \in J_f} b_j x^j, \text{ where } a_i, b_j \in \Gamma_m \setminus \{0\}.$$

We define nonnegative integers W_f, $l_{f,m}$ and $h_{f,m}$ as

$$W_f = \max\left\{ p \ \max\{w_p(i) \mid i \in I_f\}, \quad \max\{w_p(j) \mid j \in J_f\} \right\},$$

$$l_{f,m} = \left\lceil \frac{m}{W_f} \right\rceil - 1 \text{ and } h_{f,m} = \left\lfloor \frac{m}{W_f} \right\rfloor.$$

The following result is proved in [L-O].

Theorem 1. *For a non-degenerate polynomial* $f(x) \in \mathrm{GR}(p^2, m)[x]$, *we have*

$$\left| \sum_{a \in \mathbb{Z}_{p^2} \setminus p\mathbb{Z}_{p^2}} \sum_{x \in \Gamma_m} e^{2\pi i \frac{\mathrm{Tr}_m(af(x))}{p^2}} \right| \le p^{l_{f,m}+1} \left\lfloor \frac{p^{h_{f,m}} \frac{p^2-p}{2}(D_f - 1) \left\lfloor 2p^{\frac{m}{2}-h_{f,m}} \right\rfloor}{p^{l_{f,m}+1}} \right\rfloor.$$

Definition 2. *For* $1 \le D \le p^{m/2}$, *let*

$$W_D = \max\left\{ W_f : f(x) \in S(D) \setminus \{0\} \right\}, \quad l_{D,m} = \left\lceil \frac{m}{W_D} \right\rceil - 1$$

and

$$h_{D,m} = \left\lfloor \frac{m}{W_D} \right\rfloor.$$

For $n \ge 1$, the Gray map (cf. [C], [G-S], [L-B], [L-S]) Φ over $\mathbb{Z}_{p^2}^n$ is defined as follows: For $u \in \mathbb{Z}_{p^2}$ let $u = r_0(u) + pr_1(u)$ with $r_0(u), r_1(u) \in \{0, 1, \dots, p-1\}$. We denote the addition modulo p as $\oplus$. For $(u_0, u_1, \dots, u_{n-1}) \in \mathbb{Z}_{p^2}^n$, we have $\Phi(u_0, u_1, \dots, u_{n-1}) = (a_0, a_1, \dots, a_{pn-1}) \in \mathbb{F}_p^{pn}$ such that for $0 \le j \le p-1$ and $0 \le t \le n-1$, $a_{jn+t} = r_1(u_t) \oplus jr_0(u_t)$.

Definition 3. *For* $1 \le D \le p^{m/2}$, *let* $C(D)$ *be the* $\mathbb{Z}_{p^2}$-*linear code of length* p^m *defined as* $C(D) = \left\{ \left(\mathrm{Tr}_m(f(0)) + u, \mathrm{Tr}_m(f(\beta)) + u, \dots, \mathrm{Tr}_m(f(\beta^{p^m-1})) + u \right) : f(x) \in S(D) \text{ and } u \in \mathbb{Z}_{p^2} \right\}.$

Theorem 2. *For* $1 \le D \le p^{m/2}$, $\Phi(C(D))$ *is a* p-*ary code of length* p^{m+1} *of minimum distance*

$$d_{\min} \ge p^{m+1} - p^m - p^{l_{D,m}} \left\lfloor \frac{p^{h_{D,m}} \frac{p^2-p}{2}(D-1) \left\lfloor 2p^{\frac{m}{2}-h_{D,m}} \right\rfloor}{p^{l_{D,m}+1}} \right\rfloor \qquad (1)$$

and of size $|\Phi(C(D))| = p^2 \cdot p^{m\left(D - \lfloor \frac{D}{p^2} \rfloor\right)}.$

Next we consider the nonlinearity of $\Phi(C(D))$. Let T denote the set of ordered pairs $(a, b) \in \mathbb{F}_p^2$ such that $a + b \geq p$ (we identify $\mathbb{F}_p$ with $\{0, 1, \ldots, p-1\}$). Let χ denote the characteristic function of T, i.e.,

$$\chi(a, b) = \begin{cases} 1 \text{ if } (a, b) \in T, \\ 0 \text{ otherwise.} \end{cases}$$

For $\mathbf{a} = (a_1, \ldots, a_n) \in \mathbb{F}_p^n$ and $\mathbf{b} = (b_1, \ldots, b_n) \in \mathbb{F}_p^n$, we define

$$\chi(\mathbf{a}, \mathbf{b}) = (\chi(a_1, b_1), \ldots, \chi(a_n, b_n)) \in \mathbb{F}_p^n.$$

Recall that for $\boldsymbol{\alpha} = (\alpha_1, \ldots, \alpha_n) \in \mathbb{Z}_{p^2}^n$, we denote $r_0(\boldsymbol{\alpha}) = (\rho(\alpha_1), \ldots, \rho(\alpha_n)) \in \mathbb{F}_p^n$. The following lemma is found in [L-B, Theorem 4.6].

Lemma 1. *If C is a $\mathbb{Z}_{p^2}$-linear code of length n, then $\Phi(C)$ is a linear code over $\mathbb{F}_p$ if and only if, for all $\boldsymbol{\alpha}, \boldsymbol{\beta} \in C$, we have $p\chi(r_0(\boldsymbol{\alpha}), r_0(\boldsymbol{\beta})) \in C$.*

Using Lemma 1, we determine whether $\Phi(C(D))$ is linear or nonlinear in some cases.

Theorem 3. *For $1 \leq D \leq p - 1$, the code $\Phi(C(D))$ is linear. If $p \geq 3$ and $p \leq D \leq p^{m/2}/2$, then $\Phi(C(D))$ is nonlinear.*

Proof. First we prove that $\Phi(C(D))$ is linear for $D \leq p - 1$. For $\boldsymbol{\alpha}, \boldsymbol{\beta} \in C(D)$, there exist $f_1(x), f_2(x) \in S(D)$ and $u_1, u_2 \in \mathbb{Z}_{p^2}$ such that

$$\boldsymbol{\alpha} = \left(\mathrm{Tr}_m(f_1(0)) + u_1, \mathrm{Tr}_m(f_1(\beta)) + u_1, \ldots, \mathrm{Tr}_m(f_1(\beta^{p^m-1})) + u_1 \right), \text{ and}$$
$$\boldsymbol{\beta} = \left(\mathrm{Tr}_m(f_2(0)) + u_2, \mathrm{Tr}_m(f_2(\beta)) + u_2, \ldots, \mathrm{Tr}_m(f_2(\beta^{p^m-1})) + u_2 \right).$$

As $D \leq p - 1$, we have $f_1(x), f_2(x) \in pS(D)_1$. Hence

$$r_0(\boldsymbol{\alpha}) = (\rho(u_1), \ldots, \rho(u_1)), \quad r_0(\boldsymbol{\beta}) = (\rho(u_2), \ldots, \rho(u_2)) \text{ and}$$

$$p\chi(r_0(\boldsymbol{\alpha}), r_0(\boldsymbol{\beta})) = \begin{cases} (p, \ldots, p) \text{ if } \rho(u_1) + \rho(u_2) \geq p, \\ (0, \ldots, 0) \text{ if } \rho(u_1) + \rho(u_2) < p. \end{cases}$$

Since $(p, \ldots, p), (0, \ldots, 0) \in \mathbb{Z}_{p^2}^{p^m}$ are elements of $C(D)$, the proof for the case $D \leq p - 1$ is completed.

Next we consider the case $p \geq 3$ and $p \leq D \leq p^{m/2}/2 + 1$. The polynomial $f(x) = x$ belongs to $S(D)$ and hence

$$\boldsymbol{\alpha} = (\mathrm{Tr}_m(0), \mathrm{Tr}_m(\beta), \ldots, \mathrm{Tr}_m(\beta^{p^m-1})) \in C(D).$$

Clearly,

$$r_0(\boldsymbol{\alpha}) = (\mathrm{tr}_m(0), \mathrm{tr}_m(\omega), \ldots, \mathrm{tr}_m(\omega^{p^m-1})).$$

For each $a \in \mathbb{F}_p$, $\chi(a, a) = 1$ if and only if $a \geq \frac{p+1}{2}$. By the properties of the trace map tr_m, it follows that every element $a \in \mathbb{F}_p$ appears in exactly p^{m-1}

coordinates of $r_0(\alpha)$. Hence $\chi(r_0(\alpha), r_0(\alpha))$ has 1 at exactly $p^{m-1}(p-1)/2$ coordinates, and the remaining positions are 0. By (1), the minimum distance d_{min} of $\Phi(C(D))$ satisfies

$$d_{min} \geq p^{m+1} - p^m - (p-1)(D-1)p^{m/2}.$$

The distance between $\Phi(p\chi(r_0(\alpha), r_0(\alpha)))$ and the zero codeword is $p^m(p-1)/2$. For $D < p^{m/2}/2 + 1$, it is easy to see that

$$p^{m+1} - p^m - (D-1)p^{m/2} > p^m(p-1)/2.$$

Therefore $p\chi(r_0(\alpha), r_0(\alpha)) \notin C(D)$, which completes the proof.

3 p-ary Sequences with Low Correlation

For a p-ary sequence $\{s(i)\}_{i=0}^{\infty}$ and $\tau \geq 0$, the *cyclic shift* of $\{s(i)\}_{i=0}^{\infty}$ by τ is the p-ary sequence $\{s(i+\tau)\}_{i=0}^{\infty}$. Two p-ary sequences $\{s_1(i)\}_{i=0}^{\infty}$ and $\{s_2(i)\}_{i=0}^{\infty}$ are *cyclically distinct* if for each $\tau \geq 1$ neither is $\{s_1(i)\}_{i=0}^{\infty}$ the cyclic shift of $\{s_2(i)\}_{i=0}^{\infty}$ by τ nor is $\{s_2(i)\}_{i=0}^{\infty}$ the cyclic shift of $\{s_1(i)\}_{i=0}^{\infty}$ by τ.

For $n = p^m - 1$, the generalized Nechaev-Gray map (cf. [N], [L-B], [L-S]) Ψ over $\mathbb{Z}_{p^2}^n$ is defined as follows: For $u \in \mathbb{Z}_{p^2}$ let $u = r_0(u) + pr_1(u)$ with $r_0(u), r_1(u) \in \{0, 1, \ldots, p-1\}$. Recall that $\oplus$ denotes the addition modulo p. For $(u_0, u_1, \ldots, u_{n-1}) \in \mathbb{Z}_{p^2}^n$, we have $\Psi(u_0, u_1, \ldots, u_{n-1}) = (a_0, a_1, \ldots, a_{pn-1}) \in \mathbb{F}_p^{pn}$ such that for $0 \leq j \leq p-1$ and $0 \leq t \leq n-1$, $a_{jn+t} = r_1((1-p)^t u_t) \oplus jr_0((1-p)^t u_t)$. It is shown in [L-B, Corollary 3.6] that, if C is a cyclic code over $\mathbb{Z}_{p^2}$, then $\Psi(C)$ is a cyclic p-ary code.

Let $\mathcal{P}_{m,D}^1$ be the subset of $S(D) \times \mathbb{Z}_{p^2}$ defined as

$$\mathcal{P}_{m,D}^1 = \Big\{ (f(x), u) \in S(D) \times \mathbb{Z}_{p^2} : \rho(f(x)) \neq 0,$$
$$\text{and } \{\mathrm{Tr}_m(f(\beta^i))\}_{i=0}^{\infty} \text{ has period } p^m - 1 \Big\}.$$

We introduce an equivalence relation on $\mathcal{P}_{m,D}^1$: We say that $(f(x), u)$, $(g(x), v) \in \mathcal{P}_{m,D}^1$ are related if there exist $0 \leq j$, $k \leq p-1$ and $0 \leq t \leq (p^m-1)-1$ such that

$$g(x) = (1+p)^j(1-p)^t f(\beta^t x) \quad \text{and} \quad v = (1+p)^j(1-p)^t u + kp.$$

Let $\widehat{\mathcal{P}}_{m,D}^1$ be a full set of representatives of the equivalence relation. We also assume, without loss of generality, that the elements of $\widehat{\mathcal{P}}_{m,D}^1$ are of the form $(f(x), u)$ with $u \in \{0, 1, \ldots, p-1\} \subseteq \mathbb{Z}_{p^2}$. Let $\mathcal{F}_{m,D}^1$ be the family of p-ary sequences given as

$$\mathcal{F}_{m,D}^1 = \big\{ \{\Psi(\mathrm{Tr}_m(f(\beta^i)) + u)\}_{i=0}^{\infty} : (f(x), u) \in \widehat{\mathcal{P}}_{m,D}^1 \big\}.$$

Let $\mathcal{P}_{m,D}^2$ be the subset of $pS(D)_1 \times (\mathbb{Z}_{p^2} \setminus p\mathbb{Z}_{p^2})$ defined as

$$\mathcal{P}_{m,D}^2 = \Big\{ (pf(x), u) \in pS(D)_1 \times (\mathbb{Z}_{p^2} \setminus p\mathbb{Z}_{p^2}) : \{\mathrm{Tr}_m(pf(\beta^i))\}_{i=0}^{\infty} \text{ has period } p^m - 1 \Big\}.$$

We say $(pf(x), u)$, $(pg(x), v) \in \mathcal{P}^2_{m,D}$ are *cyclically related* if there exist $0 \le j \le p-1$ and $0 \le t \le (p^m - 1) - 1$ such that $pg(x) = (1+p)^j(1-p)^t pf(\beta^t x)$ and $v = (1+p)^j(1-p)^t u$. Cyclically related elements of $\mathcal{P}^2_{m,D}$ form an equivalence relation. Let $\overline{\mathcal{P}}^2_{m,D}$ denote the set of equivalence classes in $\mathcal{P}^2_{m,D}$. In fact, we can choose a full set of representatives $\widetilde{\mathcal{P}}^2_{m,D}$ of the equivalence classes in $\overline{\mathcal{P}}^2_{m,D}$ such that

$$\widetilde{\mathcal{P}}^2_{m,D} = \left\{ (pf(x), u) \in \mathcal{P}^2_{m,D} : u \in \{1, \ldots, p-1\} \subseteq (\mathbb{Z}_{p^2} \setminus p\mathbb{Z}_{p^2}) \right\}.$$

Let $\mathcal{F}^2_{m,D}$ be the family of p-ary sequences given as

$$\mathcal{F}^2_{m,D} = \left\{ \{\Psi(\mathrm{Tr}_m(pf(\beta^i)) + u)\}_{i=0}^{\infty} : (pf(x), u) \in \widetilde{\mathcal{P}}^2_{m,D} \right\}.$$

Let $\mathcal{F}_{m,D}$ be the family of p-ary sequences defined as

$$\mathcal{F}_{m,D} = \mathcal{F}^1_{m,D} \cup \mathcal{F}^2_{m,D}.$$

Theorem 4. *The families $\mathcal{F}^1_{m,D}$, $\mathcal{F}^2_{m,D}$ and $\mathcal{F}_{m,D}$ have the following properties:*

i) *The period of each sequence in $\mathcal{F}_{m,D}$ (resp. $\mathcal{F}^1_{m,D}$ and $\mathcal{F}^2_{m,D}$) is $p(p^m - 1)$.*

ii) *The sequences in $\mathcal{F}_{m,D}$ (resp. $\mathcal{F}^1_{m,D}$ and $\mathcal{F}^2_{m,D}$) are pairwise cyclically distinct.*

iii) $|\mathcal{F}^1_{m,D}| = \frac{1}{p^m - 1} \sum_{l | (p^m - 1)} \mu(l) \left\{ p^{m(\lfloor \frac{D}{l} \rfloor - \lfloor \frac{D}{p^2 l} \rfloor)} - p^{m(\lfloor \frac{D}{l} \rfloor - \lfloor \frac{D}{pl} \rfloor)} \right\}$,

$|\mathcal{F}^2_{m,D}| = \frac{p-1}{p^m - 1} \sum_{l | (p^m - 1)} \mu(l) p^{m(\lfloor \frac{D}{l} \rfloor - \lfloor \frac{D}{pl} \rfloor)}$, *and*

$|\mathcal{F}_{m,D}| = |\mathcal{F}^1_{m,D}| + |\mathcal{F}^2_{m,D}|$, *where $\mu(\cdot)$ is the Möbius function.*

iv) *For the maximal non-trivial correlation $\theta_{\max}$ of $\mathcal{F}_{m,D}$ (resp. $\mathcal{F}^1_{m,D}$ and $\mathcal{F}^2_{m,D}$), we have*

$$\theta_{\max} \le \frac{1}{p-1} p^{l_{D,m}+1} \left\lfloor \frac{p^{h_{D,m}} \frac{p^2 - p}{2}(D-1) \lfloor 2p^{\frac{m}{2} - h_{D,m}} \rfloor}{p^{l_{D,m}+1}} \right\rfloor + p.$$

Remark 1. For $p = 2$, from $\mathcal{F}^1_{m,D}$ we retrieve the family of binary sequences $Q(D)$ of [H-K, Section 8.8]. Let $\mathcal{F}^{1,0}_{m,D}$ be the subfamily of $\mathcal{F}^1_{m,D}$ defined as

$$\mathcal{F}^{1,0}_{m,D} = \left\{ \{\Psi(\mathrm{Tr}_m(f(\beta^i)))\}_{i=0}^{\infty} : (f(x), 0) \in \widehat{\mathcal{P}}^1_{m,D} \right\}.$$

Note that $\mathcal{F}^1_{m,D}$ is larger than $\mathcal{F}^{1,0}_{m,D}$ with the same upper bound on the maximal non-trivial correlation. For $p = 2$, from $\mathcal{F}^{1,0}_{m,D}$ we obtain the family of binary sequences of [S-K-H].

Remark 2. $\mathcal{F}_{m,D}$ is larger than $\mathcal{F}^1_{m,D}$ while the sequences in them have the same period and the same upper bound for their maximal non-trivial correlation in Theorem 4.

For more details of the results above we refer the reader to [L-O2].

References

[C]   C. Carlet, "$\mathbb{Z}_{2^k}$-linear codes", *IEEE Trans. Inform. Theory*, vol. 44, no. 4, pp. 1543-1547, July 1998.

[C-H]   I. Constantinescu and T. Heise, "A metric for codes over residue class rings of integers", *Prob. Inform. Transmission*, vol. 33, pp. 208-213.

[G-S]   M. Greferath and S.E. Schmidt, "Gray isometries for finite chain rings and a nonlinear ternary $(36, 3^{12}, 15)$ code", *IEEE Trans. Inform. Theory*, vol. 45, no. 7, pp. 2522-2524, November 1999.

[H-K]   T. Helleseth and P.V. Kumar, "Sequences with low correlation", in *Handbook of Coding theory Vol. I, II.* Edited by V.S. Pless and W.C. Huffman, pp. 1765-1853, North-Holland, Amsterdam, 1998.

[K-H-C]   P.V. Kumar, T. Helleseth and A.R. Calderbank, "An upper bound for Weil exponential sums over Galois rings with applications", *IEEE Trans. Inform. Theory*, vol. 41, no. 2, pp. 456-468. March 1995.

[H-K-M-S]   T. Helleseth, P.V. Kumar, O. Moreno and A.G. Shanbhag, "Improved estimates via exponential sums for the minimum distance of $\mathbb{Z}_4$-linear trace codes", *IEEE Trans. Inform. Theory*, vol. 42, no. 4, pp. 1212-1216, July 1996.

[L-B]   S. Ling and J.T. Blackford, "$\mathbb{Z}_{p^{k+1}}$-linear codes", *IEEE Trans. Inform. Theory*, vol. 48, no. 9, 2592-2605, September 2002.

[L-O]   S. Ling and F. Özbudak, "An Improvement on the bounds of Weil exponential sums over Galois rings with some application", *IEEE Trans. Inform. Theory*, vol. 50, no. 10, pp. 2529-2539, October 2004.

[L-O2]   S. Ling and F. Özbudak, "Improved p-ary codes and sequence families from Galois rings of characteristic p^2", submitted, 2004.

[L-S]   S. Ling and P. Solé, "Nonlinear p-ary sequences", *Appl. Alg. Eng. Comm. Comp.*, vol. 14, pp. 117-125, 2003.

[N]   A. A. Nechaev, "The Kerdock code in a cyclic form", *Discr. Math. Appl.*, vol. 1, pp. 365-384, 1991.

[S-K-H]   A.G. Shanbhag, P.V. Kumar and T. Helleseth, "Improved binary codes and sequence families from $\mathbb{Z}_4$-linear codes", *IEEE Trans. Inform. Theory*, vol. 42, no. 5, pp. 1582-1587, September 1996.

Quadriphase Sequences Obtained from Binary Quadratic Form Sequences

Xiaohu Tang[1], Parampalli Udaya[2], and Pingzhi Fan[1,*]

[1] Institute of Mobile Communications, Southwest Jiaotong University,
Chengdu, China
{xhutang, p.fan}@ieee.org
[2] Department of Computer Science and Software Engineering,
University of Melbourne, VIC 3010, Australia
udaya@cs.mu.oz.au

Abstract. The development of the theory of $\mathbf{Z}_4$ maximal length sequences in the last decade led to the discovery of several families of optimal quadriphase sequences. In theory, the construction uses the properties of Galois rings. In this paper, we propose a method for constructing quadriphase sequences using binary sequences based on quadratic forms. The study uses only the properties of Galois fields instead of Galois rings. We demonstrate the theory by constructing a new family of $\mathbf{Z}_4$ sequences with low correlation property.

1 Introduction

In the early 1990s, the theory of $\mathbf{Z}_4$ maximal length sequences was established, leading to the discovery of optimal quadriphase sequences meeting the Welch bound [5, 8, 2]. This family (Family $\mathcal{A}$ in [2]) comprises of $2^n + 1$ $\mathbf{Z}_4$ maximal length sequences [1] of length $2^n - 1$ and its maximum out of phase auto- and cross-correlations (C_{max}) is lower bounded by a quantity approximately equal to

---

* This work of X.H. Tang and P.Z. Fan was supported by the National Science Foundation of China (NSFC) under Grants 60302015 , the Foundation for the Author of National Excellent Doctoral Dissertation of PR China (FANEDD) under Grants 200341 and the NSFC/RGC under the Grant 60218001/N_HKUST617/02. The work of Udaya was supported by Australian Research Council(ARC) and Melbourne Research Grant Scheme (MRGS) of the University of Melbourne.

[1] The nomenclature 'maximal length' is appropriate in case of sequences over a field, because the length of any field m-sequence is the largest length possible for any sequence generated by a n-length feedback shift register. As $\mathbf{Z}_4$ is a ring, $\mathbf{Z}_4$ sequences generated by a n-length linear feedback shift register, can never attain period of $4^n - 1$. In this case the possible periods are only $2^n - 1$ and $2(2^n - 1)$. Further, any $\mathbf{Z}_4$ LFSR sequence of period $2(2^n - 1)$ is essentially an interleaved version of two appropriate linear $\mathbf{Z}_4$ sequences of period $2^n - 1$. Thus we refer 'maximal length sequences' only to the $\mathbf{Z}_4$ sequences which attain the maximal length corresponding to the residue field $\mathbf{Z}_2$.

T. Helleseth et al. (Eds.): SETA 2004, LNCS 3486, pp. 243–254, 2005.
© Springer-Verlag Berlin Heidelberg 2005

$\sqrt{N+1}$. For the same period and the same size, this value is smaller by a factor of $\sqrt{2}$, to C_{max} of any binary optimal family. Like their binary counterparts, the m-sequences over $\mathbf{Z}_4$ have trace representation as trace of successive powers of unit elements of Galois rings. A complete treatment of all such families of trace sequences over $\mathbf{Z}_4$ is given in [8] which includes the familiar families $\mathcal{A}$ and $\mathcal{B}$ given in [1, 5]. Further, this research led to the discovery of many optimal biphase sequences from $\mathbf{Z}_4$ sequences, like binary Gold-like sequences and Udaya-Siddiqi sequences [2, 7]. In 1994, Hammons et. al. developed the connection and built the fundamental theory for $\mathbf{Z}_4$ Linear codes [3].

So far, the connection between the $\mathbf{Z}_4$ sequences and the binary sequences is just one-way. Usually, we use Gray mapping or the most significant bit mapping to transform the $\mathbf{Z}_4$ sequences into binary sequences, and then investigate the properties of the binary sequences based on those of the $\mathbf{Z}_4$ sequences. In this paper, we reverse the above idea: construct $\mathbf{Z}_4$ sequences based on binary quadratic forms and then study the $\mathbf{Z}_4$ sequences by virtue of their representation using quadratic forms.

The paper is organized as follows. In Section 2, we provide basic background to present $\mathbf{Z}_4$ m-sequences and give their finite field representations. In Section 3, we define our new family of quaternary sequences and present our main results. In Section 4, we give the proof of correlation function of the new sequences making use of quadratic form techniques.

2   Quaternary m-Sequences and Connection to Binary Quadratic Forms

The properties of Galois rings play an important role in defining and determination properties of quaternary m-sequences [5, 2, 6, 7, 8]. In this section, we give definitions of quaternary m-sequences and their representation over finite fields.

Let $\mathbf{Z}_4[x]$ be the ring of all polynomials over $\mathbf{Z}_4$. A monic polynomial $f(x) \in \mathbf{Z}_4[x]$ is said to be a *primitive basic irreducible* if its projection $\mu(f(x))$

$$\mu(f(x)) \overset{\triangle}{\equiv} f(x)(\bmod 2)$$

is primitive and irreducible over $\mathbf{Z}_2[x]$.

Specifically, let $f(x) = \sum_{i=0}^{n} f_i x^i$ be a primitive basic irreducible polynomial of degree n over $\mathbf{Z}_4$, and $\mathbf{Z}_4[x]/f(x)$ denote the ring of residue classes of polynomials over $\mathbf{Z}_4$ modulo $f(x)$. It can be shown, this quotient ring is a commutative ring with identity called Galois ring, denoted as $GR(4, n)$. As a multiplicative group, the units $GR^*(4, n)$ have the following structure

$$GR^*(4, n) = G_A \otimes G_C,$$

where G_C is a cyclic group of order $2^n - 1$ and G_A is an Abelian group of order 2^n.

Let $\beta \in GR^*(4, n)$ be a generator of the cyclic group G_C, i.e.,

$$G_C = \{1, \beta, \beta^2, \ldots, \beta^{2^n-2}\}.$$

Then $\alpha = \mu(\beta)$ is a primitive root of $GF(2^n)$ with primitive polynomial $\mu(f(x))$ over $\mathbf{Z}_2$. For convenience let $\beta^\infty = \alpha^\infty = 0$.

G_A itself is a direct product of n cyclic groups, each of order 2; its elements are given by $\{1, 1 + 2\beta^i; i = 0, 1, \ldots, 2^n - 2\}$.

A trace function maps elements of $GR(4, n)$ to $\mathbf{Z}_4$, defined as

$$Tr_1^n(x) = \sum_{i=0}^{n-1} \sigma^i(x),$$

where $\sigma^i(\cdot), 0 \leq I < n$ are automorphisms of $GR(4, n)$ given by

$$\sigma^i : \sum_{j=0}^{n-1} a_j \beta^j \longrightarrow \sum_{j=0}^{n-1} a_j \beta^{j 2^i} \pmod{f(x)}, \quad a_j \in \mathbf{Z}_4.$$

Definition 1. [8, 2] $\mathbf{Z}_4$ **m-sequences:** *Let*

$$\gamma_i = \begin{cases} 1 + 2\beta^i, & 0 \leq i \leq 2^n - 2 \\ 1, & i = 2^n - 1 \\ 2, & i = 2^n \end{cases},$$

the $\mathbf{Z}_4$ maximal length sequence family (A) is defined as

$$a_i(t) = Tr_1^n(\gamma_i \beta^t), \; t = 0, 1, \ldots, 2^n - 2, i = 0, \ldots, 2^n. \tag{1}$$

Henceforth, we use $\oplus$ to denote the addition operation in $\mathbf{Z}_4$ and $+$ to denote either the addition operation in $\mathbf{Z}_2$ or the ordinary addition operation in $\mathbf{R}$.

The following theorem states the connection between the $\mathbf{Z}_4$ maximal length sequences and the binary sequences.

Theorem 1. *The sequences $a_i, 0 \leq i \leq 2^n$, given in (1) have the following representation:*

$$a_i(t) = \begin{cases} tr_1^n(\alpha^t) \oplus 2(tr_1^n(\alpha^{i+t}) + p(\alpha^t)), & 0 \leq i \leq 2^n - 2 \\ tr_1^n(\alpha^t) \oplus 2p(\alpha^t), & i = 2^n - 1 \\ 2tr_1^n(\alpha^t), & i = 2^n \end{cases} \tag{2}$$

where

$$p(x) = \begin{cases} \sum_{l=1}^{\frac{n-1}{2}} tr_1^n(x^{2^l+1}), & \text{if } n \text{ is odd} \\ \sum_{l=1}^{\frac{n}{2}-1} tr_1^n(x^{2^l+1}) + tr_1^{\frac{n}{2}}(x^{2^{\frac{n}{2}}+1}), & \text{if } n \text{ is even} \end{cases},$$

and $tr_1^n(\cdot)$ is the trace function over $\mathbf{Z}_2$ given by

$$tr_1^n(x) = \sum_{i=0}^{n-1} x^{2^i}.$$

3   Statement of Results

Theorem 1 gives connection between quaternary m-sequences and binary quadratic forms. Note that the terms $tr_1^n(\alpha^{i+t}) + p(\alpha^t)$ and $p(\alpha^t)$ in (2) are Gold-like sequences(the sequences with same correlation properties as Gold sequences but with large linear span). In this paper we replace them by the binary GKW-like sequences introduced by Kim and No [4], which is a generalization of Gold-like sequences. We can use the technique adopted here to study the Z_4 maximal length sequences defined in (1).

Definition 2. Generalized Z_4 m-sequences: *Let* $e = gcd(n, k)$ *and* $\frac{n}{e} = m$, *the new family of* Z_4 *sequences is defined as*

$$b_i(t) = \begin{cases} tr_1^n(\alpha^t) \oplus 2 \; (tr_1^n(\alpha^{i+t}) + q(\alpha^t)), & 0 \le i \le 2^n - 2 \\ tr_1^n(\alpha^t) \oplus 2 \; q(\alpha^t), & i = 2^n - 1 \\ 2 \; tr_1^n(\alpha^t), & i = 2^n \end{cases} \quad (3)$$

where

$$q(x) = \begin{cases} \sum_{l=1}^{\frac{m-1}{2}} tr_1^n(x^{2^{kl}+1}), & \text{if } m \text{ is odd} \\ \sum_{l=1}^{\frac{m}{2}-1} tr_1^n(x^{2^{kl}+1}) + tr_1^{\frac{n}{2}}(x^{2^{\frac{n}{2}}+1}), & \text{if } m \text{ is even} \end{cases}.$$

Remark 1. *When* $e = 1$, *the quadratic form* $q(x)$ *turns to be* $p(x)$ *in Theorem 1 and the corresponding* Z_4 *sequences are m-sequences (Family $\mathcal{A}$).*

In fact, $p(x)$ and $q(x)$ are quadratic forms as all the exponents of x have only two ones in their binary representation. Quadratic form techniques have been employed both in [1] and [4] to compute the correlation functions of Gold-like and GKW-like sequences, respectively. Here we apply the quadratic form techniques to the Z_4 sequences family defined in (3).

Given two quaternary sequences $a = \{a_0, a_1, \ldots, a_{N-1}\}$ and $b = \{b_0, b_1, \ldots, \ldots b_{N-1}\}$ of period N, we define the periodic cross-correlation between a and b as

$$R_{a,b}(\tau) = \sum_{i=0}^{N-1} \omega^{a_i - b_{i+\tau}}, 0 \le \tau < N, \quad (4)$$

where $\omega = \sqrt{-1}$, the fourth root of unity.

Now let S be a set of M sequences of period N. In sequence design, it is desirable that M and N are large and non trivial correlations $|R_{a,b}(\tau)| : a, b \in S, 0 < \tau < N$ or $(\tau = 0, a \ne b)$ are as small as possible.

In the next section, we prove the following result on non trivial correlations of the new family.

Theorem 2. *Let* n, e, m, k *be as defined in Definition 2 and let* $R_{a,b}(\tau)$ *denote the nontrivial correlation function of any two sequences a and b in (3) where* $0 \le \tau < 2^n - 1$. *Then the modulus value* $|R_{a,b}(\tau) + 1|$ *takes the values from the set*

Table 1. Correlation Values of the family in Example 1

Index	Corelation
1	[63]
2	$[-1\omega + 0]$
3	$[-1\omega + -8]$
4	$[7\omega + 0]$
5	$[7\omega + 8]$
6	$[-1\omega + 8]$
7	$[7\omega + -8]$
8	$[-9\omega + 0]$
9	$[-9\omega + 8]$
10	$[-9\omega + -8$

- *when m is odd:*
  - $\{0, 2^{\frac{n}{2}}\}$ *for* $e = 1$
  - $\{0, 2^{\frac{n+e-1}{2}}, 2^{\frac{n+e-2}{2}}\}$ *for* $e > 1$.
- *when m is even:*
  - $\{0, 2^{\frac{n}{2}}\}$ *for* $e = 1$
  - $\{0, 2^{\frac{n}{2}}, 2^{\frac{n+1}{2}}, 2^{\frac{n+2e-2}{2}}\}$, *for* $e > 1$.

Remark 2. *When m is odd,* θ_{max}*, the maximum non-trivial correlation values for the above quaternary family is smaller by a factor of* $\sqrt{2}$ *than for the* θ_{max} *of binary GKW sequence [4]. Similarly, when m is even,* θ_{max}*, for the quaternary family is better than the GKW family by a factor of 2.*

Example 1. *Let* $n = 6, m = 3, e = 2, k = 2$*. The sequences* $a_i, 0 \leq i \leq 64$*, of the new family are given as:*

Table 2. Correlation Values of the family in Example 2

Index	Corelation
1	[255]
2	$[-17\omega + -16]$
3	$[-1\omega + 0]$
4	$[-1\omega + 32]$
5	$[-17\omega + 16]$
6	$[15\omega + 16]$
7	$[15\omega + -16]$
8	$[15\omega + 0]$
9	$[-1\omega + -32]$
10	$[-17\omega + 0]$
11	$[-1\omega + -16]$
12	$[-1\omega + 16]$
13	$[-33\omega + 0]$
14	$[31\omega + 0]$

$$a_i(t) = \begin{cases} tr_1^6(\alpha^t) \oplus 2(tr_1^6(\alpha^{i+t}) + tr_1^6(\alpha^{5t})), & 0 \le i \le 62 \\ tr_1^6(\alpha^t) \oplus 2\,_1^6(\alpha^{5t}), & i = 63 \\ 2tr_1^6(\alpha^t), & i = 64. \end{cases} \quad (5)$$

The correlation values of the family are given in Table 1.

Example 2. *Let $n = 8, m = 4, e = 2, k = 2$. The sequences a_i, $0 \le i \le 64$, of the new family are given as:*

$$a_i(t) = \begin{cases} tr_1^8(\alpha^t) \oplus 2(tr_1^8(\alpha^{i+t}) + tr_1^8(\alpha^{5t}) + tr_1^4(\alpha^{17t})), & 0 \le i \le 255 \\ tr_1^8(\alpha^t) \oplus 2\ (tr_1^8(\alpha^{5t}) + tr_1^4(\alpha^{17t})), & i = 255 \\ 2tr_1^8(\alpha^t), & i = 256. \end{cases} \quad (6)$$

The correlation values of the family are given in Table 2.

4   Proof of Theorem 2

The last sequence in the family is isomorphic to binary m-sequence and hence its non trivial autocorrelation value is -1. We divide the other computations into two cases:

4.1   Case 1: Correlation Function of $c = b_i$ and $d = b_j$, $0 \le i, j \le 2^n - 1$

Let us consider sequences c and d over $\mathbf{Z}_4$:

$$c = tr_1^n(x) \oplus 2(tr_1^n(\eta_1 x) + q(x)),$$
$$d = tr_1^n(x) \oplus 2(tr_1^n(\eta_2 x) + q(x)),$$
$$\eta_1, \eta_2 \in \mathbf{GF}(2^n).$$

Then their correlation function at the shift τ is

$$R_{c,d}(\tau) = \sum_{x \in \mathbf{GF}(2^n)} j^{\,tr_1^n(x) \oplus 2(tr_1^n(\eta_1 x) + q(x)) \oplus 3tr(yx) \oplus 2(tr_1^n(\eta_2 yx) + q(yx))} - 1,$$

where $y = \alpha^\tau$ and $j = \sqrt{-1}$.
When $y = 1$, i.e., $\tau = 0$, the correlation function simplifies to:

$$R_{c,d}(\tau) = \sum_{x \in \mathbf{GF}(2^n)} (-1)^{tr_1^n((\eta_1 + \eta_2)x)} - 1;$$

$$= \begin{cases} 2^n - 1, & c = d, \\ -1, & \text{else.} \end{cases}$$

When $y \ne 1$, immediately we can write

$$|R_{c,d}(\tau) + 1|^2$$

$$= \sum_{x,z \in \mathbf{GF}(2^n)} j^{\,tr_1^n(x) \oplus 3tr_1^n(z) \oplus 3tr_1^n(yx) \oplus tr_1^n(yz)}$$

$$\cdot (-1)^{tr_1^n(\eta_1 x) + tr_1^n(\eta_2 yx) + tr_1^n(\eta_1 z) + tr_1^n(\eta_2 yz) + q(x) + q(yx) + q(z) + q(yz)}$$

$$= \sum_{z,w \in \mathbf{GF}(2^n)} j^{\,tr_1^n(w) \oplus 2tr_1^n(z)tr_1^n(w) \oplus 3tr_1^n(yw) \oplus 2tr_1^n(yz)tr_1^n(yw)}$$

$$\cdot (-1)^{tr_1^n(\eta_1 w) + tr_1^n(\eta_2 yw) + q(x) + q(yx) + q(z) + q(yz)} \quad w = x + z$$

$$= \sum_{z,w \in \mathbf{GF}(2^n)} j^{\,tr_1^n(w) \oplus tr_1^n(yw)} \cdot (-1)^{tr_1^n(z)tr_1^n(w) + tr_1^n(yw) + tr_1^n(yz)tr_1^n(yw)}$$

$$\cdot (-1)^{tr_1^n(\eta_1 w) + tr_1^n(\eta_2 yw) + q(z+w) + q(y(z+w)) + q(z) + q(yz)}.$$

It should be noted that $tr_1^n(w) \equiv tr_1^n(x) \oplus tr_1^n(z) \oplus 2tr_1^n(x)tr_1^n(z) = 3tr_1^n(x) \oplus tr_1^n(z) \oplus 2tr_1^n(x)tr(w) \pmod 4$ if $w = x + z$ over $\mathbf{GF}(2^n)$.

By the quadratic form technique [4], the symplectic form associated with $q(x)$ is

$$q(z+w) + q(y(z+w)) + q(z) + q(yz) + q(w) + q(yw)$$
$$= tr_1^n[z(tr_e^n(w) + ytr_e^n(yw) + w + y^2w)].$$

Define

$$f(z,w) = tr_1^n(z)tr_1^n(w) + tr_1^n(yz)tr_1^n(yw) + q(z+w) + q(y(z+w)) + q(z) +$$
$$+ q(yz) + q(w) + q(yw)$$
$$= tr_1^n[z(tr_1^n(w) + ytr_1^n(yw) + tr_e^n(w) + ytr_e^n(yw) + w + y^2w)],$$

and

$$W = \{w | tr_1^n(w) + ytr_1^n(yw) + tr_e^n(w) + ytr_e^n(yw) + w + y^2w = 0\},$$

then we get the following equation.

$$|R_{c,d}(\tau) + 1|^2$$
$$= \sum_{w \in GF(2^n)} j^{\,tr_1^n(w) \oplus tr_1^n(yw) \oplus 2(tr_1^n(yw) + tr_1^n(\eta_1 w) + tr_1^n(\eta_2 yw) + q(w) + q(yw))} \cdot FZW,$$
$$= 2^n \sum_{w \in W} j^{\,tr_1^n(w) \oplus tr_1^n(yw) \oplus 2(tr_1^n(yw) + tr_1^n(\eta_1 w) + tr_1^n(\eta_2 yw) + q(w) + q(yw))}, \qquad (7)$$

where $FZW = \sum_{z \in \mathbf{GF}(2^n)} (-1)^{f(z,w)}$. Further define

$$g(w) = tr_1^n(w) \oplus tr_1^n(yw) \oplus 2(tr_1^n(yw) + tr_1^n(\eta_1 w) + tr_1^n(\eta_2 yw) + q(w) + q(yw)).$$

Then (7) can be rewritten as

$$|R_{c,d}(\tau) + 1|^2 = 2^n \sum_{w \in W} j^{\,g(w)}. \qquad (8)$$

For any $w_1, w_2 \in W$, consider $g(w_1) \oplus g(w_2)$. By using the definition of $f(z,w)$ and W we have the following:

$g(w_1) \oplus g(w_2)$

$$= \quad tr_1^n(w_1) \oplus tr_1^n(yw_1) \oplus 2(tr_1^n(yw_1) + tr_1^n(\eta_1 w_1) + tr_1^n(\eta_2 yw_1)$$
$$+q(w_1) + q(yw_1))$$
$$\oplus tr_1^n(w_2) \oplus tr_1^n(yw_2) \oplus 2(tr_1^n(yw_2) + tr_1^n(\eta_1 w_2) +$$
$$tr_1^n(\eta_2 yw_2) + q(w) + q(yw_2))$$
$$= \quad tr_1^n(w_1 + w_2) \oplus tr_1^n(y(w_1 + w_2)) \oplus 2(tr_1^n(y(w_1 + w_2)) +$$
$$+tr_1^n(\eta_1(w_1 + w_2)) + tr_1^n(\eta_2 y(w_1 + w_2)))$$
$$\oplus 2(tr_1^n(w_1)tr_1^n(w_2) + tr_1^n(yw_1)tr_1^n(yw_2) + q(w_1) + q(yw_1) + q(w_2) + q(yw_2))$$
$$= g(w_1 + w_2).$$

This implies, in other words, that the function $g(w)$ is a linear function on W.

Suppose that the cardinality of W is $|W|$, and $k_i = |\{w | g(w) = i, \ w \in W\}|$, $i = 0, 1, 2, 3$. Obviously $k_0 \geq 1$ due to $g(0) = 0$, and $k_i \geq k_0$ if $k_i \neq 0$ from the linearity of $g(w)$. Furthermore, it follows from (8) that $k_1 = k_3$. Therefore the possibilities for k_i s are as follows:

- $k_1 = k_3 = 0$:
  - $k_2 = 0$, then we have $k_0 = |W|$.
  - $k_2 = 1$, then we have $k_0 = k_2 = \frac{|W|}{2}$.
  - $k_2 \geq 2$, then for any two elements w_1 and w_2 such that $g(w_1) = g(w_2) = 2$, we may deduce that $g(w_1 + w_2) = 0$. Consequently $k_0 \geq k_2$. That is $k_0 = k_2 = \frac{|W|}{2}$.
- $k_1 = k_3 \geq 1$: Suppose that $w \in W$ satisfies $g(w) = 1$. Then $g(0) = g(w+w) = 2g(w) = 2$ contradicts with the fact $g(0) = 0$, which implies this case never happens.

So easily we can conclude that $|R_{c,d}(\tau) + 1|^2$ is either $2^n |W|$ or 0.

The remainder task is only to determine the number of the roots of the equation over $\mathbf{GF}(2^n)$:

$$tr_1^n(w) + ytr_1^n(yw) + tr_e^n(w) + ytr_e^n(yw) + w + y^2 w = 0. \tag{9}$$

Let $tr_e^n(yw) = a$ and $tr_e^n(w) = b$, then (9) can be rewritten as

$$tr_1^e(b) + ytr_1^e(a) + b + ay + w + y^2 w = 0.$$

Since $y \neq 1$, the expression of w is

$$w = \frac{(a + tr_1^e(a))y + b + tr_1^e(b)}{1 + y^2}. \tag{10}$$

The solution of the equation depends on if m is odd or even.

m Odd: Plugging (10) into $tr_e^n(yw) = a$ and $tr_e^n(w) = b$, we have

$$(a + tr_1^e(a) + b + tr_1^e(b))X^2 + (b + tr_1^e(b))X = tr_1^e(a), \tag{11}$$
$$(a + tr_1^e(a) + b + tr_1^e(b))X^2 + (a + tr_1^e(a))X = b, \tag{12}$$

where $X = tr_e^n(\frac{1}{1+y})$. There are three different cases.

1) If $X = 0$, then $tr_1^e(a) = 0$ and $b = 0$. The number of solutions is 2^{e-1} ;

2) If $X = 1$, then $a = 0$ and $tr_1^e(b) = 0$. The number of solutions is 2^{e-1} ;

3) If $X = c \in GF(2^e)/\{0,1\}$, then

$$a + tr_1^e(a) + b + tr_1^e(b) = \frac{b + tr_1^e(a)}{c}. \tag{13}$$

Replacing $a + tr_1^e(a) + b + tr_1^e(b)$ in (11) and (12) with the right-hand side of (13), we have

$$c(tr_1^e(a) + tr_1^e(b)) = tr_1^e(a), \tag{14}$$
$$c(a + b) = b. \tag{15}$$

From (14), it is clear that $tr_1^e(a) + tr_1^e(b) = 0$ and $tr_1^e(a) = 0$ due to the assumption of $c \in \mathbf{GF}(2^e)/\{0,1\}$. It follows that $tr_1^e(a) = tr_1^e(b) = 0$. On the other hand, we obtain from (15)

$$a = \frac{c+1}{c}b. \tag{16}$$

Using the trace function on both side of (16), we require

$$tr_1^e(\frac{b}{c}) = 0.$$

Substituting (16) into (10), we get

$$w = \frac{\frac{c+1}{c}by + b}{1 + y^2}.$$

There are only 2^{e-2} solutions of w since $tr_1^e(b) = 0$ and $tr_1^e(\frac{b}{c}) = 0$.
Thus, collecting all the cases, we conclude that the correlation value $|R_{c,d}(\tau) + 1|$ is

 – either 0 or $2^{\frac{n}{2}}$ when $e = 1$,
 – either 0 or $2^{\frac{n+e-1}{2}}$ or $2^{\frac{n+e-2}{2}}$ when $e > 1$.

Therefore the maximum non-trivial correlation is $2^{\frac{n+e-1}{2}}$, which is smaller by a factor of $\sqrt{2}$ than the binary GKW-like sequences.

m **Even:** Plugging (10) into $tr_e^n(yw) = a$ and $tr_e^n(w) = b$, we have

$$(a + tr_1^e(a) + b + tr_1^e(b))X^2 + (b + tr_1^e(b))X = a, \tag{17}$$
$$(a + tr_1^e(a) + b + tr_1^e(b))X^2 + (a + tr_1^e(a))X = b, \tag{18}$$

where $X = tr_e^n(\frac{1}{1+y})$. Again we have three different cases.

1) If $X = 0$, then $b = 0$ and $a = 0$, and then $w = 0$. The number of solutions is 1 ;

2) If $X = 1$, then $tr_1^e(a) = 0$ and $tr_1^e(b) = 0$, and then $w = \frac{ay+b}{1+y^2}$. The number of solutions is 2^{2e-2} ;

3) If $X = c \in \mathbf{GF}(2^e)/\{0,1\}$, then

$$a + tr_1^e(a) + b + tr_1^e(b) = \frac{b+a}{c}. \tag{19}$$

Replacing $a + tr_1^e(a) + b + tr_1^e(b)$ in (17) and (18) with the right-hand side of (19), we have

$$a = \frac{c}{c+1} tr_1^e(b), \tag{20}$$

$$b = \frac{c}{c+1} tr_1^e(a). \tag{21}$$

It is easy to check that the number of solutions to w is one or two depending on whether $tr_1^e(\frac{c}{c+1}) = 1$ or not.

Thus, combining all the cases, we conclude that the correlation value $|R_{c,d}(\tau)+1|$ is

- either 0 or $2^{\frac{n}{2}}$ when $e = 1$,
- either 0 or $2^{\frac{n}{2}}$ or $2^{\frac{n+1}{2}}$ or $2^{\frac{n+2e-2}{2}}$ when $e > 1$.

Then the maximum non-trivial correlation is $2^{\frac{n+2e-2}{2}}$, which is smaller by a factor of 2 than the binary GKW-like sequences.

4.2   Case 2: Correlation Function of $c = b_{2^n}$ and $d = b_i$, $0 \leq i \leq 2^n - 1$

We will only need to consider this case, since the results equally apply to the case when $c = b_i$ and $d = b_{2^n}$, $0 \leq i \leq 2^n - 1$. Let us consider the last sequence d and another sequence c over Z_4:

$$c = tr_1^n(x) \oplus 2(tr_1^n(\eta_1 x) + q(x)), \quad d = 2tr_1^n(x), \quad \eta_1 \in \mathbf{GF}(2^n).$$

Then their correlation function at the shift τ is

$$R_{c,d}(\tau) = \sum_{x \in \mathbf{GF}(2^n)} j^{\,tr_1^n(x) \oplus 2(tr_1^n(\eta_1 x)+q(x)) \oplus 2tr_1^n(yx)} - 1,$$

where $y = \alpha^\tau$ and $j = \sqrt{-1}$.

Immediately, we have

$$\begin{aligned}
&|R_{c,d}(\tau) + 1|^2 \\
&= \sum_{x,z \in \mathbf{GF}(2^n)} j^{\,tr_1^n(x) \oplus 3tr_1^n(z)} \cdot (-1)^{tr_1^n(\eta_1 x)+tr_1^n(yx)+tr_1^n(\eta_1 z)+tr_1^n(yz)+q(x)+q(z)}
\end{aligned}$$

$$= \sum_{z,w \in \mathbf{GF}(2^n)} j^{\,tr_1^n(w) \oplus 2tr_1^n(z)tr_1^n(w)} \cdot (-1)^{tr_1^n(\eta_1 w)+tr_1^n(yw)+q(x)+q(z)} \quad w = x+z$$

$$= \sum_{z,w \in \mathbf{GF}(2^n)} j^{\,tr_1^n(w)} \cdot (-1)^{tr_1^n(z)tr_1^n(w)+tr_1^n(\eta_1 w)+tr_1^n(yw)+q(z+w)+q(z)}.$$

Also from [4], the symplectic form associated with $q(x)$ is

$$q(z+w) + q(z) + q(w)$$
$$= tr_1^n[z(tr_e^n(w) + w)].$$

Define $f(z,w) = tr_1^n[z(tr_1^n(w) + tr_e^n(w) + w)]$, then

$$|R_{c,d}(\tau) + 1|^2$$
$$= \sum_{w \in \mathbf{GF}(2^n)} j^{\,tr_1^n(w) \oplus 2(tr_1^n(\eta_1 w)+tr_1^n(yw)+q(w))} \cdot \sum_{z \in \mathbf{GF}(2^n)} (-1)^{f(z,w)}.$$

Similar to the last section, $|R_{c,d}(\tau) + 1|^2$ takes on $2^n|W|$ or 0 where $W = \{w | tr_1^n(w) + tr_e^n(w) + w = 0\}$.

The remainder is to determine the number of the roots to the equation

$$tr_1^n(w) + tr_e^n(w) + w = 0. \tag{22}$$

Obviously $w \in \mathbf{GF}(2^e)$.

If m is odd, then $tr_1^n(w) = tr_1^e(w)$ and $tr_e^n(w) = w$. So

$$tr_1^n(w) + tr_e^n(w) + w = tr_1^e(w) = 0, \tag{23}$$

it follows that the number of roots is 2^{e-1}.

If m is even, then $tr_1^n(w) = tr_e^n(w) = 0$. Hence

$$tr_1^n(w) + tr_e^n(w) + w = w = 0, \tag{24}$$

and it follows that the number of roots is 1.

Therefore the number of roots is either 2^{e-1} or 1 depending on whether m is odd or even. Then the non-trivial correlation value $|R_{c,d}(\tau) + 1|$ takes on 0 or $2^{\frac{n+e-1}{2}}$ when m is odd; 0 or $2^{\frac{n}{2}}$ when m is even.

References

1. Boztas S., Kumar, P.V.:Binary sequences with Gold-like correlation but larger linear span. IEEE Trans. Inform. Theory, Vol. 40. (1994) 532-537.
2. Boztas S., Hammons R., and Kumar P.V. : 4-Phase Sequences with Near-Optimum Correlation Properties, IEEE Trans. on Inform. Theory, Vol. IT-38, 1992, 1101-1113.
3. Hammons, R., Kumar, P.V., Calderbank,A.N., Sloane, N.J.A, Solé P.: The $\mathbb{Z}_4$-Linearity of Kerdock, Preparata, Goethals and Related Codes. IEEE Trans. Inform. Theory, Vol. 40. (1994) 301-319.

4. Kim S.H., No, J.S.: New families of binary sequences with low crosscorrelation property. IEEE Trans. Inform. Theory, Vol. 49. (2003) 3059-3065.
5. P. Solé. A Quaternary Cyclic Code, and a Familly of Quadriphase Sequences with low Correlation Properties. *Lect. Notes Computer Science*, 388:193–201, 1989.
6. P. Udaya and M. U. Siddiqi. Large Linear Complexity Sequences over Z4 for Quadriphase Modulated Communication Systems having Good Correlation Properties. *IEEE International Symposium on Information Theory, Budapest, Hungary, June 23-29, 1991.*
7. Udaya P., Siddiqi, M.U.: Optimal biphase sequences with large linear complexity derived from sequences over Z4. IEEE Trans. Inform. Theory, Vol.42. (1996) 206-216.
8. P. Udaya and M. U. Siddiqi. Optimal and Suboptimal Quadriphase Sequences Derived from Maximal Length Sequences over Z_4. *Journal of AAECC*, 9:161–191, 1998.

New Families of p-Ary Sequences from Quadratic Form with Low Correlation and Large Linear Span

Xiaohu Tang[1], Parampalli Udaya[2], and Pingzhi Fan[1],[*]

[1] Institute of Mobile Communications, Southwest Jiaotong University,
Chengdu, China
{xhutang, p.fan}@ieee.org
[2] Department of Computer Science and Software Engineering,
University of Melbourne, VIC 3010, Australia
udaya@cs.mu.oz.au

Abstract. The quadratic form technique has many applications in sequence design including determining certain exponential sums. A key idea in this technique is to transform the problem of computing exponential sums into determining the weights of certain quadratic forms. We use this idea to define two new families of non-binary sequences and determine their correlation properties. The new families of sequences possess low correlation and large linear complexity properties.

1 Introduction

We assume that p is an odd prime, and $n = em$. Let $q = p^e$, For simplicity, denote $\mathbf{GF}(p^e)$ as $\mathbf{GF}(q)$ and $\mathbf{GF}(p^n)$ as $\mathbf{GF}(q^m)$. Let $Tr_e^n(x)$ (and respectively $Tr_1^e(x)$) be the trace mapping from $\mathbf{GF}(q^m)$ into the subfield $\mathbf{GF}(q)$ (and respectively from $\mathbf{GF}(q)$ into $\mathbf{GF}(p)$) given by

$$Tr_e^n(x) = \sum_{i=0}^{m-1} x^{p^{ei}}, \quad Tr_1^e(x) = \sum_{i=0}^{e-1} x^{p^i}.$$

Given two sequences $a = \{a_0, a_1, \ldots, a_{N-1}\}$ and $b = \{b_0, b_1, \ldots, b_{N-1}\}$ of period N, we define the periodic cross-correlation between a and b as

$$R_{a,b}(\tau) = \sum_{i=0}^{N-1} \omega^{a_i - b_{i+\tau}}, 0 \leq \tau < N, \tag{1}$$

where ω is the pth root of unity given by $\omega = e^{j2\pi/p}, j = \sqrt{-1}$.

---

[*] This work of X.H. Tang and P.Z. Fan was supported by the National Science Foundation of China (NSFC) under Grants 60302015, the Foundation for the Author of National Excellent Doctoral Dissertation of PR China (FANEDD) under Grants 200341 and the NSFC/RGC under the Grant 60218001/N_HKUST617/02. The work of Udaya was supported by Australian Research Council(ARC) and Melbourne Research Grant Scheme (MRGS) of the University of Melbourne.

T. Helleseth et al. (Eds.): SETA 2004, LNCS 3486, pp. 255–265, 2005.
© Springer-Verlag Berlin Heidelberg 2005

Now let S be a set of M sequences of period N. For practical applications, it is desirable that M and N are large and non trivial correlations $|R_{a,b}(\tau)| : a, b \in S, 0 < \tau < N$ or $(\tau = 0, a \neq b)$ are as small as possible.

For many families of sequences defined over Galois fields, the computation of correlations in (1) is related to determining the following exponential sum

$$S(F) = \sum_{x \in \mathbf{GF}(p^n)} \omega^{F(x)},$$

where ω is the pth root of unity given by $\omega = e^{j2\pi/p}, j = \sqrt{-1}$ and $F(x)$ is a function over $\mathbf{GF}(p^n)$ with values in $\mathbf{GF}(p)$.

It is an extremely hard problem to determine the exponential sum for an arbitrary function $F(x)$. However there are several bounds like the Weil bound, Deligne bound, etc, [6] which give a reasonable estimate for the sum. In many instances, these bounds turn out to be trivial (i.e. the bound exceeds the number of terms). The problem of determining the exponential sum is tractable if $F(x)$ is of some special form. For example, if $F(x)$ is derived from a quadratic form, we may even get the exact value of the exponential sum.

This idea used for sequence design seems to have been first adopted by Trachtenberg [8], and later by Helleseth [1]. Recently, the quadratic form technique has been employed to construct many p-ary sequence families with good correlation properties [4, 5, 7].

In this paper, we take

$$F(x) = Tr_1^e(p(x) + Tr_e^n(yx)) \tag{2}$$

where $p(x)$ is a quadratic form over $\mathbf{GF}(q)$, and obtain two families of sequences with low correlation and large linear span.

The paper is organized as follows. In Section II, we give a general background on quadratic forms and give a key theorem to compute the exponential sum defined using quadratic forms and trace functions. In Section III, we define two new families of sequences and state our main results. We provide proofs to our results in Sections IV and V.

2   Quadratic Forms and Exponential Sums

In this section, we give necessary background on quadratic forms.

Definition 1. *Let* $x = \sum_{i=1}^m x_i\alpha_i$ *where* $x_i \in \mathbf{GF}(q)$ *and* $\alpha_i, i = 1, 2, \ldots, m,$ *is a basis for* $\mathbf{GF}(q^m)$ *over* $\mathbf{GF}(q)$. *Then the function* $p(x)$ *is a quadratic form over* $\mathbf{GF}(q)$ *if it can be expressed as*

$$p(x) = p(\sum_{i=1}^m x_i\alpha_i) = \sum_{i=1}^m \sum_{j=1}^m b_{i,j} x_i x_j,$$

where $b_{i,j} \in \mathbf{GF}(q)$, *that is* $p(x)$ *is a homogeneous polynomial of degree 2 in ring* $\mathbf{GF}(q)[x_1, x_2, \ldots, x_m]$.

The quadratic form in odd characteristic has been well analyzed in the literature [6, 3]. Here we follow the treatment in [3]. It should be noted that the rank of the quadratic form is the minimum number of variables required to represent the function under the nonsingular coordinate transformations. Then, any quadratic form of rank r can be transferred to one of the following three canonical forms [3].

Lemma 1. *For any quadratic form $p(x)$ in $\mathbf{GF}(q^m)$, if the rank of $p(x)$ is r, then $p(x)$ is equivalent to one of*

Type I: $B_r(x)$;
Type II: $B_{r-1}(x) + \mu x_r^2$
Type III: $B_{r-2}(x) + x_{r-1} - \lambda x_r^2$,

where $B_r(x) = x_1 x_2 + x_3 x_4 + \cdots + x_{r-1} x_r$, $\mu \in \{1, \lambda\}$ and λ is a fixed nonsquare in $\mathbf{GF}(q)$. For any element $\zeta \in \mathbf{GF}(q)$, the number of solutions to the equation $p(x) = \zeta$ is

Type I: $q^{m-1} + \upsilon(\zeta)q^{m-r/2-1}$;
Type II: $q^{m-1} + \eta(\zeta\mu)q^{m-(r+1)/2}$;
Type III: $q^{m-1} - \upsilon(\zeta)q^{m-r/2-1}$;

where $\upsilon(x)$ and $\eta(x)$ are functions in $\mathbf{GF}(q)$ respectively given by

$$\upsilon(x) = \begin{cases} -1, & \text{if } x \neq 0, \\ q-1, & \text{if } x = 0, \end{cases} \quad \text{and} \quad \eta(x) = \begin{cases} 0, & \text{if } x = 0, \\ 1, & \text{if } x \text{ is a square}, \\ -1, & \text{otherwise.} \end{cases}$$

Lemma 1 is useful to determine the distribution of the range of the quadratic form when the rank of the form is known. We use this Lemma to prove the following theorem on exponential sums.

Theorem 1. *Let $p(x)$ be a quadratic form over $GF(q)$ of rank r and define the exponential sum*

$$R = \sum_{x \in \mathbf{GF}(p^n)} \omega^{Tr_1^e(p(x)+Tr_e^n(yx))}. \tag{3}$$

Then, $|R|$ can take either 0 or $p^{n-(re/2)}$, where $|x|$ means modulus value of x.

Proof. Because of Lemma 1 and the fact that Tr is a linear function, there exists a nonsingular transformation such that $f(x) = p(x) + Tr_e^n(ax)$ can be transformed into one of the following two representatives:

$$Q_r(x) + \sum_{i=r+1}^{m} a_i x_i \quad \text{or} \quad Q_r(x) + c,$$

where $Q_r(x)$ be one of the canonical forms: Type I, II or III given in Lemma 1, and c is a constant.

Consider when $f(x)$ is of the first type. Given an element $\zeta \in \mathbf{GF}(q)$, if we choose any values for $x_1, x_2, \ldots, x_r$, we can always find a unique $x_{r+1}, x_{r+2}, \ldots, x_m$, that gives a solution to the equation $Q_r(x) + \sum_{i=r+1}^{m} a_i x_i = \zeta$. This equation always has p^{n-e} solutions. Then,

$$|R| = |p^{n-e} \sum_{\zeta \in \mathbf{GF}(q)} \omega^{Tr_1^e(\zeta)}| = 0.$$

Now consider when $f(x)$ is of the second type. Here we need to divide the computation into two cases according to the parity of r. Also let $\mathbf{GF}(q)^*$ denote the multiplicative group of $\mathbf{GF}(q)$.

1) r even: Then $Q_r(x)$ has to be of Type I or III. Here we only consider Type I since the case with Type III is similar.

From Lemma 1,

$$|R_a| = \left| \sum_{x \in \mathbf{GF}(p^n)} \omega^{Tr_1^e(Q_r(x))} \right|$$

$$= \left| p^{n-e} + (p^e - 1)p^{n-re/2-e} + (p^{n-e} - p^{n-re/2-e}) \sum_{\zeta \in \mathbf{GF}(q)^*} \omega^{Tr_1^e(\zeta)} \right|$$

$$= \left| (p^{n-e} - p^{n-re/2-e}) \sum_{\zeta \in \mathbf{GF}(q)} \omega^{Tr_1^e(\zeta)} + p^{n-re/2} \right|$$

$$= p^{n-re/2}.$$

2) r odd: Then $Q_r(x)$ has to be of Type II. Also from Lemma 1,

$$|R_a| = \left| \sum_{x \in \mathbf{GF}(q)} \omega^{Tr_1^e(Q_r(x))} \right|$$

$$= \left| p^{n-e} \sum_{\zeta \in \mathbf{GF}(q)} \omega^{Tr_1^e(\zeta)} + p^{n-(r+1)e/2} \sum_{\zeta \in \mathbf{GF}(q)^*} \eta(\zeta\mu)\omega^{Tr_1^e(\zeta)} \right|$$

$$= p^{n-(r+1)e/2} \left| \sum_{\zeta \in \mathbf{GF}(q)^*} \eta(\zeta)\omega^{Tr_1^e(\zeta)} \right|.$$

Since $\eta(\zeta)$ is a nontrivial multiplicative character and $\omega^{Tr_1^e(\zeta)}$ is a nontrivial additive character, using (Theorem 5.11, [6]), we get

$$|R_a| = p^{n-re/2}.$$

The above theorem assures that quadratic form $p(x)$ with large rank r helps us to define sequences with low correlations. Hence the main task is to find quadratic forms with large rank. The rank is related to the dimension of the vector subspace W in $\mathbf{GF}(q^m)$, i.e.,

$$W = \{w \in \mathbf{GF}(q^m) : p(x + w) = p(x) \text{ for all } x \in \mathbf{GF}(q^m)\}. \qquad (4)$$

More precisely, $r = m - \dim(W)$.

3   Statement of Results

In this section we define two new families of sequences using quadratic forms. From now on throughout this paper, n, m, e and k are integers such that $n = me$ and $gcd(n, k) = e$.

Definition 2. *When m is odd, a family of nonbinary sequences of period $p^n - 1$ with family size $p^n + 1$ is defined as*

$$\mathcal{A} = \{a_i(t) \mid 0 \leq i \leq p^n, 0 \leq t < p^n - 1\} \tag{5}$$

$$
a_i(t) = \begin{cases}
Tr_1^n(\alpha^{t+i} + \lambda\delta\alpha^{2t} + \lambda \sum_{l=1}^{(m-1)/2} \alpha^{t(p^{kl}+1)}), & 0 \leq i < (p^n - 1)/2, \\
Tr_1^n(\alpha^{t+i-(p^n-1)/2} + \delta\alpha^{2t} + \sum_{l=1}^{(m-1)/2} \alpha^{t(p^{kl}+1)}), \\
\quad (p^n - 1)/2 \leq i < p^n - 1, \\
Tr_1^n(\delta\alpha^{2t} + \sum_{l=1}^{(m-1)/2} \alpha^{t(p^{kl}+1)}), & i = p^n - 1, \\
Tr_1^n(\alpha^t), & i = p^n,
\end{cases}
$$

where $\delta \neq (1/2) \in \mathbf{GF}(p^n)$.

Definition 3. *When m is even, a family of nonbinary sequences of period $p^n - 1$ with family size $(p^n + 3)/2$ is defined as*

$$\mathcal{B} = \{b_i(t) \mid 0 \leq i \leq (p^n - 1)/2, 0 \leq t < p^n - 1\} \tag{6}$$

$$
b_i(t) = \begin{cases}
Tr_1^n(\alpha^{t+i} + \delta\alpha^{2t} + \sum_{l=1}^{m/2-1} \alpha^{t(p^{kl}+1)}) + Tr_1^{n/2}(\alpha^{t(p^{n/2}+1)}), \\
\quad 0 \leq i < (p^n - 1)/2, \\
Tr_1^n(\delta\alpha^{2t} + \sum_{l=1}^{m/2-1} \alpha^{t(p^{kl}+1)}) + Tr_1^{n/2}(\alpha^{t(p^{n/2}+1)}), & i = (p^n - 1)/2, \\
Tr_1^n(\alpha^t), & i = (p^n + 1)/2,
\end{cases}
$$

where $\delta \neq (1/2) \in \mathbf{GF}(p^n)$.

In the following sections, we will show that the correlation function of the sequences in new families can be transformed to an exponential sum similar to (3).

1) For the sequences in family $\mathcal{A}$, the quadratic form in the sum is of the form:

$$p(x) = Tr_e^n(\delta x^2 + \sum_{l=1}^{(m-1)/2} x^{p^{kl}+1} - \gamma\delta(yx)^2 - \gamma \sum_{l=1}^{(m-1)/2} (yx)^{p^{kl}+1}),$$

where $\gamma \in \{0, 1, \lambda, 1/\lambda\}$, and

Table 1. p-ary sequences from quadratic forms

Family	n	size	Maximal Correlation	Maximum Linear Span	Comments
TH	$n = me, m\ odd$ $gcd(n,k) = e$	$p^n + 1$	$1 + p^{(n+e)/2}$	$2n$	$[8, 1]$
KM	$n = me, m\ odd$ $gcd(n,k) = e$	$p^n + 1$	$1 + p^{n/2}$	$2n$	Optimal [4]
JKNH	$n = me, m\ odd$ $gcd(n,k) = e$	$p^n + 1$	$1 + p^{n/2}$	$n(m+3)/2$	Optimal [5]
TH-Like	$n = me, m\ odd$ $gcd(n,k) = e$	$p^n + 1$	$1 + p^{(n+e)/2}$	$n(m+1)/2$	[7]
$\mathcal{A}$	$n = me, m\ odd$ $gcd(n,k) = e$	$p^n + 1$	$1 + p^{(n+2e)/2}$	$n(m+3)/2$	new family
$\mathcal{B}$	$n = me, m\ even$ $gcd(n,k) = e$	$(p^n + 3)/2$	$1 + p^{(n+2e)/2}$	$n(m+3)/2$	new family

2) For sequences in family $\mathcal{B}$, the quadratic form in the sum is of the form:

$$p(x) = Tr_e^n(\delta x^2 + \sum_{l=1}^{m/2-1} x^{p^{kl}+1} - \gamma\delta(yx)^2 - \gamma \sum_{l=1}^{(m-1)/2} (yx)^{p^{kl}+1})$$
$$+ Tr_e^{n/2}(x^{p^{n/2}+1} - \gamma(yx)^{p^{n/2}+1}),$$

where $\gamma \in \{0, 1\}$.

In the next two sections we will determine the rank of the above two quadratic forms which take values m or $m - 1$ or $m - 2$. Then using Theorem 1, we prove the following two main results of the paper.

Theorem 2. *Let $R_{i,j}(\tau)$ denote the nontrivial correlation function of sequences a_i and a_j in (5), $0 \le \tau < p^n - 1$. Then the modulus value $|R_{i,j}(\tau) + 1|$ takes the values from the set $\{0, p^{n/2}, p^{(n+e)/2}, p^{(n+2e)/2}\}$.*

Theorem 3. *Let $R_{i,j}(\tau)$ denote the nontrivial correlation function of sequences b_i and b_j in (6), $0 \le \tau < p^n - 1$. Then the modulus value $|R_{i,j}(\tau) + 1|$ takes the values from the set $\{0, p^{n/2}, p^{(n+e)/2}, p^{(n+2e)/2}\}$.*

It has been proven in [2] that, if a sequence can be expressed as a polynomial, then its linear span is just the number of the distinct powers of x of the polynomial. Then straightforwardly the linear spans for the sequence families $\mathcal{A}$ and $\mathcal{B}$ are given in the following two theorems.

Theorem 4. *Let $LS(a_i)$ be linear span of the sequences $a_i \in \mathcal{A}$. Then the linear span of the sequences a_i is*

$$LS(a_i) = \begin{cases} n(m+3)/2, & if\ 0 \le i < p^n - 1, \\ n(m+1)/2, & if\ i = p^n - 1, \\ n, & if\ i = p^n. \end{cases}$$

Theorem 5. *Let $LS(b_i)$ be linear span of the sequences $b_i \in \mathcal{B}$. Then the linear span of the sequences b_i is*

$$LS(b_i) = \begin{cases} n(m+3)/2, & \text{if } 0 \le i < (p^n-1)/2, \\ n(m+1)/2, & \text{if } i = (p^n-1)/2, \\ n, & \text{if } i = (p^n+1)/2. \end{cases}$$

As a comparison, the known p-ary sequences families from quadratic form will be illustrated in Table 1.

4   Proof of Theorem 2

Our main task is to determine the rank of the quadratic form related to Family $\mathcal{A}$. Define

$$c_i = \begin{cases} \alpha^i, & \text{if } 0 \le i < (p^n-1)/2 \\ \alpha^{i-(p^n-1)/2}, & \text{if } (p^n-1)/2 \le i < p^n-1 \\ 0, & \text{if } i = p^n-1 \\ 1, & \text{if } i = p^n \end{cases}.$$

For a given shift parameter τ, $0 \le \tau < p^n - 1$, let $y = \alpha^\tau$. Then the correlation function between $a_i(t)$ and $a_j(t)$ is
$$R_{i,j}(\tau) =$$

$$\sum_{x \in \mathbf{GF}(p^n)} \omega^{Tr_1^n(\gamma_i(\delta x^2 + \sum_{l=1}^{(m-1)/2} x^{p^{kl}+1}) - \gamma_j(\delta(yx)^2 + \sum_{l=1}^{(m-1)/2}(yx)^{p^{kl}+1}) + (c_i - c_j y)x)} - 1,$$

where $\gamma_i \in \{1, \lambda\}, 0 \le i \le p^n - 1$ and $\gamma_{p^n} = 0$.

When $i = j = p^n$, it is clear that

$$R_{i,j}(\tau) = \begin{cases} p^n - 1, & \tau = 0 \\ -1, & \text{otherwise} \end{cases}$$

as the sequence $a_i(t)$ is an m-sequence over $\mathbf{GF}(p)$.

For other cases, without loss of generality assume that $\gamma_i \ne 0$. Then the correlation function is written as

$$R_{i,j}(\tau) = \sum_{x \in \mathbf{GF}(p^n)} \omega^{Tr_1^n(\gamma_i(\delta x^2 + \sum_{l=1}^{(m-1)/2} x^{p^{kl}+1} - \gamma(\delta(yx)^2 + \sum_{l=1}^{(m-1)/2}(yx)^{p^{kl}+1})))}$$

$$\cdot \omega^{Tr_1^n(c_i - c_j y)x)} - 1, \tag{7}$$

where $\gamma = \frac{\gamma_j}{\gamma_i} \in \{0, 1, \lambda, 1/\lambda\}$. From Theorem 1, the correlation function is determined by the rank of the quadratic form

$$q(x) = \gamma_i Tr_e^n(\delta x^2 + \sum_{l=1}^{(m-1)/2} x^{p^{kl}+1} - \gamma\delta(yx)^2 - \gamma\sum_{l=1}^{(m-1)/2}(yx)^{p^{kl}+1}),$$

or equivalently the rank of the quadratic form

$$p(x) = Tr_e^n(\delta x^2 + \sum_{l=1}^{(m-1)/2} x^{p^{kl}+1} - \gamma\delta(yx)^2 - \gamma\sum_{l=1}^{(m-1)/2}(yx)^{p^{kl}+1}).$$

In view of (4), it is sufficient to determine the dimension of the space ($\mathcal{W}$) satisfied by w in

$$Tr_e^n(\delta(w+x)^2 + \sum_{l=1}^{(m-1)/2}(w+x)^{p^{kl}+1} - \gamma\delta(yw+yx)^2 - \gamma\sum_{l=1}^{(m-1)/2}(yw+yx)^{p^{kl}+1})$$

$$= Tr_e^n(\delta x^2 + \sum_{l=1}^{(m-1)/2} x^{p^{kl}+1} - \gamma\delta(yx)^2 - \gamma\sum_{l=1}^{(m-1)/2}(yx)^{p^{kl}+1}) \quad \text{for all } x \in \mathbf{GF}(p^n).$$

This is equivalent to solving:

$$Tr_e^n(x(2\delta w + \sum_{l=1}^{(m-1)/2}(w^{p^{kl}} + w^{p^{-kl}}) - 2\gamma\delta y^2 w - \gamma y\sum_{l=1}^{(m-1)/2}((yw)^{p^{kl}} +$$

$$(yw)^{p^{-kl}})) + \delta w^2 + \sum_{l=1}^{(m-1)/2} w^{p^{kl}+1} - \gamma\delta(yw)^2 - \gamma\sum_{l=1}^{(m-1)/2}(yw)^{p^{kl}+1}$$

$$= Tr_e^n(x(Tr_e^n(w) - \gamma y Tr_e^n(yw) + (2\delta-1)w - \gamma(2\delta-1)y^2 w) \quad (8)$$

$$+ \delta w^2 + \sum_{l=1}^{(m-1)/2} w^{p^{kl}+1} - \gamma\delta(yw)^2 - \gamma\sum_{l=1}^{(m-1)/2}(yw)^{p^{kl}+1})$$

$$= 0$$

If (8) holds for all $x \in \mathbf{GF}(p^n)$, Then we need to have

$$Tr_e^n(w) - \gamma y Tr_e^n(yw) + (2\delta-1)w - \gamma(2\delta-1)y^2 w = 0, \quad (9)$$

and

$$Tr_e^n(\delta w^2 + \sum_{l=1}^{(m-1)/2} w^{p^{kl}+1} - \gamma\delta(yw)^2 - \gamma\sum_{l=1}^{(m-1)/2}(yw)^{p^{kl}+1}) = 0. \quad (10)$$

In fact from (10),

$$2Tr_e^n(\delta w^2 + \sum_{l=1}^{(m-1)/2} w^{p^{kl}+1} - \gamma\delta(yw)^2 - \gamma\sum_{l=1}^{(m-1)/2}(yw)^{p^{kl}+1})$$

$$= Tr_e^n(w(Tr_e^n(w) - \gamma y Tr_e^n(yw) + (2\delta-1)w - \gamma(2\delta-1)y^2 w))$$

$$= 0,$$

which means (10) holds only if (9) is true. Then we only need to consider(9), which could be classified into two cases:

1) $(2\delta - 1) - \gamma(2\delta - 1)y^2 = 0$.

By hypothesis $\delta \neq (1/2)$, clearly $\gamma y^2 = 1$. And it is impossible that $\lambda y^2 = 1$ and $\lambda^{-1}y^2 = 1$ for all $y \in GF(p^n)$, since λ is a nonsquare. So we have $\gamma = 1$ and $y = 1$. Accordingly, the correlation function (7) is given by

$$R_{i,j}(0) = \sum_{x \in \mathbf{GF}(p^n)} \omega^{Tr_1^n((c_i - c_j)x)} - 1.$$

Since $\gamma = \frac{\gamma_j}{\gamma_i} = 1$, this indicates that $0 \leq i, j < (p^n - 1)/2$ or $(p^n - 1)/2 \leq i, j < p^n - 1$. Then it follows that $c_i \neq c_j$ unless $i = j$ from the definition of c_i. Thus,

$$R_{i,j}(0) = \begin{cases} p^n - 1, i = j, \\ -1, \quad \text{otherwise.} \end{cases}$$

2) $(2\delta - 1) - \gamma(2\delta - 1)y^2 \neq 0$.

Let $Tr_e^n(w) = a$ and $Tr_e^n(yw) = b$, then from (9),

$$w = \frac{b\gamma y - a}{c - c\gamma y^2},$$

where $c = 2\delta - 1$.

To satisfy $Tr_e^n(w) = a$ and $Tr_e^n(yw) = b$, we obtain

$$Tr_e^n(w) = Tr_e^n(\frac{b\gamma y - a}{c - c\gamma y^2}) = a,$$

and

$$Tr_e^n(yw) = Tr_e^n(\frac{b\gamma y^2 - ay}{c - c\gamma y^2}) = b.$$

That is

$$b \cdot Tr_e^n(\frac{\gamma y}{c - c\gamma y^2}) - a \cdot [Tr_e^n(\frac{1}{c - c\gamma y^2}) + 1] = 0, \tag{11}$$

and

$$b \cdot [Tr_e^n(\frac{\gamma y^2}{c - c\gamma y^2}) - 1] - a \cdot Tr_e^n(\frac{y}{c - c\gamma y^2}) = 0. \tag{12}$$

If

$$\triangle = \begin{vmatrix} Tr_e^n(\frac{\gamma y}{c - c\gamma y^2}) & Tr_e^n(\frac{1}{c - c\gamma y^2}) + 1 \\ Tr_e^n(\frac{\gamma y^2}{c - c\gamma y^2}) - 1 & Tr_e^n(\frac{y}{c - c\gamma y^2}) \end{vmatrix} \neq 0,$$

then the only solution to (11) and (12) is zero. This implies $\mathcal{W}$ is a zero space, i.e., the dimension of $\mathcal{W}$ is zero. Otherwise, if $\triangle = 0$, there are q solutions to

(11) and (12), i.e., the dimension of $\mathcal{W}$ is 1, unless the following three equations hold simultaneously,

$$Tr_e^n(\frac{\gamma y}{c - c\gamma y^2}) = 0, \tag{13}$$

$$Tr_e^n(\frac{1}{c - c\gamma y^2}) + 1 = 0, \tag{14}$$

$$Tr_e^n(\frac{\gamma y^2}{c - c\gamma y^2}) - 1 = 0. \tag{15}$$

Apparently $y \notin \mathbf{GF}(q)$ otherwise (13) contradicts with (15). Then

$$b_1\gamma y - a_1 \neq b_2\gamma y - a_2 \text{ for all } (a_1, b_1) \neq (a_2, b_2) \in \mathbf{GF}(q)^2.$$

There are q^2 solution to (11) and (12), that is the dimension of $\mathcal{W}$ is 2.

Thus the rank of quadratic form $p(x)$ is m or $m - 1$ or $m - 2$. Now applying Theorem 1, we prove the result.

5   Proof of Theorem 3

As in the last section, the correlation function can be related to an exponential sum involving a quadratic form. In this case, it is sufficient to determine the rank of the following quadratic form

$$p(x) = Tr_e^n(\delta x^2 + \sum_{l=1}^{m/2-1} x^{p^{kl}+1} - \gamma\delta(yx)^2 - \gamma \sum_{l=1}^{m/2-1} (yx)^{p^{kl}+1})$$
$$+ Tr_e^{n/2}(x^{p^{n/2}+1} - \gamma(yx)^{p^{n/2}+1}).$$

where $\gamma \in \{0, 1\}$.

That is we need to investigate the number of the elements in the vector subspace $\mathcal{W}$. This means that we need to determine the dimension of the vector subspace satisfied by w in the following:

$$Tr_e^n(\delta(w + x)^2 + \sum_{l=1}^{m/2-1} (w + x)^{p^{kl}+1} - \gamma\delta(yw + yx)^2 - \gamma \sum_{l=1}^{m/2-1} (yw + yx)^{p^{kl}+1})$$
$$+ Tr_e^{n/2}((w + x)^{p^{n/2}+1} - \gamma(yw + yx)^{p^{n/2}+1})$$
$$= Tr_e^n(\delta x^2 + \sum_{l=1}^{m/2-1} x^{p^{kl}+1} - \gamma\delta(yx)^2 - \gamma \sum_{l=1}^{m/2-1} (yx)^{p^{kl}+1})$$
$$+ Tr_e^{n/2}(x^{p^{n/2}+1} - \gamma(yx)^{p^{n/2}+1}) \quad \text{for all } x \in \mathbf{GF}(p^n).$$

This is equivalent to solving:

$$Tr_e^n(w) - \gamma y Tr_e^n(yw) + (2\delta - 1)w - \gamma(2\delta - 1)y^2 w = 0, \qquad (16)$$

and

$$Tr_e^n(\delta w^2 + \sum_{l=1}^{m/2-1} w^{p^{kl}+1} - \gamma\delta(yw)^2 - \gamma \sum_{l=1}^{m/2-1} (yw)^{p^{kl}+1})$$

$$+ Tr_e^{n/2}(w^{p^{n/2}+1} - \gamma(yw)^{p^{n/2}+1}) \qquad (17)$$

$$= 0.$$

Essentially from (17),

$$2Tr_e^n(\delta w^2 + \sum_{l=1}^{m/2-1} w^{p^{kl}+1} - \gamma\delta(yw)^2 - \gamma \sum_{l=1}^{m/2-1} (yw)^{p^{kl}+1})$$

$$+ Tr_e^{n/2}(w^{p^{n/2}+1} - \gamma(yw)^{p^{n/2}+1})$$

$$= Tr_e^n(w(Tr_e^n(w) - \gamma y Tr_e^n(yw) + (2\delta - 1)w - \gamma(2\delta - 1)y^2 w))$$

$$= 0,$$

which means (17) holds provided that (16) is true. Then we only need to consider (16), which is as same as (9). Hence the rank of quadratic form $p(x)$ is m or $m - 1$ or $m - 2$. Now applying Theorem 1, we prove the result.

References

1. Helleseth,T.: Some results about the cross-correlation function between two maximal linear sequences. Discrete Math., Vol. 16. (1976) 209-232.
2. E. L. Key, "An analysis of the structure and complexity of nonlin-ear binary sequence generators," IEEE Trans. Inform. Theory, Vol. 22. (1976) 732-736.
3. Klapper, A.:Cross-correlations of geometric sequences in odd characteristic. Des. Codes, and Cryptogr., Vol. 11. (1997) 289-305.
4. Kumar, P.V., Moreno, O.: Prime-phase sequences with periodic correlation properties better than binary sequences. IEEE Trans. Inform.Theory, Vol. 37. (1991) 603-616.
5. Jang, J.W., Kim, Y.S., No, J.S., Helleseth, T.: New family of p-ary sequences with ideal autocorrelation and large linear span. IEEE Trans. Inform.Theory, Vol 50(2004) 1839-1844.
6. Lidl R., Niederreiter, H.: Finite Fields, in Encyclopedia of Mathematics, Vol. 20, Cambridge University Press, Cambridge (1983)
7. Tang, X.H., Udaya P., Fan, P.Z.: A new family of nonbinary sequences with three level correlation property and large linear span. accepted by IEEE Trans on Information Theory, to appear.
8. Trachtenberg, H. M.: On the crosscorrelation functions of maximal linear recurring sequences. Ph.D. Thesis, Univ. of Southern California (1970)

On the Distribution of Some New Explicit Nonlinear Congruential Pseudorandom Numbers

Harald Niederreiter[1] and Arne Winterhof[2]

[1] Department of Mathematics, National University of Singapore,
2 Science Drive 2, Singapore 117543, Republic of Singapore
nied@math.nus.edu.sg

[2] Johann Radon Institute for Computational and Applied Mathematics,
Austrian Academy of Sciences,
Altenbergerstr. 69, 4040 Linz, Austria
arne.winterhof@oeaw.ac.at

Abstract. Nonlinear methods are attractive alternatives to the linear congruential method for pseudorandom number generation. We introduce a new particularly attractive explicit nonlinear congruential method and present nontrivial results on the distribution of pseudorandom numbers generated by this method over the full period and in parts of the period. The proofs are based on new bounds on certain exponential sums over finite fields.

Keywords: Pseudorandom numbers, Nonlinear method, Discrepancy.

1 Introduction

Let $\mathbb{F}_p = \{0, 1, \ldots, p-1\}$ be the finite field of prime order $p \geq 3$. Further let $\eta \in \mathbb{F}_p^*$ be an element of multiplicative order $T \geq 2$. For a given polynomial $f(X) \in \mathbb{F}_p[X]$ of positive degree D we generate a sequence $\gamma_0, \gamma_1, \ldots$ of elements of $\mathbb{F}_p$ by

$$\gamma_n = f(\eta^n) \qquad \text{for } n = 0, 1, \ldots. \qquad (1)$$

This sequence is purely periodic with least period t for some $t|T$. We may restrict ourselves to the case where $t = T$ and $D < T$. If $D < T$, then we have $t = T$ if and only if, for all proper divisors d of T, the polynomial $f(X)$ is not of the form $f(X) = g(X^{T/d})$ with a polynomial $g(X) \in \mathbb{F}_p[X]$. For example, this is guaranteed if T is a prime or if $f(X)$ is a permutation polynomial of $\mathbb{F}_p$ (or more generally, $f(X)$ is injective on the group generated by η).

We study exponential sums over $\mathbb{F}_p$ which in the simplest case are of the form

$$\sum_{n=0}^{N-1} \chi(\gamma_n) \qquad \text{for } 1 \leq N \leq T,$$

where χ is a nontrivial additive character of $\mathbb{F}_p$. Upper bounds for these exponential sums are then applied to the analysis of a new nonlinear method for

T. Helleseth et al. (Eds.): SETA 2004, LNCS 3486, pp. 266–274, 2005.

pseudorandom number generation. This new method is defined as follows. We derive *explicit nonlinear congruential pseudorandom numbers of period T* in the interval $[0, 1)$ by putting

$$y_n = \gamma_n/p, \quad n = 0, 1, \ldots.$$

After some auxiliary results in Section 2 we prove some new bounds for complete and incomplete exponential sums over finite fields in Section 3 which allow us to give nontrivial results on the distribution of sequences of explicit nonlinear congruential pseudorandom numbers of period T. The application to explicit nonlinear congruential pseudorandom numbers is presented in Section 4.

Similar results on a different family of explicit nonlinear congruential pseudorandom numbers of period p were obtained in [13].

2   Auxiliary Results

We recall Weil's bound on additive character sums (see [8–Theorem 5.38], [19–Chapter II, Theorem 2E]).

Lemma 1. *Let χ be a nontrivial additive character of $\mathbb{F}_p$ and g be a nonconstant polynomial over $\mathbb{F}_p$. Then we have*

$$\left| \sum_{\xi \in \mathbb{F}_p} \chi(g(\xi)) \right| \leq (\deg(g) - 1)p^{1/2}.$$

For the following analog on hybrid character sums see [19–Chapter II, Theorem 2G].

Lemma 2. *Let χ be a nontrivial additive character and ψ a nontrivial multiplicative character of $\mathbb{F}_p$ and g a nonconstant polynomial over $\mathbb{F}_p$. Then we have*

$$\left| \sum_{\xi \in \mathbb{F}_p} \chi(g(\xi))\psi(\xi) \right| \leq \deg(g)p^{1/2}.$$

Lemma 3. *Let $\gamma_0, \gamma_1, \ldots$ be a sequence of the form (1). If $\mu_0, \mu_1, \ldots, \mu_{s-1} \in \mathbb{F}_p$ and*

$$\sum_{i=0}^{s-1} \mu_i \gamma_{n+i} = c, \quad 0 \leq n \leq T - 1,$$

for some $c \in \mathbb{F}_p$, then either

$$\mu_0 = \mu_1 = \ldots = \mu_{s-1} = 0$$

or

$$s \geq \mathrm{w}(f),$$

where $\mathrm{w}(f)$ denotes the weight of $f(X)$, i. e., the number of nonzero coefficients of $f(X)$.

Proof. We assume that not all μ_i are zero and denote by j the largest index with $\mu_j \neq 0$ (so $0 \leq j \leq s-1$). Then we have

$$\sum_{i=0}^{j} \mu_i \gamma_{n+i} = c, \quad 0 \leq n \leq T-1, \tag{2}$$

and

$$\sum_{i=1}^{j+1} \mu_{i-1} \gamma_{n+i} = c, \quad 0 \leq n \leq T-1. \tag{3}$$

Subtracting (3) from (2) yields

$$\mu_0 \gamma_n + \sum_{i=1}^{j} (\mu_i - \mu_{i-1}) \gamma_{n+i} - \mu_j \gamma_{n+j+1} = 0, \quad 0 \leq n \leq T-1.$$

Hence, $j+1$ is at least as large as the linear complexity L of the sequence $\gamma_0, \gamma_1, \ldots$, i.e., the order of the shortest linear recurrence relation over $\mathbb{F}_p$

$$\gamma_{n+L} = \sum_{i=0}^{L-1} \sigma_i \gamma_{n+i}, \quad 0 \leq n \leq T-1,$$

satisfied by the sequence. Lemma 3 follows from the well-known result

$$L = \mathrm{w}(f) \tag{4}$$

of Blahut [1]. We refer to [7–Section 6.8] for a proof. □

Put $e_T(z) = \exp(2\pi i z/T)$.

Lemma 4. *For any integer $1 \leq N \leq T$ we have*

$$\sum_{u=1}^{T-1} \left| \sum_{n=0}^{N-1} e_T(un) \right| \leq T \left(\frac{4}{\pi^2} \log T + 0.8 \right).$$

Proof. We have

$$\sum_{u=1}^{T-1} \left| \sum_{n=0}^{N-1} e_T(un) \right| = \sum_{u=1}^{T-1} \left| \frac{\sin(\pi N u/T)}{\sin(\pi u/T)} \right|$$

$$\leq \frac{4}{\pi^2} T \log T + 0.38T + 0.608 + 0.116 \frac{\gcd(N,T)^2}{T}$$

by [2–Theorem 1]. □

3   Bounds for Exponential Sums

Let $\gamma_0, \gamma_1, \ldots$ be the sequence of elements of $\mathbb{F}_p$ generated by (1). For a nontrivial additive character χ of $\mathbb{F}_p$, for $\mu_0, \mu_1, \ldots, \mu_{s-1} \in \mathbb{F}_p$, and for an integer N with $1 \leq N \leq T$ we consider the exponential sums

$$S_N = \sum_{n=0}^{N-1} \chi \left(\sum_{i=0}^{s-1} \mu_i \gamma_{n+i} \right).$$

Theorem 1. *Let* $1 \leq s < \mathrm{w}(f)$ *and suppose that* $\mu_0, \mu_1, \ldots, \mu_{s-1} \in \mathbb{F}_p$ *are not all 0. Then we have*

$$|S_T| \leq \left(D - \frac{T}{p-1} \right) p^{1/2} + \frac{T}{p-1}.$$

Proof. We have

$$|S_T| = \left| \sum_{n=0}^{T-1} \chi \left(\sum_{i=0}^{s-1} \mu_i f(\eta^{n+i}) \right) \right|$$

$$= \frac{T}{p-1} \left| \sum_{\xi \in \mathbb{F}_p^*} \chi \left(\sum_{i=0}^{s-1} \mu_i f(\eta^i \xi^{(p-1)/T}) \right) \right|.$$

Since at least one μ_i is nonzero and $s < \mathrm{w}(f)$, Lemma 3 implies that

$$\sum_{i=0}^{s-1} \mu_i f(\eta^i X^{(p-1)/T})$$

is not constant and the result follows by Lemma 1.   □

Theorem 2. *Let* $1 \leq s < \mathrm{w}(f)$ *and suppose that* $\mu_0, \mu_1, \ldots, \mu_{s-1} \in \mathbb{F}_p$ *are not all 0. Then we have*

$$|S_N| < Dp^{1/2} \left(\frac{4}{\pi^2} \log T + 1.8 \right) \qquad \text{for } 1 \leq N < T.$$

Proof. With $\sigma_n = \sum_{i=0}^{s-1} \mu_i \gamma_{n+i}$ we have

$$S_N = \sum_{n=0}^{T-1} \chi(\sigma_n) \sum_{t=0}^{N-1} \frac{1}{T} \sum_{u=0}^{T-1} e_T(u(n-t))$$

$$= \frac{1}{T} \sum_{u=0}^{T-1} \left(\sum_{t=0}^{N-1} e_T(-ut) \right) \left(\sum_{n=0}^{T-1} \chi(\sigma_n) e_T(un) \right)$$

$$= \frac{N}{T} \sum_{n=0}^{T-1} \chi(\sigma_n) + \frac{1}{T} \sum_{u=1}^{T-1} \left(\sum_{t=0}^{N-1} e_T(-ut) \right) \left(\sum_{n=0}^{T-1} \chi(\sigma_n) e_T(un) \right),$$

and so

$$|S_N| \leq \frac{N}{T}|S_T| + \frac{1}{T}\sum_{u=1}^{T-1}\left|\sum_{t=0}^{N-1}e_T(ut)\right|\left|\sum_{n=0}^{T-1}\chi(\sigma_n)e_T(un)\right|.$$

For $1 \leq u \leq T-1$ we define the nontrivial multiplicative character ψ_u of $\mathbb{F}_p$ by

$$\psi_u(\vartheta^n) = e_T(un), \quad 0 \leq n \leq p-2,$$

with a primitive element ϑ of $\mathbb{F}_p$. Then we have

$$\left|\sum_{n=0}^{T-1}\chi(\sigma_n)e_T(un)\right| = \frac{T}{p-1}\left|\sum_{\xi\in\mathbb{F}_p^*}\chi\left(\sum_{i=0}^{s-1}\mu_i f(\eta^i\xi^{(p-1)/T})\right)\psi_u(\xi)\right|$$

$$\leq Dp^{1/2}$$

by Lemma 2. Lemma 4 yields

$$\sum_{u=1}^{T-1}\left|\sum_{t=0}^{N-1}e_T(ut)\right|\left|\sum_{n=0}^{T-1}\chi(\sigma_n)e_T(un)\right| \leq Dp^{1/2}\sum_{u=1}^{T-1}\left|\sum_{t=0}^{N-1}e_T(ut)\right|$$

$$\leq Dp^{1/2}T\left(\frac{4}{\pi^2}\log T + 0.8\right).$$

Hence we obtain by Theorem 1,

$$|S_N| \leq \frac{N}{T}\left(\left(D - \frac{T}{p-1}\right)p^{1/2} + \frac{T}{p-1}\right)$$

$$+ Dp^{1/2}\left(\frac{4}{\pi^2}\log T + 0.8\right).$$

Simple calculations yield the theorem. $\qquad\square$

4   Discrepancy Bound

We use the bounds for exponential sums obtained in Theorems 1 and 2 to derive results on the distribution of sequences of explicit nonlinear congruential pseudorandom numbers of period T over the full period and in parts of the period.

Let $\gamma_0/p, \gamma_1/p, \ldots$ be a sequence of explicit nonlinear congruential pseudorandom numbers of least period $T \geq 2$ obtained from (1) with a polynomial $f(X)$ of degree $D \geq 1$. For any integer $1 \leq N \leq T$ we define the s-dimensional (extreme) discrepancy

$$D_s(N) = \sup_J\left|\frac{A_N(J)}{N} - V(J)\right|,$$

where the supremum is extended over all subintervals J of $[0,1)^s$, $A_N(J)$ is the number of points

$$(\gamma_n/p, \ldots, \gamma_{n+s-1}/p) \in [0,1)^s, \quad 0 \le n \le N-1,$$

falling into J, and $V(J)$ denotes the s-dimensional volume of J.

In the following we establish an upper bound for $D_s(N)$.

Theorem 3. *For any fixed integer $1 \le s < \mathrm{w}(f)$, the s-dimensional discrepancy $D_s(N)$ satisfies*

$$D_s(N) < 1 - \left(1 - \frac{1}{p}\right)^s + \frac{Dp^{1/2}}{N}\left(\frac{4}{\pi^2}\log T + 1.8\right)\left(\frac{4}{\pi^2}\log p + 1.72\right)^s$$

for $1 \le N < T$ and

$$D_s(T) \le 1 - \left(1 - \frac{1}{p}\right)^s + \left(\left(D - \frac{T}{p-1}\right)\frac{p^{1/2}}{T} + \frac{1}{p-1}\right)\left(\frac{4}{\pi^2}\log p + 1.72\right)^s.$$

Proof. By a general discrepancy bound in [14–Corollary 3.11] we obtain

$$D_s(N) \le 1 - \left(1 - \frac{1}{p}\right)^s + \frac{B}{N}\left(\frac{4}{\pi^2}\log p + 1.72\right)^s,$$

where B is the maximum over all $(\mu_0, \ldots, \mu_{s-1}) \in \mathbb{F}_p^s \setminus (0, \ldots, 0)$ of the exponential sums S_N. The result follows from Theorems 1 and 2. $\qquad\square$

5 Final Remarks

For $1 \le D \le T-1$ with $\gcd(D, p-1) = 1$ and $a, b \in \mathbb{F}_p^*$, the polynomial

$$f(X) = a(X+b)^D = a\sum_{i=0}^{D}\binom{D}{i}b^{D-i}X^i \tag{5}$$

of weight $D+1$ is a permutation polynomial of $\mathbb{F}_p$, and so the sequence (1) has least period T. It has linear complexity $D+1$ by (4). Therefore and by [17–Section 2] it passes the D-dimensional lattice test introduced by Marsaglia (see [9]). In contrast to sequences defined with a general polynomial of large weight, it can be rather efficiently generated.

Theorem 3 is nontrivial only if D is at most of the order of magnitude $Tp^{-1/2}(\log p)^{-s}$. However, for polynomials of the form (5) with D close to $p-2$ (in case $T = p-1$), or more generally for rational functions of the form

$$f(X) = a(X+b)^{-d}$$

(with the convention $0^{-1} = 0$) with small d, we can obtain similar results using the following analogs of Lemmas 1 and 2 for rational functions which can be found in [12, 18].

Lemma 5. *Let χ be a nontrivial additive character of $\mathbb{F}_p$ and let f/g be a rational function over $\mathbb{F}_p$. Let v be the number of distinct roots of the polynomial g in the algebraic closure $\overline{\mathbb{F}_p}$ of $\mathbb{F}_p$. Suppose that f/g is not constant. Then*

$$\left| \sum_{\substack{\xi \in \mathbb{F}_p \\ g(\xi) \neq 0}} \chi\left(\frac{f(\xi)}{g(\xi)}\right) \right| \leq (\max(\deg(f), \deg(g)) + v^* - 2)p^{1/2} + \delta,$$

where $v^* = v$ and $\delta = 1$ if $\deg(f) \leq \deg(g)$, and $v^* = v+1$ and $\delta = 0$ otherwise.

Lemma 6. *Let χ be a nontrivial additive character and ψ a nontrivial multiplicative character of $\mathbb{F}_p$ and let f/g be a rational function over $\mathbb{F}_p$. Let v be the number of distinct roots of the polynomial g in the algebraic closure $\overline{\mathbb{F}_p}$ of $\mathbb{F}_p$. Then*

$$\left| \sum_{\substack{\xi \in \mathbb{F}_p^* \\ g(\xi) \neq 0}} \chi\left(\frac{f(\xi)}{g(\xi)}\right) \psi(\xi) \right| \leq (\max(\deg(f), \deg(g)) + v^* - 1)p^{1/2},$$

where $v^* = v$ if $\deg(f) \leq \deg(g)$, and $v^* = v + 1$ otherwise.

The particularly interesting case $d = 1$ is investigated in [20]. In this case we have the following main character sum bound.

Theorem 4. *If $\mu_1, \mu_2, \ldots, \mu_s$ are not all 0, then we have*

$$|S_N| < s\left(2p^{1/2} + 1\right)\left(\frac{4}{\pi^2} \log T + 1.8\right) \text{ for } 1 \leq N < T.$$

These *inversive generators* have also desirable structural properties (see [3, 4, 10]).

Mordell [11] established the bound

$$\left| \sum_{\xi \in \mathbb{F}_p} e_p(f(\xi)) \right| \leq (k_1 k_2 \cdots k_w \gcd(p - 1, k_1, k_2, \ldots, k_w))^{1/2w} p^{1-1/2w}$$

for polynomials of the type

$$f(X) = c_1 X^{k_1} + \cdots + c_w X^{k_w}, \quad 1 \leq k_1 < \ldots < k_w < p - 1, \ p \nmid c_1 \cdots c_w.$$

This bound is nontrivial for a restricted set of polynomials of large degree and can be used to obtain nontrivial discrepancy bounds for these particular polynomials.

For the p-periodic sequences $\gamma_0, \gamma_1, \ldots$ defined by

$$\gamma_n = f(n) \quad \text{for } n = 0, 1, \ldots$$

we have discrepancy bounds of the same order of magnitude as in Theorem 3 for all dimensions s with $2 \leq s \leq \deg(f)$ (see [13]). For the analogous results on the inversive sequence

$$\gamma_n = (an + b)^{-1} \quad \text{for } n = 0, 1, \ldots$$

see [5]. Appropriate bounds for corresponding sequences over arbitrary finite fields were obtained in [16].

Recursively defined generators

$$\gamma_{n+1} = f(\gamma_n) \quad \text{for } n = 0, 1, \ldots$$

with some initial value u_0 were investigated in [15]. However, the results are much weaker than for the explicitly defined sequences. For the particular case of inversive sequences

$$\gamma_{n+1} = a\gamma_n^{-1} + b \quad \text{for } n = 0, 1, \ldots$$

much better results were proven in [6]. The character sum bounds are of the order of magnitude $O(N^{1/2}p^{1/4})$ (vs. $O(p^{1/2}\log p)$ in Theorem 4 or in [5]). The method of [6, 15] can also be applied to explicit generators yielding character sum bounds of the order of magnitude $O(N^{1/2}p^{1/4})$.

Acknowledgments

Parts of this paper were written during a visit of the second author to the National University of Singapore. He wishes to thank the Department of Mathematics for the hospitality. The second author is supported by the Austrian Academy of Sciences and by the Austrian Science Fund (FWF) grant S8313. The research of the first author is partially supported by the grant R-394-000-011-422 with Temasek Laboratories in Singapore.

References

1. Blahut, R.E.: Theory and Practice of Error Control Codes. Addison-Wesley, Reading, MA (1983)
2. Cochrane, T.: On a trigonometric inequality of Vinogradov. J. Number Theory **27** (1987) 9–16
3. Dorfer, G., Winterhof, A.: Lattice structure and linear complexity profile of nonlinear pseudorandom number generators. Appl. Algebra Engrg. Comm. Comput. **13** (2003) 499–508
4. Dorfer, G., Winterhof, A.: Lattice structure of nonlinear pseudorandom number generators in parts of the period. In: Niederreiter, H. (ed.): Monte Carlo and Quasi-Monte Carlo Methods 2002. Springer-Verlag, Berlin (2004) 199–211
5. Eichenauer-Herrmann, J.: Statistical independence of a new class of inversive congruential pseudorandom numbers. Math. Comp. **60** (1993) 375–384

6. Gutierrez, J., Niederreiter, H., Shparlinski, I.: On the multidimensional distribution of inversive congruential pseudorandom numbers in parts of the period. Monatsh. Math. **129** (2000) 31–36
7. Jungnickel, D.: Finite Fields: Structure and Arithmetics. Bibliographisches Institut, Mannheim (1993)
8. Lidl, R., Niederreiter, H.: Finite Fields. Cambridge University Press, Cambridge (1997)
9. Marsaglia, G.: The structure of linear congruential sequences. In: Zaremba, S.K. (ed.): Applications of Number Theory to Numerical Analysis. Academic Press, New York (1972) 249–285
10. Meidl, W., Winterhof, A.: On the linear complexity profile of some new explicit inversive pseudorandom numbers. J. Complexity **20** (2004) 350–355
11. Mordell, L.J.: On a sum analogous to a Gauss's sum. Quart. J. Math. **3** (1932) 161–167
12. Moreno, C.J., Moreno, O.: Exponential sums and Goppa codes: I. Proc. Amer. Math. Soc. **111** (1991) 523–531
13. Niederreiter, H.: Statistical independence of nonlinear congruential pseudorandom numbers. Monatsh. Math. **106** (1988) 149–159
14. Niederreiter, H.: Random Number Generation and Quasi-Monte Carlo Methods. SIAM, Philadelphia (1992)
15. Niederreiter, H., Shparlinski, I.: On the distribution and lattice structure of nonlinear congruential pseudorandom numbers. Finite Fields Appl. **5** (1999) 246–253
16. Niederreiter, H., Winterhof, A.: Incomplete exponential sums over finite fields and their applications to new inversive pseudorandom number generators. Acta Arith. **93** (2000) 387–399
17. Niederreiter, H., Winterhof, A.: Lattice structure and linear complexity of nonlinear pseudorandom numbers. Appl. Algebra Engrg. Comm. Comput. **13** (2002) 319–326
18. Perel'muter, G.I.: Estimate of a sum along an algebraic curve (Russian). Mat. Zametki **5** (1969) 373–380; Engl. transl. Math. Notes **5** (1969) 223–227
19. Schmidt, W.M.: Equations over Finite Fields. Lecture Notes in Mathematics, Vol. 536. Springer, Berlin (1976)
20. Winterhof, A.: On the distribution of some new explicit inversive pseudorandom numbers and vectors. In: Niederreiter, H., Talay, D. (eds.): Monte Carlo and Quasi-Monte Carlo Methods 2004, Springer, Berlin, to appear

Distribution of r-Patterns in the Most Significant Bit of a Maximum Length Sequence over $\mathbb{Z}_{2^l}$

Patrick Solé[1] and Dmitrii Zinoviev[2]

[1] CNRS-I3S, ESSI, Route des Colles,
06 903 Sophia Antipolis, France
`ps@essi.fr`
[2] Institute for Problems of Information Transmission,
Russian Academy of Sciences, Bol'shoi Karetnyi,
19, GSP-4, Moscow, 101447, Russia
`dzinov@iitp.ru`

Abstract. The number of subwords of length r and of given value within a period of a sequence in the title is shown to be close to equidistribution. Important tools in the proof are a higher order correlation and Galois ring character sum estimates.

Keywords: Galois Rings, Maximum Length Sequences, Most Significant Bit, Higher Order Correlation.

1 Introduction

The analogues of M-sequences over $\mathbb{Z}_{2^l}$ were introduced by Dai in 1992 [1], under the name ML-sequence over rings. In [4] the distribution properties of the most significant bit of these ML-sequences were investigated. The frequency of zeroes (resp. ones) in a period of such a binary sequence was shown to be close to $1/2$, up to a bound of order square root of the period. These bounds were improved (by roughly removing the factor of order 2^l from the bounding constant) in [10]. In [2] the frequency of subwords of length r in a period was shown to be close to $1/2^r$. The present contribution [1] improves on the latter bound. Essential tools are the higher order correlation and the character sums estimates of [7].

2 Preliminaries

Let $R = GR(2^l, m)$ denote the Galois ring of characteristic 2^l with 2^{lm} elements. Let ξ be an element in $GR(2^l, m)$ that generates the Teichmüller set $\mathcal{T}$

---

[1] The paper has been written under the partial financial support of the Russian fund for fundamental research (under project No. 03 - 01 - 00098).

T. Helleseth et al. (Eds.): SETA 2004, LNCS 3486, pp. 275–281, 2005.

of $GR(2^l, m)$. Specifically, let $\mathcal{T} = \{0, 1, \xi, \xi^2, \ldots, \xi^{2^m-2}\}$ and $\mathcal{T}^* = \{1, \xi, \xi^2, \ldots, \xi^{2^m-2}\}$. The 2-adic expansion of $x \in GR(2^l, m)$ is given by

$$x = x_0 + 2x_1 + \cdots + 2^{l-1}x_{l-1},$$

where $x_0, x_1, \ldots, x_{l-1} \in \mathcal{T}$. The Frobenius operator F is defined for such an x as

$$F(x_0 + 2x_1 + \cdots + 2^{l-1}x_{l-1}) = x_0^2 + 2x_1^2 + \cdots + 2^{l-1}x_{l-1}^2,$$

and the trace Tr, from $GR(2^l, m)$ down to $\mathbb{Z}_{2^l}$, as

$$\text{Tr}(x) := \sum_{j=0}^{m-1} F^j(x).$$

We also define another trace tr from $\mathbb{F}_{2^m}$ down to $\mathbb{F}_2$ as

$$\text{tr}(x) := \sum_{j=0}^{m-1} x^{2^j}.$$

Throughout this note, we let $n = 2^m$ and $R^* = R \backslash 2R$. Let $\gamma = \xi(1+2\lambda) \in R$, where $\xi \in \mathcal{T}$ and $\lambda \in R^*$. Assume $1 + 2\lambda$ is of order 2^{l-1}. Since ξ is of order $2^m - 1$ then γ is an element of order $N = 2^{l-1}(2^m - 1)$. Following [4-Lemma 2], we define the sequence:

$$S_{l,m} := \{(\text{Tr}(\alpha\gamma^t))_{t=0}^{N-1} \mid \alpha \in R^*\}. \tag{1}$$

Let MSB : $\mathbb{Z}_{2^l}^n \to \mathbb{Z}_2^n$ be the most-significant-bit map, i.e.,

$$\text{MSB}(x_0 + 2x_1 + \ldots + 2^{l-1}x_{l-1}) := x_{l-1}.$$

3   The r-th Moments

Let l be a positive integer (without loss of generality we assume that $l \geq 4$) and $\omega = e^{2\pi i/2^l}$ be a primitive 2^l-th root of 1 in $\mathbb{C}$. Let ψ_k be the additive character of $\mathbb{Z}_{2^l}$ such that

$$\psi_k(x) = \omega^{kx}.$$

Let $\mu : \mathbb{Z}_{2^l} \to \{\pm 1\}$ be the mapping $\mu(t) = (-1)^c$, where c is the most significant bit of $t \in \mathbb{Z}_{2^l}$, i.e. it maps $0, 1, \ldots, 2^{l-1} - 1$ to $+1$ and $2^{l-1}, 2^{l-1} + 1, \ldots, 2^l - 1$ to -1. Our goal is to express this map as a linear combination of characters. Recall the Fourier transformation formula on $\mathbb{Z}_{2^l}$:

$$\mu = \sum_{j=0}^{2^l-1} \mu_j \psi_j, \quad \text{where } \mu_j = \frac{1}{2^l} \sum_{x=0}^{2^l-1} \mu(x)\psi_j(-x). \tag{2}$$

For all $\beta \in R = GR(2^l, m)$, we denote by Ψ_β the character

$$\Psi_\beta : R \to \mathbb{C}^*, \quad x \mapsto \omega^{\text{Tr}(\beta x)}.$$

Note that for the defined above ψ_k and Ψ_β, we have:

$$\psi_k(\mathrm{Tr}(\beta x)) = \Psi_{\beta k}(x). \tag{3}$$

The following lemma follows from [7].

Lemma 1. *For all $\lambda \in R$, $\lambda \neq 0$, we have:*

$$\left| \sum_{x \in \mathcal{T}} \Psi_\lambda(x) \right| \leq (2^{l-1} - 1)\sqrt{2^m}.$$

Proof. We restate [7–Theorem 1] for the special Galois Ring of concern here. Let $f(X)$ denote a polynomial in $R[X]$ and let

$$f(X) = F_0(X) + 2F_1(X) + \ldots + 2^{l-1}F_{l-1}(X)$$

denote its 2-adic expansion. Let d_i be the degree in X of F_i. Let ψ be an arbitrary additive character of R, and set D_f to be the *weighted degree* of f, defined as

$$D_f = \max(d_0 2^{l-1}, d_1 2^{l-2}, \ldots, d_{l-1}).$$

With the above notation, we have (under mild technical conditions) the bound

$$\left| \sum_{x \in \mathcal{T}} \psi(f(x)) \right| \leq (D_f - 1)2^{m/2}.$$

See [7] for details. The result follows upon considering a linear f so that $d_0 = d_1 = \ldots = d_{l-1} = 1$, and D_f being equal to 2^{l-1}. □

We now have the following results on the r-th moment function of the binary sequence $(c_t)_{t \in \mathbb{N}}$, which is the image of $S_{l,m}$ defined by (1) under the MSB map.

First, we need the following technical lemma:

Lemma 2. *Let $\gamma = \xi(1 + 2\lambda) \in R$, where $\xi \in \mathcal{T}^*$, $\lambda \in R^* = R \backslash 2R$, and $1 + 2\lambda \in R$ is an element of order 2^{l-1}. Set $N = 2^{l-1}(2^m - 1)$. Then, for any $0 \neq \beta \in R$, we have:*

$$\sum_{j=0}^{N-1} \Psi_\beta(\gamma^j) = \sum_{j=0}^{2^{l-1}-1} \left(\sum_{x \in \mathcal{T}^*} \Psi_{\beta(1+2\lambda)^j}(x) \right).$$

Proof. Since $(2^{l-1}, 2^m - 1) = 1$, as j ranges over $\{0, 1, \ldots, N-1\}$, the set of ordered pairs

$$\{(j \,(\mathrm{mod}\, 2^{l-1}), j \,(\mathrm{mod}\, 2^m - 1))\}$$

runs over all pairs (j_1, j_2), where $j_1 \in \{0, 1, \ldots, 2^{l-1}-1\}$ and $j_2 \in \{0, 1, \ldots, 2^m - 2\}$. Thus the set

$$\{\gamma^j; \ j = 0, 1, \ldots, N-1\} = \{\xi^j(1 + 2\lambda)^j; \ j = 0, 1, \ldots, N-1\}.$$

is equal to the cartesian product of sets:

$$\{(1 + 2\lambda)^{j_1}; \ j_1 = 0, 1, \ldots, 2^{l-1} - 1\} \times \{\xi^{j_2}; \ j_2 = 0, 1, \ldots, 2^m - 2\} \tag{4}$$

From (4) we obtain:

$$\sum_{j=0}^{N-1} \Psi_\beta(\gamma^j) = \sum_{j_1=0}^{2^{l-1}-1} \sum_{j_2=0}^{2^m-2} \Psi_\beta((1+2\lambda)^{j_1}\xi^{j_2}),$$

whereas observing that

$$\Psi_\beta((1+2\lambda)^{j_1}\xi^{j_2}) = \Psi_{\beta'}(\xi^{j_2}),$$

where $\beta' = \beta(1+2\lambda)^{j_1}$, yields:

$$\sum_{j_1=0}^{2^{l-1}-1} \sum_{j_2=0}^{2^m-2} \Psi_{\beta(1+2\lambda)^{j_1}}(\xi^{j_2}).$$

The Lemma follows.   □

We now proceed to bound the r-th moments, by bounding first the correlation of order r.

Theorem 3. *With notation as above, and for all pair-wise distinct phase shifts* $\tau_1, \tau_2,... \tau_r \in [1, 2^m - 2]$ *(r is a positive integer), let*

$$\Theta(\tau_1, ..., \tau_r) = \sum_{t=0}^{N-1} (-1)^{c_{t+\tau_1}+c_{t+\tau_2}+...+c_{t+\tau_r}},$$

where $c_t = MSB(Tr(\alpha\gamma^t))$. *We then have (for $l \geq 4$) the bound :*

$$|\Theta(\tau_1, ..., \tau_r)| \leq \left(\frac{2l}{\pi}\ln(2)+1\right)^r 2^{l-1}[(2^{l-1}-1)\sqrt{2^m}+1] \approx C_l\sqrt{2^m}.$$

Here C_l is a constant in l of order $l^r 2^{2l}$.

Proof.   Again let $\gamma = \xi(1+2\lambda) \in R$, where $\xi \in T$ is a generator of the Teichmüller set and $\lambda \in R^*$. As we have $c_t = MSB(Tr(\alpha\gamma^t))$, and by (2), we obtain that $(-1)^{c_t}$ is equal to:

$$\mu(Tr(\alpha\gamma^t)) = \sum_{j=0}^{2^l-1} \mu_j \psi_j(Tr(\alpha\gamma^t)) = \sum_{j=0}^{2^l-1} \mu_j \Psi_{\alpha j}(\gamma^t).$$

Changing the order of summation, we obtain that:

$$\Theta(\tau_1, ..., \tau_r) = \sum_{j_1=0}^{2^l-1} ... \sum_{j_r=0}^{2^l-1} \mu_{j_1}...\mu_{j_r} \sum_{t=0}^{N-1} \Psi_\beta(\gamma^t).$$

Here $\beta = \alpha(j_1\gamma^{\tau_1} + ... + j_r\gamma^{\tau_r})$ and $j_1\gamma^{\tau_1} + ... + j_r\gamma^{\tau_r} \neq 0$. Applying Corollary 7.4 of [9] (for $l \geq 4$), we have:

$$\sum_{j_1=0}^{2^l-1} ... \sum_{j_r=0}^{2^l-1} |\mu_{j_1}...\mu_{j_r}| = \left(\sum_{j=0}^{2^l-1} |\mu_j|\right)^r \leq \left(\frac{2l}{\pi}\ln(2)+1\right)^r. \tag{5}$$

Applying Lemma 2, we have:

$$\sum_{j=0}^{N-1} \Psi_\beta(\gamma^j) = \sum_{j=0}^{2^{l-1}-1} \left(\sum_{x \in \mathcal{T}} \Psi_{\beta(1+2\lambda)^j}(x) \right),$$

where $0 \neq \beta(1+2\lambda)^j \in R$ so that each sum over x can be estimated using Lemma 1. Thus, we have:

$$\left| \sum_{j=0}^{N-1} \Psi_\beta(\gamma^j) \right| \leq 2^{l-1}[(2^{l-1}-1)\sqrt{2^m} + 1]. \tag{6}$$

Combining (5) with (6) the Lemma follows. □

4 Distribution of r-Patterns

In this note our main goal is to study the distribution of elements in the binary sequence $(c_t)_{t \in \mathbb{N}}$, which is the image of $S_{l,m}$ defined by (1) under the MSB map.

For a positive integer r, fix $\tau_1, \ldots, \tau_r \in [1, \ldots, 2^m - 2]$ and let $v = (v_1, ..., v_r) \in \mathbb{Z}_2^r$. Then define $N(v)$ to be the number of integers $t \in [0, N-1]$ such that

$$c_{t+\tau_i} = v_i, \ 1 \leq i \leq r.$$

If $u = (u_1, ..., u_r) \in \mathbb{Z}_2^r$ and $v = (v_1, ..., v_r) \in \mathbb{Z}_2^r$, let

$$\langle u \cdot v \rangle = \sum_{i=1}^{r} u_i v_i.$$

The main result is the following estimate:

Theorem 4. *With notation as above, the bound :*

$$\left| \frac{N(v)}{N} - \frac{1}{2^r} \right| \leq \left(\frac{2l}{\pi} \ln(2) + 1 \right)^r 2^{(l-1)} \sqrt{2^{-m}},$$

where $N = 2^{l-1}(2^m - 1)$.

Proof. For any $t \in [0, N-1]$, let $\mathbf{c}_t = (c_{t+\tau_1}, ..., c_{t+\tau_r}) \in \mathbb{Z}_2^r$. Let $u = (u_1, \ldots, u_r) \in \mathbb{Z}_2^r$. Then

$$\langle u \cdot \mathbf{c}_t \rangle = \sum_{i=1}^{r} u_i c_{t+\tau_i}$$

and by definition of $N(v)$ we have

$$S(u) = \sum_{t=0}^{N-1} (-1)^{\langle u \cdot \mathbf{c}_t \rangle} = \sum_{v \in \mathbb{Z}_2^r} (-1)^{\langle u \cdot v \rangle} N(v).$$

Thus we have

$$N(v) = \frac{1}{2^r} \sum_{u \in \mathbb{Z}_2^r} (-1)^{\langle u \cdot v \rangle} S(u).$$

Note that $S(0) = N$, and we obtain

$$N(v) = \frac{1}{2^r} \left(N + \sum_{0 \neq u \in \mathbb{Z}_2^r} (-1)^{\langle u \cdot v \rangle} S(u) \right).$$

It implies

$$\left| \frac{N(v)}{N} - \frac{1}{2^r} \right| = \frac{1}{N} \left| \sum_{0 \neq u \in \mathbb{Z}_2^r} (-1)^{\langle u \cdot v \rangle} S(u) \right|$$

$$< \frac{2^r}{N} \times \max_{0 \neq u \in \mathbb{Z}_2^r} \{ |S(u)| \}.$$

For any non-zero vector $u = (u_1, \ldots, u_r) \in \mathbb{Z}_2^r$, we have

$$\langle u \cdot \mathbf{c}_t \rangle = \sum_{i=1}^r u_i c_{t+\tau_i} = \sum_{i \in I} c_{t+\tau_i},$$

where $I = \{ i \in [1,r] : u_i = 1 \}$. Let I be $\{ i_1, \ldots, i_s \}$ then

$$S(u) = \Theta(\tau_{i_1}, \ldots, \tau_{i_s}),$$

where $s \leq r$, thus we can apply Theorem 3 to obtain the bound

$$\max_{0 \neq u \in \mathbb{Z}_2^r} \{ |S(u)| \} \leq \left(\frac{2l}{\pi} \ln(2) + 1 \right)^r 2^{l-1} [(2^{l-1} - 1)\sqrt{2^m} + 1].$$

Substituting this estimate into (7) and using that $N = 2^{l-1}(2^m - 1)$ the Theorem follows. □

References

1. Zong-Duo Dai, Binary sequences derived from ML-sequences over rings I: period and minimal polynomial, J. Cryptology (1992) 5:193–507.
2. Zong-Duo Dai, Ye Dingfeng, Wang Ping, Fang Genxi, Distribution of r−patterns in the highest level of p−adic sequences over Galois Rings, Golomb Symposium, USC 2002.
3. H. Davenport, *Multiplicative Number Theory*, GTM 74, Springer Verlag (2000).
4. S. Fan, W. Han, Random properties of the highest level sequences of primitive Sequences over $\mathbb{Z}_{2^e}$, IEEE Trans. Inform. Theory vol.IT-**49**,(2003) 1553–1557.
5. T. Helleseth, P.V. Kumar, Sequences with low Correlation, in *Handbook of Coding Theory*, vol. II, V.S. Pless, W.C. Huffman, eds, North Holland (1998), 1765–1853.
6. P.V. Kumar, T. Helleseth, An expansion of the coordinates of the trace function over Galois rings, AAECC **8**, 353–361, (1997).

7. P.V. Kumar, T. Helleseth and A.R. Calderbank, An upper bound for Weil exponential sums over Galois rings and applications, IEEE Trans. Inform. Theory, vol. IT-**41**, pp. 456–468, May 1995.

8. J. Lahtonen, On the odd and the aperiodic correlation properties of the binary Kasami sequences, IEEE Trans. Inform. Theory, vol.IT-**41**, pp. 1506–1508, September 1995.

9. J. Lahtonen, S. Ling, P. Solé, D. Zinoviev, $\mathbb{Z}_8$-Kerdock codes and pseudo-random binary sequences, Journal of Complexity, Volume 20, Issues 2-3, April-June 2004, Pages 318-330.

10. P. Solé, D. Zinoviev,The most significant bit of maximum length sequences over $\mathbb{Z}_{2^l}$ autocorrection and imbalance, IEEE IT, in press (2004).

Algebraic Feedback Shift Registers Based on Function Fields

Andrew Klapper

University of Kentucky, Department of Computer Science,
779 A Anderson Hall, Lexington, KY 40506-0046, USA
klapper@cs.uky.edu
http://www.cs.uky.edu/~klapper/

Abstract. We study algebraic feedback shift registers (AFSRs) based on quotients of polynomial rings in several variables over a finite field. These registers are natural generalizations of linear feedback shift registers. We describe conditions under which such AFSRs produce sequences with various ideal randomness properties. We also show that there is an efficient algorithm which, given a prefix of a sequence, synthesizes a minimal such AFSR that outputs the sequence.

1   Introduction

Linear feedback shift registers (LFSRs) are useful for many applications, including cryptography, coding theory, CDMA, radar ranging, and pseudo-Monte Carlo simulation. There is a large body of literature on these simple devices for generating pseudorandom sequences. Naturally, many variations and generalizations of LFSRs have been studied. Algebraic feedback shift registers (AFSRs) are a very general class of sequence generators that include LFSRs as a special case [8]. Each class of AFSRs is based on a choice of an algebraic ring R and a parameter $r \in R$. In the case of LFSRs R is $F[x]$, the ring of polynomials in one variable x over a finite field F, and $r = x$. In previous work we have studied AFSRs over the integers [7] and certain finite extensions of the integers [3, 5, 9]. More recently [4] we have studied AFSRs over $F[x]$ when $r \neq x$. In this paper we generalize this last work further and study AFSRs over polynomial rings of transcendence degree one over finite fields. We call this the function field case since, in the language of algebraic geometry, the field of fractions of such a ring is the function field of an algebraic curve. We describe the basic properties (such as periodicity) of the resulting sequences.

Two aspects of LFSRs that make them especially interesting are the statistical randomness of maximum period LFSR sequences (*m-sequences*) and the existence of the Berlekamp-Massey algorithm. This algorithm efficiently solves the register synthesis problem: given a prefix of a sequence, find a smallest LFSR that outputs the sequence. This algorithm plays a role in both cryptanalysis and error correction. It has been generalized to the setting of codes defined over algebraic curves, or *algebraic geometry codes*, where it is used to solve the so called

T. Helleseth et al. (Eds.): SETA 2004, LNCS 3486, pp. 282–297, 2005.

"key equation" [12, 13, 14], and it has been generalized to the setting of FCSRs and some more general AFSRs [9]. There is, however, no known generalization of the Berlekamp-Massey algorithm that works for all classes of AFSRs, although there is a general approach that works if there is a reasonable analog of degree [9]. We show here that there is such an analog in the function field case, resulting in a solution to the register synthesis problem for AFSRs based on function fields. We characterize the maximal period sequences for a given length of AFSR and show that they have certain good statistical properties — uniform distribution of small subsequences, the run property, and ideal autocorrelations. We also compare them to Blackburn's classification of sequences [1] with the shift and add (SAA) property and show that all sequences with the SAA property and uniform distributions are maximal period sequences over function fields.

A class of AFSRs depends on a choice of ring R, $r \in R$, and $S \subseteq R$, a complete set of representatives modulo r. An AFSR is determined by $q_0, q_1, \cdots, q_k \in R$, q_0 invertible modulo r. States are tuples $(a_0, a_1, \cdots, a_{k-1}; m)$ with $a_i \in S$ and $m \in R$. The element m is called the *extra memory*. The AFSR changes states as follows. There are unique $a_k \in S$ and $m' \in R$ so that

$$q_0 a_k + rm' = m + \sum_{i=1}^{k} q_i a_{k-i}. \tag{1}$$

Then the new state is $(a_1, a_2, \cdots, a_k; m')$. An LFSR is an AFSR with $R = F[x]$ for some field F, $r = x$, $S = F$, all $q_i \in S$, and $q_0 = 1$. AFSRs are analyzed in terms of coefficient sequences of r-adic numbers $\sum_{i=0}^{\infty} a_i r^i$, $a_i \in S$. The set of r-adic numbers is denoted by R_r. R_r is a ring, and such an r-adic number is invertible in R_r if and only if a_0 is invertible modulo r. The r-adic number $\alpha(A, r) = \sum_{i=0}^{\infty} a_i r^i$ associated with the output sequence $A = a_0, a_1, \cdots$ is in fact a rational element u/q where $u \in R$ and $q = \sum_{i=1}^{k} q_i r^i - q_0$. In some cases maximal period AFSR sequences, or ℓ-sequences, share many desirable properties with m-sequences [8]. We have shown, for example, that when R is $\mathbf{Z}$ or $\mathbf{Z}[\sqrt{N}]$ for some integer N, then ℓ-sequences have excellent randomness properties, similar to those of m-sequences [4, 5, 7].

One can also consider rings R with nonzero characteristic. Previously we showed that if $R = F[x]$ and $R/(r)$ is not a prime field, ℓ-sequences are distinct from m-sequences but have similar statistical properties: the distributions of subsequences in these ℓ-sequences are as uniform as possible, their distributions of lengths of runs matches the expectation and they have the shift and add property [4]. Thus, with an appropriate definition, they have ideal autocorrelations. In this paper we describe conditions under which the same randomness properties hold in the higher genus case.

It has been shown by Blackburn that every sequence of period $p^{hk} - 1$ over an extension F_{p^h} of F_p with the shift and add property is the F_p-linear image of successive powers of a primitive element in $F_{p^{hk}}$ [1]. Moreover, the sequences that also have uniform distributions can be identified within this classification. The advantage of the approach described here is that it leads to more efficient implementations of generators of the sequences than Blackburn's description.

2   Setting and Hypotheses

In this section we describe the general setting for the sequences we are studying and various conditions that may need to hold to obtain sequences with good properties.

Let $p \in \mathbf{Z}$ be prime, $h > 0 \in \mathbf{Z}$, and I and r be an ideal and an element in $\mathbf{F}_{p^h}[x_1, \cdots, x_n]$. Assume that $R = \mathbf{F}_{p^h}[x_1, \cdots, x_n]/I$ has transcendence degree 1 over $\mathbf{F}_{p^h}$ and that $K = R/(r)$ is finite. Then $R/(r)$ is a vector space over $\mathbf{F}_{p^h}$ so its cardinality is a power of p^h, say p^{he}. If $n = 1$, $r = x_1$, and $I = (0)$, then the AFSRs we obtain are exactly the LFSRs.

One goal is to obtain conditions under which the output from an AFSR based on these ingredients is statistically random. In order to construct AFSRs with good statistical properties we need a "well structured" complete set of representatives S for R modulo r.

Hypothesis H1: S is closed under addition and contains F_{p^h}.

It is straightforward to see that such sets S exist in abundance (the F_{p^h}-span of a lift of a basis containing 1 from $R/(r)$ to R with 1 lifted to 1). Any S that satisfies H1 is closed under multiplication by F_p, but possibly not under multiplication by any larger field. In general we can represent any element of R as an r-adic element with coefficients in S, but in order that we get good randomness properties we need to be able to represent the elements of R finitely.

Hypothesis H2: If $v \in R$ then for some $\ell \in \mathbf{Z}$ and $v_0, \cdots, v_\ell \in S$,

$$v = \sum_{i=0}^{\ell} v_i r^i. \tag{2}$$

Since r is not a zero divisor, the representation in equation (2) is unique if $v_\ell \neq 0$. Indeed, suppose that

$$\sum_{i=0}^{m} u_i r^i = \sum_{i=0}^{\ell} v_i r^i$$

for some $u_i, v_i \in S$ with $u_m \neq 0 \neq v_\ell$ and $(u_0, u_1, \cdots) \neq (v_0, v_1, \cdots)$ and let ℓ be the minimal integer so that such a pair of representations exists. Reading this equation modulo r we see that $u_0 = v_0$, so by subtraction we may assume that $u_0 = v_0 = 0$. But the fact that r is not a zero divisor then implies that

$$\sum_{i=0}^{m-1} u_{i+1} r^i = \sum_{i=0}^{\ell-1} v_{i+1} r^i$$

and $(u_1, u_2, \cdots) \neq (v_1, v_2, \cdots)$. This contradicts the minimality of ℓ.

We say that the ℓ in Hypothesis H2 is the *r-degree of v*, $\ell = \deg_r(v)$.

Lemma 1. *If Hypotheses H1 and H2 hold, then for all $u, v \in R$, $\deg_r(u + v) \leq \max(\deg_r(u), \deg_r(v))$. Let $a = \max\{\deg(st : s, t \in S\}$. Then for all $u, v \in R$,*

$$\deg_r(uv) \leq \deg_r(u) + \deg_r(v) + a.$$

To prove randomness properties we need an additional hypothesis. Let V_q denote the set of elements u of R such that v/q has a periodic r-adic expansion.

Hypothesis H3: The elements of V_q are distinct modulo r^k.

3   Periodicity

Let $A = a_0, a_1, \cdots$ be the output from an AFSR based on R, r, S with connection element

$$q = \sum_{i=1}^{k} q_i r^i - q_0, q_i \in S,$$

with q_0 invertible modulo r. It follows that

$$\sum_{i=0}^{\infty} a_i r^i = \frac{u}{q}$$

for some $u \in R$. We call this a rational representation of A. Any left shift of A can be generated by the same AFSR (with a different initial state), so also has a rational representation with denominator q. Our first task is to analyze the period of A. We need one fact from the general theory of AFSRs (which was misstated in the original paper and is correctly stated here).

Theorem 1. *([8]) Let A be periodic. Let U denote the set of elements $v \in R$ such that v/q is a rational representation of a shift of A. Suppose no two elements of U are congruent modulo q and let V be a complete set of representatives modulo q containing U. Then*

$$a_i = q_0^{-1}(wr^{-i} \pmod{q}) \pmod{r}, \tag{3}$$

for some $w \in R/(q)$.

By equation (3) we mean first find the multiplicative inverse δ of the image of r in $R/(q)$. Raise δ to the ith power and multiply by w. Then lift the result to an element of V. Reduce the result modulo r to an element of $R/(r)$. Finally, multiply that element by the inverse of the image of q_0 in $R/(r)$.

Let V_q denote the set of elements u of R such that v/q has a periodic r-adic expansion.

Corollary 1. *If A is periodic and no two elements of V_q are congruent modulo q, then*

$$a_i = q_0^{-1}(wr^{-i} \pmod{q}) \pmod{r}$$

for some $w \in R/(q)$.

In our case the stronger condition in the corollary holds.

Theorem 2. *A is eventually periodic. If S satisfies Hypotheses H1 and H2, then A's eventual period is a divisor of the multiplicative order of r modulo q. If $R/(q)$ is an integral domain, then the period equals the multiplicative order of r modulo q. In general, for a given q there is at least one periodic sequence with connection element q whose period is the multiplicative order of r modulo q.*

Proof: To see that A is periodic, it suffices to show that the r-degree of the extra memory of the AFSR is bounded, for then there are finitely many distinct states of the AFSR. Eventually the state repeats and from then on the output is periodic.

Suppose that at some point the AFSR is in state

$$(a_j, a_{j+1}, \cdots, a_{j+k-1}; m)$$

with $m = \sum_{i=0}^{\ell} m_i r^i$ and $m_0, \cdots, m_\ell \in S$. Also, suppose that the maximal r-degree of the product of two elements of S is d. Then the carry m' of the next state satisfies

$$rm' = m + \sum_{i=1}^{k-1} q_i a_{j+k-i} - q_0 a_{j+k}.$$

The right hand side is divisible by r and its r-degree is at most $\max\{\ell, d\}$, so the r-degree of m' is at most $\max\{\ell - 1, d - 1\}$. Thus the r-degree of the carry decreases monotonically until it is less than d, and then remains less than d forever.

To describe the eventual period, it suffices to consider strictly periodic sequences since q is also a connection element of every shift of A. We claim that no two elements of V_q are congruent modulo q. One consequence of Hypothesis H1 is that the r-adic sum of two periodic sequences is the term-wise sum, so is also periodic. Thus to prove our claim it suffices to show that no nonzero element of V_q is divisible by q. Suppose to the contrary that $uq \in V_q$. Then $u = uq/q$ has a periodic r-ary expansion. But this contradicts the fact that any element of R has a unique r-adic expansion.

Now consider the series of numerators $u_0, u_1, \cdots$ of the rational representations of the r-adic elements associated with the shifts of A. The period is the least t such that $u_t = u_0$. By the argument in the preceding paragraph, this is equivalent to $u_t \equiv u_0 \pmod q$. For every i,

$$\frac{u_{i-1}}{q} = a_{i-1} + r\frac{u_i}{q},$$

and so $u_{i-1} = qa_{i-1} + ru_i$. Therefore $u_i \equiv r^{-1} u_{i-1} \equiv r^{-i} u_0 \pmod q$ by induction. Thus $u_t \equiv u_0 \pmod q$ if and only if $r^t u_0 \equiv u_0 \pmod q$, which is equivalent to $(r^t - 1)u_0 \equiv 0 \pmod q$.

If $R/(q)$ is an integral domain, then this says that i is the multiplicative order of r modulo q. If $R/(q)$ is not an integral domain, then it implies that i is a divisor of the multiplicative order of r modulo q.

Finally, consider the coefficient sequence of the r-adic expansion of $1/q$. This sequence may not be periodic, but it is eventually periodic, for some j the its shift by j positions is periodic. This shift has a rational representation u/q, and by the above argument, $u \equiv r^{-j} \pmod q$. In particular, u is invertible modulo q, so $(r^i - 1)u \equiv 0 \pmod q$ if and only if $r^i \equiv 1 \pmod q$. Thus in this case the period equals the multiplicative order of r modulo q.   □

Corollary 2. (To the proof.) *A has an exponential representation.*

4 ℓ-Sequences and Randomness

Let A be an AFSR of sequence of the type described in Section 2. The period A is largest if $R/(q)$ is a field and r is primitive. Then A has period $p^{hg} - 1$ for some g. A is a *punctured de Bruijn sequence of span k* if each nonzero sequence B of length k in a period of A occurs once and the all-zero sequence of length k does not occur. To produce punctured de Bruijn sequences over $\mathbf{F}_{p^{he}}$ we want the period to be of the form $p^{hek} - 1$. Thus $hg = hek$, so $g = ek$.

Definition 1. *Let $r, q \in R$, $q = \sum_{i=1}^{k} q_i r^i - q_0$ with $q_k \neq 0$, $|R/(r)| = p^{he}$, $S \subseteq R$, and suppose Hypotheses H1 and H2 hold. Let $|R/(q)| = p^{hg}$, and let $R/(q)$ be a field. Let $A = (a_0, a_1, \cdots)$ be the coefficient sequence of the r-adic expansion of a rational function u/q such that $u \neq 0 \in R$ (or equivalently, that is the nonzero periodic output sequence from an AFSR with connection element q). Then A is an (r, q)-adic ℓ-sequence if A is periodic with period $p^{hg} - 1$. That is, if r is primitive modulo q.*

Theorem 3. *Suppose A is an (r, q)-adic ℓ-sequence, Hypotheses H1, H2, and H3 hold, and $q_k \in \mathbf{F}_{p^h}$. Then the following hold.*

1. *A is a punctured de Bruijn sequence. Thus the number of occurrences of a sequence B of length $m \leq k$ in a period of A is $p^{he(k-m)}$ if $B \neq (0, \cdots, 0)$ and is $p^{he(k-m)} - 1$ otherwise.*
2. *A has the shift and add property*
3. *A is balanced.*
4. *A has the run property.*
5. *A has ideal autocorrelations. (Care is needed in defining autocorrelations over non-prime fields.)*

Detailed definitions of these properties may be found in Golomb's book [2].

Proof: Properties (3), (4) and (5) follow from properties (1) and (2).

Since $q_k = F_{p^{he}}$ and Hypothesis H2 holds, we have

$$|R/(q)| = |S|^k = p^{hek}.$$

Thus A has period $p^{hek} - 1$. The various shifts of A plus the all-zero sequence give p^{hek} periodic sequences corresponding to elements u/q. Thus,

$$|V_q| \geq p^{hek}.$$

We have seen that the elements of V_q are distinct modulo q, so

$$|V_q| \leq p^{hek}.$$

Thus,

$$|V_q| = p^{hek} = |R/(r^k)|.$$

By Hypothesis H3, the elements of V_q are distinct modulo r^k, so V_q is a complete set of representatives modulo r^k.

The set of occurrences in A of a block B of k elements corresponds to the set of shifts of A that begin with B. By the above, every nonzero u/q with $u \in V_q$ occurs as a shift of A, so the set of occurrences in A of B corresponds to the set of nonzero u/q, $u \in V_q$, that begin with B. We claim that the u/q are distinct modulo r^k, so that each nonzero B occurs once. Suppose not, so that $u/q \equiv v/q \pmod{r^k}$ for some $u \neq v \in V_q$. Then $u \equiv v \pmod{r^k}$ since q is invertible modulo r, and hence also modulo r^k. But by Hypothesis H3 the elements of V_q are distinct modulo r^k. It follows that A is a punctured de Bruijn sequence.

Furthermore, if $u, v \in V_q$, then $u + v \in V_q$. The shifts of A account for all the u/q with $u \neq 0 \in V_q$ so the SAA property follows.   □

5   ℓ-Sequences and Blackburn Sequences

Blackburn [1] showed that a sequence $A = a_0, a_1, \cdots$ of period $p^{ek} - 1$ over F_{p^e} has the shift and add (SAA) property if and only if there is a primitive element $\alpha \in F_{p^{ek}}$ and a surjective F_p-linear function $T : F_{p^{ek}} \to F_{p^e}$ such that $a_i = T(\alpha^i)$ for all $i \geq 0$. We call sequences realized this way *Blackburn sequences*.

Theorem 4. *Let A be a Blackburn sequence that is a punctured de Bruijn sequence. Then A is an (r, q)-adic ℓ-sequence over F_p.*

This section is devoted to a proof of this theorem.

Corollary 3. *Every punctured de Bruijn sequence with the SAA property is an (r, q)-adic ℓ-sequence over F_p.*

Let $A = a_0, a_1, \cdots$ be a sequence with $a_i = T(\alpha^i)$ with T an F_p-linear function from $F_{p^{ek}}$ to F_{p^e} and α primitive in $F_{p^{ek}}$. We first find necessary and sufficient conditions for A to be a punctured de Bruijn sequence. If $\beta_0, \cdots, \beta_{e-1}$ is a basis of F_{p^e} over F_p, then there are F_p-linear functions $T_i : F_{p^{ek}} \to F_p$, $i = 0, \cdots, e-1$, such that $T = \sum_{i=0}^{e-1} \beta_i T_i$.

Lemma 2. *([10–p. 56]) If $f : F_{p^n} \to F_p$ is F_p-linear, then there is a constant $u \in F_{p^n}$ such that*

$$f(x) = Tr_p^{p^n}(ux).$$

Thus we have

$$T_j(x) = Tr_p^{p^{ek}}(u_j x) \text{ with } u_j \in F_{p^{ek}}, i = 0, \cdots, e-1.$$

As was pointed out by a referee of an earlier paper [4], this makes it possible to characterize the sequences that have the SAA property and uniform distributions. We include a proof since, to our knowledge, this fact has not been described in the literature.

Theorem 5. *Let A have the SAA property. Then A is a punctured de Bruijn sequence if and only if*

$$V = \{u_j \alpha^i : 0 \leq j < k, 0 \leq i < e\}$$

is a basis for $F_{p^{ek}}$ over F_p.

Proof: The sequence A is a punctured de Bruijn sequence if and only if each nonzero k-tuple of elements of F_{p^e} occurs exactly once in each period of A, and the zero k-tuple does not occur. Since the period of A is $p^{ek} - 1$, this is equivalent to each such k-tuple occurring at most once in A, and the zero k-tuple not occurring. Since A has the SAA property, it is equivalent simply to the zero k-tuple not occurring. Indeed, if any k-tuple occurs twice, then we can shift A by the distance between the two occurrences, then subtract A (the same as adding A $p - 1$ times) from this shift to obtain an occurrence of the all zero k-tuple.

The all-zero k-tuple occurs if and only if for some n we have $a_n = a_{n+1} = \cdots = a_{n+k-1} = 0$. That is,

$$Tr_p^{p^{ek}}(u_j \alpha^{i+n}) = 0$$

for $0 \leq j < k$ and $0 \leq i < e$. The set

$$\alpha^n V = \{u_j \alpha^{i+n} : 0 \leq j < k, 0 \leq i < e\}$$

is a basis for $F_{p^{ek}}$ over F_p if and only if V is. A linear function is zero on a basis if and only if it is identically zero. But the trace function is not identically zero. Thus, if V is a basis, then A is a punctured de Bruijn sequence.

Conversely, if

$$\sum_{i=0}^{e-1}\sum_{j=0}^{k-1} c_{ij} u_j \alpha^i = 0$$

with each c_{ij} in F_p and not all zero, then for any n,

$$\sum_{i=0}^{e-1}\sum_{j=0}^{k-1} c_{ij} Tr_p^{p^{ek}}(u_j \alpha^{i+n}) = 0.$$

That is, the F_p-coordinates of all k-tuples satisfy a common linear relation. Hence not all nonzero values of k-tuples can occur and A is not a punctured de Bruijn sequence. □

Our goal now is to realize a shift of A as a maximum period AFSR sequence over a function field, i.e., an ℓ-sequence. That is, we want to find a ring $R = F_p[z_1, \cdots, z_n]/I$ with I an ideal, an element $r \in R$, a subset $S \subseteq R$, and an element $q \in R$ so that A is the output from an AFSR based on R, r, S with connection element q. That is,

$$a_i = q_0^{-1}(br^i \pmod{q}) \pmod{r},$$

and equivalently, $\sum_{i=0}^{\infty} a_i r^i = u/q$ in the ring R_r of r-adic elements over R, for some $u \in R$.

We achieve this as follows. For any F_p-linear function f, let K_f denote the kernel of f. We construct an appropriate R together with functions $\Gamma : R \to F_{p^{ek}}$ and $\Delta : R \to F_{p^e}$ so that $K_\Gamma = (q), K_\Delta = (r)$, and $\Delta(s) = qT(c\Gamma(s))$ for $s \in S$ and some constant $c \neq 0 \in F_{p^{ek}}$.

Let $f(x)$ be the minimal polynomial of $r = \alpha^{-1}$ over F_{p^e} so that

$$f(x) = \sum_{i=0}^{k} f_i x^i, f_i \in F_{p^e}.$$

We have $f_k = 1 \in F_p$ and

$$f_0 = r^{(p^{ek}-1)/(p^e-1)},$$

since f_0 is the product of the Galois conjugates r^{p^j}, $j = 0, \cdots, k-1$. Let $\beta = f_0$. Then β is primitive in F_{p^e} and $1, \beta, \cdots, \beta^{e-1}$ is a basis of F_{p^e} over F_p.

We have

$$f_i = \sum_{j=0}^{e-1} f_{ij} \beta^j, f_{ij} \in F_p.$$

Thus we can write

$$f(x) = \sum_{j=0}^{e-1} (\sum_{i=0}^{k} f_{ij} x^i) \beta^j = \sum_{j=0}^{e-1} z_j(x) \beta^j.$$

The polynomial $z_j(x) \in F_p[x]$ has degree at most k. In particular, if $e \geq 2$, then $z_j(x)$ does not have r as a root unless $z_j(x)$ is identically zero.

Note that $f_k = 1$, which implies that $z_0(x)$ has degree k and is nonzero. All other $z_i(x)$ have degree at most $k - 1$. Since $f_0 = \beta$, $z_1(x)$ has constant term 1, so is nonzero, and all other $z_i(x)$ have constant term 0.

Lemma 3. *There exist* $c, \gamma_0, \cdots, \gamma_{e-1} \in F_{p^{ek}}$ *so that*

a. $\sum_{j=0}^{e-1} z_j(r)\gamma_j = 0$;
b. $\gamma_0 = 1$;
c. $T(c\gamma_1) = 1$; *and*
d. $T(c\gamma_0), \cdots, T(c\gamma_{e-1})$ *are linearly independent over* F_p.

Proof: See Appendix A. □

Suppose conditions (a), (b), (c), and (d) hold. By conditions (b) and (d) we have $T(c) = T(c\gamma_0) \neq 0$. We define $\Gamma(y_i) = \gamma_i$ and $\Gamma(x) = r$. We also define $\Delta(y_i) = \delta_i = T(c)^{-1}T(c\gamma_i)$ for $i = 0, \cdots e - 1$, and $\Delta(x) = 0$. We then extend these to ring homomorphisms. Let I be the intersections of the kernels of Γ and Δ. The functions Γ and Δ induce functions on $R = F_p[x, y_0, \cdots, y_{e-1}]/I$ for which we shall use the same names. It follows from condition (a) that

$$\sum_{i=0}^{k} \sum_{j=0}^{e-1} f_{ij} \gamma_j r^i = 0.$$

Let

$$S = \{\sum_{i=0}^{e-1} s_i y_i : s_i \in F_p\}.$$

Let

$$q_i = \sum_{j=0}^{e-1} f_{ij} y_j \in S,$$

so that

$$\Gamma(q_i) = \sum_{j=0}^{e-1} f_{ij} \gamma_j.$$

Let $q = \sum_{i=0}^{k} q_i x^i$. Then $\Gamma(q) = 0$. We have $\Gamma(y_1) = f_0$ so $q_0 = y_1$ and $\Delta(q) = \Delta(q_0) = T(c)^{-1}T(c\gamma_1) = T(c)^{-1}$ by condition (c). Also, $f_k = 1$ so $\sum_{j=0}^{e-1} f_{ij}\gamma_j = 1$ and $q_k = 1$.

Lemma 4. *Let* $c, \gamma_0, \cdots, \gamma_{e-1}$ *satisfy the conditions of Lemma 3. If* R, S, *and* q *are as above, then Hypotheses H1, H2, and H3 hold.*

Proof: See Appendix B. □

The sequence generated by an AFSR based on R, x, and S with connection element q is given by $b_i = q_0^{-1}(x^{-i} \pmod{q}) \pmod{x}$ which really means

$$b_i = \Delta(q_0)^{-1} \cdot \Delta(\Gamma_S^{-1}(\Gamma(x)^{-i})))$$
$$= \Delta(q_0)^{-1} \cdot \Delta(\Gamma_S^{-1}(r^{-i}))$$
$$= T(c) \cdot \Delta(\Gamma_S^{-1}(\alpha^i)),$$

where Γ_S is the restriction of Γ to S. On the other hand $a_i = T(c\alpha^i)$ so we want to see that

$$T(c\alpha^i) = T(c) \cdot \Delta(\Gamma_S^{-1}(\alpha^i))$$

for every i. The powers of α are precisely the images of the nonzero elements of S under Γ, so it suffices to show that for every $y \in S$ we have

$$T(c\Gamma(y)) = T(c) \cdot \Delta(y). \tag{4}$$

Since T, Γ, and Δ are F_p-linear, it suffices to see that equation (4) holds for y in an F_p-basis for S, such as $\{y_0, y_1, \cdots, y_{e-1}\}$. That is, it suffices to see that $T(c\gamma_i) = T(c)\delta_i$. This holds by the definition of δ_i. This completes the proof of Theorem 4.

6   Rational Approximation

In this section we consider the problem of finding an AFSR over R, r, S that generates a sequence A. This is equivalent [8] to finding $u, q \in R$ such that $\alpha(A, r) = u/q$, so we define the Rational Approximation Problem for AFSRs over R, r, and S as:

Rational Approximation over R, r, and S
Instance: A prefix of sequence A of elements of S.
Problem: Find elements u and $q \in R$ so that $\alpha(A, r) = u/q$.

We may approach problem with successive *rational approximations*: For each i, find $u_i, q_i \in R$ with $\alpha(A, r) \equiv u_i/q_i \pmod{r^i}$. Such an algorithm *converges* after T steps if $\alpha(A, r) = u_T/q_T$. We measure the quality of the algorithm in terms of the smallest T after which it converges, with the smallest such T expressed as a function of the size of the smallest AFSR that outputs A and we measure the quality in terms of the time complexity expressed as a function of T. In the Berlekamp-Massey algorithm (rational approximation for LFSRs), when the previous approximation fails at the next stage, a new approximation is formed by adding a multiple of a (carefully chosen) earlier approximation. If λ is the size of the smallest LFSR that outputs A, then the algorithm converges in 2λ steps with time complexity $O(T^2)$ [11].

One might try the same approach with AFSRs over more general rings. But if there are carries when approximations are added, then the proof of convergence breaks down. Xu and the author overcame this in a general setting [9]. The key idea is to produce a new approximation that works for several new terms. To make this work, we need an *index function* $\phi : R \to \mathbf{Z} \cup \{-\infty\}$ so that the following holds.:

Property 1. There are non-negative integers $a, b \in \mathbf{Z}$ such that

1. $\phi(0) = -\infty$ and $\phi(z) \geq 0$ if $z \neq 0$;
2. for all $z, y \in R$ we have $\phi(zy) \leq \phi(z) + \phi(y) + a$;
3. for all $z, y \in R$, we have $\phi(z \pm y) \leq \max\{\phi(z), \phi(y)\} + b$; and
4. for all $z \in R$ and $n \geq 0 \in \mathbf{Z}$, we have $\phi(r^n z) = n + \phi(z)$.

We define $-\infty + c = -\infty$ for every integer c. Let $n_\phi = \max\{\phi(z) : z \in S\} \cup \{\phi(1)\}$. For a pair $x, y \in R$ we define $\Phi(x, y) = \max(\phi(x), \phi(y))$.

If an AFSR over R and r has connection number $q = \sum_{i=0}^{k} q_i r^i$ with $q_i \in T$ and produces output sequence A with $\alpha = \alpha(A, r) = u/q$, then $\phi(q)$ and $\phi(u)$ are bounded by affine functions of k and $\phi(m)$, where m is the initial memory.

Generally $\phi(m)$ measures the storage needed for m. From this and an expression for u in terms of q and the initial state, we show that $\lambda = \max(\phi(u), \phi(q))$ is at most linear in the size of the AFSR. If we bound the execution time of a rational approximation algorithm in terms of λ, then we will have bounded the execution time in terms of the size of the AFSR. If A is an infinite sequence of elements of S and λ is the minimal value of $\Phi(u, q)$ over all pairs u, q with $\alpha(A, r) = u/q$, then we say λ is the *r-adic complexity* of A.

We also need a subset P of R so that the following holds.

Property 2. There are $c > d \geq 0 \in \mathbf{Z}$ such that

1. if $s \in P$, then r^c does not divide s;
2. if $z, y \in R$, then there exist $s, t \in P$ such that $r^c | sz + ty$; and
3. if $z, y \in R$ and $s, t \in P$, then $\phi(sz + ty) \leq \max\{\phi(z), \phi(y)\} + d$.

We call P an *interpolation set*. Let M_P be the cost of finding s, t in Property 2.2.

Theorem 6. *[9] Let ϕ be an index function and let P be an interpolation set for R and r. Then there is a register synthesis algorithm for AFSRs over R, r, S with time complexity $O(M_P T^2)$ such that if λ is the r-adic complexity of a sequence A, then given a prefix of*

$$T > \frac{2c}{c-d}\lambda + \frac{c(2(a+b) + c + b\lceil \log(c) \rceil + n_\phi)}{c-d} + 1 \in O(\lambda)$$

symbols of A, the algorithm produces an AFSR that outputs A.

Suppose

$$R = F_{p^h}[x_1, \cdots, x_n]/I = F_{p^h}[\overline{x}]/I$$

and $r \in R$, so that $R/(r)$ is finite. Thus $|R/(r)| = r = p^{hg}$ for some $g \in \mathbf{Z}$. Let S be a set of representatives for R modulo r. We want to construct an index function and interpolation set. Define $\phi(v) = \deg_r(v)$ and $\phi(0) = -\infty$. Then Property 1 holds with $a = \max\{\deg_r(uv) : u, v \in S\}$ and $b = 0$.

If $a = 0$, then our AFSRs are LFSRs, so we assume that $a \geq 1$. Let

$$P = \{\sum_{i=0}^{a} s_i r^i : s_i \in S, s_a \in F_{p^h}, (s_0, \cdots, s_a) \neq (0,0)\},$$

$c = 2a$, and $d = 2a - 1$ so $0 \leq d < c$. Part 1 of Property 2 is immediate.

If

$$s = \sum_{i=0}^{a} s_i r^i \in P \text{ and } z = \sum_{i=0}^{n} z_i r^i$$

with $z_i \in S$, then $\phi(s_j z_i) \leq a$ and $\phi(s_a z_i) \leq 0$, so that $\phi(sz) \leq n + 2a - 1$. Thus if $s, t \in P$ and $z, y \in R$, then $\phi(sz + ty) \leq \Phi(sz, ty) \leq \Phi(z, y) + 2a - 1 = d..$ Thus part 3 of Property 2 holds.

Next let $z, y \in R$. Let

$$\mu : (P \cup \{(0,0)\})^2 \to R/(r^c)$$

be defined by $\mu(s, t) = sz + ty \pmod{r^c}$. Then μ is an F_{p^h}-linear map from a set of cardinality

$$(|P| + 1)^2 = |S|^{2a}|F_{p^h}|^2 = p^{2h(ea+1)}$$

to a set of cardinality

$$|S|^c = p^{2hea}.$$

The former set is larger, so μ has a nontrivial kernel. That is, there exist $s, t \in P$, not both zero, such that r^c divides $sz + ty$. This proves part 2 of Property 2. It follows that we have a rational approximation algorithm in this setting.

Theorem 7. *Let S be a complete set of representatives modulo r satisfying Hypotheses H1 and H2. Let $a = \max\{\deg_r(uv) : u, v \in S\}$. Then there is a register synthesis algorithm for AFSRs over R, r, S with time complexity $O(M_P T^2)$, such that if λ is the r-adic complexity of a sequence A, then the algorithm produces an AFSR that generates the sequence if it is given a prefix of $T > 4a\lambda + 8a^2 + 1 \in O(\lambda)$ symbols of the sequence.*

For cryptography this gives us another security test that stream ciphers must pass. For coding theory this may give us new classes of algebraic geometry codes with efficient decoding algorithms, but this is a subject for future research.

References

1. S. Blackburn: A Note on Sequences with the Shift and Add Property. Designs, Codes, and Crypt. **9** (1996) pp. 251-256.
2. S. Golomb: Shift Register Sequences. Aegean Park Press, Laguna Hills CA (1982).
3. M. Goresky and A. Klapper: Periodicity and Correlations of of d-FCSR Sequences. Designs, Codes, and Crypt. **33** (2004) 123-148.
4. M. Goresky and A. Klapper: Polynomial pseudo-noise sequences based on algebraic feedback shift registers, under review.
5. A. Klapper: Distributional properties of d-FCSR sequences. J. Complexity **20** (2004) 305-317.
6. A. Klapper, Pseudonoise Sequences Based on Algebraic Function Fields, presented at ISIT 2004, Chicago, USA.
7. A. Klapper and M. Goresky: Feedback Shift Registers, Combiners with Memory, and 2-Adic Span. J. Cryptology **10** (1997) 111-147.
8. A. Klapper and J. Xu: Algebraic feedback shift registers, Theoretical Comp. Sci. **226** (1999) 61-93.
9. A. Klapper and J. Xu: Register synthesis for algebraic feedback shift registers based on non-primes. Designs, Codes, and Crypt. **31** (2004) 227-25.
10. R. Lidl and H. Niederreiter: Finite Fields, Encycl. Math. Appl. **20**. Addision Wesley, Reading, MA (1983).
11. J. Massey: Shift-register synthesis and BCH decoding. IEEE Trans. Info. Thy. **IT-15** (1969) 122-127.
12. M. O'Sullivan: Decoding of codes defined by a single point on a curve. IEEE Trans. Info. Thy. **41** (1995) 1709-1719.
13. S. Porter, B.-Z. Shen, and R. Pellikaan: Decoding geometric Goppa codes using an extra place. IEEE Trans. Info. Thy. **38** (1992) 1663-1676.
14. S. Sakata, H. Jensen, and T. Hoholdt: Generalized Berlekamp-Massey decoding of algebraic geometry codes up to half the Feng-Rao bound. IEEE Trans. Info. Thy. **41** (1995) 1762-1768.

A   Proof of Lemma 3

Suppose that

$$\frac{z_1(r)}{z_0(r)} \cdot K_T \subseteq K_T + F_p T^{-1}(1).$$

The right hand side equals $K_T + F_p v$ for any element v with $T(v) = 1$. Let $u \in F_{p^e} - F_p$ and define $T'(y) = uT(y)$. Then T' is also F_p-linear and $K_{T'} = K_T$. However

$$\frac{z_1(r)}{z_0(r)} \cdot K_{T'} \not\subseteq K_{T'} + F_p T'^{-1}(1).$$

Now suppose we can show that any sequence $A = a_0, a_1, \cdots$ generated by Blackburn's method using primitive element r^{-1} and linear function T' in fact is an ℓ-sequence, say based on ring R, $x \in R$, and set of representatives S, with connection element $q \in R$, and satisfying Hypotheses H1, H2, and H3. That is, $\sum_{i=0}^{\infty} a_i x^i = z/q$ for some $z \in R$.

Lemma 5. *The sequence $A' = u^{-1}a_0, u^{-1}a_1, \cdots$ is an ℓ-sequence based on R, $x \in R$, S, and connection integer q satisfying Hypotheses H1, H2, and H3.*

Proof: Let $\hat{u} \in S$ reduce to u modulo x. Let $S' = \hat{u}^{-1}S$. Then S' is an F_p-vector space and contains F_p. Thus Hypothesis H1 holds. Hypothesis H2 holds, since any $v \in R$ can be written $v = \hat{u}^{-1}v'$ for some $v \in R$. If moreover $w \in R$, then we can write

$$\hat{u}w = \sum_{i=0}^{t} w_i x^i,$$

with $w_i \in S$ so

$$w = \sum_{i=0}^{t} \hat{u}^{-1}w_i x^i.$$

Since $\sum_{i=0}^{\infty} a_i x^i = z/q$, we also have

$$\sum_{i=0}^{\infty} \hat{u}^{-1}a_i x^i = \frac{\hat{u}^{-1}z}{q}.$$

It follows that A' is an ℓ-sequence. Hypothesis H3 holds by similar reasoning. □

Thus we may assume from here on that

$$\frac{z_1(r)}{z_0(r)} \cdot K_T \not\subseteq K_T + F_p T^{-1}(1). \tag{5}$$

We claim that we can pick $\gamma_1, \gamma_2, \cdots, \gamma_{e-1}$ so that $T(\gamma_1) = 1$, $T(\gamma_1), \cdots, T(\gamma_{e-1})$ are linearly independent over F_p, and

$$\frac{z_1(r)}{z_0(r)}K_T \not\subseteq K_T + \sum_{j=1}^{e-1} F_p \gamma_j. \tag{6}$$

This is possible: To see this, first pick γ_1 arbitrarily so that $T(\gamma_1) = 1$. Then by equation (5) there exists $\kappa \in K_T$ so that $(z_1(r)/z_0(r))\kappa \notin K_T + F_p T^{-1}(1)$. Finally, we can pick $\gamma_2, \cdots, \gamma_{e-1}$ so that

$$T(\frac{z_1(r)}{z_0(r)}\kappa), 1, T(\gamma_2), \cdots, T(\gamma_{e-1})$$

are F_p-linearly independent.

In particular, we have $\kappa \in K_T$ with

$$\frac{z_1(r)}{z_0(r)}\kappa \notin K_T + \sum_{j=1}^{e-1} F_p \gamma_j.$$

Let

$$\gamma_0 = -\sum_{j=1}^{e-1} \frac{z_j(r)}{z_0(r)}\gamma_j.$$

Thus $\sum_{j=0}^{e-1} z_j(r)\gamma_j = 0$.

Suppose that $T(\gamma_0), \cdots, T(\gamma_{e-1})$ are linearly dependent over F_p. Then we have $T(\gamma_0) = \sum_{j=1}^{e-1} b_j T(\gamma_j)$ for some $b_j \in F_p$. Let

$$\gamma_j' = \begin{cases} \gamma_j + \frac{z_j(r)}{z_0(r)}\kappa & \text{if } j = 0 \\ \gamma_j - \kappa & \text{if } j = i \\ \gamma_j & \text{otherwise.} \end{cases}$$

Then $T(\gamma_1') = 1$, $T(\gamma_1'), \cdots, T(\gamma_{e-1}')$ are linearly independent over F_p, and $\sum_{j=0}^{e-1} z_j(r)\gamma_j' = 0$. Suppose that $T(\gamma_0'), \cdots, T(\gamma_{e-1}')$ are linearly dependent over F_p. Then we have $T(\gamma_0') = \sum_{j=1}^{e-1} c_j \gamma_j'$ for some $c_j \in F_p$. Since $T(\gamma_j') = T(\gamma_j)$ for $j \geq 1$, it follows that

$$T(\frac{z_j(r)}{z_0(r)}\kappa) = \sum_{j=1}^{e-1}(c_j - b_j)T(\gamma_j).$$

This contradicts equation (6) and proves the following lemma.

Lemma 6. *There exist $\gamma_0, \gamma_1, \cdots, \gamma_{e-1} \in R$ so that $T(\gamma_1) = 1$, $\sum_{j=0}^{e-1} z_j(r)\gamma_j = 0$, and the images $T(\gamma_0), T(\gamma_1), \cdots, T(\gamma_{e-1})$ are linearly independent over F_p.*

Now let $c = \gamma_0$. For $j = 0, \cdots, e-1$, let $\gamma_j' = c^{-1}\gamma_j$. Then

1. $\displaystyle\sum_{j=0}^{e-1} z_j(r)\gamma_j' = 0$;

2. $\gamma_0' = 1$;

3. $T(c\gamma_1') = 1$; and

4. $T(c\gamma_0'), T(c\gamma_1'), \cdots, T(c\gamma_{e-1}')$ are linearly independent over F_p.

This completes the proof of Lemma 3.

B   Proof of Lemma 4

That $F_p \subseteq S$ follows from (2), and the closure under addition is immediate from the definition. Thus Hypothesis H1 holds.

We have $\Delta(y_0) = 1$ and $\Gamma(y_0) = 1$, so $y_0 - 1 \in I$. We know from (4) that the set of δ_j spans F_{p^e} over F_p, so every product $\delta_i \delta_j$ can be written uniquely as

$$\delta_i \delta_j = \sum_{\ell=0}^{e-1} v_{ij\ell} \delta_\ell.$$

Therefore

$$y_i y_j - \sum_{\ell=0}^{e-1} v_{ij\ell} y_\ell \in K_\Delta.$$

If

$$y_i y_j - \sum_{\ell=0}^{e-1} v_{ij\ell} y_\ell \in K_\Gamma,$$

then it is in I. Otherwise we have

$$\Gamma\left(y_i y_j - \sum_{\ell=0}^{e-1} v_{ij\ell} y_\ell\right) = \alpha^t$$

for some t. Thus

$$y_i y_j - \sum_{\ell=0}^{e-1} v_{ij\ell} y_\ell - x^t \in I.$$

In particular, every element of R can be written as a finite sum $\sum_{i=0}^{s} u_i x^i$ with $u_i \in S$. This implies that every element of $R/(q)$ can be written in this form with $s < k$. Thus $|R/(q)| \leq p^{ek}$. But the image of R in $F_{p^{ek}}$ is all of $F_{p^{ek}}$ and is a quotient of $R/(q)$. Hence $|R/(q)| = p^{ek}$ and $(q) = K_\Gamma$ in R. Thus Hypothesis H2 holds.

Suppose that Hypothesis H3 is false. Then there are distinct elements $u, v \in R$ such that u/q and v/q have periodic x-adic expansions and $u \equiv v \pmod{q}$. Since the termwise difference of two periodic sequences is periodic and corresponds to the difference of the corresponding x-adic elements, we can assume that $u \in R$ with $u = wx^k$. But then the x-adic expansion of u/q is x^k times the x-adic expansion of w/q, hence has k consecutive zeros. But this is false for a punctured de Bruijn sequence. Thus Hypothesis H3 holds.

New LFSR-Based Cryptosystems and the Trace Discrete Log Problem (Trace-DLP)

Kenneth J. Giuliani[1] and Guang Gong[2]

[1] Dept. of Combinatorics and Optimization,
University of Waterloo,
Waterloo, ON, Canada, N2L 3G1
kjgiulia@cacr.math.uwaterloo.ca
[2] Dept. of Electrical and Computer Engineering,
University of Waterloo,
Waterloo, ON, Canada, N2L 3G1
ggong@calliope.uwaterloo.ca

Abstract. In order to reduce key sizes and bandwidth, cryptographic systems have been proposed using minimal polynomials to represent finite field elements. These systems are essentially equivalent to systems based on characteristic sequences generated by a linear feedback shift register (LFSR). We propose a general class of LFSR-based key agreement and signature schemes based on n-th order characteristic sequences. These schemes have the advantage that they do not require as much bandwidth as their counterparts based on finite fields. In particular, we present a signature scheme based on a new computational problem, the *Trace Discrete Logarithm Problem (Trace-DLP)*. The Trace-DLP and its variants are discussed and their relationship with well-known finite field-based computational problems is examined. In addition, we prove the equivalence between several LFSR-based computational problems and their finite field-based counterparts.

1 Introduction

A good portion of public-key cryptography is based upon finite fields. Some of the most notable examples are Diffie-Hellman key agreement [1] and the Digital Signature Standard [13]. However, since field sizes must be chosen large enough to avoid the so-called "index-calculus" attacks, finite field elements normally require a large amount of bits in order to represent them. For applications where bandwidth is limited, this is undesirable.

As a result, several cryptosystems have been proposed which reduce the representation of finite field elements. Examples of such systems are LUC [9, 16], GH [6], and XTR [7]. These systems reduce representations of finite field elements by representing them with the coefficients of their minimal polynomials. Due to the Newton Identity, these methods are essentially the same as systems based on n-th order characteristic sequences generated by linear feedback shift registers (LFSR's). In particular, LUC can be considered as a second-order sequence, while GH and XTR can be considered as third-order sequences. We also

T. Helleseth et al. (Eds.): SETA 2004, LNCS 3486, pp. 298–312, 2005.

note that fifth-order sequences have also been proposed [14, 3, 4] and that Niederreiter [10, 11, 12] has proposed encryption and key agreement schemes based on general n-th order LFSR sequences.

This paper proposes schemes based on n-th order *characteristic* sequences generated by an LFSR. In particular, we propose general key agreement and signature schemes where the sizes of signatures and keys are directly related to the sizes of the representations of elements. In addition, we present new computational problems, namely the *Trace-Discrete Logarithm Problem (Trace-DLP)* and its variants, on which the security of our signature schemes is based. We present a thorough discussion of these computational problems and tie them to more well-known problems. In particular, we prove the equivalence of several LFSR-based computational problems to their counterparts based on finite fields.

This paper is presented as follows. Section 2 describes the connection between finite fields and characteristic sequences which lead to the reduced representations produced in Section 3. Section 4 lists the operations required to be able to perform the necessary computation for our cryptographic schemes, which are given in Section 5. Section 6 gives a thorough discussion of the computational problems based on finite fields and sequences. Section 7 summarizes and suggests some areas for future study.

2   LFSR's, Characteristic Sequences, and Minimal Polynomials

We first draw the connection between LFSR sequences and finite fields through the use of characteristic sequences and minimal polynomials.

Consider the sequence generated by a linear feedback shift register (LFSR) of order n over $GF(q)$ where q is a prime power. This sequence is given by the recurrence

$$s_{k+n} = a_1 s_{k+n-1} - a_2 s_{k+n-2} + \cdots + (-1)^{n+1} a_n s_k$$

for all $k \geq 0$ where $a_1, \ldots, a_n$ are elements of $GF(q)$.

Let

$$f(x) = x^n - a_1 x^{n-1} + a_2 x^{n-2} - \cdots + (-1)^n a_n \tag{1}$$

and suppose that this polynomial is irreducible over $GF(q)$ with γ a root of f in $GF(q^n)^*$. Also, let $\bar{s}_i = (s_i, s_{i+1}, \ldots, s_{i+n-1})$ be the i-th state of the LFSR sequence. By choosing our initial state $\bar{s}_0$ in a special way, namely $s_i = Tr(\gamma^i)$ for $i = 0, \ldots, n-1$ where Tr is the trace map from $GF(q^n)$ to $GF(q)$, we ensure that $s_k = Tr(\gamma^k)$ for all $k \geq 0$.

The sequence is strictly periodic with period, say, Q. We may then define $s_k = s_{Q+k}$ for all $k \leq 0$. Hence, we may consider the sequence $\{s_k\}$ with indices running over all integers.

A sequence defined in this fashion will be called the *n-th order characteristic sequence* over $GF(q)$ generated by γ or $f(x)$. It is well-known that the following is true.

Proposition 1. *Let $\{s_i\}$ be an n-th order characteristic sequence over $GF(q)$ generated by $\gamma \in GF(q^n)^*$. Then the period of $\{s_i\}$ is equal to the order of γ.*

Characteristic sequences are closely related to minimal polynomials of finite field elements by the Newton Identity.

Lemma 1 (Newton Identity). *Let $\gamma \in GF(q^n)^*$ and $t_i = Tr(\gamma^i)$, $0 \leq i < n$. Then*

$$\prod_{j=0}^{n-1}(x - \gamma^{q^j}) = x^n + \sum_{j=1}^{n}(-1)^j b_j x^{n-j}$$

where for $0 < j \leq n$, the b_j's is defined recursively by the relation

$$t_j - t_{j-1}b_1 + \cdots + (-1)^j j b_j = 0 \tag{2}$$

For all integers k, define the minimal polynomial of γ^k over $GF(q)$ to be

$$f_{\gamma^k}(x) = x^n - a_{1,k}x^{n-1} + a_{2,k}x^{n-2} - \cdots + (-1)^{n-1}a_{n-1,k}x + (-1)^n a_{n,k}$$

The Newton Identity tells us that for any $m \in \{1, \ldots, n\}$, we can efficiently determine the set $\{a_{1,k}, a_{2,k}, \ldots, a_{m,k}\}$ from $\{s_k, s_{2k}, \ldots, s_{mk}\}$ and vice-versa.

Let us now specialize to the case where $n \geq 2$ and γ, and hence γ^k for all k, has an order dividing $q^{n-1} + q^{n-1} + \cdots + q + 1$. Now

$$a_{i,k} = \sum_{0 \leq j_1 < j_2 < \cdots < j_i \leq n-1} \gamma^{k(q^{j_1} + q^{j_2} + \cdots + q^{j_i})}$$

When $i = n$, this becomes $a_{n,k} = \gamma^{k(1+q+q^2+\cdots+q^{n-1})} = 1$. Also, for any $1 \leq i \leq n-1$, we have that

$$a_{i,k} = \sum_{0 \leq j_1 < j_2 < \cdots < j_i \leq n-1} \gamma^{k(q^{j_1} + q^{j_2} + \cdots + q^{j_i})}$$

$$= \sum_{0 \leq j_1 < j_2 < \cdots < j_{n-i} \leq n-1} \gamma^{-k(q^{j_1} + q^{j_2} + \cdots + q^{j_{n-i}})}$$

$$= a_{n-i,-k} \tag{3}$$

Remark 1. In the previous discussion, it is not necessary that $f(x)$ be irreducible. We may loosen this restriction by saying that $f(x)$ is a polynomial whose roots are conjugate over $GF(q)$. The previous and following analysis will still hold.

3   Reducing Representations of Finite Field Elements

Let $\gamma \in GF(q^n)^*$. Ordinarily, for any integer k, γ^k would require $n \log q$ bits for its representation, normally in a polynomial basis over $GF(q)$. Our goal in this section is to obtain a smaller representation of γ^k. Again, suppose that $n \geq 2$

and that γ has an order dividing $q^{n-1} + q^{n-2} + \cdots q + 1$. Then the minimal polynomial of γ^k would be

$$f_{\gamma^k}(x) = x^n - a_{1,k}x^{n-1} + a_{2,k}x^{n-2} - \cdots + (-1)^{n-1}a_{n-1,k}x + (-1)^n$$

since the constant term is $a_{n,k} = 1$ as shown in the previous section. Thus, at the tradeoff of representing of giving γ^k the same representation as its conjugates $\{\gamma^k, \gamma^{kq}, \ldots, \gamma^{kq^{n-1}}\}$, we may represent γ^k by the set $(a_{1,k}, a_{2,k}, \ldots, a_{n-1,k})$, a total of $(n-1)\log q$ bits. For our purposes, we will accept this tradeoff. Observe that we are now using $\frac{n-1}{n}$ as many bits as in the ordinary case.

We now describe two special cases where we obtain an even shorter representation. Suppose $q = p^2$ and n is even. Let $\gamma \in GF(p^{2n})^*$ have an order dividing $p^n + 1$. Then $\gamma^{kp^n} = \gamma^{-k}$ from which we see that for each $1 \leq i \leq n-1$,

$$a_{n-i,k} = a_{i,-k} = \sum_{0 \leq j_1 < j_2 < \cdots < j_i \leq n-1} \gamma^{-k(q^{j_1} + q^{j_2} + \cdots + q^{j_i})}$$

$$= \sum_{0 \leq j_1 < j_2 < \cdots < j_i \leq n-1} \gamma^{kp^n(q^{j_1} + q^{j_2} + \cdots + q^{j_i})}$$

$$= a_{i,k}^{p^n} = a_{i,k}$$

where the first equality was established in (3) and the last equality follows from the fact that n is even and $a_{i,k}^{p^2} = a_{i,k}$ since $a_{i,k} \in GF(p^2)$. Hence, we may represent γ^k (and its conjugates) by the set $(a_{1,k}, \ldots, a_{n/2,k})$ which requires $\frac{n}{2}\log q$ bits to represent. This now requires $\frac{1}{2}$ as many bits as in the ordinary case.

Finally, suppose $q = p^2$ and n is odd. Let $\gamma \in GF(p^{2n})^*$ have an order dividing $p^{n-1} - p^{n-2} + \cdots - p + 1$. We again have that $\gamma^{kp^n} = \gamma^{-k}$ and for each $1 \leq i \leq n-1$, we get the similar result

$$a_{n-i,k} = a_{i,-k} = a_{i,k}^{p^n} = a_{i,k}^p$$

where the last equality is established from the fact that n is odd and $a_{i,k}^{p^2} = a_{i,k}$. Hence, we may represent γ^k (and its conjugates) by the set $(a_{1,k}, \ldots, a_{(n-1)/2,k})$ which requires $\frac{n-1}{2}\log q$ bits to represent. This now requires $\frac{n-1}{2n}$ as many bits as in the ordinary case.

To consider all three cases concurrently, we shall define r by

$$r = \begin{cases} n-1 & \text{for general } q \text{ and } n \\ n/2 & \text{if } q = p^2 \text{ and } n \text{ is even} \\ (n-1)/2 & \text{if } q = p^2 \text{ and } n \text{ is odd} \end{cases}$$

and define the set $A_k = (s_k, s_{2k}, \ldots, s_{rk})$. Observe that from the Newton Identity, we can recover the minimal polynomial coefficients $\{a_{1,k}, a_{2,k}, \ldots, a_{r,k}\}$ and hence the entire minimal polynomial f_{γ^k} from A_k.

Example 1. Let $n = 2$, $q = p$ where p is prime, and the order Q of $\gamma \in GF(p^2)^*$ divides $p + 1$. Then $A_k = (s_k)$ which needs only $\log p$ bits for representation, only $\frac{1}{2}$ as many as in the ordinary case. This is the basis of the LUC [9, 16] cryptosystem. Hence, LUC may be viewed as using second-order characteristic sequences over $GF(p)$.

Example 2. Let $n = 3$, $q = p$ where p is prime, and the order Q of $\gamma \in GF(p^2)^*$ divides p^2+p+1. Then $A_k = (s_k, s_{2k})$ which needs only $2 \log p$ bits for representation, only $\frac{2}{3}$ as many as in the ordinary case. This is the basis of the GH [6] cryptosystem. Hence, GH makes use third-order characteristic sequences over $GF(p)$.

Example 3. Let $n = 3$, $q = p^2$ where p is prime, and the order Q of $\gamma \in GF(p^2)^*$ divides $p^2 - p + 1$. Then $A_k = (s_k)$ which needs only $2 \log p$ bits for representation, only $\frac{1}{3}$ as many as in the ordinary case. This is the basis of the XTR [6] cryptosystem. Hence, XTR may be viewed as using third-order characteristic sequences over $GF(p^2)$.

Example 4. Let $n = 5$, $q = p$ where p is prime, and the order Q of $\gamma \in GF(p^2)^*$ divides $p^4 + p^3 + p^2 + p + 1$. Then $A_k = (s_k, s_{2k}, s_{3k}, s_{4k})$ which needs only $8 \log p$ bits for representation, only $\frac{4}{5}$ as many as in the ordinary case. This is the basis of the cryptosystem proposed by Giuliani and Gong [3]. Hence, this system makes use of fifth-order characteristic sequences over $GF(p)$.

Example 5. Let $n = 5$, $q = p^2$ where p is prime, and the order Q of $\gamma \in GF(p^2)^*$ divides $p^4 - p^3 + p^2 - p + 1$. Then $A_k = (s_k, s_{2k})$ which needs only $4 \log p$ bits for representation, only $\frac{2}{5}$ as many as in the ordinary case. This is the basis of the cryptosystem proposed by Quoos and Mjølsnes [14] and Giuliani and Gong [3, 4]. Hence, may be viewed as using fifth-order characteristic sequences over $GF(p^2)$.

4   Operations to Calculate Sequence Terms

There are two main operations which will be needed for our schemes in the next section. We describe them here.

Sequence Operation I (SO1): Given A_k and an integer l, where $0 \leq k, l < Q$, compute A_{kl}.

Sequence Operation II (SO2): Given states $\bar{s}_k$ and $\bar{s}_l$ for some $0 \leq k, l < Q$, compute $\bar{s}_{k+l}$.

SO1 can be performed efficiently by the algorithm due to Fiduccia [2], while SO2 can be done efficiently from the theory of shift register sequences [5]. We will now work toward detailing these procedures, starting first with some background.

Define the $n \times n$ matrix

$$
C = \begin{bmatrix}
0 & 0 & 0 & \cdots & 0 & 0 & (-1)^{n+1} \\
1 & 0 & 0 & \cdots & 0 & 0 & (-1)^n a_{n-1} \\
0 & 1 & 0 & \cdots & 0 & 0 & (-1)^{n-1} a_{n-2} \\
\vdots & \vdots & \vdots & \ddots & \vdots & \vdots & \vdots \\
0 & 0 & 0 & \cdots & 0 & 0 & a_3 \\
0 & 0 & 0 & \cdots & 1 & 0 & -a_2 \\
0 & 0 & 0 & \cdots & 0 & 1 & a_1
\end{bmatrix}
$$

The determinant of C is 1, which means that C is nonsingular. If we now consider the i-th state $\overline{s}_i = (s_i, \ldots, s_{i+n-1})$ as a row vector, then we have that

$$\overline{s}_{i+1} = \overline{s}_i C$$

for all $i \in \mathbb{Z}$. We can iteratively apply this relation to get

$$\overline{s}_{i+k} = \overline{s}_i C^k \tag{4}$$

for all $i, k \in \mathbb{Z}$ with the convention that $C^0 = I_n$, the $n \times n$ identity matrix. For each $i \in \mathbb{Z}$, construct the matrix M_i whose rows consist of the states $\overline{s}_i, \ldots, \overline{s}_{i+n-1}$, that is

$$M_i = \begin{bmatrix} \overline{s}_i \\ \overline{s}_{i+1} \\ \cdots \\ \overline{s}_{i+n-1} \end{bmatrix}$$

Then from (4), we get

$$M_{i+k} = M_i C^k \tag{5}$$

for all $i, k \in \mathbb{Z}$. Note that since C is nonsingular, the M_i's all have the same rank. If the M_i's were singular, we would have a relation of the form

$$\overline{s}_{i+l} = d_1 \overline{s}_{i+l-1} + \cdots + d_l \overline{s}_i$$

for all $i \in \mathbb{Z}$ and some l with $1 \leq l < n$, and we could consider this as an l-th order sequence over $GF(q)$. For our purposes, we will assume that the M_i's are nonsingular.

To perform SO2, we simply calculate $C^k = M_0^{-1} M_k$ from (5), whence we can calculate $\overline{s}_{k+l}$ from (4).

To perform SO1, we populate C with the coefficients $a_{1,k}, \ldots, a_{n-1,k}$ in place of $a_1, \ldots, a_{n-1}$ and use the well-known Cayley-Hamilton theorem.

Theorem 1 (Cayley-Hamilton). *A matrix satisfies its characteristic polynomial. That is, if $F(x)$ is the characteristic polynomial of a matrix C, then $F(C)$ is the zero matrix.*

It is easily seen that the characteristic polynomial of C is $f_{\gamma^k}(x)$. We now define $\overline{s}_{k,l} = (s_{kl}, s_{k(l+1)}, \ldots, s_{k(l+n-1)})$. This is exactly the l-th state of the n-th order characteristic sequence generated by γ^k. We can obtain the states $\overline{s}_{k,l}, \overline{s}_{k,2l}, \ldots, \overline{s}_{k,rl}$ by Fiduccia's algorithm.

Algorithm 1. Fiduccia

INPUT: C, $f_{\gamma^k}(x)$, $\overline{s}_{k,1}$, l
OUTPUT: $\overline{s}_{k,l}$.
1. $w \leftarrow \lfloor \log_2 l \rfloor$.
2. $l_i \in \{0, 1\}$ are such that $l = \sum_{i=0}^{w} l_i 2^i$.
3. $R(x) \leftarrow x$.
4. for i from $w - 2$ down to 0 do
    4.1 $R(x) \leftarrow R(x) \cdot R(x) \bmod f_{\gamma^k}(x)$.
    4.2 if $l_i = 1$, $R(x) \leftarrow R(x) \cdot x \bmod f_{\gamma^k}(x)$.
5. $D \leftarrow R(C)$.
6. $\overline{s}_{k,l} \leftarrow \overline{s}_{k,0} D$.

---

5   Cryptographic Schemes

In this section, we list some cryptographic schemes using finite fields and LFSR sequences. The first two examples are finite field-based key agreement and signature schemes which are well-known in cryptography. The three LFSR-based schemes which follow it are analogues of these systems.

5.1   Finite Field Diffie-Hellman Key Agreement

Diffie and Hellman [1] proposed the following key agreement scheme in their seminal paper.

Domain Parameters: q, n, Q, γ.

Alice:

1. Chooses random private key l, $0 \leq l < Q$.
2. Computes public key γ^l and transmits this to Bob.
3. Receives Bob's public key γ^k.
4. Computes the shared secret $(\gamma^k)^l = \gamma^{kl}$.

Bob: Performs the symmetric operation.

Alice and Bob then both have the shared secret γ^{kl}.

5.2   Finite Field ElGamal Signature Scheme

The following signature scheme has served as the basis for the Digital Signature Standard (DSS) [13]. To be consistent with notation from the DSS, we shall write g in place of γ.

Domain parameters: q, n, Q, g, Q prime.
Private Key: w, $0 \leq w < Q$.
Public Key: $h = g^w$.

Signature Generation:

1. Hash the message M to obtain $H(M)$.
2. Choose random k and compute $r = g^k$.
3. Calculate $s \equiv k^{-1}(H(M) + wr) \pmod{Q}$.
4. The signature is (r, s).

Signature Verification:

1. Compute $u = r^s$.
2. Compute $v = g^{H(M)} h^r$.
3. If $u = v$ accept, otherwise reject.

5.3   LFSR-Based Diffie-Hellman Key Agreement

Niederreiter first proposed a Diffie-Hellman key agreement scheme using general LFSR sequences in [11, 12]. We propose a Diffie-Hellman scheme specifically using characteristic sequences with reduced representations. This has the benefit that it requires Alice and Bob to transmit fewer bits to obtain the shared secret.

Domain Parameters: q, n, Q, A_1.

Alice:

1. Chooses random private key l, $0 \leq l < Q$.
2. Computes public key A_l from A_1 and l using SO1 and transmits this to Bob.
3. Receives Bob's public key A_k.
4. Computes the shared secret A_{kl} from A_k and l using SO1.

Bob: Performs the symmetric operation.

Observe that both Alice and Bob transmit sets of the form A_k which are only $r \log q$ bits.

5.4   General LFSR-Based ElGamal Signature Scheme

Domain parameters: q, n, Q, A_1, Q prime.
Private Key: w, $0 \leq w < Q$.
Public Key: $\overline{s}_w, \overline{s}_{2w}, \ldots, \overline{s}_{mw}$ where $1 \leq m \leq r$.

Signature Generation:

1. Hash the message M to obtain $H(M)$.
2. Choose random k and compute A_k using SO1.
3. Let $r = s_k$ obtained from A_k.
4. Calculate $s \equiv k^{-1}(H(M) + wr) \pmod{Q}$.
5. The signature is (A_k, s).

Signature Verification:

1. Compute r and $l = H(M)r^{-1} \pmod{Q}$.
2. Compute first $\overline{s}_l, \overline{s}_{2l}, \ldots, \overline{s}_{ml}$ from A_1 and l using SO1.
3. Compute $u = (s_{w+l}, s_{2(w+l)}, \ldots, s_{m(w+l)})$ from the public key and the previous step using SO2.
4. Compute $A_{ksr^{-1}}, A_{2ksr^{-1}}, \ldots, A_{mksr^{-1}}$ from A_k and sr^{-1} using SO1 and let $v = (s_{ksr^{-1}}, s_{2ksr^{-1}}, \ldots, s_{mksr^{-1}})$.
5. If $u = v$ accept, otherwise reject.

The signature consists of A_k which is $r \log q$ bits in size and s which is $\log Q$ bits. However, the public key is $mn \log q$ bits which is quite large. We can reduce this to $n \log q$ bits by taking $m = 1$. The public key would then be the same size as a finite field element in canonical representation. However, the signature would still use a reduced representation. The specific signature for $m = 1$ is listed as follows.

5.5   Specific LFSR-Based ElGamal Signature Scheme

Domain parameters: q, n, Q, A_1, Q prime.
Private Key: $w, 0 \leq w < Q$.
Public Key: $\bar{s}_w$.

Signature Generation:

1. Hash the message M to obtain $H(M)$.
2. Choose random k and compute A_k using SO1.
3. Let $r = s_k$ obtained from A_k.
4. Calculate $s \equiv k^{-1}(H(M) + wr) \pmod{Q}$.
5. The signature is (A_k, s).

Signature Verification:

1. Compute r and $l = H(M)r^{-1} \pmod{Q}$.
2. Compute first $\bar{s}_l$ from A_1 and l using SO1.
3. Compute $u = s_{w+l}$ from the public key and the previous step using SO2.
4. Compute $A_{ksr^{-1}}$ from A_k and sr^{-1} using SO1 and let $v = s_{ksr^{-1}}$.
5. If $u = v$ accept, otherwise reject.

Observe that signature generation is unchanged. Only signature verification has been modified.

6   Computational Complexity Problems

Let us examine the computational complexity problems relevant to the schemes of the previous section and discuss their relations to more well-known problems.

For a problem $\mathcal{A}$, we write $\mathcal{A} \in P$ if it can be solved in probabilistic polynomial time in its inputs. For two problems $\mathcal{A}$ and $\mathcal{B}$, we shall write $\mathcal{A} \leq_P \mathcal{B}$ if $\mathcal{A}$ can be solved in probabilistic polynomial time with polynomially many queries to an oracle solving $\mathcal{B}$. We also write $\mathcal{A} =_P \mathcal{B}$ if both $\mathcal{A} \leq_P \mathcal{B}$ and $\mathcal{B} \leq_P \mathcal{A}$.

6.1   LFSR-Related Problems

Let us first state some traditional finite field problems.

Definition 1. *The **Discrete Logarithm Problem (DLP)** is, given $\beta \in \langle \gamma \rangle$, to find l such that $\beta = \gamma^l$.*

Definition 2. *The **Diffie-Hellman Problem (DHP)** is, given γ along with γ^k and γ^l, to determine γ^{kl}.*

Definition 3. *The **Decisional Diffie-Hellman Problem (DDHP)** is, given γ along with $\gamma^k, \gamma^l, \gamma^{kl}, \gamma^c$ where c is chosen randomly so that $\gamma^c \neq \gamma^{kl}$, to determine which one of γ^{kl} or γ^c is the solution to the DHP with $\gamma, \gamma^k, \gamma^l$.*

We now define the analogous complexity problems involving LFSR's. We shall refer to them as *LFSR-based* problems.

Definition 4. *The* **LFSR-Based Discrete Logarithm Problem (LFSR-DLP)** *is, given A_1 and A_l, to find l.*

Definition 5. *The* **LFSR-Based Diffie-Hellman Problem (LFSR-DHP)** *is, given A_1 along with A_k and A_l, to determine A_{kl}.*

The key agreement scheme in Section 5.3 is based upon the LFSR-DHP.

Definition 6. *The* **LFSR-Based Decisional Diffie-Hellman Problem (LFSR-DDHP)** *is, given A_1 along with A_k, A_l, A_{kl}, A_c where c is chosen randomly so that $A_c \neq A_{kl}$, to determine which one of A_{kl} or A_c is the solution to the LFSR-DHP with input A_1, A_k, A_l.*

In [17], it was essentially proven that the LFSR-DLP is computationally equivalent to the DLP. We show this proof and prove the analogous for the Diffie-Hellman and Decisional Diffie-Hellman problems below.

Theorem 2.   *1. DLP $=_P$ LFSR-DLP.*
2. DHP $=_P$ LFSR-DHP.
3. DDHP $=_P$ LFSR-DDHP.

Proof. In the course of this proof, we will repeatedly transfer from sets of the form A_k to the minimal polynomial γ^k. This can be done by using the Newton Identity. We will also need to find a root of f_{γ^k}. This can be done efficiently using a root-finding algorithm such as the ones due to Rabin [15] or van Oorschot and Vanstone [18].

1) Given an instance A_1, A_k of the LFSR-DLP, we find roots γ and β of the respective minimal polynomials. Then the discrete log l where $\beta = \gamma^l$ is a solution to the LFSR-DLP.

Conversely, if $\beta = \gamma^l$, then we can find the respective minimal polynomials of γ and β and obtain the sets A_1 and A_l. Solving the LFSR-DLP would give an integer k such that $l = kq^i$ for some $i = 0, \ldots, n-1$. A quick check will tell us which is the correct l. This proves the first assertion.

2) Given an instance A_1, A_k, A_l of the LFSR-DHP, we can again find the respective polynomial roots $\gamma, \gamma^{kq^i}, \gamma^{lq^j}$ where $0 \leq i, j < n$. Solving the DHP with these three inputs gives $\gamma^{klq^{i+j}}$ whose minimal polynomial yields the set $A_{klq^{i+j}} = A_{kl}$.

Conversely, suppose we have an instance $\gamma, \gamma^k, \gamma^l$ of the DHP. Converting to minimal polynomial representations, we solve the LFSR-DHP with instances A_1, A_k, A_l and A_1, A_{k+1}, A_l to get A_{kl} and A_{kl+l} respectively. Finding roots gives us γ^{klq^i} and $\gamma^{(kl+l)q^j}$ where $0 \leq i, j < n$. We now calculate $\gamma^{(kl+l)q^j - l}$. Finding the value of j such that this is equal to γ^{klq^i} for some i gives the solution to the DHP. This proves the second assertion.

3) To prove the third assertion, we note that a solution to the DHP with input $\gamma, \gamma^k, \gamma^l$ is γ^c if and only if A_c is a solution to the LFSR-DHP with input A_1, A_k, A_l and A_{c+l} is a solution to the LFSR-DHP with input A_1, A_{k+1}, A_l.

Hence, these 2 LFSR-DDHP oracle queries would give the correct decision for the DHP.

Conversely, A_c is a solution to the LFSR-DHP with input A_1, A_k, A_l if and only if γ^{cq^i} is a solution to the DHP with input $\gamma, \gamma^k, \gamma^l$ for some $i = 0, \ldots, n-1$. This can be ascertained with at most n queries to the DDHP oracle.   □

We now turn our attention to representations using states. We can define the following computational problems.

Definition 7. *The **State-Based Discrete Logarithm Problem (S-DLP)** is, given $\bar{s}_1$ and $\bar{s}_l$, to determine l.*

Definition 8. *The **State-Based Diffie-Hellman Problem (S-DHP)** is, given $\bar{s}_1$ along with $\bar{s}_k$ and $\bar{s}_l$, to determine $\bar{s}_{kl}$.*

Definition 9. *The **State-Based Decisional Diffie-Hellman Problem (S-DDHP)** is, given $\bar{s}_1$ along with $\bar{s}_k, \bar{s}_l, \bar{s}_{kl}, \bar{s}_c$ where c is chosen randomly so that $\bar{s}_c \neq \bar{s}_{kl}$, to determine which one of $\bar{s}_{kl}$ or $\bar{s}_c$ is the solution to the S-DHP with input $\bar{s}_1, \bar{s}_k, \bar{s}_l$.*

Lemma 2. *1. DLP $\leq_P$ S-DLP.*
2. DHP $\leq_P$ S-DHP.
3. DDHP $\leq_P$ S-DDHP.

Proof. This lemma follows immediately from the fact that given γ and γ^k, we can calculate $\bar{s}_k$ from the relation

$$s_{k+i} = Tr(\gamma^{k+i}) = Tr(\gamma^k(\gamma)^i)$$

and that Tr is an efficiently computable function.

Lemma 3. *1. S-DLP $\leq_P$ LFSR-DLP.*
2. S-DHP $\leq_P$ LFSR-DHP.
3. S-DDHP $\leq_P$ LFSR-DDHP.

Proof. Given a state $\bar{s}_k = (s_k, \ldots, s_{k+n-1})$, we can use SO2 to calculate $s_{2k}, s_{3k}, \ldots, s_{rk}$ and hence get A_k. Following the proof of Theorem 2, the result follows.

Theorem 3. *1. DLP $=_P$ S-DLP.*
2. DHP $=_P$ S-DHP.
3. DDHP $=_P$ S-DDHP.

Proof. This follows immediately from Lemmas 2 and 3 and Theorem 2.

6.2   Trace-Related Complexity Problems

Let us now formally define problems related to the security of our signature scheme. The security of the signature scheme listed in Section 5.5 is based on the following problem.

Definition 10. *The **Trace Discrete Log Problem (Trace-DLP)** is the problem of finding, given an element $t \in GF(q)$, an index l such that $Tr(\gamma^l) = t$, or determining that there is no such index.*

The security of the more general signature scheme in Section 5.4 is based on this problem.

Definition 11. *The m-**Trace Discrete Logarithm Problem** (m-**Trace-DLP**) is the problem, given elements $t_1, \ldots, t_m$, of finding an integer l such that $Tr(\gamma^{li}) = t_i$ for all $i = 1, \ldots, m$, or determining that there is no such index.*

If this problem were tractable, then we could forge signatures by performing the following algorithm.

Algorithm 2.

INPUT: Signature parameters, public key, message M.
OUTPUT: Valid signature.

1. Perform the first three steps of signature generation as indicated.
2. Use SO1 to calculate $(\overline{s}_{h(M)r-1}, \ldots, \overline{s}_{mh(M)r-1})$.
3. Use SO2 to compute $(s_{h(M)r-1+w}, \ldots, s_{m(H(M)r-1+w)})$.
4. Solve the m-Trace-DLP with $t_i = s_{i(H(M)r-1+w)}$ to get l.
5. The forged signature is (A_k, s) with $s = rlk^{-1} \pmod{Q}$.

The following problems are related to the Trace-DLP problems and will aid is in their examination.

Definition 12. *The **Trace Inverse Problem (TraceInv)** is the problem, given $t \in GF(q)$, of finding an element $\beta \in \langle\gamma\rangle$ such that $Tr(\beta) = t$, or determining that no such element exists.*

Definition 13. *The m-**Trace Inverse Problem** (m-**TraceInv**) is the problem, given $t_1, \ldots, t_m \in GF(q)$, of finding an element $\beta \in \langle\gamma\rangle$ such that $Tr(\beta^i) = t_i$ for $i = 1, \ldots, m$, or determining that no such element exists.*

6.3   The Complexity of Trace-Related Problems

Let us now try to examine how feasible the m-Trace-DLP is, how this changes for different values of m, and whether or not we can relate it to the DLP.

Intuitively, it would seem at first glance that the m_2-Trace-DLP should be at least as difficult as the m_1-Trace-DLP if $m_1 < m_2$. For given $t_1, \ldots, t_{m_2}$ and an element $\beta \in \langle\gamma\rangle$ which solves the m_2-Trace-DLP, β would also solve the m_1-Trace-DLP with input $t_1, \ldots, t_{m_1}$. In addition, there may be elements $\beta \in \langle\gamma\rangle$ which solve the m_1-Trace-DLP , but not the m_2-Trace-DLP with these inputs.

However, when trying to make an actual reduction argument, we run into a problem. Suppose we have an oracle to solve the m_2-Trace-DLP and we wish to solve the m_1-Trace-DLP (again with $m_1 < m_2$) with input $t_1, \ldots, t_{m_1}$. In order to use the oracle, we must extend by including elements $t_{m_1+1}, \ldots, t_{m_2}$ in the oracle call. But it is unclear how to choose these elements. If we choose them at random, then there may not be a solution to the m_2-Trace-DLP with this input, even though there is a solution to the m_1-Trace-DLP with the truncated input.

We now discuss the relations amongst the problems presented in the previous two subsections. We begin by first relating the Trace-DLP and the DLP. We can use TraceInv to aid in this connection.

Lemma 4. m-*TraceInv* $\leq_P m$-*Trace-DLP*.

Proof. Given γ and $t_1, \ldots, t_m$, use the oracle for the m-Trace-DLP to find l. Then set $\beta = \gamma^l$.

This theorem is useful for making the following association.

Theorem 4. *For* $m_1 < m_2$, *if* m_1-*TraceInv* $\in P$, *then* m_1-*Trace-DLP* $\leq_P m_2$-*Trace-DLP*.

Proof. Let $t_1, \ldots, t_m \in GF(q)$. Since m_1-TraceInv can be solved in polynomial time, we can find $\beta = \gamma^l$ such that $Tr(\beta^i) = t_i$ for $i = 1, \ldots, m_1$. We then set $t_i = Tr(\beta^i)$ for $i = m_1+1, \ldots, m_2$. Now present $t_1, \ldots, t_{m_2}$ to the oracle to solve the m_2-Trace-DLP to get l which solves the m_1-Trace-DLP.

For the moment, let us assume that m-TraceInv $\in P$. Then the larger we choose m, the more difficult the m-Trace-DLP becomes. But does it get increasingly more difficult with every incrementation or is there a limit to how hard it can get? This is answered by the following theorem.

Theorem 5. r-*Trace-DLP* $=_P$ *DLP*.

Proof. The r-Trace-DLP is exactly the LFSR-DLP, which by Theorem 2 is computationally equivalent to the DLP.

Corollary 1. *For all* $m \geq r$, m-*Trace-DLP* $=_P r$-*Trace-DLP*.

Proof. Clearly, r-Trace-DLP $\leq_P m$-Trace-DLP since given A_k, we can uniquely calculate s_{ik} from A_k for all $i > r$. But for any instance of the m-Trace-DLP, the solution of the truncated instance to the r-Trace-DLP would be a solution to the m-Trace-DLP.

The question now becomes, how difficult is m-TraceInv? Let us examine just the TraceInv. Given an element $t \in GF(q)$, it is actually quite simple to find an element $\beta \in GF(q^n)$ such that $Tr(\beta) = t$. In fact $\beta = \frac{1}{n}t$ is a preimage of t provided that n is coprime to q. However, when we want our preimage to be in a (relatively small) subgroup of $GF(q^n)^*$, it becomes much more difficult to find preimages. In general, this problem is still open to the knowledge of this author, and appears difficult to solve. But in some instances, TraceInv and m-TraceInv can be solved in polynomial time with high probability.

Proposition 2. *Suppose that γ has order approximately q^r. Then m-TraceInv $\in P$.*

Proof. We may assume that $m \leq r$, since if $m > r$, we may truncate any input $t_1, \ldots, t_m$ to $t_1, \ldots, t_r$ and solve r-TraceInv with this input. Given a solution β, we simply check that $t_i = Tr(\beta^i)$ for $i = r + 1, \ldots, m$. Note that any other solution to the r-TraceInv with this input would be conjugate to β and thus give the same traces for all of their powers.

Let $t_1, \ldots, t_m \in GF(q)$. We wish to find one of its solutions $\beta \in \langle \gamma \rangle$, if one exists.

Let A be the m-tuple of elements $t_1, \ldots, t_m$ in $GF(q)$ by choosing $t_{m+1}, \ldots, t_r$ at random in $GF(q)$ such that t_i takes the place of s_{ik} in the representation within A. If $A = A_k$ for some k, we would be able to construct its corresponding minimal polynomial f_{γ^k} over $GF(q)$. Finding a root β of f_{γ^k} would give $\beta = \gamma^{kq^j}$, with $Tr(\beta^i) = t_i$ for $i = 1, \ldots, m$, solving the m-TraceInv Problem.

Given a candidate polynomial, we need only find a root β and then check that β has order dividing Q. These are efficient operations.

Thus, we need only determine the probability of success if we choose our set A in this fashion. Since the order of γ is $\sim q^r$ and each minimal polynomial has n roots, there are approximately, q^r/n polynomials which would give success. But there are q^r possible r-tuples of elements in $GF(q)$. Thus, the probability that a randomly chosen tuple represents the coefficents of a minimal polynomial is approximately $q^r/nq^r = 1/n$. Hence, repeating this trial polynomially many times in n, we are likely to succeed with high probability. If no valid β is produced after a small number of trials, then with high probability, there is no such β.

Remark 2. We note that if γ has order instead approximately q^c for some $c < d$. Then the probability for success would be $1/nq^{d-c}$. It would then take exponentially many attempts to achieve a non-negligible probability.

7   Conclusions and Discussion

We have presented a new general class of cryptosystems based on characteristic sequences generated by LFSR's. We have proposed a signature scheme based on the Trace-DLP and its variants which takes advantage of the compact representations finite field elements. The complexity of the Trace-DLP and related problems was examined. We also have proven the equivalence of the LFSR-based and State-based sequence problems with their counterparts based on finite fields.

To the knowledge of the authors, this is the first time complexity problems involving traces have been proposed. This is an area which deserves more study. It would also be very nice to find an LFSR-based signature scheme in which both the public key and the signature can be represented by reduced representations. Finally, it would be nice to develop other applications dependent upon the Trace-DLP and TraceInv problems.

References

1. Diffie, W., Hellman, M.E.: New Directions in Cryptography. IEEE Trans. IT. **22** (1976) 644–654.
2. Fiduccia, C.M.: An Efficient Formula for Linear Recurrences. SIAM J. Comput. **14** (1985) 106–112.
3. Giuliani, K., Gong, G.: Analogues to the Gong-Harn and XTR Cryptosystems. Combinatorics and Optimization Research Report CORR 2003-34, University of Waterloo (2003).
4. Giuliani, K., Gong, G.: Efficient Key Agreement and Signature Schemes Using Compact Representations in $GF(p^{10})$. In: Proceedings of the 2004 IEEE International Symposium on Information Theory - ISIT 2004. Chicago (2004) 13–13.
5. Golomb, S. W.: Shift Register Sequences. Holden-Day, San Francisco (1967).
6. Gong, G., Harn, L.: Public-Key Cryptosystems Based on Cubic Finite Field Extensions. IEEE Trans. IT. **24** (1999) 2601–2605.
7. Lenstra, A., Verheul, E.: The XTR Public Key System. In: Advances in Cryptology – Crypto 2000. Lecture Notes In Computer Science, Vol. 1880. Springer-Verlag, Berlin Heidelberg New York (2000) 1–19.
8. Lidl, N., Niederreiter, H.: Finite Fields. Addison-Wesley, Reading (1983).
9. Müller, W. B., Nobauer, R.: Cryptanalysis of the Dickson scheme. In: Advances in Cryptology – Eurocrypt 1985. Lecture Notes In Computer Science, Vol. 219. Springer-Verlag, Berlin Heidelberg New York (1986) 50–61.
10. Niederreiter, H.: A Public-Key Cryptosystem Based on Shift-Register Sequences. In: Advances in Cryptology – Eurocrypt 1985. Lecture Notes In Computer Science, Vol. 219. Springer-Verlag, Berlin Heidelberg New York (1986) 35–39.
11. Niederreiter, H.: Some New Cryptosystems Based on Feedback Shift Register Sequences. Math. J. Okayama Univ. **30** (1988) 121-149.
12. Niederreiter, H.: Finite Fields and Cryptology. In: Finite Fields, Coding Theory, and Advances in Communications and Computing. M. Dekker, New York (1993) 359–373.
13. National Institute of Standards (NIST): Digital Signature Standard. U. S. Government Standard. FIPS-186. (1994).
14. Quoos, L., Mjølsnes, S. F.: Public Key Systems Based on Finite Field Extensions of Degree Five. Presented at Fq7 conference (2003).
15. Rabin, M.: Probabilistic Algorithms in Finite Fields. SIAM J. Comput. **9** (1980) 273–280.
16. Smith, P., Skinner, C.: A Public-Key Cryptosystem and a Digital Signature System Based on the Lucas Function Analogue to Discrete Logarithms. In: Advances in Cryptology – Asiacrypt '94. Lecture Notes In Computer Science, Vol. 917. Springer-Verlag, Berlin Heidelberg New York (1994) 357–364.
17. Tan, C.-H., Yi, X., Siew, C.-K.: On the n-th Order Shift Register Based Discrete Logarithm. IEICE Trans. Fundamentals. **E86-A** (2003) 1213–1216.
18. van Oorschot, P. C.,Vanstone, S. A.: A Geometric Approach to Root Finding in $GF(q^m)$. IEEE Trans. IT. **35** (1989) 444–453.

Cryptanalysis of a Particular Case of Klimov-Shamir Pseudo-Random Generator

Vincent Bénony[1], François Recher[2], Éric Wegrzynowski[1],
and Caroline Fontaine[1,3]

[1] USTL-LIFL/IRCICA,
Cité Scientifique, Bâtiment M3,
59655 Villeneuve d'Ascq cedex – France
{Vincent.Benony, Eric.Wegrzynowski, Caroline.Fontaine}@lifl.fr
[2] USTL-Laboratoire Paul Painlevé, Bâtiment M2,
59655 Villeneuve d'Ascq cedex – France
Francois.Recher@math.univ-lille1.fr
[3] Also with CNRS

Abstract. T-functions have been introduced by Shamir and Klimov in [1]. Those functions can be used in order to design a new class of stream ciphers. We present in this paper an algorithm which can retrieve the internal state of a particular class of pseudo-random generators based on T-functions. This algorithm has time complexity of $O(2^{\frac{n}{4}})$ and has memory complexity of $O(n \log_2 n)$ for pseudo random generators which put out the $n/2$ most significants bits of their internal state at each time clock, n being the length of the internal state of the pseudo-random generator.

1 Introduction

Designing pseudo-random generators (PRGs) is of major interest in cryptography. A PRG can be viewed as the composition of a deterministic automaton (whose internal state is evolving according to its transition function f), and a filtering function, which computes the output of the PRG from the internal state of the automaton. Our work concerns a recent kind of automaton, proposed by Klimov and Shamir in 2002 [1]. The authors discussed its suitability for cryptography in [2], presenting two attacks, which enable an opponent to derive the internal state of the automaton from consecutive outputs of the PRG. We present here an improvement of these attacks. This does not discredit the choice of the automaton itself, but shows that the filtering function has to be chosen very carefully.

2 Presentation of Klimov and Shamir's Scheme

Klimov and Shamir proposed in [1] a new method to design invertible mappings: those mappings are a composition of primitive functions, which can be found on

T. Helleseth et al. (Eds.): SETA 2004, LNCS 3486, pp. 313–322, 2005.

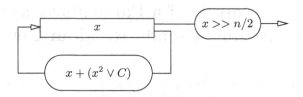

Fig. 1. Klimov and Shamir proposal

any recent micropocessor. The interest of using such functions is the efficiency of their software implementation.

The Automaton: we will only recall the properties we need to present our result; the reader can refer to [1, 2] for more details. The internal state at time t is an integer, $0 \leq x_t < 2^n$, whose binary expansion, also denoted by x_t, is expressed with exactly n bits. Given a function $f : \mathbb{N} \to \mathbb{N}$, the transition from a state x_t to a state x_{t+1} is defined by $x_t \mapsto x_{t+1} = f(x_t) \mod 2^n$. We will use the same notation as in [2], and denote by $[x_t]_i$ the i-th component of x_t; $[x_t]_0$ is the least significant bit, and $[x_t]_{n-1}$ the most significant one. When the index t is not needed, we omit it. Klimov and Shamir study *triangular functions (T-functions)*, that is, functions f such that the bit $[f(x)]_i$ only depends on the bits $[x]_0, \ldots, [x]_i$. They propose a suitable subset of invertible T-functions f, which ensure that their iteration will define a single cycle, of maximal length 2^n.

Hence, they focus on a particular case of such functions, taking:

$$f(x) = x + (x^2 \vee C) \mod 2^n$$

where the addition and multiplication are the usual ones on integers, $\vee$ denotes the bitwise logic OR operator, and C is a constant integer satisfying $[C]_0 = [C]_2 = 1$.

The Filtering Function: in [2], the authors propose to put out the $m \ll n$ higher bits. In [1], they mention the really good statistical behavior of the particular case $m = n/2$, arguing that it "is better than for some of the AES candidates themselves!". Hence, $m = n/2$ is a very interesting case to study and we will focus on it in this paper (see Figure 1).

The Two Attacks: Klimov and Shamir propose two attacks in [2], requiring only the output of the PRG. They have data and memory complexity of $O(2^{\frac{n}{2}})$ for the first one, and $O(2^{\frac{n}{3}})$ for the second one. These attacks can be applied for any value of $m \leq n/2$.

We present here an improved attack for the case $m = n/2$, whose data complexity is $O(2^{\frac{n}{4}})$ and memory complexity is $O(n \log_2 n)$.

3   Definitions and Facts

Our attack aims at retrieving the internal state of a PRG in the particular case where C satisfies $\log_2 C \leq n/4$, by observing on average $O(2^{\frac{n}{4}})$ consecutive outputs of the PRG. In [1], the authors propose to consider the particular case $C = 5$ which can ensure a single cycle. Let us take a look at some particular states which could give us useful information: states where the half-higher of the half-lower bits are all zeros, that is, states x_t such that

$$[x_t]_{(\frac{n}{2}-1)...\frac{n}{4}} = 0 \iff x_t \mod 2^{n/2} < 2^{n/4}. \tag{1}$$

Let us introduce some more notation : let u_t be the output of the PRG at time t, and v_t the unknown part of the internal state:

$$u_t = [x_t]_{(n-1)...\frac{n}{2}} \text{ and } v_t = [x_t]_{(\frac{n}{2}-1)...0}$$

Now, we can state some simple lemmas:

Lemma 1. $\forall x, y \in \mathbb{N}$, we have $\max(x, y) \leq x \vee y \leq x + y$.

Proof. The inequality $\max(x, y) \leq x \vee y$ is obvious, since $x \leq x \vee y$ and $y \leq x \vee y$.

We will now prove the second inequality by induction on the smallest length of the binary expansion of x and y, denoted by $L \in \mathbb{N}$.

Let $L = 0$. This means that x or y is equal to 0, and the result states easily.

Now, suppose that the inequality holds for any smallest length less than or equal to $L \leq k$; we will show that it still holds for a smallest length equal to $L + 1$. Suppose that the smallest length of the binary expansions of x and y is equal to $L + 1$. Two cases have to be considered, according to the occurence of a carry when computing $x + y$.

1. No carry: in this case, $x \vee y = x + y$.
2. A carry occurs; without loss of generality, let us suppose that the first carry occurs when adding bits at position $0 \leq i \leq L$. Let $x = x_h 2^i + 2^{i-1} + x_\ell$ and $y = y_h 2^i + 2^{i-1} + y_\ell$, with $x_\ell, y_\ell < 2^{i-1}$. We can state

$$\begin{aligned} x + y &= (x_h + y_h + 1)2^i + (x_\ell \vee y_\ell) \\ &\geq (x_h + y_h)2^i + (x_\ell \vee y_\ell) \\ &\geq (x_h \vee y_h)2^i + (x_\ell \vee y_\ell) \qquad \text{by induction hypothesis.} \end{aligned}$$

Since the right part of the last inequality is in fact $x \vee y$, this concludes the proof. ◇

Lemma 2. Let $x = u2^{n/2} + v$, where $u < 2^{n/2}$ and $v < 2^{n/4}$. Let $x' = f(x) \mod 2^n = u'2^{n/2} + v'$, where $u' < 2^{n/2}$ and $v' < 2^{n/2}$. Then, we have:

$$u' = u(2v + 1) \mod 2^{n/2}. \tag{2}$$

Proof. From Lemma 1 and $v < 2^{n/4}$, we can write:

$$v + (v^2 \vee C) \leq 2^{n/2} - 2^{n/4} + C.$$

But, as $\log_2 C \leq n/4$, we have:

$$v + (v^2 \vee C) < 2^{n/2}.$$

Hence,

$$
\begin{array}{r|c|c|c|}
x = & u & 0 & v \\
\end{array}
$$

$$
\begin{array}{r|c|c|}
x^2 \vee C = & 2uv & v^2 \vee C \\
\end{array}
$$

$$
\begin{array}{r|c|c|}
x + (x^2 \vee C) = & u(2v+1) & v + (v^2 \vee C) \\
\end{array}
$$

which concludes the proof.   ◇

Lemma 3. *Let x and x' be two integers defined as in Lemma 2; then, u and u' are divisible by the same powers of two.*

Proof. Let k be the number of zeros at the end of u. So, $u = 2^k \tilde{u}$, where $\tilde{u}$ is an odd number. Then $u' = 2^k \tilde{u}(2v+1) \mod 2^{n/2}$. As $2v+1$ is an odd number too, the product $\tilde{u}(2v+1)$ is odd, and u' has exactly k zeroes at its end.   ◇

4   The Attack

We now focus on the effective realization. From Lemma 2, we can see that if we have an internal state whose half-higher of half-lower bits are all zeros, then it is possible to retrieve this whole state from consecutive outputs. Lemma 3 gives us a necessary condition on the PRG outputs in order to detect if two consecutive outputs are really interesting for the attack. We will try to solve Equation (2) (in v) for all pairs (for $u = u_t$, $u' = u_{t+1}$) of consecutive outputs having the same divisibility by powers of two. If solutions exist, this equation can be solved by using the *extended Euclid's algorithm* to determine v as a candidate value for v_t; we can remark here that if the computed value is greater than or equal to $2^{n/4}$, then it is sure that the attack missed, because we are only trying to recover states for which $v_t < 2^{n/4}$; we can also remark that if the computed value is smaller that $2^{n/4}$, then it may be the right one, or not (see below). This computation is achieved by finding a and b such that:

$$
2^{n/2} a + u_t b = 2^s, \text{ where } 2^s = \gcd(2^{n/2}, u_t).
$$

Once we have found them, we have :

$$
v_t = \frac{b u_{t+1} - 2^s}{2^{s+1}} \mod 2^{\frac{n}{2} - s}. \tag{3}
$$

Unfortunately, not all these states can be used for the attack, as some will give a wrong computed value for v_t. Indeed, from Equation (3), the number of solutions for v_t may be greater than one if $s \geq n/4$. Considering such cases should change the memory complexity of our algorithm, so we will omit them for the moment.

The major problem for the attacker is to be sure that the attack succeeded after the computations. In order to check it, a new PRG will be designed using the computed state. Looking for a divergence between this simulated PRG and the outputs of the attacked one will solve the problem.

If we denote by c_t the state he computed/guessed from the outputs u_t and u_{t+1}, we have $c_t = x_t + \varepsilon$ where ε is the error he made. If $\varepsilon = 0$, then we can say that the attack was successful; if $\varepsilon \neq 0$, we can estimate the divergence which will be introduced between the simulated PRG and the real one.

Let us take a look at the evolution of these two PRGs. After one iteration of the PRG, the internal state of the real PRG will be

$$x_{t+1} = x_t + (x_t^2 \vee C)$$

and the internal state of the simulated PRG will be

$$c_{t+1} = c_t + (c_t^2 \vee C) = x_t + \varepsilon + [(x_t^2 + \varepsilon^2 + 2\varepsilon x_t) \vee C]$$

which is roughly equal to

$$x_t + (x_t^2 \vee C) + E(\varepsilon) = x_{t+1} + E(\varepsilon)$$

where the error $E(\varepsilon)$ is a polynomial of degree 2 in ε.

This error will grow quadratically at each iteration of the PRG. We can approximate the error's evolution by saying that the error's length will double at each iteration; so, it will need $\log_2(n/2)$ iterations before the error disturbs the $n/2$ most significant bits of the internal state.

We may want to know how many consecutive outputs the attack needs to succeed; as $2^{3n/4}$ states satisfy Relation (1), the average distance between a randomly chosen state and one of those states is $2^{n/4}$, $i.e.$ we will have to produce $2^{n/4}$ outputs of the PRG on average in order to reach a "good state".

Hence, a simple method to determine if the computed state is the good one is to compare $\log_2(n/2)$ real outputs of the PRG with $\log_2(n/2)$ outputs of the simulated PRG. The verification process has data complexity $O(n \log_2(n))$.

5   An Example

In order to be more clear, we will now describe a scenario for the attack. Let us consider, as an example, a PRG with an internal state of 16 bits. This is a really small PRG, but this will illustrate the attack.

The first step of our attack is to find a pair of consecutive outputs which verify the condition on divisibility, that is, two outputs which have the same number of zeros at the end of their binary expansion. Let have a look at Figure 2.

The first pair of consecutive outputs (left column) which meets this condition is:10010110, 01001110. The attacker will suppose that the unknown part of the internal state has $n/4 = 4$ zeros as its half most significant bits. We can see here, that this assumption is false, but, the attacker has no way to know this, $a\ priori$.

So, the attacker has to solve the equation exposed in Lemma 2:

$$u' = u(2v + 1) \mod 2^{n/2}$$
$$78 = 150(2v + 1) \mod 2^8$$
$$v = 10$$

Internal states ($n = 16$)

High parts   Low parts
(outputs)   (hidden)

01011101   01010000
10010110   01010101
01001110   10010010
10011001   11010111
01001100   01101100
10011010   00000001
11001110   00000110

⋮   ⋮

Fig. 2. Failed attack

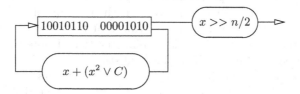

Real PRG :   01001110, 10011001, 01001100, ...
Simulated PRG :   01001110, 00100010, 11001000, ...

Fig. 3. Failed attack - verification

The computed value satisfies $v \leq 2^{n/4}$, so it may be the right one. But the attacker has to check it. To do so, he will create a new PRG which will have as internal state, the value he just computed (see Figure 3). He can be sure that if he did not compute the right value, the output of this new PRG and the output of the real PRG will diverge after at most $\log_2(n/2) = 3$ iterations! Here, we can see that the third output is not the same than the real one, so the attacker states he did not compute the right value for v_t.

Now, let us have a look at another pair of consecutive outputs which meets the condition on divisibility. The two states are now 10011010 and 11001110 (see Figure 4).

The attacker tries to solve the equation:

$$u' = u(2v + 1) \mod 2^{n/2}$$
$$206 = 154(2v + 1) \mod 2^8$$
$$v = 1$$

The computed value satisfies $v \leq 2^{n/4}$, so it is perhaps the right one. But, as previously, he has to check it (see Figure 5).

Internal states $(n = 16)$

High parts | Low parts
(outputs) | (hidden)

01011101 | 01010000
10010110 | 01010101
01001110 | 10010010
10011001 | 11010111
01001100 | 01101100
10011010 | 00000001
11001110 | 00000110

⋮   ⋮

Fig. 4. Successfull attack

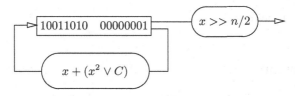

Real PRG : $11001110, 01110110, 00100001, \ldots$
Simulated PRG : $11001110, 01110110, 00100001, \ldots$

Fig. 5. Successfull attack - verification

This time, $\log_2(n/2) = 3$ consecutives output of the new PRG coincide with the real ones, so, the attacker can be sure that this v was the good one.

6   The Algorithm

Here is a pseudo-code version of the used algorithm:

```
crack(u_t, u_{t+1})  :=
    s := log_2(gcd(u_t, 2^{n/2}))
    s' := log_2(gcd(u_{t+1}, 2^{n/2}))
    if u_t ≠ 0 and s = s' then
        find a and b, b ≥ 0, an integer solution of 2^{n/2}a + u_t b = 2^s
        v_t := (bu_{t+1}-2^s)/(2^{s+1})  mod 2^{n/2 -s}, and x_t := v_t + 2^{n/2}u_t
        if v_t < 2^{n/4} then return x_t else return false
    else
        return false
```

```
attack(u) :=
```
$n :=$ `the length of the internal state`
$l := \log_2 n$
$t := 0$
$seeds := \{\mathtt{false}, \mathtt{false}, \ldots, \mathtt{false}\}$, `containing` l `elements`
$success := \{0, 0, \ldots, 0\}$, `containing` l `elements`
```
do
```
$\qquad seeds_{0\ldots(l-2)} := seeds_{1\ldots(l-1)}$
$\qquad success_{0\ldots(l-2)} := success_{1\ldots(l-1)}$
$\qquad seeds_{l-1} := \mathtt{crack}(u_t, u_{t+1})$
$\qquad success_{l-1} := 0$
$\qquad$`for all elements` $seeds_j$ `of` $seeds$ `do`
$\qquad\qquad$`if` $seeds_j \neq \mathtt{false}$ `then` $seeds_j := f(seeds_j)$
$\qquad\qquad$`if` $seeds_j = u_{t+1}$ `then` $success_j := success_j + 1$
$\qquad$`if` $success_0 = l$ `we are done!`
$\qquad t := t + 1$
```
loop until success
```

7   Discussion

We saw in previous sections that the time complexity of our attack is $O(2^{\frac{n}{4}})$. This time complexity only measures in average the number of PRG outputs we need to retrieve the internal state. Indeed, in such attacks the aim is to retrieve the internal state of the PRG with the least possible outputs. Notice that it does not measure the time consuming of calculation itself.

From this point of view, we can improve our attack by considering consecutive outputs of the form :

$$u_t = \boxed{\quad \ldots 1 \mid 0 \quad}$$
$$u_{t+1} = \boxed{\quad \ldots 1 \mid \ldots \quad}$$

(we do not impose divisibility condition on u_t and u_{t+1}) with the further assumption on the internal state :

$$[x_t]_{(\frac{n}{2}-1)\ldots\frac{n}{4}+\frac{s}{2}} = 0 \iff x_t \mod 2^{n/2} < 2^{n/4+s/2}$$

where $s = \log_2(\gcd(u_t, 2^{n/2}))$ is supposed to be even.

This is an extension of conditions imposed in the attack. Namely, objects can be considered as parts of higher dimension objects : $n + 2s$ where some bits are ignored ($2s$ higher bits).

We can thus apply the attack in this context with the following consecutive truncated outputs :

$$\tilde{u}_t = \frac{u_t}{2^s}, \qquad \tilde{u}_{t+1} = \frac{u_{t+1} - (u_{t+1} \mod 2^s)}{2^s}$$

and solve the equation :

$$\tilde{u}_t(2\tilde{x}_t + 1) = \tilde{u}_{t+1} \mod 2^{\frac{n}{2}-s}$$

to find solutions in $\tilde{x}_t$ such that $x_t \mod 2^{n/2} < 2^{n/4+s/2}$.

The equation may have several many solutions.

Clearly this new attack needs more calculation since half pairs $(\tilde{u}_t, \tilde{u}_{t+1})$ verify the required hypothesis.

Let us calculate the time complexity of this new attack. The number of internal states that can be used for fixed s is :

for $s = 0$: | ... | ... | 0 | ... | : $2^{\frac{3n}{4}}$

for $s = 2$: | ... | ... 0 0 | 0 | 1 | ... | : $2^{\frac{3n}{4}-2}$

for $s = 4$: | ... | ... 0 0 0 0 | 0 | 1 . | ... | : $2^{\frac{3n}{4}-3}$

$\vdots$

for $s = n/2$: | 0 | 0 | 1 ... | ... | : $2^{\frac{3n}{4}-\left(\frac{n}{4}+1\right)}$

So the total number of states is :

$$2^{\frac{3n}{4}} + \sum_{\tilde{s}=1}^{n/4} 2^{\frac{3n}{4}-(\tilde{s}+1)} = 2^{\frac{3n}{4}} + 2^{\frac{3n}{4}-1} - 2^{\frac{n}{2}-1}.$$

So asymptotically, the number of internal states that can be used in the new attack is 50% greater than in the previous one.

8   Conclusion

The attack has been implemented in $MAPLE$, and the algorithm's time complexity is $O(2^{\frac{n}{4}})$, whereas its memory complexity is $O(n \log_2 n)$; this is better than the attacks exposed in [2]. As an example, the average time in order to retrieve the internal state of a 64 bits long PRG is 10 seconds on a 2.4Ghz processor. It would be possible to retrieve the internal state of a 128 bits long PRG by using 2^{32} outputs of this PRG.

There are many possibilities to prevent this attack:

- by choosing a constant C which satisfies $\log_2 C > n/4$; this may introduce the $\vee$ function into equations, and make the resolution more difficult;
- by using another filtering function to produce the output from the internal state; this could significantly reduce the data rate of the PRG, but it would also increase its security.

Those precautions could prevent the attacker from knowing a significant part of the internal state and, consequently, from deducing the content of the internal state.

References

1. A. Klimov and A. Shamir. A new class of invertible mappings. In *CHES 2002*, number 2523 in Lecture Notes in Computer Science, pages 470–483. Springer-Verlag, 2002.
2. A. Klimov and A. Shamir. Cryptographic applications of T-functions. In *Selected Areas in Cryptography – SAC 2003*, Lecture Notes in Computer Science. Springer-Verlag.

Generating Functions Associated with Random Binary Sequences Consisting of Runs of Lengths 1 and 2[*]

Vladimir B. Balakirsky

Institute for Experimental Mathematics,
Ellernstr. 29, 45326 Essen, Germany
v.b.balakirsky@tue.nl

Abstract. We associate a generating function of two formal variables with a given binary sequence and a generating function of three formal variables with a given pair of binary sequences. The first function gives information about all subsequences of the sequence and the second function gives information about all common subsequences of the pair of sequences. It is shown that, in many cases, these functions can be easily found, which is of interest for various applications such as reconstruction of sequences, pattern recognition, data transmission over channels with deletions, etc. [1], [2], [3]. This conclusion is demonstrated for random sequences chosen from a completely randomized probabilistic ensemble of binary sequences and from the ensemble of random sequences consisting of runs of lengths 1 and 2. The results show that the latter ensemble can be considered as a very good candidate for the ensemble of random codes capable of correcting deletion errors.

1 Generating Functions Associated with Binary Sequences

Let $\mathbf{v} = (v_1, v_2, \ldots)$ be an infinite binary sequence and let $\mathbf{v}^t = (v_1, \ldots, v_t)$ for all $t \geq 1$. For all $r \geq 1$, let $\mathbf{v}[r] = (0^r 1^r)^*$ denote the sequence obtained by concatenating runs of length r with the first bit equal to 0. In particular,

$$\mathbf{v}[1] \triangleq (01)^*, \quad \mathbf{v}[2] \triangleq (0011)^*.$$

Given a sequence $\mathbf{v}$ and a vector $\mathbf{u}^\ell \in \{0,1\}^\ell$, introduce $\mathrm{Del}(\mathbf{v}|\mathbf{u}^\ell) \in \{0, 1, \ldots\}$ as the smallest integer d such that $\mathbf{u}^\ell$ is a subsequence of the string $\mathbf{v}^{\ell+d}$. Thus, $\mathrm{Del}(\mathbf{v}|\mathbf{u}^\ell) = d$ implies that one can delete d components of the vector $\mathbf{v}^{\ell+d-1}$ and attach component $v_{\ell+d}$ to the result in order to obtain $\mathbf{u}^\ell$, and this procedure is not possible if d decreases. The total number of subsequences of length ℓ of the

---

[*] This work was supported by the Philips Research Laboratories (The Netherlands) and DFG (Germany).

T. Helleseth et al. (Eds.): SETA 2004, LNCS 3486, pp. 323–338, 2005.

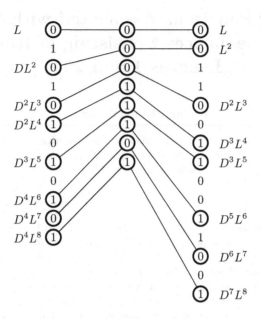

Fig. 1. Mappings $\ell \to \ell + \mathrm{Del}(\mathbf{v}|\mathbf{u}^\ell)$, $\ell = 1, \ldots, 8$, for sequences $\mathbf{u}^8 = 00011101$ and $\mathbf{v} = \mathbf{v}[1], \mathbf{v}[2]$. The generating functions of the mapping of the sequence $\mathbf{u}^8$ to the sequences $\mathbf{v}[1]$ and $\mathbf{v}[2]$ are equal to $D^4 L^8$ and $D^7 L^8$, respectively. The generating function of the mapping of the sequence $\mathbf{u}^8$ to the pair of sequences $(\mathbf{v}[1], \mathbf{v}[2])$ is equal to $D_1^4 D_2^7 L^8$

string $\mathbf{v}^t$, where $t \geq \ell$, is equal to the number of vectors $\mathbf{u}^\ell$ with $\mathrm{Del}(\mathbf{v}|\mathbf{u}^\ell) \leq t - \ell$. This number can be expressed as $n_{\mathbf{v}}(0, \ell) + \ldots + n_{\mathbf{v}}(t - \ell, \ell)$, where

$$n_{\mathbf{v}}(d, \ell) \stackrel{\triangle}{=} \left| \left\{ \mathbf{u}^\ell : \mathrm{Del}(\mathbf{v}|\mathbf{u}^\ell) = d \right\} \right| \tag{1}$$

for all $d, \ell \geq 0$. We will assume that $n_{\mathbf{v}}(0, 0) = 1$ and $n_{\mathbf{v}}(d, 0) = 0$ for all $d \neq 0$.

Given two sequences, $\mathbf{v}_1$ and $\mathbf{v}_2$, let

$$n_{\mathbf{v}_1, \mathbf{v}_2}(d_1, d_2, \ell) \stackrel{\triangle}{=} \left| \left\{ \mathbf{u}^\ell : \left(\mathrm{Del}(\mathbf{v}_1|\mathbf{u}^\ell), \mathrm{Del}(\mathbf{v}_2|\mathbf{u}^\ell) \right) = (d_1, d_2) \right\} \right| \tag{2}$$

for all $d_1, d_2, \ell \geq 0$. We will assume that $n_{\mathbf{v}}(0, 0, 0) = 1$ and $n_{\mathbf{v}}(d_1, d_2, 0) = 0$ for all $(d_1, d_2) \neq (0, 0)$. All vectors $\mathbf{u}^\ell$ counted in $n_{\mathbf{v}_1, \mathbf{v}_2}(d_1, d_2, \ell)$ are common subsequences of $\mathbf{v}_1^{\ell+d_1}$ and $\mathbf{v}_2^{\ell+d_2}$. If $\mathbf{v}_1^{t_1}$ and $\mathbf{v}_2^{t_2}$ are two given strings, then the length of their longest common subsequence is equal to the largest ℓ such that $n_{\mathbf{v}_1, \mathbf{v}_2}(d_1, d_2, \ell) > 0$ for some $d_1 \leq t_1 - \ell$, $d_2 \leq t_2 - \ell$.

Notice that different vectors $(j_1, \ldots, j_\ell)$ such that $j_1 < j_2 < \ldots < j_\ell = \ell + \mathrm{Del}(\mathbf{v}|\mathbf{u}^\ell)$ and $u_i = v_{j_i}$ for all $i = 1, \ldots, \ell$ may exist. However $(\mathrm{Del}(\mathbf{v}|\mathbf{u}^1), \ldots, \mathrm{Del}(\mathbf{v}|\mathbf{u}^\ell))$ is a unique vector, and it is constructed by the following greedy algorithm : (a) set $i = j = 1$; (b) find the smallest $d \geq 0$ such that $u_i = v_{j+d}$ and set $\mathrm{Del}(\mathbf{v}|\mathbf{u}^i) = d$; (c) increase i by 1 and j by d; (d) if $i \leq \ell$, then go to (b). Notation above is illustrated in Fig. 1 and Table 1.

Table 1. Representation of vectors of length $\ell = 1, 2, 3$ as subsequences of the sequences $\mathbf{v}[1]$ and $\mathbf{v}[2]$, where $d_1 = \mathrm{Del}(\mathbf{v}[1]|\mathbf{u}^\ell)$ and $d_2 = \mathrm{Del}(\mathbf{v}[2]|\mathbf{u}^\ell)$. Components of the vectors $\mathbf{v}^{\ell+d_1}[1]$ and $\mathbf{v}^{\ell+d_2}[2]$ that are mapped to components of the vector $\mathbf{u}^\ell$ are given in the bold font

ℓ	$\mathbf{u}^\ell$	d_1	$\mathbf{v}^{\ell+d_1}[1]$	d_2	$\mathbf{v}^{\ell+d_2}[2]$
1	0	0	**0**	0	**0**
	1	1	0 **1**	2	0 0 **1**
2	0 0	1	0 **1 0**	0	**0 0**
	0 1	0	**0 1**	1	0 **0 1**
	1 0	1	0 **1 0**	3	0 0 **1 1** 0
	1 1	2	0 1 **0 1**	2	0 0 **1 1**
3	0 0 0	2	0 **1 0 1** 0	2	0 0 **1 1** 0
	0 0 1	1	0 **1 0 1**	0	**0 0 1**
	0 1 0	0	**0 1 0**	2	0 0 **1 1** 0
	0 1 1	1	0 **1 0 1**	1	0 **0 1 1**
	1 0 0	2	0 **1 0 1** 0	3	0 0 **1 1** 0 0
	1 0 1	1	0 **1 0 1**	4	0 0 **1 1** 0 0 **1**
	1 1 0	2	0 1 **0 1 0**	2	0 0 **1 1** 0
	1 1 1	3	0 1 0 **1 0 1**	4	0 0 **1 1** 0 0 **1**

Given a sequence $\mathbf{v}$ and a pair of sequences $(\mathbf{v}_1, \mathbf{v}_2)$, the integers defined in (1), (2) can be specified by the generating functions

$$N_\mathbf{v}(D, L) \overset{\triangle}{=} \sum_{d,\ell \geq 0} n_\mathbf{v}(d, \ell) D^d L^\ell, \tag{3}$$

$$N_{\mathbf{v}_1, \mathbf{v}_2}(D_1, D_2, L) \overset{\triangle}{=} \sum_{d_1, d_2, \ell \geq 0} n_{\mathbf{v}_1, \mathbf{v}_2}(d_1, d_2, \ell) D_1^{d_1} D_2^{d_2} L^\ell, \tag{4}$$

where D, D_1, D_2, L are formal variables.

We will consider two ensembles of binary sequences. Let the ensemble of "completely random" binary sequences consist of all sequences whose components are i.i.d. random variables taking values 0 and 1 with probability $1/2$. Denote

$$\overline{n_{1/2}}(d, \ell) \overset{\triangle}{=} 2^{-(\ell+d)} \sum_{\mathbf{v}^{\ell+d}} n_{\mathbf{v}^{\ell+d}}(d, \ell), \tag{5}$$

$$\overline{n_{1/2}}(d_1, d_2, \ell) \overset{\triangle}{=} 2^{-(2\ell+d_1+d_2)} \sum_{\mathbf{v}_1^{\ell+d_1}, \mathbf{v}_2^{\ell+d_2}} n_{\mathbf{v}_1^{\ell+d_1}, \mathbf{v}_2^{\ell+d_2}}(d_1, d_2, \ell). \tag{6}$$

Introduce also the $[1_p 2_q]$ probabilistic ensemble of binary sequences consisting of runs of lengths 1 and 2 and having the first bit equal to 0, $p \in [0, 1]$ is fixed and $q = 1 - p$. Let $v_{-1} = v_0 = 1$. For all $t \geq 1$, associate the probability

$$\Omega_{1_p 2_q}(\mathbf{v}^t) \triangleq \prod_{j=1}^{t} \omega_{1_p 2_q}(v_j | v_{j-1}, v_{j-2})$$

with the binary string $\mathbf{v}^t$, where

$$\begin{bmatrix} \omega_{1_p 2_q}(0|0,0) & \omega_{1_p 2_q}(1|0,0) \\ \omega_{1_p 2_q}(0|0,1) & \omega_{1_p 2_q}(1|0,1) \\ \omega_{1_p 2_q}(0|1,0) & \omega_{1_p 2_q}(1|1,0) \\ \omega_{1_p 2_q}(0|1,1) & \omega_{1_p 2_q}(1|1,1) \end{bmatrix} \triangleq \begin{bmatrix} 0 & 1 \\ q & p \\ p & q \\ 1 & 0 \end{bmatrix}.$$

If $\Omega_{1_p 2_q}(\mathbf{v}^t) > 0$, then we say that $\mathbf{v}^t$ belongs to the $[1_p 2_q]$ ensemble. Let us denote

$$\overline{n_{1_p 2_q}}(d, \ell) \triangleq \sum_{\mathbf{v}} \Omega_{1_p 2_q}(\mathbf{v}) n_{\mathbf{v}}(d, \ell), \tag{7}$$

$$\overline{n_{1_p 2_q}}(d_1, d_2, \ell) \triangleq \sum_{\mathbf{v}_1, \mathbf{v}_2} \Omega_{1_p 2_q}(\mathbf{v}_1) \Omega_{1_p 2_q}(\mathbf{v}_2) n_{\mathbf{v}_1, \mathbf{v}_2}(d_1, d_2, \ell), \tag{8}$$

where $\mathbf{v} = \mathbf{v}^{\ell+d}$, $\mathbf{v}_1 = \mathbf{v}_1^{\ell+d_1}$, $\mathbf{v}_2 = \mathbf{v}_2^{\ell+d_2}$, and introduce the generating functions

$$\overline{N_{1_p 2_q}}(D, L) \triangleq \sum_{d, \ell \geq 0} \overline{n_{1_p 2_q}}(d, \ell) D^d L^\ell, \tag{9}$$

$$\overline{N_{1_p 2_q}}(D_1, D_2, L) \triangleq \sum_{d_1, d_2, \ell \geq 0} \overline{n_{1_p 2_q}}(d_1, d_2, \ell) D_1^{d_1} D_2^{d_2} L^\ell. \tag{10}$$

In the present correspondence we show that coefficients defined in (5)–(8) can be easily found. Moreover, the method of their computing is rather general and it can be also successfully used for computing coefficients of generating functions defined in (3), (4) for arbitrary sequences $\mathbf{v}$ and arbitrary pairs of sequences $(\mathbf{v}_1, \mathbf{v}_2)$. As a result, one gets data needed for the study of problems related to structure of sequences and their applications. The role of ensembles of random sequences introduced above is explained by the point that formalization is very simple in this case and that the properties of the obtained generating functions $\overline{N_{1_p 2_q}}(D, L)$ and $\overline{N_{1_p 2_q}}(D_1, D_2, L)$ can be interesting for constructing efficient random block codes capable of correcting deletion errors.

2   Coefficients of Generating Functions Associated with "Completely Random" Binary Sequences

Lemma 1. *For all $d, d_1, d_2, \ell \geq 0$,*

$$\overline{n_{1/2}}(d, \ell) = \binom{\ell + d - 1}{d} 2^{-d}, \tag{11}$$

$$\overline{n_{1/2}}(d_1, d_2, \ell) = \binom{\ell + d_1 - 1}{d_1} \binom{\ell + d_2 - 1}{d_2} 2^{-(d_1 + d_2 + \ell)}, \tag{12}$$

where $\overline{n_{1/2}}(d, \ell)$, $\overline{n_{1/2}}(d_1, d_2, \ell)$ are defined in (5), (6), and

$$\overline{E_{1/2}}(\delta) \triangleq \lim_{\ell \to \infty} \frac{1}{\ell} \log \overline{n_{1/2}}(\ell\delta, \ell) = (1+\delta)h\left(\frac{\delta}{1+\delta}\right) - \delta, \qquad (13)$$

$$\overline{E_{1/2}}(\delta_1, \delta_2) \triangleq \lim_{\ell \to \infty} \frac{1}{\ell} \log \overline{n_{1/2}}(\ell\delta_1, \ell\delta_2, \ell) = \overline{E_{1/2}}(\delta_1) + \overline{E_{1/2}}(\delta_2) - 1, \quad (14)$$

$h(x) \triangleq -x \log x - (1-x)\log(1-x)$, $x \in (0,1)$, is the binary entropy function and $\delta, \delta_1, \delta_2 \geq 0$. Thus, $\overline{E_{1/2}}(\delta) < \overline{E_{1/2}}(1) = 1$ for all $\delta \neq 1$ and $\overline{E_{1/2}}(\delta) > 0$ for all $\delta \in (0, \delta_0)$, where $\delta_0 = 3.404\ldots$ is solution to the equation $h(\delta/(1+\delta)) = \delta/(1+\delta)$. Furthermore, $|\overline{E_{1/2}}(\delta)| < \infty$ for all $\delta \geq 0$.

Proof: Given a vector $\mathbf{u}^\ell$, one can construct all strings $\mathbf{v}^{\ell+d}$ with $\mathrm{Del}(\mathbf{v}|\mathbf{u}^\ell) = d$ by the following algorithm. Choose ℓ positions $j_1 < j_2 < \ldots < j_\ell$ in such a way that $j_1, \ldots, j_\ell \in \{1, \ldots, \ell + d\}$ and $j_\ell = \ell + d$. Set $v_{j_i} = u_i$ for all $i = 1, \ldots, \ell$. For all $i \in \{1, \ldots, \ell - 1\}$ such that $j_{i+1} > j_i + 1$, set $v_j = 1 \oplus u_{j_{i+1}}$ for all $j \in \{j_i + 1, \ldots, j_{i+1} - 1\}$. Thus,

$$\overline{n_{1/2}}(d, \ell) = 2^{-(\ell+d)} \left| \left\{ (\mathbf{u}^\ell, \mathbf{v}^{\ell+d}) : \mathrm{Del}(\mathbf{v}^{\ell+d}|\mathbf{u}^\ell) = d \right\} \right|$$

$$= 2^{-(\ell+d)} \sum_{\mathbf{u}^\ell} \left| \left\{ \mathbf{v}^{\ell+d} : \mathrm{Del}(\mathbf{v}^{\ell+d}|\mathbf{u}^\ell) = d \right\} \right|$$

$$= 2^{-d} \binom{\ell+d-1}{\ell-1},$$

and (11) follows. Similarly,

$$\overline{n_{1/2}}(d_1, d_2, \ell) = 2^{-(2\ell+d_1+d_2)} \sum_{\mathbf{u}^\ell} \binom{\ell+d_1-1}{d_1} \binom{\ell+d_2-1}{d_2},$$

and (12) follows. The asymptotic formulas (13), (14) readily follow from (11), (12) and Stirling's approximation for the binomial coefficients.

The counting argument used in the proof of Lemma 1 is based on the fact that the cardinality of the set of vectors $\mathbf{v}^{\ell+d}$ such that $\mathrm{Del}(\mathbf{v}^{\ell+d}|\mathbf{u}^\ell) = d$ does not depend on $\mathbf{u}^\ell$. For example, if $d = \ell = 2$, we construct 3 strings of length $\ell + d = 4$ having a fixed vector $\mathbf{u}^\ell$ as a subsequence, and $\overline{n_{1/2}}(2,2) = 3 \cdot 2^\ell \cdot 2^{-(\ell+d)} = 3/4$. Namely, if $\mathbf{u}^2 = 00$, then $\mathbf{v}^4 \in \{0110, 1010, 1100\}$; if $\mathbf{u}^2 = 01$, then $\mathbf{v}^4 \in \{0001, 1001, 1101\}$; if $\mathbf{u}^2 = 10$, then $\mathbf{v}^4 \in \{1110, 0110, 0010\}$; if $\mathbf{u}^2 = 11$, then $\mathbf{v}^4 \in \{1001, 0101, 0011\}$. Similarly, if $d_1 = d_2 = \ell$, then there are 3^2 pairs of strings of length 4 having a fixed vector $\mathbf{u}^2$ as a common subsequence, and $\overline{n_{1/2}}(2,2,2) = (3/4)^2 \cdot 2^{-4}$.

Exponents $\overline{E_{1/2}}(\delta)$ and $\overline{E_{1/2}}(\delta, \delta)$ will be shown in Fig. 6 and Fig. 7 together with the corresponding exponents obtained for the $[1_p 2_q]$ ensemble.

3   Generating Functions Associated with Random Binary Sequences Consisting of Runs of Lengths 1 and 2

3.1   Computing the Generating Function $N_{\mathbf{v}[r]}(D, L)$

Let us introduce the r-run representation of a given vector $\mathbf{u}^\ell$ by the formula

$$\mathbf{u}^\ell = (0^{r_1}, 1^{r_2}, \dots, 0^{r_{k-1}}, 1^{r_k}), \tag{15}$$

where parameters k and $r_1, \dots, r_k$ are uniquely determined by (15) and the conditions :

$$\begin{cases} r_1, \dots, r_k \in \{0, \dots, r\}, \\ r_1 + \dots + r_k = \ell, \\ r_i = 0 \Rightarrow r_{i-1} = r, \quad i = 1, \dots, k \end{cases} \tag{16}$$

(we assume that $0^0, 1^0$ are the 0-length vectors and set $r_0 \overset{\triangle}{=} r$). For example, if $r = 3$, then $111110100001 = (0^0, 1^3, 0^0, 1^2, 0^1, 1^1, 0^3, 1^0, 0^1, 1^1)$.

The r-run representation is unique and

$$\mathrm{Del}(\mathbf{v}[r]|\mathbf{u}^\ell) = \sum_{i=1}^{k-1}(r - r_i) \tag{17}$$

for all $\ell = r_1 + \dots + r_{k-1} + 1, \dots, r_1 + \dots + r_k$. Therefore, enumeration of all possible numbers k and $r_1, \dots, r_k$ satisfying (16) gives coefficients of the generating function $N_{\mathbf{v}[r]}(D, L)$. Let us sequentially form a vector $\mathbf{r} = (r_1, r_2, \dots)$ using the machine with 3 possible states : S, S_+, and F (see Fig. 2). If the current state of the machine is S, then the last integer included in the vector $\mathbf{r}$ is equal to r. In this case, the next integer has to be chosen from the set $\{0, \dots, r\}$. If this integer is equal to r, then the machine stays in state S, otherwise it moves to state S_+ where 0 cannot be chosen as the next integer. State F is introduced to describe all possible ways of terminating the process.

One can easily see that the assignment of transfer functions associated with the edges of the transition diagram in Fig. 2 and the definition of the transfer function of a path as the product of transfer functions of edges of that path result in a conclusion that the generating function $N_{\mathbf{v}[r]}(D, L)$ is equal to the sum of transfer functions of all paths leading to state F. Therefore

$$N_{\mathbf{v}[r]}(D, L) = N^{(S)} + (N^{(S)} + N^{(S_+)}) \sum_{j=1}^{r-1} L^j,$$

where $N^{(s)} = N^{(s)}(D, L)$, $s = S, S_+$, are solutions to the system of linear equations

$$\begin{bmatrix} 1 - L^r & -L^r \\ -D^r - \sum_{j=1}^{r-1} D^{r-j} L^j & 1 - \sum_{j=1}^{r-1} D^{r-j} L^j \end{bmatrix} \begin{bmatrix} N^{(S)} \\ N^{(S_+)} \end{bmatrix} = \begin{bmatrix} 1 \\ 0 \end{bmatrix}.$$

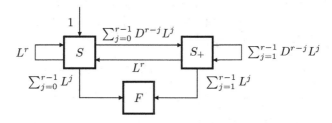

Fig. 2. The 3-state machine introduced to compute the number of subsequences of the sequence $\mathbf{v}[r]$

Table 2. Transitions of the 3-state machine at a 0-run of length $r = 3$ represented as results of transformations of input vectors

s	Type(s)	Input	s'	Type(s')	$h_r(s, s')$
S or S_+	0 0 0 1	000	S	0 0 0 1	L^3
S or S_+	0 0 0 1	001	S_+	0 0 0 1	DL^2
S or S_+	0 0 0 1	01	S_+	0 0 0 1	D^2L
S	0 0 0 1	1	S_+	0 0 0 1	D^3

The method of counting described above seems to be very general. In particular, computing the generating function $N_\mathbf{v}(D, L)$ for an arbitrary sequence $\mathbf{v}$ can be also organized using the formula (17) with the parameter r depending on i in such a way that r is equal to the length of a run of the sequence $\mathbf{v}$ containing the $(r_1 + \ldots + r_{i-1} + 1)$-st bit of this sequence. If $\mathbf{v}$ is a sequence of period T, then this approach brings a machine having at most $2T + 1$ states, and the solution has be easily found. To extend this method to the problem of computing the joint generating function for two given sequences, let us also introduce a representation of the transitions of the machine in Fig. 2 as results of transformations of input vectors belonging to the $(r + 1)$-element set consisting of the vector 0^r and vectors $(0^i, 1)$, $i = 0, \ldots, r - 1$, for the 0-run and consisting of the vector 1^r and vectors $(1^i, 0)$, $i = 0, \ldots, r - 1$, for the 1-run. This representation can be easily understood from the numerical example given in Table 2, where $h_r(s, s')$ denotes the transfer function associated with the corresponding transformation. We will use the method above for computing the generating functions of random binary sequences with the restricted lengths of all runs.

3.2   Computing the Generating Function $N_{\mathbf{v}[1],\mathbf{v}[2]}(D_1, D_2, L)$

The suggested way of computing the generating function $N_{\mathbf{v}[1],\mathbf{v}[2]}(D_1, D_2, L)$ is an extension of considerations of the previous subsection. This extension should be clear from Fig. 3 and Table 3. As a result,

$$N_{\mathbf{v}[1],\mathbf{v}[2]}(D_1, D_2, L) = N^{(\mathbf{S})} + (N^{(\mathbf{S})} + N^{(\mathbf{S_+})})L,$$

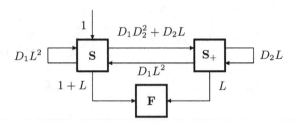

Fig. 3. The 3-state machine introduced to compute the number of common subsequences of a pair of sequences $(\mathbf{v}[1], \mathbf{v}[2])$

Table 3. Transitions of the 3-state machine at a pair of 0-runs of the pair of sequences $(\mathbf{v}[1], \mathbf{v}[2])$ represented as results of transformations of input vectors

s	Type(s)	Input	s′	Type(s′)	$h_{1,2}(\mathbf{s}, \mathbf{s}')$
S or S$_+$	$\underline{0}\,1\,0\ 1$ $\underline{0}\,0\,1\ 1$	0 0	S	$0\,1\,0\ \underline{1}$ $0\,0\,1\ \underline{1}$	$D_1 L^2$
S or S$_+$	$\underline{0}\,1\,0\ 1$ $\underline{0}\,0\,1\ 1$	0 1	S$_+$	$0\,1\,\underline{0}\ 1$ $0\,0\,1\ \underline{1}$	$D_2 L$
S	$0\,1\,0\ 1$ $\underline{0}\,0\,1\ 1$	1	S$_+$	$0\,1\,\underline{0}\ 1$ $0\,0\,\underline{1}\ 1$	$D_1 D_2^2$
S or S$_+$	$\underline{0}\,1\,0\ 1$ $\underline{0}\,0\,1\ 1$	0	F	$0\,1\,0\ 1$ $0\,0\,1\ 1$	L
S	$0\,1\,0\ 1$ $\underline{0}\,0\,1\ 1$	∅	F	$0\,1\,0\ 1$ $0\,0\,1\ 1$	1

where $N^{(\mathbf{s})} = N^{(\mathbf{s})}(D_1, D_2, L)$, $\mathbf{s} = \mathbf{S}, \mathbf{S}_+$, are solutions to the system of linear equations

$$\begin{bmatrix} 1 - D_1 L^2 & -D_1 L^2 \\ -D_1 D_2^2 - D_2 L & 1 - D_2 L \end{bmatrix} \begin{bmatrix} N^{(\mathbf{S})} \\ N^{(\mathbf{S}_+)} \end{bmatrix} = \begin{bmatrix} 1 \\ 0 \end{bmatrix}.$$

3.3   Computing the Generating Functions Associated with Random Binary Sequences Consisting of Runs of Lengths 1 and 2

Note that the $[1_1 2_0]$ and $[1_0 2_1]$ ensembles contain only sequences $\mathbf{v}[1]$ and $\mathbf{v}[2]$, respectively. Notation above allows us to present a simple proof of the important known result [4], which says that the sum $\sum_{d \leq \ell\delta} n_{\mathbf{v}}(d, \ell)$, being interpreted as the size of "a ball" of radius $\ell\delta$ having the center $\mathbf{v}^{\ell(1+\delta)}$, is bounded from above by the size of a corresponding ball having the center $\mathbf{v}^{\ell(1+\delta)}[1]$. However, the statement that, for all sequences belonging to the $[1_p 2_q]$ ensemble, the size of a ball is bounded from below by the size of a corresponding ball having the center $\mathbf{v}^{\ell(1+\delta)}[2]$ is not true in general.

Lemma 2. *For all sequences* **v** *and all* $\ell \geq 1$, $\delta \in [0, 1]$,

$$\sum_{d \leq \ell\delta} n_{\mathbf{v}}(d, \ell) \leq \sum_{d \leq \ell\delta} n_{\mathbf{v}[1]}(d, \ell). \tag{18}$$

Proof: Given a vector $\mathbf{u}^\ell$ with $\text{Del}(\mathbf{v}|\mathbf{u}^\ell) = d \leq \ell\delta$, let $\mathbf{\Delta}_\ell = (\Delta_1, \ldots, \Delta_\ell)$, where $\Delta_i \overset{\triangle}{=} \text{Del}(\mathbf{v}|\mathbf{u}^i) - \text{Del}(\mathbf{v}|\mathbf{u}^{i-1})$ for all $i = 1, \ldots, \ell$ and $\text{Del}(\mathbf{v}|\mathbf{u}^0) \overset{\triangle}{=} 0$. Construct the vector $\mathbf{\Delta}'_\ell = (\Delta'_1, \ldots, \Delta'_\ell)$ whose components are defined as $\Delta'_i = \min\{\Delta_i, 1\}$, $i = 1, \ldots, \ell$. Then $\Delta'_1 \leq \Delta_1, \ldots, \Delta'_\ell \leq \Delta_\ell$ and

$$\Delta'_1 + \ldots + \Delta'_\ell \leq \Delta_1 + \ldots + \Delta_\ell. \tag{19}$$

Different vectors $\mathbf{u}^\ell$ specify different vectors $\hat{\mathbf{u}}^\ell$ by the following rule : set $\hat{u}_i = v_{j_i}[1]$, where $j_i = i + \Delta'_1 + \ldots + \Delta'_{i-1}$ for all $i = 1, \ldots, \ell$. Since $\text{Del}(\mathbf{v}[1]|\hat{\mathbf{u}}^\ell) = \Delta'_1 + \ldots + \Delta'_\ell$, this construction and inequality (19) prove (18).

Theorem 1. *Let the polynomials* $N^{(s)} = N^{(s)}(D, L)$, $s = S, S_+$, *and* $N^{(\mathbf{s})} = N^{(\mathbf{s})}(D_1, D_2, L)$, $\mathbf{s} = \mathbf{S}, \mathbf{S}_+$, *be defined as solutions to systems of linear equations*

$$\begin{bmatrix} 1-a & -a \\ -b-c & 1-c \end{bmatrix} \begin{bmatrix} N^{(S)} \\ N^{(S_+)} \end{bmatrix} = \begin{bmatrix} 1 \\ 0 \end{bmatrix}, \quad \begin{bmatrix} 1-a_2 & -a_2 \\ -b_2-c_2 & 1-c_2 \end{bmatrix} \begin{bmatrix} N^{(\mathbf{S})} \\ N^{(\mathbf{S}_+)} \end{bmatrix} = \begin{bmatrix} 1 \\ 0 \end{bmatrix},$$

where

$$a \overset{\triangle}{=} pL + qL^2, \quad b \overset{\triangle}{=} (p+qD)D, \quad c \overset{\triangle}{=} qDL,$$

and

$$a_2 \overset{\triangle}{=} p^2 L + \left(q^2 + pq(p+qD_1)D_1 + pq(p+qD_2)D_2\right)L^2,$$

$$b_2 \overset{\triangle}{=} \left(p^2 + pq(D_1 + D_2) + q^2 D_1 D_2\right)D_1 D_2,$$

$$c_2 \overset{\triangle}{=} \left(pq(D_1 + D_2) + q^2 D_1 D_2\right)L.$$

Then the generating functions $\overline{N_{1_p 2_q}}(D, L)$ *and* $\overline{N_{1_p 2_q}}(D_1, D_2, L)$ *defined in* (7)−(10) *can be expressed as*

$$\overline{N_{1_p 2_q}}(D, L) = N^{(S)} + (N^{(S)} + N^{(S_+)})qL$$

$$= \left(1 + q(1+b)L - c\right)\sum_{k \geq 0}\left[\overline{Z_{1_p 2_q}}(D, L)\right]^k,$$

$$\overline{N_{1_p 2_q}}(D_1, D_2, L) = N^{(\mathbf{S})} + (N^{(\mathbf{S})} + N^{(\mathbf{S}_+)})(1-p^2)L$$

$$= \left(1 + (1-p^2)(1+b_2)L - c_2\right)\sum_{k \geq 0}\left[\overline{Z_{1_p 2_q}}(D_1, D_2, L)\right]^k,$$

where

$$\overline{Z_{1_p 2_q}}(D, L) \overset{\triangle}{=} a + c + ab, \quad \overline{Z_{1_p 2_q}}(D_1, D_2, L) \overset{\triangle}{=} a_2 + c_2 + a_2 b_2. \tag{20}$$

Proof: To find the generating function $\overline{N}_{1_p 2_q}(D, L)$ we use the approach developed for computing the function $N_{\mathbf{v}[r]}(D, L)$ and introduce the 3-state machine whose transition diagram is given in Fig. 4. The transfer functions associated with edges are obtained as linear combinations of transfer functions associated with corresponding edges at the diagrams of machines for $N_{\mathbf{v}[1]}(D, L)$ and $N_{\mathbf{v}[2]}(D, L)$ taken with coefficients p and q, respectively.

The way of computing the generating function $\overline{N}_{1_p 2_q}(D_1, D_2, L)$ can be also easily understood, since it is based on the transition diagram in Fig. 5, which is constructed as an extension of the diagrams in Fig. 2 and Fig. 3. As soon as the machine enters states $\mathbf{S}$ or $\mathbf{S}_+$, we choose lengths of current runs in $\mathbf{v}_1$ and $\mathbf{v}_2$. By construction, both runs are either 0-runs or 1-runs. If these runs have equal lengths, then we assign the transfer functions to edges according to Fig. 2 for $r = 1$ with the weight p^2 and for $r = 2$ with the weight q^2. If the length of the run in $\mathbf{v}_1$ is equal to 1 and the length of the run in $\mathbf{v}_2$ is equal to 2, then we take transfer functions from Fig. 3 with the weight pq. In the case when lengths of runs are 2 and 1, then the procedure is the same with the exchanged indices of variables D_1 and D_2.

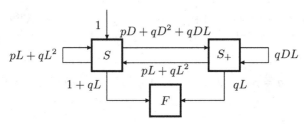

Fig. 4. The 3-state machine introduced to compute the expected number of subsequences of random sequences belonging to the $[1_p 2_q]$ ensemble

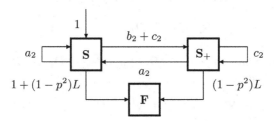

Fig. 5. The 3-state machine introduced to compute the expected number of common subsequences of a pair of random sequences belonging to the $[1_p 2_q]$ ensemble

Note that the result of Theorem 1 implies

$$N_{\mathbf{v}[1]}(D, L) = \sum_{k \geq 0} \left(L + DL\right)^k = \sum_{\ell \geq 0} \sum_{d=0}^{\ell} \binom{\ell}{d} D^d L^\ell, \tag{21}$$

$$N_{\mathbf{v}[2]}(D, L) = \left(1 + (1 - D + D^2)L\right) \sum_{k \geq 0} \left(DL + L^2 + D^2 L^2\right)^k, \tag{22}$$

$$N_{\mathbf{v}[1],\mathbf{v}[2]}(D_1, D_2, L) = \left(1 + (1 - D_2 + D_1 D_2^2)L\right)$$

$$\cdot \sum_{k \geq 0} \left(D_2 L + D_1 L^2 + D_1^2 D_2^2 L^2\right)^k. \quad (23)$$

One can easily check that these equalities agree with initial parts of the generating functions given in Table 1. Notice also that

$$N_{\mathbf{v}[1],\mathbf{v}[2]}(D, 1, L) = N_{\mathbf{v}[1]}(D, L), \quad N_{\mathbf{v}[1],\mathbf{v}[2]}(1, D, L) = N_{\mathbf{v}[2]}(D, L).$$

3.4 Asymptotic Behavior of the Expected Number of Subsequences of Random Binary Sequences Consisting of Runs of Lengths 1 and 2

Let

$$\overline{E_{1_p 2_q}}(\delta) \triangleq \limsup_{\ell \to \infty} \frac{1}{\ell} \log \overline{n_{1_p 2_q}}(\ell\delta, \ell),$$

$$\overline{E_{1_p 2_q}}(\delta_1, \delta_2) \triangleq \limsup_{\ell \to \infty} \frac{1}{\ell} \log \overline{n_{1_p 2_q}}(\ell\delta_1, \ell\delta_2, \ell),$$

where $\delta, \delta_1, \delta_2 \in [0, f]$ are fixed, $f = 1$ if $p \in \{0, 1\}$, and $f = 2$ if $p \in (0, 1)$. The solution

$$E_{\mathbf{v}[1]}(\delta) \triangleq \overline{E_{1_1 2_0}}(\delta) = h(\delta)$$

can be readily found from (21) and Stirling's approximation for the binomial coefficients. However, in general case, (20) bring polynomials such that one is supposed to compute the sum of all coefficients of monomials $D^{\ell\delta} L^\ell$ and $D_1^{\ell\delta_1} D_2^{\ell\delta_2} L^\ell$ obtained from $[\overline{Z_{1_p 2_q}}(D, L)]^k$ and $[\overline{Z_{1_p 2_q}}(D_1, D_2, L)]^k$, $k = 0, 1, \ldots$ Nevertheless, we have closed formulas by the statement below.

Theorem 2. *1. The function $\overline{E_{1_p 2_q}}(\delta)$ can be expressed as*

$$\overline{E_{1_p 2_q}}(\delta) = -\xi\delta - \lambda, \quad (24)$$

where ξ and λ are reals determined by the equations

$$\overline{Z_{1_p 2_q}}(2^\xi, 2^\lambda) = 1, \quad \frac{\partial \overline{Z_{1_p 2_q}}(2^\xi, 2^\lambda)/\partial \xi}{\partial \overline{Z_{1_p 2_q}}(2^\xi, 2^\lambda)/\partial \lambda} = \delta. \quad (25)$$

2. The function $\overline{E_{1_p 2_q}}(\delta_1, \delta_2)$ can be expressed as

$$\overline{E_{1_p 2_q}}(\delta_1, \delta_2) = -\xi_1\delta_1 - \xi_2\delta_2 - \lambda, \quad (26)$$

where ξ_1, ξ_2, and λ are reals determined by the equations

$$\overline{Z_{1_p 2_q}}(2^{\xi_1}, 2^{\xi_2}, 2^\lambda) = 1$$

and

$$\frac{\partial \overline{Z_{1_p 2_q}}(2^{\xi_1}, 2^{\xi_2}, 2^\lambda)/\partial \xi_1}{\partial \overline{Z_{1_p 2_q}}(2^{\xi_1}, 2^{\xi_2}, 2^\lambda)/\partial \lambda} = \delta_1, \quad \frac{\partial \overline{Z_{1_p 2_q}}(2^{\xi_1}, 2^{\xi_2}, 2^\lambda)/\partial \xi_2}{\partial \overline{Z_{1_p 2_q}}(2^{\xi_1}, 2^{\xi_2}, 2^\lambda)/\partial \lambda} = \delta_2.$$

Proof: We use a general statement given below. The proof is included in the Appendix.

Lemma 3. *Suppose that $\mathcal{I}$ is a finite set of pairs of non–negative integers and that a polynomial*

$$Z(D, L) = \sum_{(i,j)\in\mathcal{I}} z(i, j) D^i L^j$$

is given in such a way that $z(i, j) > 0$ for all $(i, j) \in \mathcal{I}$. Let $\mathcal{Z}(\delta)$ be the set of pairs of reals (ξ, λ) such that

$$Z(2^\xi, 2^\lambda) = 1, \quad \frac{\partial Z(2^\xi, 2^\lambda)/\partial\xi}{\partial Z(2^\xi, 2^\lambda)/\partial\lambda} = \delta. \tag{27}$$

Then

$$|\mathcal{Z}(\delta)| \geq 1 \Longrightarrow E(\delta|Z) = \max_{(\xi,\lambda)\in\mathcal{Z}(\delta)}\left[-\xi\delta - \lambda\right],$$

$$\mathcal{Z}(\delta) = \emptyset \Longrightarrow E(\delta|Z) = -\infty,$$

where

$$E(\delta|Z) \triangleq \lim_{\ell\to\infty} \frac{1}{\ell} \log \sum_{k\geq 0} \mathrm{Coef}_{(\ell\delta,\ell)}\left[Z^k(D, L)\right],$$

where $\mathrm{Coef}_{(\ell\delta,\ell)}\left[Z^k(D, L)\right]$ is the coefficient of the monomial $D^{\ell\delta}L^\ell$ in the representation

$$Z^k(D, L) = \sum_{d,\ell\geq 0} \mathrm{Coef}_{(d,\ell)}\left[Z^k(D, L)\right] D^d L^\ell.$$

One can easily check that the set $\mathcal{Z}(\delta)$ constructed for the polynomial $\overline{Z_{1_p 2_q}}(\delta)$ is either an empty set or it contains only one pair (ξ, λ), and the first claim of Theorem 2 follows. The second claim is proven similarly with the use of a generalized version of Lemma 3.

Conditions (25) imply that parameters ξ and λ in (24) are reals determined by the equations

$$(P_1(2^\xi) + P_2(2^\xi)2^\lambda)2^\lambda = 1, \quad \frac{P_1'(2^\xi) + P_2'(2^\xi)2^\lambda}{P_1(2^\xi) + 2P_2(2^\xi)2^\lambda} = \delta \tag{28}$$

where

$$\begin{bmatrix} P_1(D) & P_2(D) \\ P_1'(D) & P_2'(D) \end{bmatrix} = \begin{bmatrix} p + (p^2 + q)D + pqD^2 & q + pqD + q^2D^2 \\ (p^2 + q)D + 2pqD^2 & pqD + 2q^2D^2 \end{bmatrix}.$$

Hence, the parameter λ can be expressed as

$$\lambda = \log\frac{\sqrt{1 + 4P_2(2^\xi)/P_1^2(2^\xi)} - 1}{2P_2(2^\xi)/P_1(2^\xi)}$$

In particular, if $p \in \{0,1\}$, then

$$\xi = 0 \Longrightarrow E_{\mathbf{v}[1]}(1/2) = E_{\mathbf{v}[2]}(2/3) = 1;$$
$$\xi \to \infty \Longrightarrow \delta \to 1^-, \; E_{\mathbf{v}[1]}(\delta) \to 0, E_{\mathbf{v}[2]}(\delta) \to \log(\sqrt{5}-1) - 1;$$
$$\delta > 1 \Longrightarrow E_{\mathbf{v}[1]}(\delta) = E_{\mathbf{v}[2]}(\delta) = -\infty.$$

Some properties of the function $\overline{E_{1_p2_q}}(\delta)$ for $p \in (0,1)$ are stated below.

Lemma 4. *If $p \in (0,1)$, then*

$$|\overline{E_{1_p2_q}}(\delta)| < \infty, \quad for \; all \; \delta < 2,$$
$$\lim_{\delta \to 2^-} \overline{E_{1_p2_q}}(\delta) = \log(pq),$$
$$\overline{E_{1_p2_q}}(\delta) = -\infty, \quad for \; all \; \delta > 2.$$

Proof: If $\xi \to \infty$, then

$$P_2(2^\xi)/P_1(2^\xi) \to q/p, \;\; P_2(2^\xi)/P_1^2(2^\xi) \to (1/p^2)2^{-2\xi}, \;\; \lambda = \to \frac{1}{pq}2^{-2\xi},$$

because $\sqrt{1+\varepsilon} - 1 = \varepsilon/\sqrt{1+\varepsilon} + O(\varepsilon^2)$ by Taylor power series. As a result, $\delta \to 2^-$ and $-\xi\delta - \lambda \to \log(pq)$.

Examples of exponents $\overline{E_{1_p2_q}}(\delta)$ are shown in Fig. 6. Notice that

$$\max_{p\in[0,1]} \delta_1(p) = 7/2 - 2\sqrt{2}, \;\; p^* \overset{\triangle}{=} \arg \max_{p\in[0,1]} \delta_1(p) = 3 - 2\sqrt{2},$$

where δ_1 is defined by the equation $\overline{E_{1_p2_q}}(\delta_1) = 1$, i.e., if $\delta = 1$ and a sequence $\mathbf{v}$ belongs to the $[1_p2_q]$ ensemble, then the sum at the left–hand side of (18) is bounded from below by the corresponding sum for a random sequence belonging to the ensemble $[1_{p^*}2_{1-p^*}]$.

A further discussion of properties of the exponents $\overline{E_{1_p2_q}}(\delta)$ and $\overline{E_{1_p2_q}}(\delta_1,\delta_2)$ is postponed to another paper. In the present correspondence we only indicate interesting points of the exponent $\overline{E_{1_p2_q}}(\delta)$ in Table 4 and show the function $\overline{E_{1_p2_q}}(\delta,\delta)$ in Fig. 7 for $p = 1/2$.

3.5 Appendix: Proof of Lemma 3

Let $\mathcal{A}(\delta)$ be a collection of probability distributions $\boldsymbol{\alpha} = \left(\alpha(i,j) \in [0,1], (i,j) \in \mathcal{I}\right)$ such that

$$\sum_{(i,j)\in\mathcal{I}} i\alpha(i,j) = \delta, \quad \sum_{(i,j)\in\mathcal{I}} j\alpha(i,j) = 1. \tag{29}$$

For a fixed $\ell \geq 1$, let $\mathcal{A}^{(\ell)}(\delta) \in \mathcal{A}(\delta)$ be a finite set consisting of probability distributions with $\ell\alpha(i,j) \in \{0,\ldots,\ell\}$ for all $(i,j) \in \mathcal{I}$. Then

$$|\mathcal{A}^{(\ell)}(\delta)| \leq (\ell+1)^{|\mathcal{I}|},$$

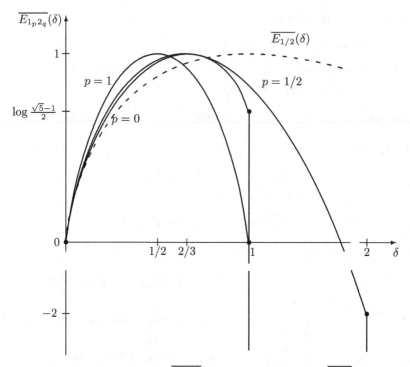

Fig. 6. Exponents $\overline{E_{1_p 2_q}}(\delta)$ for $p = 0, 1/2, 1$ and $\overline{E_{1/2}}(\delta)$

because each entry of $\alpha \in \mathcal{A}^{(\ell)}(\delta)$ can take at most $\ell + 1$ values. Furthermore, conditions $z(i,j) > 0$, $(i,j) \in \mathcal{I}$, imply an existence of at most

$$\sum_{(i,j)\in\mathcal{I}} \left[1 + \min\left\{ \frac{\ell\delta}{i}, \frac{\ell}{j} \right\} \right] \le (\ell+1)|\mathcal{I}|$$

different values of k such that $\mathrm{Coef}_{(\ell\delta,\ell)}[Z^k(D,L)] > 0$. Since each value of k can be obtained as $\ell \sum_{(i,j)\in\mathcal{I}} \alpha(i,j)$ for some $\alpha \in \mathcal{A}^{(\ell)}(\delta)$ and *vice versa* : for any fixed k one can find the corresponding probability distribution α,

$$2^{\ell F_\ell(Z)} \le \sum_{k \ge 0} \mathrm{Coef}_{(\ell\delta,\ell)}\left[Z^k(D,L) \right] \le |\mathcal{I}|(\ell+1)^{1+|\mathcal{I}|} 2^{\ell F_\ell(Z)},$$

where

$$F_\ell(Z) \stackrel{\triangle}{=} \frac{1}{\ell} \log \max_{\alpha\in\mathcal{A}(\delta)} \frac{\left(\ell\sum_{(i,j)\in\mathcal{I}} \alpha(i,j)\right)!}{\prod_{(i,j)\in\mathcal{I}}(\ell\alpha(i,j))!} \prod_{(i,j)\in\mathcal{I}} \left(z(i,j) \right)^{\ell\alpha(i,j)},$$

Using Stirling's approximation formula for the factorial, we write

$$F_\ell(Z) = \max_{\alpha\in\mathcal{A}(\delta)} \Phi(\alpha) + O\left(\frac{\log\ell}{\ell} \right),$$

Table 4. Values of the parameters δ_1, δ_0, $\delta_{-\infty}$ defined by the equations $\overline{E_{1_p 2_q}}(\delta_1) = 1$, $\overline{E_{1_p 2_q}}(\delta_0) = 0$, $\delta_0 > 0$, and $\overline{E_{1_p 2_q}}(\delta_{-\infty}) = -\infty$. The corresponding values of the function $\overline{E_{1/2}}(\delta)$ are given in the last line

p	δ_1	δ_0	$\delta_{-\infty}$	$\overline{E_{1_p 2_q}}(1)$	$\overline{E_{1_p 2_q}}(2^-)$
1	1/2	1	1	0	$-\infty$
0.9	0.545	1.267	2	0.595	-3.474
0.8	0.582	1.378	2	0.716	-2.644
0.7	0.611	1.445	2	0.778	-2.252
0.6	0.633	1.484	2	0.813	-2.059
0.5	0.650	1.501	2	0.832	-2
0.4	0.662	1.497	2	0.841	-2.059
0.3	0.669	1.473	2	0.840	-2.252
0.2	0.671	1.425	2	0.829	-2.644
0.1	0.671	1.342	2	0.802	-3.474
0	2/3		1	0.694	$-\infty$
	1	3.404	∞	1	0.755

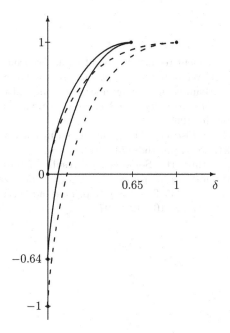

Fig. 7. Increasing parts of the functions $\overline{E_{1_p 2_q}}(\delta)$, $\overline{E_{1_p 2_q}}(\delta, \delta)$ for $p = 1/2$ (solid lines) and $\overline{E_{1/2}}(\delta)$, $\overline{E_{1/2}}(\delta, \delta)$ (dashed lines)

where

$$\Phi(\alpha) = \sum_{(i,j)\in\mathcal{I}} \alpha(i,j) \log \frac{z(i,j) \sum_{(i',j')\in\mathcal{I}} \alpha(i',j')}{\alpha(i,j)}$$

One can check that $\Phi(\alpha)$ is a convex up function of $\alpha(i,j)$ for all $(i,j) \in \mathcal{I}$. Therefore, using the Lagrange multipliers method, we conclude that if $\alpha \in \mathcal{A}(\delta)$ maximizes $\Phi(\alpha)$, then

$$\frac{\partial \Phi_{\xi,\lambda}(\alpha)}{\partial \alpha(i,j)} = 0, \text{ for all } (i,j) \in \mathcal{I},$$

where ξ, λ are reals chosen in such a way that conditions (29) are satisfied and

$$\Phi_{\xi,\lambda}(\alpha) \overset{\triangle}{=} \sum_{(i,j)\in\mathcal{I}} \alpha(i,j) \log \frac{z(i,j) \sum_{(i',j')\in\mathcal{I}} \alpha(i',j')}{\alpha(i,j)}$$

$$+ \xi \Big(\sum_{(i,j)\in\mathcal{I}} i\alpha(i,j) - \delta \Big) + \lambda \Big(\sum_{(i,j)\in\mathcal{I}} j\alpha(i,j) - 1 \Big).$$

Hence

$$\frac{\alpha(i,j)}{\sum_{(i',j')\in\mathcal{I}} \alpha(i',j')} = z(i,j) 2^{\xi i + \lambda j}.$$

and conditions (29) can be equivalently introduced as equations (27).

References

1. V. I. Levenshtein, "Efficient reconstruction of sequences from their subsequences or supersequences", *J. Combinatorial Theory, Ser. A*, vol. 93, pp. 310–332, 2001.
2. D. S. Hirschberg, "Bounds on the number of string subsequences", in : *Proc. 10th Symp. on Combinatorial Pattern Matching*, Warwick, UK, LNCS, vol. 1645, Springer–Verlag, Berlin, 1999.
3. R. A. Wagner and M. J. Fischer, "The string–to–string correction problem", *Journal of the ACM*, vol. 21, no. 1, pp. 168–173, 1974.
4. L. Calabi and W. E. Harnett, "Some general results of coding theory with applications to the study of codes for the correction of synchronization errors", *Inform. Control*, vol. 15, no. 3, pp. 235–249, 1969. Reprinted in : W. E. Harnett (ed.), *Foundation of Coding Theory*, pp. 107–121, 1974.

Multi-continued Fraction Algorithm and Generalized B-M Algorithm over F_2*

Zongduo Dai[1], Xiutao Feng[1], and Junhui Yang[2]

[1] State Key Lab of Information Security,
(Graduate School of Chinese Academy of Sciences), Beijing, China, 100049
yangdai@public.bta.net.cn, fengxt@mails.gscas.ac.cn
[2] Software Institute, Chinese Academy of Sciences, Beijing, China, 100049

Abstract. It is shown that the generalized Berlekamp-Massey algorithm (GBMA, in short) for solving the linear synthesis problem of a multi-sequence $\underline{r}$ over F_2 can be obtained naturally from a special form of the multi-continued fraction algorithm, called the multi-strict continued fraction algorithm (m-SCFA, in short). Moreover, the discrepancy sequence in acting GBMA on $\underline{r}$ is expressed explicitly by the data associated to the multi-strict continued fraction expansion $C(\underline{r})$ which is obtained by applying m-SCFA on $\underline{r}$. As a consequence, a 1-1 correspondence between multi-sequences of any given length and certain multi-strict continued fractions is established.

1 Introduction

Continued fraction algorithm (CFA) [1, 2, 3] for a single power series over a field F is a powerful tool in dealing with the optimal rational approximation problem. The CFA was applied [1] to solve the linear synthesis problem of a single sequence over F. It was shown [2] that the CFA is virtually equivalent to Berlekamp's algorithm (BMA, in short) [4, 5] in dealing with the linear synthesis problem of a single sequence. Later it was shown [6] further that an iterative algorithm obtained naturally from the CFA is exactly the same as the BMA when F is the binary field F_2.

Recently, a multi-continued fraction algorithm (m-CFA, in short), as a generalization of the CFA, is introduced in dealing with the optimal rational approximation problem of the multi-formal Laurent series (including power series).

In this work it is shown that the generalized Berlekamp-Massey algorithm[7, 8] (GBMA, in short) for solving the linear synthesis problem of a multi-sequence $\underline{r}$ over F_2 can be also obtained naturally from a special form of the multi-continued fraction algorithm, called the multi-strict continued fraction algorithm (m-SCFA, in short). Moreover, the discrepancy sequence in acting GBMA on $\underline{r}$ is expressed explicitly by the data associated to the multi-strict continued fraction expansion

---

* This work is partly supported by NSFC (Grant No. 60173016), and the National 973 Project (Grant No. 1999035804).

T. Helleseth et al. (Eds.): SETA 2004, LNCS 3486, pp. 339–354, 2005.

$C(\underline{r})$ which is obtained by applying m-SCFA on $\underline{r}$. As a consequence, a 1-1 correspondence between multi-sequences of any given length and certain multi-strict continued fractions is established.

2   Multi-strict Continued Fraction Algorithm (m-SCFA)

Let Z be the ring of integers, F_2 the binary field, $F_2[z]$ the polynomial ring over F_2, $F_2((z^{-1})) = \{\sum_{i \geq t} a_i z^{-i} | t \in Z, a_i \in F_2\}$ the Laurent series field over F_2, m a positive integer, $F_2[z]^m$ and $F_2((z^{-1}))^m$ the column vector space over $F_2[z]$ and $F_2((z^{-1}))$ of dimension m respectively, and Z_m the integer set $\{1, 2, \cdots, m\}$.

A linear order on $Z_m \times Z$ is defined as follows: for any two elements (h, v) and (h', v') in $Z_m \times Z$, we define $(h, v) < (h', v')$ if $v < v'$, or $v = v'$ and $h < h'$. We denote $(m, n)^+ = (1, n + 1)$, and $(j, n)^+ = (j + 1, n)$ if $j < m$. For each $(j, i) \in Z_m \times Z$, we call $z^{-i}\underline{e}_j$ a *monomial*, which will be denoted by $\underline{m}^{(j,i)}$ occasionally, where $\underline{e}_j$ is the j-th standard basis element, and

$$z^{-i}\underline{e}_j = \begin{pmatrix} 0 \\ \vdots \\ 0 \\ z^{-i} \\ 0 \\ \vdots \\ 0 \end{pmatrix} \longleftarrow j \,.$$

We denote by $\mathcal{M}$ the set of all possible monomials. Then, any non-zero element $\underline{r} = (r_1, \cdots r_j, \cdots, r_m)^\tau$ in $F_2((z^{-1}))^m$ can be expressed uniquely as

$$\underline{r} = \sum_{(h,v) \leq (j,i) \in Z_m \times Z} c_{j,i}(\underline{r}) z^{-i}\underline{e}_j, \quad c_{j,i}(\underline{r}) \in F_2$$

for some $(h, v) \in Z_m \times Z$, which is called the *monomial decomposition* of $\underline{r}$, where τ means transpose. We call $c_{j,i}(\underline{r})$ the (j, i)-th *coefficient* of $\underline{r}$; we say $z^{-i}\underline{e}_j$ belongs to $\underline{r}$ and denote $z^{-i}\underline{e}_j \in \underline{r}$ if $c_{j,i}(\underline{r}) \neq 0$.

Define $Iv(\underline{r}) = (h, v)$ and $v(\underline{r}) = v$, where $(h, v) = \min\{(j, n) \in Z_m \times Z | z^{-n}\underline{e}_j \in \underline{r}\}$. We call $Iv(\underline{r})$ and $v(\underline{r})$ the *indexed valuation* and *valuation* of $\underline{r}$ respectively. By convention, $Iv(\underline{0}) = (1, \infty)$, $v(\underline{0}) = \infty$, where $\underline{0}$ is the m-tuple made of all zeros.

Fact 1. *Let* $\alpha, \beta \in F_2((z^{-1}))^m$. *Then*

1. $Iv(\alpha) = (1, \infty) \Leftrightarrow \alpha = 0$.
2. *If* $Iv(\alpha) = (h, v)$, *then* $Iv(r\alpha) = (h, v + v(r))$ *for any* $r \in F_2((z^{-1}))$.
3. $Iv(\alpha + \beta) \geq \min\{Iv(\alpha), Iv(\beta)\}$, *and the equality holds true if and only if* $Iv(\alpha) \neq Iv(\beta)$.   □

For any finite sum $A = \sum_{(j,i) \in \mathcal{M}} c_{j,i} z^{-i} \underline{e}_j \in F_2((z^{-1}))$, we define $Supp^+(A) = \max\{(j,i) \mid c_{j,i} \neq 0\}$, which is the maximal indexed valuation of the monomials which belong to A.

For any element $\beta = \sum b_i z^{-i}$ in $F_2((z^{-1}))$, denote $\lfloor \beta \rfloor = \sum_{i \leq 0} b_i z^{-i}$ and $\{\beta\} = \sum_{i \geq 1} b_i z^{-i}$. For any element $\underline{r} = (r_1, r_2 \cdots, r_m)^\tau \in F_2((z^{-1}))^m$, denote $\lfloor \underline{r} \rfloor = (\lfloor r_1 \rfloor, \lfloor r_2 \rfloor, \cdots, \lfloor r_m \rfloor)^\tau$ and $\{\underline{r}\} = (\{r_1\}, \{r_2\}, \cdots, \{r_m\})^\tau$. Denote by $Diag.(\lambda_1, \cdots, \lambda_k, \cdots, \lambda_m)$ the diagonal matrix of order m with the k-th diagonal element being equal to λ_k, where $\lambda_k \in F_2((z^{-1}))$ and $1 \leq k \leq m$.

Multi-strict Continued Fraction Algorithm (m-SCFA) [9, 10]: Given $\underline{0} \neq \underline{r} \in F_2((z^{-1}))^m$ with $v(\underline{r}) > 0$. Initially, set $\underline{a}_0 = \underline{0} \in F_2[z]^m$, $\Delta_{-1} = I_m$, and $\alpha_0 = \beta_0 = \underline{r}$. Suppose $[\underline{a}_0, h_1, \underline{a}_1, \cdots, h_{k-1}, \underline{a}_{k-1}]$,

$$\Delta_{k-2} = Diag.(z^{-c_{k-1,1}}, \cdots, z^{-c_{k-1,j}}, \cdots, z^{-c_{k-1,m}}),$$

α_{k-1} and $\beta_{k-1} = (\beta_{k-1,1}, \cdots, \beta_{k-1,j}, \cdots, \beta_{k-1,m})^\tau \in F_2((z^{-1}))^m$ have been defined for an integer $k \geq 1$, where I_m is the identity matrix of order m, $h_k \in Z_m$, $\underline{a}_k \in F_2[z]^m$, $c_{k-1,j} \in Z$, Δ_{k-2} is a diagonal matrix of order m. If $\alpha_{k-1} \neq \underline{0}$, then the computations for the k-th round are defined by the following steps:

1. Take $(h_k, c_k) = Iv(\Delta_{k-2}\alpha_{k-1}) \in Z_m \times Z$.
2. Take an $m \times m$ diagonal matrix

$$\Delta_{k-1} = Diag.(z^{-c_{k,1}}, \cdots, z^{-c_{k,j}}, \cdots, z^{-c_{k,m}}),$$

where

$$c_{k,j} = \begin{cases} c_{k-1,j} & \text{if } j \neq h_k, \\ c_k & \text{if } j = h_k. \end{cases}$$

3. Take $\rho_k = (\rho_{k,1}, \cdots, \rho_{k,j}, \cdots, \rho_{k,m})^\tau \in F_2((z^{-1}))^m$, where

$$\rho_{k,j} = \begin{cases} \frac{\beta_{k-1,j}}{\beta_{k-1,h_k}} & \text{if } j \neq h_k, \\ \frac{1}{\beta_{k-1,h_k}} & \text{if } j = h_k. \end{cases}$$

4. Take $\alpha_k = \{\rho_k\}$ and $A_k = \lfloor \rho_k \rfloor$.
5. Let $A_k = \sum_{\underline{m} \in \mathcal{M}} c_{\underline{m}}(A_k)\underline{m}$ be the monomial decomposition of the polynomial tuple A_k, $c_{\underline{m}}(A_k) \in F_2$. Take $\beta_k = \rho_k - \underline{a}_k$, where

$$\underline{a}_k = \sum_{\underline{m} \in A_k, Iv(\Delta_{k-1}\underline{m}) < Iv(\Delta_{k-1}\alpha_k)} c_{\underline{m}}(A_k) \, \underline{m}.$$

6. Take $\mu = k$ and the algorithm terminates if $\alpha_k = \underline{0}$; otherwise, go to the $(k+1)$-th round.

Denote $\mu = \infty$ if the above procedure never terminates. As a result of the m-SCFA, we get an expansion form

$$C = C(\underline{r}) = \begin{bmatrix} h_1, & h_2, & \cdots, & h_k, & \cdots \\ \underline{0}, & \underline{a}_1, & \underline{a}_2, & \cdots, & \underline{a}_k, & \cdots \end{bmatrix}, \quad 1 \leq k \leq \mu, \ \mu = \infty \ or \ \mu < \infty, \quad (1)$$

which is called the multi-strict continued fraction expansion (m-SCFE, in short) of $\underline{r}$. We call m the dimension of C and μ the length of C.

Associated to $C(\underline{r})$, let a_{k,h_k} be the h_k-th component of $\underline{a}_k$, and

$$t_k = \deg(a_{k,h_k}),$$

$$v_{k,j} = \sum_{1 \le i \le k, h_i = j} t_i, \ v_{0,j} = 0, \ v_k = v_{k,h_k},$$

$$d_k = \sum_{1 \le i \le k} t_i,$$

$$n_k = d_k + v_{k-1,h_k},$$

$$(h_{\mu+1}, v_{\mu+1}) = (1, \infty), \ if \ \mu < \infty,$$

$$n_{\mu+1} = \infty \ if \ \mu < \infty,$$

$$D_k = Diag.(z^{v_{k,1}}, \cdots, z^{v_{k,j}}, \cdots, z^{v_{k,m}}), D_0 = I_m,$$

where $\deg(a_{k,h_k})$ denotes the degree of the polynomial a_{k,h_k}, $1 \le k \le \mu$ and $1 \le j \le m$. In order to show these parameters are associated to $C(\underline{r})$, we also write $\mu = \mu(C)$, $d_k = d_k(C)$, $D_k = D_k(C)$ and so on. It is known [9, 10] that $C(\underline{r})$ satisfies the following three conditions: for $1 \le k \le \mu$,

1. $t_k \ge 1$. In particular, $a_{k,h_k} \ne 0$.
2. $Iv(D_k\underline{a}_k) = (h_k, v_{k-1,h_k})$.
3. $Supp^+(D_k\underline{a}_k) < (h_{k+1}, v_{k+1})$.

As a consequence, the parameters satisfies the following conditions:

$$(h_k, v_{k-1,h_k}) < (h_{k+1}, v_{k+1}), \ 1 \le k \le \mu,$$

or equivalently,

$$(h_k, n_k) < (h_{k+1}, n_{k+1}), \ 1 \le k \le \mu.$$

Conversely, any expansion of the form (1) is called a multi-strict continued fraction (m-SCF, in short) if it satisfies the above three conditions.

3   m-SCFA and GBMA over F_2

We restate the linear synthesis problem by the language of formal Laurent series. An infinite sequence $r = \{a_i\}_{i \ge 1}$ of elements a_i in F_2 is identified with the Laurent series whose valuation is larger than 0: $r = \sum_{i \ge 1} a_i z^{-i} \in F_2((z^{-1}))$; and the n-prefix $(a_1, a_2, \cdots, a_n)$ of r, which is a sequence of length n and denoted by $r^{(n)}$, is identified with the element $\sum_{1 \le i \le n} a_i z^{-i}$ in $F_2((z^{-1}))$. Correspondingly, an m-tuple $\underline{r} = (r_1, \cdots r_j, \cdots, r_m)^\tau$ of infinite sequences over F_2 is considered as an element in $F_2((z^{-1}))^m$ whose valuation is larger than 0, that is, the symbol r_j is considered as a power series; the (j, n)-prefix $(r_1^{(n)}, \cdots, r_j^{(n)}, r_{j+1}^{(n-1)}, \cdots r_m^{(n-1)})$ of the multi-sequence $\underline{r}$, considered as a multi-sequence of length (j, n) and denoted by $\underline{r}^{(j,n)}$, is identified with the element $\sum_{1 \le i \le j} r_i^{(n)} \underline{e}_i + \sum_{j < i \le m} r_i^{(n-1)} \underline{e}_i$.

Now some well-known concepts [12] can be restated as follows: a minimal polynomial of $\underline{r}^{(j,n)}$ is a polynomial f with the minimal degree such that $Iv(f\underline{r}) > (j, n - \deg(f))$; the linear complexity of $\underline{r}^{(j,n)}$, denote by $L_{\underline{r}}(j, n)$, is the degree of the minimal polynomials of $\underline{r}^{(j,n)}$; the (j, n)-th discrepancy of $f(z)$ on $\underline{r}$, denoted by $\delta_{j,n}(f; \underline{r})$, is the $(j, n - \deg(f))$-th coefficient of $f\underline{r}$. We call $\{g_{j,n}\}_{(j,n)\geq(1,1)}$ a minimal polynomial profile of $\underline{r}$ and $\{\delta_{(j,n)+}(g_{j,n}; \underline{r})\}_{(j,n)\geq(1,1)}$ the corresponding discrepancy sequence of $\underline{r}$, which will be identified with $\sum_{(1,1)\leq(j,n)} \delta_{(j,n)+}(g_{j,n}; \underline{r})$ $\underline{m}^{(j,n)+}$, if $g_{j,n}$ is a minimal polynomial of $\underline{r}^{(j,n)}$ for each (j, n). The linear synthesis problem is mainly to find a minimal polynomial profile of $\underline{r}$.

In the sequel we fix a non-zero multi-sequence $\underline{r} \in F_2((z^{-1}))^m$, and keep the notation $C(\underline{r})$ for it.

Associated to $C(\underline{r})$, let q_k be the $(m+1, m+1)$-th element of the matrix B_k, $0 \leq k \leq \mu$, where B_k are square matrices of order $(m+1)$ over $F_2[z]$ and defined iteratively as below:

$$B_0 = I_{m+1}, \ B_k = B_{k-1}E_{h_k}A(\underline{a}_k), \ A(\underline{a}_k) = \begin{pmatrix} I_m & \underline{a}_k \\ 0 & 1 \end{pmatrix}, \ k \geq 1,$$

where E_h is the matrix of order $(m + 1)$ which comes by exchanging the h-th column and the $(m + 1)$-th column of the identity matrix I_{m+1}, i.e.,

$$E_h = \begin{pmatrix} I & 0 & 0 & 0 \\ 0 & 0 & 0 & 1 \\ 0 & 0 & I & 0 \\ 0 & 1 & 0 & 0 \end{pmatrix} \begin{matrix} \\ \longleftarrow h \\ \\ \longleftarrow m+1 \end{matrix}$$

Denote

$$N_k = \begin{cases} \{(j, n) | (1, 1) \leq (j, n) < (h_1, n_1)\} \text{ for } k = 0, \\ \{(j, n) | (h_k, n_k) \leq (j, n) < (h_{k+1}, n_{k+1})\} \text{ for } 1 \leq k \leq \mu. \end{cases}$$

Then we have $Z_m \times Z^+ = \dot\bigcup_{0\leq k\leq\mu} N_k$, where Z^+ denotes the set made of all positive integers, and $\dot\bigcup$ denotes the disjoint union. A minimal polynomial profile of $\underline{r}$ is obtained directly from the m-SCFA as shown by the following theorem.

Theorem 1. [10] $\deg(q_k) = d_k$, and q_k is a minimal polynomial of $\underline{r}^{(j,n)}$ for each $(j, n) \in N_k$. As a consequence, denote $g_{j,n} = q_k$ for $(j, n) \in N_k$, then $\{g_{j,n}\}$ is a minimal polynomial profile of $\underline{r}$, and the corresponding discrepancy sequence is expressed directly by the data associated to $C(\underline{r})$ as follows:

$$\sum_{(j,n)\geq(m,0)} \delta_{(j,n)+}(g_{j,n}; \underline{r}) = \sum_{k\geq1} z^{-n_k}\underline{e}_{h_k},$$

where we make convention that $g_{m,0} = 1$.

We will show in the following Theorem 2 that another minimal polynomial profile $\{f_{j,n}\}$ of $\underline{r}$, which is different to that given by Theorem 1, is constructed naturally from $C(\underline{r})$, and an iterative algorithm for getting them is also obtained naturally from $C(\underline{r})$.

Let $\underline{a}_k = \sum_{1 \leq i \leq \mu_k} \underline{m}_{k,i}$ be the monomial decomposition of $\underline{a}_k$, where the monomials $\underline{m}_{k,i}$ are ordered in the way such that they satisfy the inequality:

$$Iv(D_k \underline{m}_{k,i}) < Iv(D_k \underline{m}_{k,i+1}).$$

We denote

$$Iv(D_k \underline{m}_{k,i}) = (j_{k,i}, v_{k-1,h_k} + x_{k,i}) \in Z_m \times Z, \ 1 \leq k \leq \mu, \ 1 \leq i \leq \mu_k.$$

Then

$$(j_{k,i}, v_{k-1,h_k} + x_{k,i}) < (j_{k,i+1}, v_{k-1,h_k} + x_{k,i+1}), \ 1 \leq i < \mu_k,$$
$$(j_{k,\mu_k}, v_{k-1,h_k} + x_{k,\mu_k}) = Supp^+(D_k \underline{a}_k) < (h_{k+1}, v_{k+1}),$$

or equivalently,

$$(j_{k,i}, n_k + x_{k,i}) < (j_{k,i+1}, n_k + x_{k,i+1}), \ 1 \leq i < \mu_k,$$
$$(j_{k,\mu_k}, n_k + x_{k,i}) = Supp^+(D_k \underline{a}_k) < (h_{k+1}, n_{k+1}).$$

Define

$$f_{j,n} = \begin{cases} q_0 = 1, \ if \ (j,n) \in N_0, \\ q_{k,i}, \quad if \ (j,n) \in N_{k,i}, \end{cases}$$

where $q_{k,i}$ is the last component of the column vector

$$B_{k-1} E_{h_k} \begin{pmatrix} \underline{a}_{k,i} \\ 1 \end{pmatrix}, \quad \underline{a}_{k,i} = \sum_{1 \leq t \leq i} \underline{m}_{k,t}$$

and

$$N_{k,i} = \begin{cases} \{(j,n)|(j_{k,i}, n_k + x_{k,i}) \leq (j,n) < (j_{k,i+1}, n_k + x_{k,i+1})\} \ if \ 1 \leq i < \mu_k, \\ \{(j,n)|(j_{k,i}, n_k + x_{k,\mu_k}) \leq (j,n) < (h_{k+1}, n_{k+1})\} \ if \ i = \mu_k. \end{cases}$$

It is clear that

$$N_k = \dot{\cup}_{1 \leq i \leq \mu_k} N_{k,i}.$$

In order to show an iterative relation among $f_{j,n}$, we let

$$A(k) = \begin{pmatrix} A(k;1) \\ \vdots \\ A(k;h) \\ \vdots \\ A(k;m) \end{pmatrix}$$

be an array of size $m \times 2$, whose h-th row, denoted by $A(k;h)$, is defined as

$$A(k;h) = (Q_{k-1,h}, v_{k,h}) \in F_2[z] \times Z, \ 0 \le k \le \mu,$$

where $Q_{k-1,h}$ is the $(m+1,h)$-th element of the matrix B_k. Define

$$A_{j,n} = A(k), \ \forall \ (j,n) \in N_k, \ 0 \le k \le \mu,$$

and the h-th row of $A_{j,n}$ will be denoted by $A_{j,n}(h)$.

Theorem 2. 1. $\{f_{j,n}\}_{(j,n) \ge (1,1)}$ is a minimal polynomial of $\underline{r}$.
2. The polynomials $f_{j,n}$ can be obtained iteratively according to the following iterative algorithm:
Initially, set $f_{m,0} = 1$, $\deg(f_{m,0}) = 0$ and $A_{m,0} = A(0)$.
Assume $f_{j,n}$, $\deg(f_{j,n})$ and $A_{j,n}$ have been obtained for some $(j,n) \ge (m,0)$.
Denote $d = \deg(f_{j,n})$, $A_{j,n}(h) = (g_h, w_h) \in F_2[z] \times Z$ for $h \in Z_m$, $(j^*, n^*) = (j,n)^+$, $d^* = \deg(f_{j^*,n^*})$. Compute $\delta = \delta_{(j,n)+}(f_{j,n}; \underline{r})$. Then
(a) If $\delta = 0$, then $(f_{j^*,n^*}, d^*, A_{j^*,n^*}) = (f_{j,n}, d, A_{j,n})$.
(b) If $\delta = 1$ and $n^* - d > w_{j^*}$, then

$$f_{j^*,n^*} = f_{j,n} z^{n^*-d-w_{j^*}} + g_{j^*},$$
$$d^* = n^* - w_{j^*},$$
$$A_{j^*,n^*}(h) = \begin{cases} A_{j,n}(h) & \text{if } h \ne j^*, \\ (f_{j,n}, n^* - d) & \text{if } h = j^*. \end{cases}$$

(c) If $\delta = 1$ and $n^* - d \le w_{j^*}$, then

$$f_{j^*,n^*} = f_{j,n} + g_{j^*} z^{w_{j^*}-n^*+d},$$
$$d^* = d,$$
$$A_{j^*,n^*} = A_{j,n}.$$

The Theorem 2 will be proved in the next sections. Comparing the iterative algorithm given in Theorem 2 with the GBMA [7, 8, 13], we get the following corollary.

Corollary 1. Acting on $\underline{r}$, the iterative algorithm obtained from m-SCFA as shown in Theorem 2 is the same as the generalized Berlekamp-Massey Algorithm; and $\overline{C(\underline{r})} = \sum_{1 \le k \le \mu} z^{-d_k} D_k \underline{a}_k$ is the corresponding discrepancy sequence, expressed explicitly by data associated to $C(\underline{r})$. □

Denote by **R** the set of all possible multi-infinite sequences over F_2 of dimension m. For any multi-infinite sequence $\underline{r}$ over F_2 of dimension m, denote

$$\overline{C(\underline{r})} = \sum_{1 \le k \le \mu} z^{-d_k} D_k \underline{a}_k.$$

The following corollary is clear.

Corollary 2. The mapping $\underline{r} \mapsto \overline{C(\underline{r})}$ is an isometry [14] in the sense that it is injective from **R** onto **R**, and preserves distance in the sense that:

$$Iv(\underline{r} - \underline{r}') = Iv(\overline{C(\underline{r})} - \overline{C(\underline{r}')}) \ \forall \underline{r} \in \mathbf{R}, \ \underline{r}' \in \mathbf{R}.$$ □

Let $\mathbf{R}^{(h,n)}(m)$ be the set of all multi-sequences of dimension m and length (h, n); and let $\mathbf{C}^{(h,n)}(m)$ be the set of all possible multi-strict continued fractions C of dimension m and of finite length satisfying the condition $Supp^+(D_\mu \underline{a}_\mu) \leq (h, n - d_\mu)$, where we assume C is of the form (1), $\mu = \mu(C)$, $D_\mu = D_\mu(C)$ and $d_\mu = d_\mu(C)$ are parameters associated to C as defined in section 2, and $d_\mu(C)$ will be called the d-value of C, denoted simply by $d(C)$.

For any given multi-strict continued fraction of the form (1), the (h, n)-segment of C, denoted by $C^{(h,n)}$, is defined as

$$C^{(h,n)} = \begin{bmatrix} h_1, & h_2, & \cdots, & h_k, & \cdots, & h_{w-1}, & h_w \\ 0, & \underline{a}_1, & \underline{a}_2, & \cdots, & \underline{a}_k, & \cdots, & \underline{a}_{w-1}, & \underline{a}_w^* \end{bmatrix},$$

where w is an integer such that $0 < w \leq \mu$ and $(h_w, n_w) \leq (h, n) < (h_{w+1}, n_{w+1})$ and $\underline{a}_w^*$ is the sum of all monomials $\underline{m}$ which belong to $\underline{a}_w$ such that $Iv(D_w \underline{m}) \leq (h, n - d_w)$, where all these parameters h_w, n_w, d_w and D_w are associated to C.

The following theorem is proved based on Theorem 2, here we omit its proof, as the space is limited.

Theorem 3. *The map $\underline{r} \mapsto C(\underline{r})^{(h,n)}$ is a 1-1 correspondence from $\mathbf{R}^{(h,n)}(m)$ onto $\mathbf{C}^{(h,n)}(m)$. Moreover, the linear complexity of $\underline{r}$, denoted by $L_{\underline{r}}$, equals to $d(C(\underline{r})^{(h,n)})$. In particular, if we denote $\mathbf{R}_{n,d}(m) = \{\underline{r} \in \mathbf{R}^{(m,n)}(m) \mid L_{\underline{r}} = d\}$, and $\mathbf{C}_{n,d}(m) = \{C \in \mathbf{C}^{(m,n)}(m) \mid d(C) = d\}$, then $|\mathbf{R}_{n,d}(m)| = |\mathbf{C}_{n,d}(m)|$.*

4   Proof of Theorem 2

Before proving Theorem 2, we recall some properties of $C(\underline{r})$ which are developed in [10, 11], and give some lemmas as preparations.

For $1 \leq k \leq \mu$ and $1 \leq j \leq m$, define $l(k, j) = k_0$ if there exists an integer $k_0 \geq 1$ such that $h_{k_0} = j$ and $h_i \neq j$ for all $k_0 < i \leq k$, and $l(k, j) = 0$ otherwise. It is clear that $l(k, j)$ is a function associated to $C(\underline{r})$ and defined on the set $[1, \mu] \times Z_m$, where $[1, \mu] = \{k \in Z \mid 1 \leq k \leq \mu\}$.

Denote

$$\underline{r}_k = (-I_m \ \underline{r}) \begin{pmatrix} \underline{p}_k \\ q_k \end{pmatrix} = \underline{r}q_k - \underline{p}_k; \tag{2}$$

and denote by $\underline{P}_{k-1,j}(\in F_2[z]^m)$, $Q_{k-1,j}(\in F_2[z])$ and $R_{k-1,j}(\in F_2((z^{-1}))^m)$ the j-th column of P_{k-1}, Q_{k-1} and R_{k-1} for $1 \leq j \leq m$, respectively. It is clear that

$$(-I_m \ \underline{r})B_k = (-R_{k-1} \ \underline{r}_k), \tag{3}$$

$$(-I_m \ \underline{r}) \begin{pmatrix} P_{k-1,j} \\ Q_{k-1,j} \end{pmatrix} = -R_{k-1,j}. \tag{4}$$

Proposition 1. *[10, 11]*

1. For $1 \leq k \leq \mu$, we have

$$\begin{cases} l(k,j) < k \ \ if \ j \neq h_k, \\ h_{l(k,j)} = j \ \ if \ l(k,j) > 0, \\ v_{k,j} = v_{l(k,j)}, \ \ \ and \\ v_{k,j} = 0 \Leftrightarrow l(k,j) = 0. \end{cases}$$

2. For $1 \leq k \leq \mu$, we have
  (a)

$$B_{k-1}E_{h_k} = \begin{pmatrix} P_{k-1} \ P_{k-2,h_k} \\ Q_{k-1} \ Q_{k-2,h_k} \end{pmatrix}.$$

  (b)

$$\begin{pmatrix} P_{k-1,j} \\ Q_{k-1,j} \end{pmatrix} = \begin{cases} \begin{pmatrix} P_{k-2,j} \\ Q_{k-2,j} \end{pmatrix} & if \ j \neq h_k, \\ \begin{pmatrix} \underline{p}_{k-1} \\ q_{k-1} \end{pmatrix} & if \ j = h_k; \end{cases}$$

  and

$$\begin{pmatrix} P_{k-1,j} \\ Q_{k-1,j} \end{pmatrix} = \begin{cases} \begin{pmatrix} \underline{p}_{l(k,j)-1} \\ q_{l(k,j)-1} \end{pmatrix} & if \ l(k,j) \geq 1, \\ \begin{pmatrix} \underline{e}_j \\ 0 \end{pmatrix} & if \ l(k,j) = 0. \end{cases}$$

  As a consequence, $l(k,j) \geq 1$ if $Q_{k-1,j} \neq 0$.
3. For $1 \leq k \leq \mu$, we have
  (a)

$$(-I_m \ \underline{r})B_{k-1}E_{h_k} = (-R_{k-1} \ - R_{k-2,h_k}).$$

  (b)

$$-R_{k-1,j} = \begin{cases} -R_{k-2,j} & if \ j \neq h_k, \\ \underline{r}_{k-1} & if \ j = h_k; \end{cases}$$

  and

$$R_{k-1,j} = \begin{cases} R_{l(k,j)-1,j} = -\underline{r}_{l(k,j)-1} & if \ l(k,j) \geq 1, \\ \underline{e}_j & if \ l(k,j) = 0. \end{cases}$$

4. For $0 \leq k \leq \mu$, we have
  (a) $c_{k,j} = v_{k,j}$ and $\Delta_{k-1} = D_k$ for $0 \leq k \leq \mu$.
  (b) $Iv(R_{k-1,j}) = (j, v_{k,j})$ and $Iv(\underline{r}_k) = (h_{k+1}, v_{k+1})$.
  (c) The matrix R_{k-1} is invertible. In particular, $R_{k-1,j} \neq 0$ for $1 \leq j \leq m$.
  (d) $Iv(R_{k-1}\alpha) = Iv(\Delta_{k-1}\alpha) = Iv(D_k\alpha)$ for any $\alpha \in F_2((z^{-1}))^m$.   □

Lemma 1. *Let* $1 \leq k \leq \mu$. *Then*

1. $\underline{m}_{k,i} = \underline{e}_{j_{k,i}} z^{v_{k,j_{k,i}} - v_{k-1,h_k} - x_{k,i}}$ *for* $1 \leq i \leq \mu_k$. *In particular,* $v_{k,j_{k,i}} - v_{k-1,h_k} - x_{k,i} \geq 0$.
2. *Denote* $\underline{a}_{k,i}^+ = \sum_{i \leq t \leq \mu_k} \underline{m}_{k,t}$ *for* $1 \leq i \leq \mu_k$. *Then*

$$Iv(R_{k-1}\underline{a}_{k,i}^+) = Iv(D_k\underline{m}_{k,i}).$$

Proof. 1. Write simply $\underline{m}_{k,i} = z^x \underline{e}_h$, then $D_k\underline{m}_{k,i} = z^{-v_{k,h}+x}\underline{e}_h$, thus

$$(j_{k,i}, v + x_{k,i}) = Iv(D_k\underline{m}_{k,i}) = Iv(z^{-v_{k,h}+x}\underline{e}_h) = (h, v_{k,h} - x),$$

which leads to this item.

2. Note that

$$R_{k-1}\underline{a}_{k,i+1}^+ = R_{k-1}\underline{a}_{k,i}^+ - R_{k-1}\underline{m}_{k,i}$$

and

$$Iv(R_{k-1}\underline{m}_{k,i}) = Iv(D_k\underline{m}_{k,i}) < Iv(D_k\underline{m}_{k,i+1})$$
$$\leq Iv(D_k\underline{a}_{k,i+1}^+) = Iv(R_{k-1}\underline{a}_{k,i+1}^+),$$

we see

$$Iv(R_{k-1}\underline{a}_{k,i}^+) = Iv(-R_{k-1}\underline{m}_{k,i}). \qquad \square$$

Lemma 2. *Let* $1 \leq k \leq \mu$. *Then*

1.

$$q_{k,1} = z^{t_k} q_{k-1} + Q_{k-2,h_k},$$
$$q_{k,i} = q_{k,i-1} + z^{v_{k,j_{k,i}} - v_{k-1,h_k} - x_{k,i}} Q_{k-1,j_{k,i}}, \quad 2 \leq i \leq \mu_k.$$

2. $A(0; h) = (0, 0, -1)$ *for* $1 \leq h \leq m$; *and* $A(k; h) = A(k - 1; h)$ *if* $h \neq h_k$.

Proof. 1. We have $q_{k,i} = Q_{k-1}\underline{a}_{k,i} + Q_{k-2,h_k}$. Write simply $v = v_{k-1,h_k}$. Note that $\underline{m}_{k,i} = \underline{e}_{j_{k,i}} z^{v_{k,j_{k,i}} - v - x_{k,i}}$, we see $Q_{k-1}\underline{m}_{k,i} = Q_{k-1,j_{k,i}} z^{v_{k,j_{k,i}} - v - x_{k,i}}$. In particular, $Q_{k-1}\underline{m}_{k,1} = Q_{k-1,h_k} z^{t_k}$, since $(j_{k,1}, x_{k,1}) = (h_k, 0)$ and $v_{k,h_k} - v = t_k$. Then

$$q_{k,1} = Q_{k-1}\underline{m}_{k,1} + Q_{k-2,h_k} = z^{t_k} Q_{k-1,h_k} + Q_{k-2,h_k}$$
$$= z^{t_k} q_{k-1} + Q_{k-2,h_k}.$$

For $2 \leq i \leq \mu_k$, we have

$$q_{k,i} = Q_{k-1}(\underline{a}_{k,i-1} + \underline{m}_{k,i}) + Q_{k-2,h_k} = q_{k,i-1} + Q_{k-1}\underline{m}_{k,i}$$
$$= q_{k,i-1} + z^{v_{k,j_{k,i}} - v - x_{k,i}} Q_{k-1,j_{k,i}}.$$

2. It comes from definitions and Proposition 1. $\qquad \square$

Lemma 3. $\deg(q_{k,i}) = d_k$ *for* $1 \leq k \leq \mu$ *and* $1 \leq i \leq \mu_k$.

Proof. Write simply $v_{k-1,h_k} = v$. Note that $\underline{a}_k = \underline{a}_{k,i} + \sum_{i < t \leq \mu_k} \underline{m}_{k,t}$, we have

$$q_{k,i} = (0_{1 \times m} \ 1) B_{k-1} E_{h_k} \begin{pmatrix} \underline{a}_{k,i} \\ 1 \end{pmatrix}$$

$$= (Q_{k-1}, Q_{k-2,h_k}) \begin{pmatrix} \underline{a}_k - \sum_{i < t \leq \mu_k} \underline{m}_{k,t} \\ 1 \end{pmatrix}$$

$$= q_k - \sum_{i < t \leq \mu_k} Q_{k-1} \underline{m}_{k,t}$$

$$= q_k - \sum_{i < t \leq \mu_k, Q_{k-1,j_{k,t}} \neq 0} Q_{k-1,j_{k,t}} z^{v_{k,j_{k,t}} - v - x_{k,t}}.$$

Denote $d(k,t) = \deg(Q_{k-1,j_{k,t}} z^{v_{k,j_{k,t}} - v - x_{k,t}})$. For any given $t > i$, write simply $s = j_{k,t}$. Then

$$d(k,t) = \deg(Q_{k-1,s}) + v_{k,s} - v - x_{k,t} = d_{l(k,s)-1} + v_{l(k,s)} - v - x_{k,t}$$
$$= n_{l(k,s)} - v - x_{k,t},$$

thus

$$d_k - d(k,t) = d_k - n_{l(k,s)} + v + x_{k,t} = n_k + x_{k,t} - n_{l(k,s)}.$$

Note that $t > i \geq 1$ and

$$(s, n_{l(k,s)}) = (h_{l(k,s)}, n_{l(k,s)}) \leq (h_k, n_k) = (j_{k,1}, n_k + x_{k,1})$$
$$< (j_{k,t}, n_k + x_{k,t}) = (s, n_k + x_{k,t}),$$

we get $n_{l(k,s)} < n_k + x_{k,t}$, hence $d_k - d(k,t) = n_k + x_{k,t} - n_{l(k,s)} > 0$.   □

In the sequel, we denote

$$\underline{r}_{k,i} = \underline{r} q_{k,i} - \underline{p}_{k,i}, \tag{5}$$

$$N_{k,i}^{(0)} = (j_{k,i}, n_k + x_{k,i}) \tag{6}$$

for $1 \leq k \leq \mu$, $1 \leq i \leq \mu_k$ and $N_{\mu+1,1}^{(0)} = (1, \infty)$ if $\mu < \infty$. It is clear that

$$\begin{cases} \underline{a}_{k,\mu_k} = \underline{a}_k, \\ (\underline{p}_{k,\mu_k}, q_{k,\mu_k}) = (\underline{p}_k, q_k), \\ \underline{r}_{k,\mu_k} = \underline{r}_k; \end{cases} \tag{7}$$

and

$$(h_k, n_k) = N_{k,1}^{(0)} < N_{k,2}^{(0)} < \cdots < N_{k,\mu_k}^{(0)} < N_{k+1,1}^{(0)}. \tag{8}$$

Denote by $c_{j,n}(\alpha)$ the (j,n)-th coefficient of α for any $\alpha \in F_2((z^{-1}))^m$. Then

Lemma 4. *1. $\delta_{j,n}(q_0; \underline{r}) = c_{j,n}(\underline{r})$ for all $(1,1) \leq (j,n)$.*
2. Let $1 \leq k \leq \mu$ and $1 \leq i \leq \mu_k$. Then

$$\delta_{j,n}(q_{k,i}; \underline{r}) = c_{j,n-d_k}(\underline{r}_{k,i}) \text{ if } N_{k,i}^{(0)} < (j,n).$$

In particular, $\delta_{j,n}(q_k; \underline{r}) = c_{j,n-d_k}(\underline{r}_k)$ if $N_{k,\mu_k}^{(0)} < (j,n)$.

Proof. 1. It is easy to check, based on the fact that $q_0 = 1$.
2. We need only prove $c_{j,n-d_k}(\underline{rq}_{k,i}) = c_{j,n-d_k}(\underline{r}_{k,i})$ for $N_{k,i}^{(0)} < (j,n)$, since $\deg(q_{k,i}) = d_k$ and $\delta_{j,n}(q_{k,i}; \underline{r}) = c_{j,n-d_k}(\underline{rq}_{k,i})$. Note that $\underline{r}_{k,i} = \underline{rq}_{k,i} - \underline{p}_{k,i}$ and $(j_{k,i}, n_k + x_{k,i}) = N_{k,i}^{(0)}$, it is enough to prove $c_{j,n-d_k}(\underline{p}_{k,i}) = 0$ for $(j_{k,i}, n_k + x_{k,i}) < (j,n)$, or equivalently,

$$Supp^+(\underline{p}_{k,i}) \leq (j_{k,i}, n_k + x_{k,i} - d_k) = (j_{k,i}, v_{k-1,h_k} + x_{k,i}). \tag{9}$$

Note that $\underline{p}_{k,i} \in F_2[z]^m$, we see $Supp^+(\underline{p}_{k,i}) \leq (m,0) < (j_{k,i}, v_{k-1,h_k} + x_{k,i})$ if $v_{k-1,h_k} + x_{k,i} > 0$. Now we may assume $v_{k-1,h_k} + x_{k,i} = 0$. With this assumption, we have $v_{k-1,h_k} = x_{k,i} = 0$. Then we need only prove $Supp^+(\underline{p}_{k,i}) \leq (j_{k,i}, 0)$. Note that $(h_k, 0) = (j_{k,1}, x_{k,1}) < (j_{k,2}, x_{k,2}) < \cdots < (j_{k,i}, x_{k,i})$, we get

$$\begin{cases} x_{k,1} = x_{k,2} = \cdots = x_{k,i} = 0, \\ h_k = j_{k,1} < j_{k,2} < \cdots < j_{k,i}, \\ \underline{a}_{k,i} = \sum_{1 \leq s \leq i} z^{v_{k,j_{k,s}}} \underline{e}_{j_{k,s}}, \\ P_{k-2,h_k} = \underline{e}_{h_k} \text{ (since } v_{k-1,h_k} = 0). \end{cases}$$

Therefore, we have

$$\underline{p}_{k,i} = (P_{k-1}, P_{k-2,h_k}) \binom{\underline{a}_{k,i}}{1} = \underline{e}_{h_k} + \sum_{1 \leq s \leq i} P_{k-1,j_{k,s}} z^{v_{k,j_{k,s}}}.$$

Note that $Supp^+(\underline{e}_{h_k}) = (h_k, 0) < (j_{k,i}, 0)$, it is enough to prove

$$Supp^+(P_{k-1,j_{k,s}} z^{v_{k,j_{k,s}}}) \leq (j_{k,i}, 0) \tag{10}$$

for those s such that $1 \leq s \leq i$ and $P_{k-1,j_{k,s}} \neq \underline{0}$. For the case $v_{k,h_{j_{k,i}}} > 0$, we see $Supp^+(P_{k-1,j_{k,s}} z^{v_{k,j_{k,s}}}) \leq (m, -v_{k,j_{k,s}}) < (j_{k,i}, 0)$, since $Supp^+(P_{k,j_{k,i}}) \leq (m,0)$. For the case $v_{k,h_{j_{k,i}}} = 0$, we have $l(k, j_{k,i}) = 0$, then $P_{k-1,j_{k,s}} = \underline{e}_{j_{k,s}}$, hence, $Supp^+(P_{k-1,j_{k,s}} z^{v_{k,j_{k,s}}}) = Supp^+(\underline{e}_{j_{k,s}}) = (j_{k,s}, 0) \leq (j_{k,i}, 0)$. □

Lemma 5. *1. For $1 \leq k \leq \mu$, and $1 < i \leq \mu_k$, we have*
(a) $Iv(\underline{r}_{k,i-1}) = (j_{k,i}, v_{k-1,h_k} + x_{k,i}) = (j_{k,i}, n_k + x_{k,i} - d_k)$.
(b) $c_{j^*,n^*-d_k}(\underline{r}_{k,i-1}) = \begin{cases} 0 \text{ if } (j^*, n^*) < N_{k,i}^{(0)}, \\ 1 \text{ if } (j^*, n^*) = N_{k,i}^{(0)}. \end{cases}$

2. For $1 \le k \le \mu$, we have

 (a) $Iv(\underline{r}_{k-1}) = (h_k, v_k) = (h_k, n_k - d_{k-1})$.

 (b) $c_{j^*,n^*-d_{k-1}}(\underline{r}_{k-1}) = \begin{cases} 0 & \text{if } (j^*, n^*) < N_{k,1}^{(0)}, \\ 1 & \text{if } (j^*, n^*) = N_{k,1}^{(0)}. \end{cases}$

Proof. We have

$$\underline{r}_{k,i} = \begin{pmatrix} -I & \underline{r} \end{pmatrix} \begin{pmatrix} \underline{p}_{k,i} \\ q_{k,i} \end{pmatrix} = \begin{pmatrix} -I & \underline{r} \end{pmatrix} B_{k-1} E_{h_k} \begin{pmatrix} \underline{a}_{k,i} \\ 1 \end{pmatrix}$$

$$= \begin{pmatrix} -R_{k-1} & -R_{k-2,h_k} \end{pmatrix} \begin{pmatrix} \underline{a}_{k,i} \\ 1 \end{pmatrix} = -R_{k-1}\underline{a}_{k,i} - R_{k-2,h_k},$$

and then

$$\underline{r}_k = \underline{r}_{k,\mu_k} = -R_{k-1}\underline{a}_{k,\mu_k} - R_{k-2,h_k} = -R_{k-1}\underline{a}_k - R_{k-2,h_k}.$$

1. Note that

$$\underline{r}_k = -R_{k-1}(\underline{a}_{k,i-1} + \underline{a}_{k,i}^+) - R_{k-2,h_k} = \underline{r}_{k,i-1} - R_{k-1}\underline{a}_{k,i}^+$$

and

$$Iv(-R_{k-1}\underline{a}_{k,i}^+) = Iv(D_k\underline{m}_{k,i}) \le Iv(D_k\underline{m}_{k,\mu_k})$$
$$\le Supp^+(D_k\underline{a}_k) < (h_{k+1}, v_{k+1}) = Iv(\underline{r}_k),$$

where $\underline{a}_{k,i}^+$ is defined as in Lemma 1, we get

$$Ld(\underline{r}_{k,i-1}) + Ld(-R_{k-1}\underline{a}_{k,i}^+) = 0.$$

Then $Iv(\underline{r}_{k,i-1}) = Iv(R_{k-1}\underline{a}_{k,i}^+)$, which together with Lemma 1 lead to $Iv(\underline{r}_{k,i-1}) = Iv(D_k\underline{m}_{k,i}) = (j_{k,i}, v_{k-1,h_k} + x_{k,i})$. The part (b) is an easy consequence of (a).

2. The part (a) is known from Proposition 1, and (b) is an easy consequence of (a). □

For the sake of convenience, we denote $N_{k,\mu_k+1}^{(0)} = N_{k+1,1}^{(0)}$ for $1 \le k \le \mu$.

Lemma 6. *Let* $1 \le k \le \mu$. *Then*

1. $\delta_{j,n}(q_0; \underline{r}) = \begin{cases} 0 & \text{if } (j, n) < N_{1,1}^{(0)}, \\ 1 & \text{if } (j, n) = N_{1,1}^{(0)}. \end{cases}$

2. $\delta_{j,n}(q_{k,i}; \underline{r}) = \begin{cases} 0 & \text{if } N_{k,i}^{(0)} < (j, n) < N_{k,i+1}^{(0)}, \\ 1 & \text{if } (j, n) = N_{k,i+1}^{(0)}. \end{cases}$

Proof. It is an easy consequence of Lemma 4 and Lemma 5. □

The following lemma is prepared for transferring the relation between $q_{k,i}$ and $q_{k,i-1}$, which is given in Lemma 2, to that between $f_{(j,n)+}$ and $f_{j,n}$.

Lemma 7. Let $(j, n) \geq (m, 0)$. Denote $(j^*, n^*) = (j, n)^+$, $\delta = \delta_{j^*, n^*}(f_{j,n}, \underline{r})$, $d = \deg f_{j,n}$, $d^* = \deg f_{j^*, n^*}$. Denote by $A_{j,n}(h) = (g_h, w_h)^\tau$ the h-th column of $A_{j,n}$ for $1 \leq h \leq m$. Then

1. If $(j, n)^+ \notin \cup_{1 \leq k \leq \mu}\{N_{k,i}^{(0)} | 1 \leq i \leq \mu_k\}$, then $\delta = 0$ and

$$(f_{j^*, n^*}, d^*, A_{j^*, n^*}) = (f_{j,n}, d, A_{j,n}).$$

2. If $(j, n)^+ = N_{k,1}^{(0)}$ for some $1 \leq k \leq \mu$, then

$$\delta \neq 0,$$
$$(j^*, n^*) = (h_k, n_k),$$
$$(f_{j,n}, f_{j^*, n^*}) = (q_{k-1}, q_{k,1}),$$
$$(d, d^*) = (d_{k-1}, d_k),$$
$$(A_{j,n}, A_{j^*, n^*}) = (A(k-1), A(k)),$$
$$(f_{j,n}, n^* - d, -\delta)^\tau = A(k; h_k),$$
$$n^* - d - w_{j^*} = t_k.$$

3. If $(j, n)^+ = N_{k,i}^{(0)}$ for some $1 \leq k \leq \mu$ and $1 < i \leq \mu_k$, then

$$\delta \neq 0,$$
$$(j^*, n^*) = (j_{k,i}, n_k + x_{k,i}),$$
$$(f_{j,n}, f_{j^*, n^*}) = (q_{k,i-1}, q_{k,i}),$$
$$(d, d^*) = (d_k, d_k),$$
$$(A_{j,n}, A_{j^*, n^*}) = (A(k), A(k)),$$
$$n^* - d - w_{j^*} = -(v_{k,j_{k,i}} - v_{k-1,h_k} - x_{k,i}) \leq 0.$$

4. $\delta \neq 0$ if and only if $(j, n)^+ = N_{k,i}^{(0)}$ for some $1 \leq k \leq \mu$ and $1 \leq i \leq \mu_k$. Moreover, If $\delta \neq 0$, then $(j, n)^+ = N_{k,1}^{(0)}$ if and only if $n^* - d > w_{j^*}$; and $(j, n)^+ = N_{k,i}^{(0)}$ with $i > 1$ if and only if $n^* - d \leq w_{j^*}$.

Proof. In this proof we use Lemma 1, Lemma 3 and Lemma 6 frequently.

1. We prove it for the cases $(j, n)^+ \in N_0$ and $(j, n)^+ \notin N_0$ separately. If $(j, n)^+ \in N_0$, then $(j, n) = (m, 0)$ or $(j, n) \in N_0$, $f_{j^*, n^*} = q_0 = f_{j,n}$, $\delta = \delta_{(j,n)^+}(q_0; \underline{r}) = 0$, $d^* = \deg(q_0) = d$, $A_{j^*, n^*} = A(0) = A_{j,n}$. If $(j, n)^+ \notin N_0$, then $N_{k,i}^{(0)} \leq (j, n) < (j, n)^+ \in N_{k,i}$ for some $1 \leq k \leq \mu$ and $1 \leq i \leq \mu_k$. Then $f_{j,n} = q_{k,i} = f_{j^*, n^*}$, $\delta = \delta_{j^*, n^*}(q_{k,i}; \underline{r}) = 0$, $d = \deg(f_{j,n}) = \deg(q_{k,i}) = \deg(f_{j^*, n^*}) = d^*$, $A_{j^*, n^*} = A(k) = A_{j,n}$.

2. Under the assumption we have $(j, n) \in N_{k-1, \mu_{k-1}}$, $f_{j,n} = q_{k-1}$ and $(j^*, n^*) = (h_k, n_k)$. Then $f_{j^*, n^*} = q_{k,1}$, $\delta = \delta_{j^*, n^*}(q_{k-1}, \underline{r}) = \delta_{h_k, n_k}(q_{k-1}, \underline{r}) = 1$, $d^* = \deg(q_{k,1}) = d_k$, $d = \deg(q_{k-1}) = d_{k-1}$, $A_{j,n} = A(k-1)$, $A_{j^*, n^*} = A(k)$, $(g_{j^*}, w_{j^*})^\tau = A_{j,n}(j^*) = A(k-1; h_k) = (Q_{k-2,h_k}, v_{k-1,h_k})^\tau$. Note that $A(k; h_k) = (Q_{k-1,h_k}, v_{k,h_k})$, $Q_{k-1,h_k} = q_{k-1} = f_{j,n}$, $d = \deg(f_{j,n}) = \deg(q_{k-1}) = d_{k-1}$ and $v_{k,h_k} = n_k - d_{k-1} = n^* - d$, we get $A(k; h_k) = (f_{j,n}, n^* - d)$. Then $n^* - d - w_{j^*} = v_{k,h_k} - v_{k-1,h_k} = t_k$.

3. Under the assumption we have $(j,n) \in N_{k,i-1}$, $(j^*,n^*) = N_{k,i}^{(0)} = (j_{k,i}, n_k + x_{k,i})$, $f_{j,n} = q_{k,i-1}$, $f_{j^*,n^*} = q_{k,i}$, $d = \deg(q_{k,i-1}) = d_k = \deg(q_{k,i}) = d^*$, $\delta = \delta_{j^*,n^*}(q_{k,i-1}; \underline{r}) = 1$. Note that $(g_{j^*}, w_{j^*})^{\tau} = A_{j,n}(j^*) = A(k; j^*) = (Q_{k-1,j^*}, v_{k,j^*})^{\tau}$, we see

$$n^* - d - w_{j^*} = n_k + x_{k,i} - d_k - v_{k,j^*} = -(v_{k,j_{k,i}} - v_{k-1,h_k} - x_{k,i}) \leq 0.$$

4. It is an easy consequence of the above items. $\qquad\square$

Proof of Theorem 2. In this proof we use Lemma 2, Lemma 3 and Lemma 7 frequently.

1. If $(j,n) \in N_0$, then $f_{j,n} = q_0$, hence $f_{j,n}$ is a minimal polynomial of $\underline{r}^{(j,n)}$ from Proposition 1. Now we assume $(j,n) \in N_k$ with $1 \leq k \leq \mu$. W.l.o.g., we may assume $(j,n) \in N_{k,i}$ for some i: $1 \leq i \leq \mu_k$, hence, $f_{j,n} = q_{k,i}$. We see q_k is a minimal polynomial of $\underline{r}^{(j,n)}$, and the linear complexity $L(j,n)$ of $\underline{r}^{(j,n)}$ is $\deg(q_k) = d_k$ from Proposition 1. If $i = \mu_k$, we have $f_{j,n} = q_{k,\mu_k} = q_k$, so $f_{j,n}$ is a minimal polynomial of $\underline{r}^{(j,n)}$. If $i < \mu_k$, note that $\deg(q_{k,i}) = d_k$, it is enough to prove $q_{k,i}$ is a characteristic polynomial of $\underline{r}^{(j,n)}$. Note that $\underline{r}_{k,i} = \underline{r}q_{k,i} - \underline{p}_{k,i}$ and $(j,n) < N_{k,i+1}^{(0)} = (j_{k,i+1}, n_k + x_{k,i+1})$, we see

$$Iv(\{\underline{r}q_{k,i}\}) = Iv(\{\underline{r}_{k,i}\}) \geq Iv(\underline{r}_{k,i}) = Iv(D_k \underline{m}_{k,i+1})$$
$$= (j_{k,i+1}, v_{k-1,h_k} + x_{k,i+1}) = (j_{k,i+1}, n_k - d_k + x_{k,i+1}) > (j, n - d_k),$$

hence, $q_{k,i}$ is a characteristic polynomial of $\underline{r}^{(j,n)}$.

2. (a) It is an easy consequence of Lemma 7.

(b) If $\delta \neq 0$, $n^* - d > w_{j^*}$, then $(j^*,n^*) = (j,n)^+ = N_{k,1}^{(0)} = (h_k, n_k)$. We have

$$f_{j^*,n^*} = q_{k,1} = z^{t_k} q_{k-1} + Q_{k-2,h_k} = z^{n^* - d - w_{j^*}} f_{j,n} + g_{j^*};$$
$$\deg(f_{j^*,n^*}) = d^* = d_k = d_{k-1} + t_k = d + t_k = n^* - w_{j^*};$$
$$A_{j^*,n^*}(h) = A(k;h) = A(k-1;h) = A_{j,n}(h) \text{ if } h \neq h_k = j^*;$$
$$A_{j^*,n^*}(j^*) = A(k;j^*) = A(k;h_k) = (f_{j,n}, n^* - d).$$

(c) If $\delta \neq 0$, $n^* - d \leq w_{j^*}$, then $(j^*,n^*) = (j,n)^+ = N_{k,i}^{(0)}$ with $i > 1$. We have

$$f_{j^*,n^*} = q_{k,i} = q_{k,i-1} + z^{v_{k,j_{k,i}} - v - x_{k,i}} Q_{k-1,j_{k,i}}$$
$$= f_{j,n} + z^{-n^* + d + w_{j^*}} g_{j^*};$$
$$\deg(f_{j^*,n^*}) = d^* = d_k = d;$$
$$A_{j^*,n^*} = A(k) = A_{j,n}.$$

3. Keep the notation $\underline{m}^{(j,n)} = z^{-n} \underline{e}_j$ and $\underline{m}^{N_{k,i}^{(0)}} = z^{-(n_k + x_{k,i})} \underline{e}_{j_{k,i}}$. We have

$$\underline{\delta} = \sum_{(m,0) \leq (j,n)} \delta_{(j,n)^+}(f_{j,n}; \underline{r}) \underline{m}^{(j,n)^+}$$

$$= \sum_{(j,n) \in N_0} \delta_{(j,n)^+}(q_0; \underline{r}) \underline{m}^{(j,n)^+} + \sum_{k=1}^{\mu} \sum_{i=1}^{\mu_k} \sum_{(j,n) \in N_{k,i}} \delta_{(j,n)^+}(q_{k,i}; \underline{r}) \underline{m}^{(j,n)^+}.$$

Note that

$$z^{-d_k}D_k\underline{m}_{k,i} = z^{-d_k-v_{k-1,h_k}-x_{k,i}}\underline{e}_{j_{k,i}} = z^{-n_k-x_{k,i}}\underline{e}_{j_{k,i}} = \underline{m}^{N_{k,i}^{(0)}},$$

we get

$$\delta = \begin{cases} \underline{m}^{N_{1,1}^{(0)}} + \sum_{1\le k\le\mu-1,1\le i\le\mu_k}\underline{m}^{N_{k,i+1}^{(0)}} + \sum_{1\le i<\mu_\mu}\underline{m}^{N_{\mu,i+1}^{(0)}} & if\ \mu < \infty \\ \underline{m}^{N_{1,1}^{(0)}} + \sum_{1\le k\le\mu,1\le i\le\mu_k}\underline{m}^{N_{k,i+1}^{(0)}} & if\ \mu = \infty \end{cases}$$

$$= \sum_{1\le k\le\mu,1\le i\le\mu_k}\underline{m}^{N_{k,i}^{(0)}}$$

$$= \sum_{1\le k\le\mu,1\le i\le\mu_k} z^{-d_k}D_k\underline{m}_{k,i}$$

$$= \sum_{1\le k\le\mu} z^{-d_k}D_k\underline{a}_k. \qquad \Box$$

References

1. W.H. Mills, Continued fractions and linear recurrences, Math. Comp. 29 (1975) pp.173–180.
2. L.R. Welch and R.A. Scholtz, Continued fractions and Berlekamp's algorithm, IEEE Transactions on Information Theory, Vol. IT-25, No.1, Jan. 1979, pp.19-27.
3. W.M.Schmidt, On continued fraction and diophantine approximation in power series fields, Acta Arith. 95(2000), pp.139-166.
4. E.R. Berlekamp, Algebraic coding theory, McGraw Hill, New York, 1969.
5. J.L. Massey, Shift register synthesis and BCH decoding, IEEE Transactions on Information Theory, Vol.15, No. 1, Jan. 1969, pp.122-127.
6. Z. Dai and K. Zeng, Continued fractions and Berlekamp-Massey algorithm, Advances in Cryptology-AUSCRYPT'90, Springer-Verlag LNCS 453 (1990), pp.24-31.
7. S. Sakata, Extension of Berlekamp-Massey algorithm to N dimension, Inform. And Comput. 84 (1990) pp.207-239.
8. G.L. Feng, K.K.Tzeng, A generalization of the Berlekamp-Massey algorithm for multisequence shift-register synthesis with applications to decoding cyclic codes, IEEE Transactions on Information Theory IT-37 (1991) pp.1274-1287.
9. Z. Dai, K. Wang and D. Ye, m-Continued Fraction Expansions of Multi-Laurent Series, ADVANCES IN MATHEMATICS(CHINA), 2004, Vol.33, No.2, pp.246-248.
10. Z. Dai, K. Wang and D. Ye, Multidimensional Continued Fraction and Rational Approximation, http://arxiv.org/abs/math.NT/0401141, 2004.
11. Zongduo Dai, Kunpeng Wang and Dingfeng Ye, Multi-Continued Fraction Algorithm on Multi-Formal Laurent Series, preprint, pp.1-21.
12. H. Niederreiter, Some computable complexity measures for binary sequences, in: C. Ding, T. Helleseth, H. Niederreiter(Eds.), Sequences and their applications, Springer, London, 1999, pp.67-78.
13. L. Wang, Lattice bases reduction algorithm in function fields and multisequence linear feedback shift-register synthesis, PhD thesis, Graduate school of Chinese Academy of Sciences, Beijing, 2003.
14. Michael Vielhaber, A unified view on sequence complexity measures as isometries, Proceedings (Extended Abstracts) 2004 International Conference on Sequenes and Their Applications, pp.34-38.

A New Search for Optimal Binary Arrays with Minimum Peak Sidelobe Levels[*]

Gummadi S. Ramakrishna[1] and Wai Ho Mow[2]

[1] Department of Electrical Engineering,
Indian Institute of Technology Madras, Chennai, India
g.s.ramakrishna@gmail.com
[2] Department of Electrical and Electronic Engineering,
Hong Kong University of Science and Technology,
Clear Water Bay, Hong Kong S.A.R., China
w.mow@ieee.org
http://www.ee.ust.hk/~eewhmow

Abstract. The group structure of the 2-dimensional sidelobe-invariant transformations for binary arrays is characterized. The design of an efficient exhaustive backtracking search algorithm, which exploits such group structure to reduce the search space and, in the meantime, applies the partially determined autocorrelation values as backtracking conditions, is presented. As a consequence of applying the algorithm, all optimal binary arrays with minimum peak sidelobe levels consisting of up to 49 elements are obtained and tabulated.

1 Introduction

Consider an $M \times N$ binary array $a = [a(m,n)]$ where each $a(m,n)$ is 1 or -1. Its 2-dimensional (2D) aperiodic autocorrelation function $C(j,k)$ is defined as

$$C(j,k) = \sum_m \sum_n a(m+j, n+k)a(m,n) \qquad (1)$$

Here, we take $a(m,n) = 0$ when $m \notin [1,M]$ or $n \notin [1,N]$ such that $C(j,k)$ is 0 for $|j| \geq M$ or $|k| \geq N$. We define an array to be optimal if its largest out-of-phase autocorrelation magnitude (also called the peak sidelobe level), $|C(j,k)|$, is no greater than that of any other array of the same sizes. The minmum peak sidelobe (MPS) level for a size of $M \times N$ is the peak sidelobe level of an optimal array of the specified size. An optimal array is also called a MPS array in the literature. It is further called a Barker array if the MPS level achieved is only 1. The problem of finding the MPS levels and the corresponding optimal arrays for large sizes is evidently a computationally challenging task.

[*] This work was supported by the National Science Foundation of China (NSFC) and the Research Grants Council of Hong Kong (RGC) joint research scheme with Project No. N_HKUST617/02 (No. 60218001).

T. Helleseth et al. (Eds.): SETA 2004, LNCS 3486, pp. 355–360, 2005.

Alquaddoomi and Scholtz [1] presented a potential application of MPS arrays to high resolution radar and results of exhaustive backtracking searches for optimal arrays of sizes $M \times N$ with $M \leq 8$ for $N = 2$, $M \leq 9$ for $N = 3$, and $M \leq 7$ for $N = 4$, all with MPS ≤ 3. Kutruff and Quadt [2] conducted trial-and-error searches for near-optimal binary arrays with sizes no more than 5×5 towards the purpose of applications to loudspeaker arrays. Such arrays with sizes 4×3 and 5×5 were cited in [3–Table VI] as Kutruff-Quadt codes. (A comparison with Table 2 here shows that the 5×5 array therein is actually suboptimal.) Other potential applications of MPS arrays include 2D system identication, 2D synchronization/positioning, coded aperture imaging [4] and spread spectrum image watermarking [5].

Exhaustive enumeration results for optimal binary sequences (i.e., $M = 1$) have been conducted for lengths up to 48 (see [6]). The group structure of sidelobe-invariant transformations for binary sequences and its use for efficient representation and regeneration of the class of equivalent optimal sequences were discussed in [7] and [8]. The group of 2D sidelobe-invariant transforms for binary arrays is larger and has a far more complicated structure. In this paper, the group structure of 2D sidelobe-invariant transforms is characterized in details. An efficient backtracking search algorithm exploiting some of the group properties is developed, and a new search for all 2D optimal arrays of sizes $M \times N$ with $MN \leq 49$ is accomplished.

2   The Group of 2D Sidelobe-Invariant Transformations

Let S_{MN} denote the set of all $M \times N$ binary arrays. A sidelobe-invariant transformation is defined to be a function from S_{MN} to itself that preserves the set of sidelobe levels. For the binary case, the five 2D transformations given in Table 1 are sidelobe-invariant. The final column gives the $(i, j)^{th}$ element in the transformed array in terms of $a(i, j)$, the original array.

Let G denote the group generated by the five sidelobe-invariant transformations in Table 1. The group G partitions S_{MN} into equivalence classes in which any two elements a, b are in the same equivalence class if and only if $\exists g \in G$ such that $g(a) = b$. In each equivalence class, we identify a unique array in order to reduce the search space. This can be specified as the array that gives the lexicographically (lex) smallest string formed by taking row-wise, the logarithm to base -1 of each of its elements.

Theorem 1. *The elements of the group of 2D sidelobe-invariant transformations for binary arrays can be represented by a binary 5-tuple that indicates the composition of the 5 basic transformations C, H, V, P, Q listed in Table 1. As a corollary, the order of the group is 32.*

2.1   Characterization of the Group Structure

While the order of the group of transformations is always 32, the specific group structure depends on the parity of M and N. When both M and N are odd, the

Table 1. 2D sidelobe-invariant transformations for binary arrays

Transformation	Notation	Definition
Complementation	C	$-a(i,j)$
Horizontal reversal	H	$a(i, N - j + 1)$
Vertical reversal	V	$a(M - i + 1, j)$
Alternate row complementation	P	$(-1)^{i+1}a(i,j)$
Alternate column complementation	Q	$(-1)^{j+1}a(i,j)$

non-commutative cases do not arise and we have an abelian group isomorphic to $(Z_2)^5$. For the case where M is even and N is odd, we have only one non-commutative relation, $PV \equiv CVP$. Since Q and H are part of the center of G (they commute with all elements), we see that G is isomorphic to the direct product of the group of order 8 formed by $\langle C, V, P \rangle$ and the abelian group, $Z_2 \times Z_2$ formed by $\langle H, Q \rangle$. The group formed by $\langle C, V, P \rangle$ is D_4, the dihedral group of order 8, which can be verified from the presentation relations (see [7] and [8] for a similar discussion on sequences). When N is even and M is odd, $\langle C, H, Q \rangle$ forms D_4 and $\langle V, P \rangle$ forms $(Z_2)^2$ to generate G as their direct product. For the case when M and N are both even, we have both the relations, $PV \equiv CVP$ and $QH \equiv CHQ$ occurring simultaneously. In this case, G can not be decomposed as a direct product but is a central product of two dihedral groups of order 8, $D_4 * D_4$, with C being the element amalgamated in the central product. The two dihedral groups in question can be verified as $\langle P, V, C \rangle$ and $\langle Q, H, C \rangle$ from the presentation relations. This group is also known as the "extraspecial 2-group of order 32 and type plus" (see [9] for a presentation of this group).

2.2   Identifying the Representative in an Equivalence Class

For any array a, define the transformation, N, as,

$$N(a) = C^\alpha \cdot P^\beta \cdot Q^\gamma(a) \tag{2}$$

where α, β, γ are either 0 or 1, and are determined from the array values, $a(1,1)$, $a(2,1)$ and $a(1,2)$ so as to force $a(1,1)$, $a(2,1)$ and $a(1,2)$ to 1 in the transformed array, $N(a)$. This process may be called normalization.

Theorem 2. $N(a)$ is the lex smallest element in the equivalence class of a due to the subgroup generated by $\langle C, P, Q \rangle$. As a corollary, a is the representative element of its equivalence class if and only if a is the lexicographically smallest of $\{N(a), N(H(a)), N(V(a)), N(HV(a))\}$.

This is one of the ideas incorporated in the backtracking algorithm to search only one array out of each equivalence class.

3   Design of the Backtracking Search Algorithm

The basic structure of the algorithm used is the classical backtrack using the depth first search (see [10] for a general discussion on the backtracking technique). Traversing down a path in the search tree corresponds to assigning values to the array elements in a specific order. We use backtracking conditions to avoid traversing down the paths that are certain to not result in the arrays being sought for. Two types of backtracking conditions have been used here. One type is the backtrack executed when the path can not lead to an array with the minimum peak sidelobe (see [6],[13]). The elements are filled alternately from the top left corner and the bottom right corner. The specific sequence of filling the array elements can be varied heuristically to adopt to the order that might be expected to give strong correlation backtrack constraints early in the tree traversal. The backtrack conditions are verified for every new position in the search tree (i.e. for every new assignment of an array value). Another type corresponds to the backtrack executed when the path can not lead to the representative array of an equivalence class (see [1],[12]). It can be implemented in an efficient manner based on the relevant characterization result in Section 2. The details are omitted here due to the limitation of space.

4   The Search Results

Numerous previously unknown optimal binary arrays have been found by applying our efficient backtracking search. Table 2 lists the MPS level and the total number N_{eq} of equivalence classes of the optimal binary arrays of every size $M \times N$ with $MN \leq 49$. Comparing the search results reported in [1], new optimal binary arrays are obtained for sizes $M \times N$ with $9 \leq M \leq 24$ for $N = 2$, $10 \leq M \leq 16$ for $N = 3$, $8 \leq M \leq 12$ for $N = 4$, and all tabulated sizes for $N = 5$, 6, and 7. The MPS level increases faster for the 2D than the 1D case if we compare MPS arrays and sequences with the same number of elements. While all the sizes till 48 for the 1D case have MPS = 3, the value 3 saturates earlier for the 2D case at around 30 for MN. Also, the number of arrays which achieve MPS = 4 is very large. An analogous count for the number of MPS = 4 sequences has not been exhibited previously for any length. Note that the new searches described in [13] were not exhaustive.

5   Concluding Summary

The group structure for the 2D sidelobe-invariant transforms was presented and the reduced array groups due to the presence of certain symmetry were analyzed. Efficient backtracking search algorithm that takes advantages of the group structure of sidelobe-invariant transformations was developed. As a consequence of carrying out the search, numerous new MPS arrays with up to 49 elements have been discovered.

Table 2. $M \times N$ optimal binary arrays with $MN \leq 49$

$M \times N$	MPS	N_{eq}	SSE
2×2	1	1	2
3×2	2	3	11
4×2	2	4	12
5×2	2	6	29
6×2	2	5	42
7×2	2	8	39
8×2	2	6	48
9×2	3	106	73
10×2	3	89	58
11×2	3	92	83
12×2	3	73	84
13×2	3	27	109
14×2	3	20	130
15×2	3	3	135
16×2	3	4	176
17×2	3	1	193
18×2	4	153627	190
19×2	4	123746	219
20×2	4	149635	236
21×2	4	91759	281
22×2	4	108622	294
23×2	4	48335	295
24×2	4	54668	352
3×3	2	6	17
4×3	2	7	24
5×3	2	15	41
6×3	2	4	48
7×3	3	87	77

$M \times N$	MPS	N_{eq}	SSE
8×3	2	1	80
9×3	3	24	117
10×3	3	2	168
11×3	4	245354	161
12×3	4	342207	176
13×3	4	388928	197
14×3	4	322901	252
15×3	4	299315	265
16×3	4	207554	288
4×4	2	4	40
5×4	3	104	62
6×4	3	71	84
7×4	3	12	110
8×4	3	6	128
9×4	4	464909	166
10×4	4	619256	228
11×4	4	552020	238
12×4	4	403650	420
5×5	3	46	90
6×5	3	7	162
7×5	4	461388	168
8×5	4	688470	216
9×5	4	653050	242
6×6	4	539884	162
7×6	4	647846	233
8×6	4	576041	296
7×7	4	488588	315

References

1. S. Alquaddoomi and R.A. Scholtz, "On the nonexistence of Barker arrays and related matters," *IEEE Transactions on Information Theory*, Vol. 35, Issue 5, pp. 1048–1057, Sept. 1989.

2. H. Kutruff and H. P. Quadt, "Ebene Schallstrallergruppen mit Ungebündelter Abstrahlung (Plane sound source arrays with omnidirectional radiation)," *Acoustica*, Vol. 50, pp. 273–279, 1982.

3. S. E. El-Khamy, O. A. Abdel-Alim, A. M. Rushdi, and A. H. Banah, "Code-fed omnidirectional arrays," *IEEE Journal of Oceanic Engineering*, Vol. 14, Issue 4, pp. 384–395, Oct. 1989.

4. G. K. Skinner, "Imaging with coded-aperture masks," *Nucl. Instrum. Methods*, vol. 220, pp. 33–40, 1984.

5. Andrew Z. Tirkel and Thomas E. Hall, "A unique watermark for every image." *IEEE Multimedia*, vol. 8, no. 4, pp. 30–37, 2001.

6. Marvin N. Cohen, Marshall R. Fox, and John M. Baden, "Minimum peak sidelobe pulse compression codes," *Proc. of 1990 IEEE Radar Conference*, 7-10 May 1990, pp. 633–638.

7. L. M. Fourdan, "About the enumeration of best biphase codes," *IEEE Transactions on Signal Processing*, Vol. 39, Issue 6, pp. 1427–1428, June 1991.

8. W. H. Mow, 'Comments on "About enumeration of best biphase codes"', *IEEE Trans. on signal Processing*, Vol. 42, No. 7, pp. 1876, July 1994.

9. Tara L. Smith, "Extra-special groups of order 32 as Galois groups," *Canad. J. Math.*, vol. 46, pp. 886–896, 1994.

10. S. W. Golomb and L. D. Baumert, "Backtrack programming," *Journal ACM*, vol. 12, no. 4, pp. 516–524, 1965.

11. M. J. E. Golay and D. B. A. Harris, "New search for skewsymmetric binary sequences with optimal merit factors," *IEEE Transactions on Information Theory*, Vol. 36, Issue 5, pp. 1163–1166, Sept. 1990.

12. W. H. Mow, "Enumeration techniques for best N-phase codes," *Electronics Letters*, Vol. 29, Issue 10, pp. 907–908, 13 May 1993.

13. G. E. Coxson, A. Hirschel, and M. N. Cohen, "New results on minimum-PSL binary codes," *Proc. of 2001 IEEE Radar Conference*, 1-3 May 2001, pp. 153–156.

New Constructions of Quaternary Hadamard Matrices

Ji-Woong Jang[1], Sang-Hyo Kim[1], Jong-Seon No[1], and Habong Chung[2]

[1] School of Electrical Engineering and Computer Science,
Seoul National University, Seoul 151-742, Korea
{stasera, shkim}@ccl.snu.ac.kr, jsno@snu.ac.kr
[2] School of Electronics and Electrical Engineering,
Hong-Ik University, Seoul 121-791, Korea
habchung@hongik.ac.kr

Abstract. In this paper, we propose two new construction methods for quaternary Hadamard matrices. By the first method, which is applicable for any positive integer n, we are able to construct a quaternary Hadamard matrix of order 2^n from a binary sequence with ideal autocorrelation. The second method also gives us a quaternary Hadamard matrix of order 2^n from a binary extended sequence of period $2^n - 1$, where n is a composite number.

1 Introduction

A generalized Hadamard matrix $\mathcal{H}$ of order N is an $N \times N$ matrix satisfying $\mathcal{H}\mathcal{H}^\dagger = NI_N$, where $\dagger$ denotes the conjugate transpose and I_N is the identity matrix of order N [3,8,13]. In other words, any two distinct rows of $\mathcal{H}$ are orthogonal. For this reason, Hadamard matrices have been studied for the applications in many areas such as wireless communication systems, coding theory, and signal design[1,4,14,15,16]. Hadamard matrices have strong ties to sequences. Matsufuji and Suehiro proposed the complex Hadamard matrices related to bent sequences[9]. Popovic, Suehiro, and Fan[12] proposed orthogonal sets of quaternary sequences by using quadriphase sequence family $\mathcal{A}$ by Boztas, Hammons, and Kumar[2].

In this paper, we propose two new construction methods for quaternary Hadamard matrices. By the first method, which is applicable for any positive integer n, we are able to construct a quaternary Hadamard matrix of order 2^n from a binary sequence with ideal autocorrelation. The second method also gives us a quaternary Hadamard matrix of order 2^n from a binary extended sequence of period $2^n - 1$, where n is a composite number. Before we proceed to the next section, let us clarify some terms and notations used throughout this paper.

Let F_{2^n} be the finite field with 2^n elements. Let $F_{2^n}^* = F_{2^n} \setminus \{0\}$ and $s(x)$ be a mapping from F_{2^n} to F_2 or Z_4. If we restrict the mapping $s(x)$ to $F_{2^n}^*$ and replace x by α^t, where α is a primitive element in F_{2^n}, then we can obtain a sequence $s(\alpha^t)$, $0 \le t \le 2^n - 2$, of period $2^n - 1$. Hence, for convenience, we

T. Helleseth et al. (Eds.): SETA 2004, LNCS 3486, pp. 361–372, 2005.

will use the expression 'a binary or quaternary sequence $s(\alpha^t)$ of period $2^n - 1$' interchangeably with 'a mapping $s(x)$ from F_{2^n} to F_2 or Z_4'.

For $\delta \in F_{2^n}^*$, the crosscorrelation function between two quaternary sequences $s_i(x)$ and $s_j(x)$ is defined as

$$R_{i,j}(\delta) = \sum_{x \in F_{2^n}^*} w_4^{s_i(x\delta) - s_j(x)},$$

where w_4 is a complex fourth root of unity.

Let $f(x)$ be a mapping from F_{2^n} onto F_{2^m}, where $m|n$. The function $f(x)$ is said to be *balanced* if each nonzero element of F_{2^m} appears 2^{n-m} times and zero element $2^{n-m} - 1$ times in the list $\{f(x) | x \in F_{2^n}^*\}$. A function $f(x)$ is said to be *difference-balanced* if $f(\delta x) - f(x)$ is balanced for any $\delta \in F_{2^n} \setminus \{0, 1\}$. It is easy to see that the binary sequence with difference-balance property has the ideal autocorrelation property necessarily and sufficiently.

It is not difficult to see that a variable v over Z_4 can be expressed using two binary variables v_1 and v_2 as

$$v = v_1 + 2v_2,$$

where addition is modulo 4.

Let us define two maps ϕ and ψ as

$$\phi(v) = v_1, \quad \psi(v) = v_2.$$

It can be shown that $\phi(v - w)$ and $\psi(v - w)$ of the difference $v - w$ are expressed as

$$\phi(v - w) = v_1 + w_1$$
$$\psi(v - w) = v_1 w_1 + w_1 + w_2 + v_2. \tag{1}$$

2   New Constructions of Quaternary Hadamard Matrices

In this section, we propose two constructions for quaternary Hadamard matrices from binary sequences with ideal autocorrelation.

Lemma 1. For a positive integer n, let $g(t)$ be a binary sequence of period $2^n - 1$ with ideal autocorrelation. Then for any z, $1 \le z \le 2^n - 2$, the following sequence $q_z(t)$ given by

$$q_z(t) = g(t) + 2g(t + z)$$

is balanced over Z_4.

Proof. Let $N_z(a, b)$, $a, b \in \{0, 1\}$ be the number of t such that $g(t) = a$ and $g(t + z) = b$. Since $g(t)$ has the ideal autocorrelation property, it is balanced and difference-balanced. Thus we have

$$N_z(0, 0) + N_z(0, 1) = 2^{n-1} - 1$$
$$N_z(0, 0) + N_z(1, 0) = 2^{n-1} - 1$$
$$N_z(0, 0) + N_z(1, 1) = 2^{n-1} - 1.$$

Finally, from the facts that

$$\sum_a \sum_b N_z(a, b) = 2^n - 1,$$

we can conclude that $q_z(t)$ is balanced.   □

Using the above lemma, we get the quaternary Hadamard matrices as in the following theorem.

Theorem 1. Let n be an integer and $g(t)$, $0 \leq t \leq 2^n - 2$, be a sequence of period $2^n - 1$ with ideal autocorrelation. Then the following matrix $\mathcal{H}_Q$ is a $2^n \times 2^n$ quaternary Hadamard matrix.

$$\mathcal{H}_Q = (h_{ij}), \quad 0 \leq i, j \leq 2^n - 1,$$

where h_{ij} is given as

$$h_{ij} = \begin{cases} 1, & \text{for } i = 0 \text{ or } j = 0 \\ w_4^{2g(j-1)}, & \text{for } i = 1 \text{ and } 1 \leq j \leq 2^n - 1 \\ w_4^{g(j-1)+2g(i-1+j-1)} = w_4^{q_{i-1}(j-1)}, & \text{otherwise.} \end{cases}$$

Proof. Let u_i be the ith row of $\mathcal{H}_Q$. It is clear that $u_i u_i^\dagger = 2^n$, $0 \leq i \leq 2^n - 1$. In proving the orthogonality between u_i and u_k, we should consider the following three cases.

Case 1) $i = 0$ and $1 \leq k \leq 2^n - 1$:
From Lemma 1 and balance property of $g(t)$ and $q_k(t)$, it is clear that u_0 is orthogonal to u_k, for any k, $1 \leq k \leq 2^n - 1$.

Case 2) $i = 1$, $2 \leq k \leq 2^n - 1$:
In this case, $u_1 u_k^\dagger$ is given as

$$u_1 u_k^\dagger = 1 + \sum_{t=0}^{2^n-2} w_4^{2g(t)-g(t)-2g(t+k-1)}$$

$$= 1 + \sum_{t=0}^{2^n-2} w_4^{g(t)-2g(t+k-1)}.$$

From Lemma 1, it is straightforward that $g(t) - 2g(t+k-1)$ is also balanced and thus $u_1 u_k^\dagger = 0$, i.e., u_1 is orthogonal to u_k.

Case 3) $2 \leq i < k \leq 2^n - 1$:

In this case, $u_i u_k^\dagger$ is given as

$$u_i u_k^\dagger = 1 + \sum_{t=0}^{2^n-2} w_4^{\{g(t)+2g(t+i-1)\}-\{g(t)+2g(t+k-1)\}}$$

$$= 1 + \sum_{t=0}^{2^n-2} w_4^{2(g(t+i-1)+g(t+k-1))}$$

$$= 1 + \sum_{t=0}^{2^n-2} (-1)^{g(t+i-1)+g(t+k-1)}.$$

From the difference-balance property of $g(t)$, $u_i u_k^\dagger = 0$.   □

Here is an example of an 8×8 quaternary Hadamard matrix constructed from the above theorem.

Example 1. Let α be a primitive element in F_{2^3}. Using the m-sequence $\mathrm{tr}_1^3(\alpha^t)$ of period 7, we can construct the quaternary sequences of period 7 as

$$s_0(t) = 2\mathrm{tr}_1^3(\alpha^t)$$
$$s_i(t) = \mathrm{tr}_1^3(\alpha^t) + 2\mathrm{tr}_1^3(\alpha^{t+i}), \ 1 \leq i \leq 6,$$

which gives us $\mathcal{H}_Q$

$$\mathcal{H}_Q = \begin{bmatrix} w_4^0 & w_4^0 & w_4^0 & w_4^0 & w_4^0 & w_4^0 & w_4^0 & w_4^0 \\ w_4^0 & w_4^2 & w_4^0 & w_4^0 & w_4^2 & w_4^0 & w_4^2 & w_4^2 \\ w_4^0 & w_4^1 & w_4^0 & w_4^2 & w_4^1 & w_4^2 & w_4^3 & w_4^3 \\ w_4^0 & w_4^1 & w_4^0 & w_4^2 & w_4^1 & w_4^2 & w_4^3 & w_4^3 \\ w_4^0 & w_4^1 & w_4^2 & w_4^0 & w_4^3 & w_4^2 & w_4^3 & w_4^1 \\ w_4^0 & w_4^3 & w_4^0 & w_4^2 & w_4^3 & w_4^2 & w_4^1 & w_4^1 \\ w_4^0 & w_4^1 & w_4^2 & w_4^2 & w_4^3 & w_4^0 & w_4^1 & w_4^3 \\ w_4^0 & w_4^3 & w_4^2 & w_4^2 & w_4^1 & w_4^0 & w_4^3 & w_4^1 \\ w_4^0 & w_4^3 & w_4^2 & w_4^0 & w_4^1 & w_4^2 & w_4^1 & w_4^3 \end{bmatrix}.$$

□

No, Yang, Chung, and Song constructed *extended sequences* with ideal auto-correlation from sequences of shorter period with ideal autocorrelation[11].

Theorem 2 (No, Yang, Chung, and Song[11]). Let n and m be positive integers such that $m|n$. Let $f(y)$ be the function from F_{2^m} to F_2 with difference-balance property such that $f(0) = 0$. Let r be an integer such that $\gcd(r, 2^m - 1) = 1$ and $1 \leq r \leq 2^m - 2$. Then the sequence of period $2^n - 1$ defined by

$$f([\mathrm{tr}_m^n(x)]^r)$$

has the ideal autocorrelation property.   □

Using the extended sequences, we can construct the quaternary Hadamard matrix as in the following theorem.

Theorem 3. Let n and m be integers such that $m|n$, and r be an integer such that $1 \le r \le 2^m - 2$ and $\gcd(r, 2^m - 1) = 1$. Let $T = \frac{2^n - 1}{2^m - 1}$ and $f(y)$ be the sequence from F_{2^m} to F_2 which has the balance and difference-balance properties. Let $s_i(\alpha^t)$ be defined as

$$s_0(\alpha^t) = 2f([\mathrm{tr}_m^n(\alpha^t)]^r)$$
$$s_i(\alpha^t) = f([\mathrm{tr}_m^n(\alpha^t)]^r + 2f([\mathrm{tr}_m^n(\beta^i \alpha^t)]^r), \quad 1 \le i \le 2^m - 2,$$

where $\beta = \alpha^T$ is a primitive element in F_{2^m}.

Then the following matrix $\mathcal{H}_L$ is a $2^n \times 2^n$ quaternary Hadamard matrix.

$$\mathcal{H}_L = (h_{ij}),$$

where h_{ij} is given as

$$h_{ij} = \begin{cases} 1, & \text{if } i = 0 \text{ or } j = 0 \\ w_4^{s_{\lfloor (i-1)/T \rfloor}(j-1+i_T)}, & \text{otherwise,} \end{cases}$$

where $\lfloor x \rfloor$ denotes the greatest integer not exceeding x and $i_T = (i-1) \bmod T$.
□

Proof of the above theorem requires following lemmas.

Lemma 2. Let m, e, and n be positive integers such that $n = em$. Let $q = 2^m$ and $A = \{1, \alpha, \cdots, \alpha^{T-1}\}$, where α is a primitive element in F_{2^n} and $T = \frac{q^e - 1}{q - 1}$. Let $v(x)$ be a function from F_{q^e} onto F_q with the balance and difference-balance properties. Further assume that $v(x)$ satisfies $v(yx) = yv(x)$ for any $y \in F_q$ and $x \in F_{q^e}$. For a given $\delta \in F_{q^e} \setminus F_q$, let $M_\delta(a, b)$ be the number of $x_2 \in A$ satisfying

$$v(\delta x_2) = a \quad \text{and} \quad v(x_2) = b, \quad a, b \in F_q.$$

Then, we have

$$M_\delta(0, 0) = \frac{q^{e-2} - 1}{q - 1} = \frac{2^{n-2m} - 1}{2^m - 1}$$

$$\sum_{c \in F_q^*} M_\delta(c, 0) = \sum_{c \in F_q^*} M_\delta(0, c) = q^{e-2} = 2^{n-2m}$$

$$\sum_{d \in F_q^*} M_\delta(cd, d) = q^{e-2} = 2^{n-2m}, \quad \text{for any } c \in F_q^*.$$

Proof. Let $N_\delta(a, b)$ be the number of $x \in F_{q^e}^*$ satisfying $v(\delta x) = a$ and $v(x) = b$. Let $x = x_1 x_2$, where $x_1 \in F_q$ and $x_2 \in A$. Because $v(x)$ is difference-balanced, $v(\delta x) - v(cx) = v(\delta x) - cv(x)$ is balanced for any $c \in F_q^*$ and 0 occurs $q^{e-1} - 1$ times as x varies over $F_{q^e}^*$. Thus we have

$$\sum_{a \in F_q} N_\delta(ca, a) = q^{e-1} - 1. \tag{2}$$

Since $v(x)$ is balanced, we have

$$\sum_{a\in F_q} N_\delta(a,0) = \sum_{b\in F_q} N_\delta(0,b) = q^{e-1} - 1. \tag{3}$$

Also, note that

$$\sum_{a\in F_q}\sum_{b\in F_q} N_\delta(a,b) = q^e - 1. \tag{4}$$

Now, we have

$$\sum_{a\in F_q}\sum_{b\in F_q} N_\delta(a,b) = \sum_{a\in F_q} N_\delta(a,0) + \sum_{b\in F_q} N_\delta(0,b)$$

$$-N_\delta(0,0) + \sum_{c\in F_q^*}\sum_{a\in F_q^*} N_\delta(a,ca)$$

$$= \sum_{a\in F_q} N_\delta(a,0) + \sum_{b\in F_q} N_\delta(0,b) - N_\delta(0,0)$$

$$+ \sum_{c\in F_q^*}\left\{\sum_{a\in F_q} N_\delta(a,ca) - N_\delta(0,0)\right\}. \tag{5}$$

Plugging (2), (3), and (4) into (5), we have

$$N_\delta(0,0) = q^{e-2} - 1. \tag{6}$$

From (2) and (6), we also have

$$\sum_{a\in F_q^*} N_\delta(ca,a) = \sum_{a\in F_q} N_\delta(ca,a) - N_\delta(0,0) = q^{e-2}(q-1).$$

Let $\beta = \alpha^T$. For a given x_2 such that $v(\delta x_2) = cv(x_2)$, the ordered pair $(v(\delta x), v(x)) = (x_1 v(\delta x_2), x_1 v(x_2))$ takes each value in the list

$$(c,1), (c\beta, \beta), \cdots, (c\beta^{q-2}, \beta^{q-2})$$

exactly once as x_1 varies over F_q^*. Therefore we have

$$\sum_{a\in F_q^*} N_\delta(ca,a) = (q-1) \sum_{a\in F_q^*} M_\delta(ca,a),$$

which, in turn, tells us that

$$\sum_{a\in F_q^*} M_\delta(ca,a) = q^{e-2}.$$

Similarly, we have

$$M_\delta(0,0) = \frac{N_\delta(0,0)}{q-1} = \frac{q^{e-2}-1}{q-1}$$

$$\sum_{c \in F_q^*} M_\delta(c,0) = \frac{\sum_{c \in F_q^*} N_\delta(c,0)}{q-1} = q^{e-2}$$

$$\sum_{c \in F_q^*} M_\delta(0,c) = \frac{\sum_{c \in F_q^*} N_\delta(0,c)}{q-1} = q^{e-2}.$$

$\square$

Lemma 3. Let $s(x)$ be a function from any domain B to Z_4, where $s(0) = 0$. Define two Boolean constituent functions of $s(x)$ as

$$\phi_s(x) = \phi(s(x)), \quad \psi_s(x) = \psi(s(x))$$

and their modulo-2 sum as

$$\mu_s(x) = \phi_s(x) + \psi_s(x). \tag{7}$$

Let $N_f(c)$ denote the number of occurrences of $f(x) = c$ as x varies over B. Then, we have

$$\sum_{x \in B} \omega_4^{s(x)} = (N_{\psi_s}(0) - N_{\mu_s}(1)) + j(N_{\mu_s}(1) - N_{\psi_s}(1)).$$

Proof. It is clear that

$$\sum_{x \in B} \omega_4^{s(x)} = (N_s(0) - N_s(2)) + j(N_s(1) - N_s(3))$$

and

$$N_{\psi_s}(1) = N_s(2) + N_s(3) \tag{8}$$
$$N_{\psi_s}(0) = 2^n - N_{\psi_s}(1) = N_s(0) + N_s(1) \tag{9}$$
$$N_{\mu_s}(1) = N_s(1) + N_s(2). \tag{10}$$

From (8), (9), and (10), we have

$$N_s(0) - N_s(2) = N_{\psi_s}(0) - N_{\mu_s}(1)$$
$$N_s(1) - N_s(3) = N_{\mu_s}(1) - N_{\psi_s}(1).$$

Thus we prove the lemma.

$\square$

Corollary 1. Let $s(x)$ be a function from F_{2^n} to Z_4. Then,

$$\sum_{x \in F_{2^n}} \omega_4^{s(x)} = 0$$

if and only if the functions $\psi_s(x)$ and $\mu_s(x)$ are balanced.

$\square$

Now we are ready to prove Theorem 3.

Proof of Theorem 3. Let v_i be the ith row of $\mathcal{H}_L$, $0 \le i \le 2^n - 1$. We have to show that $v_i v_k^\dagger = 0$ for all $i \ne k$. The case when $i = 0$ is simple. Since v_0 is an all one sequence, we need to show that the row sum is zero for each row $v_k, k \ne 0$. From the structure of $\mathcal{H}_L$, it is manifest that the rows v_{1+lT} through $v_{T+lT}, 0 \le l \le 2^m - 2$, are the cyclic shifts of $s_l(x)$. Also note that $s_0(x)$ is balanced since it is in fact the binary extended sequence, and $s_l(x), l \ne 0$, is also balanced from Lemma 1. Thus we have $v_0 v_k^\dagger = 0$ for all $k \ne 0$.

Now, for any nonzero i and k, $i \ne k$, $v_i v_k^\dagger$ can be expressed as

$$v_i v_k^\dagger = 1 + \sum_{t=0}^{2^n - 2} w_4^{s_{\lfloor (i-1)/T \rfloor}(t+i_T) - s_{\lfloor (k-1)/T \rfloor}(t+k_T)}$$

$$= 1 + \sum_{x \in F_{2^n}^*} w_4^{s_{i'}(\delta x) - s_{k'}(x)},$$

where $\delta = \alpha^{i_T - k_T}$, $i' = \lfloor (i-1)/T \rfloor$, and $k' = \lfloor (k-1)/T \rfloor$. For $\delta = \alpha^{i_T - k_T}$, showing that $v_i v_k^\dagger = 0$ is equivalent to showing that the crosscorrelation $R_{i',k'}(\delta)$ between $s_{i'}(x)$ and $s_{k'}(x)$ is -1.

For $a, b \in F_{2^m} \setminus F_2$, define two quaternary sequences $u_a(x)$ and $u_b(x)$ of period $2^m - 1$ as

$$u_a(x) = f(x) + 2f(ax)$$
$$u_b(x) = f(x) + 2f(bx)$$

and let $d(x, \eta) = u_a(\eta x) - u_b(x)$. Define S_{ψ_d} and S_{μ_d} as

$$S_{\psi_d} = \sum_{x \in F_{2^m}^*} \sum_{\eta \in F_{2^m}^*} (-1)^{\psi(d(x,\eta))}$$

$$S_{\mu_d} = \sum_{x \in F_{2^m}^*} \sum_{\eta \in F_{2^m}^*} (-1)^{\mu(d(x,\eta))}.$$

Then from (1) and (7), S_{ψ_d} and S_{μ_d} can be expressed as

$$S_{\psi_d} = \sum_{x \in F_{2^m}^*} \sum_{\eta \in F_{2^m}^*} (-1)^{f(\eta x)f(x) + f(x) + f(bx) + f(a\eta x)} \tag{11}$$

$$S_{\mu_d} = \sum_{x \in F_{2^m}^*} \sum_{\eta \in F_{2^m}^*} (-1)^{f(\eta x)f(x) + f(\eta x) + f(bx) + f(a\eta x)}. \tag{12}$$

Now, let $I_1(x)$ and $I_2(x)$ be the inner summation in (11),

$$\sum_{\eta \in F_{2^m}^*} (-1)^{f(\eta x)f(x)+f(x)+f(bx)+f(a\eta x)}$$

for the cases when $f(x) = 0$ and $f(x) = 1$, respectively, i.e.,

$$I_1(x) = \sum_{\eta \in F_{2^m}^*} (-1)^{f(bx)+f(a\eta x)}$$

and

$$I_2(x) = \sum_{\eta \in F_{2^m}^*} (-1)^{f(\eta x)+1+f(bx)+f(a\eta x)}.$$

Then S_{ψ_d} can be expressed as

$$S_{\psi_d} = \sum_{x \in \{x | f(x)=0, x \in F_{2^m}^*\}} I_1(x) + \sum_{x \in \{x | f(x)=1, x \in F_{2^m}^*\}} I_2(x). \tag{13}$$

The first term in (13) is computed as

$$\sum_{x \in \{x | f(x)=0, x \in F_{2^m}^*\}} I_1(x) = \sum_{x \in \{x | f(x)=0, x \in F_{2^m}^*\}} (-1)^{f(bx)} \sum_{\eta \in F_{2^m}^*} (-1)^{f(a\eta x)}$$

$$= \sum_{x \in \{x | f(x)=0, x \in F_{2^m}^*\}} (-1)^{f(bx)+1},$$

since $f(x)$ is balanced.

The second term in (13) is computed as

$$\sum_{x \in \{x | f(x)=1, x \in F_{2^m}^*\}} I_2(x) = \sum_{x \in \{x | f(x)=1, x \in F_{2^m}^*\}} (-1)^{f(bx)+1} \sum_{\eta \in F_{2^m}^*} (-1)^{f(\eta x)+f(a\eta x)}$$

$$= \sum_{x \in \{x | f(x)=1, x \in F_{2^m}^*\}} (-1)^{f(bx)}$$

since $f(x)$ is difference-balanced.
Thus, we have

$$S_{\psi_d} = \sum_{x \in \{x | f(x)=1, x \in F_{2^m}^*\}} (-1)^{f(bx)} - \sum_{x \in \{x | f(x)=0, x \in F_{2^m}^*\}} (-1)^{f(bx)}.$$

Finally, from the difference-balance property, we have

$$(f(x), f(bx)) = \begin{cases} (0,0), & 2^{m-2} - 1 \text{ times} \\ (1,0), & 2^{m-2} \text{ times} \\ (0,1), & 2^{m-2} \text{ times} \\ (1,1), & 2^{m-2} \text{ times,} \end{cases}$$

as x varies over $F_{2^m}^*$. Therefore, we have

$$S_{\psi_d} = 1.$$

In the similar way, we get $S_{\mu_d} = 1$.

Now consider two sequences

$$s_{i'}(x) = f([\mathrm{tr}_m^n(x)]^r) + 2f(a^r[\mathrm{tr}_m^n(x)]^r)$$
$$s_{k'}(x) = f([\mathrm{tr}_m^n(x)]^r) + 2f(b^r[\mathrm{tr}_m^n(x)]^r),$$

where $a = \beta^{i'}$ and $b = \beta^{k'}$ for nonzero i' and k'. Then $R_{i',k'}(\delta)$ is given by

$$R_{i',k'}(\delta) = \sum_{x \in F_{2^n}^*} \omega_4^{s_{i'}(\delta x) - s_{k'}(x)}$$

$$= \sum_{x_2 \in A} \sum_{x_1 \in F_{2^m}^*} \omega_4^{\{f(x_1^r[\mathrm{tr}_m^n(\delta x_2)]^r) + 2f(x_1^r a^r[\mathrm{tr}_m^n(\delta x_2)]^r)\}}$$

$$\cdot \omega_4^{-\{f(x_1^r[\mathrm{tr}_m^n(x_2)]^r) + 2f(x_1^r b^r[\mathrm{tr}_m^n(x_2)]^r)\}}.$$

Case 1) $i' \neq k'$ for nonzero i' and k' :

For $\delta \notin F_{2^m}$, with the replacement of $\mathrm{tr}_m^n(\delta x_2)$ by cd and $\mathrm{tr}_m^n(x_2)$ by d and also from Lemma 2, $R_{i',k'}(\delta)$ is rewritten as

$$R_{i',k'}(\delta) = \sum_{d \in F_{2^m}^*} M_\delta(cd, d) \sum_{c \in F_{2^m}^*} \sum_{x_1 \in F_{2^m}^*} \omega_4^{\{f([x_1 cd]^r) + 2f([x_1 acd]^r)\} - \{f([x_1 d]^r) + 2f([x_1 bd]^r)\}}$$

$$+ M_\delta(0, 0) \sum_{x_1 \in F_{2^m}^*} \omega_4^0$$

$$+ \sum_{c \in F_{2^m}^*} M_\delta(c, 0) \sum_{x_1 \in F_{2^m}^*} \omega_4^{\{f([x_1 c]^r) + 2f([x_1 ac]^r)\}}$$

$$+ \sum_{c \in F_{2^m}^*} M_\delta(0, c) \sum_{x_1 \in F_{2^m}^*} \omega_4^{-\{f([x_1 c]^r) + 2f([x_1 bc]^r)\}}$$

$$= 2^{n-2m} \sum_{c \in F_{2^m}^*} \sum_{x_1 \in F_{2^m}^*} \omega_4^{\{f([x_1 c]^r) + 2f([x_1 ac]^r)\} - \{f(x_1^r) + 2f([x_1 b]^r)\}}$$

$$+ \frac{2^{n-2m} - 1}{2^m - 1} \sum_{x_1 \in F_{2^m}^*} \omega_4^0$$

$$+ 2^{n-2m} \sum_{x_1 \in F_{2^m}^*} \omega_4^{\{f([x_1 c]^r) + 2f([x_1 ac]^r)\}}$$

$$+ 2^{n-2m} \sum_{x_1 \in F_{2^m}^*} \omega_4^{-\{f([x_1 c]^r) + 2f([x_1 bc]^r)\}}.$$

From Lemma 3 and the facts that $S_{\psi_d} = 1$ and $S_{\mu_d} = 1$, $R_{i',k'}(\delta)$ can be computed as

$$R_{i',k'}(\delta) = 2^{n-2m} + 2^{n-2m} - 1 + 2 \times 2^{n-2m}(-1) = -1.$$

For $\delta = 1$, we have

$$R_{i',k'}(1) = \sum_{x \in F_{2^n}^*} \omega_4^{(f([\mathrm{tr}_m^n(x)]^r) + 2f(a^r[\mathrm{tr}_m^n(x)]^r)) - (f([\mathrm{tr}_m^n(x)]^r) + 2f(b^r[\mathrm{tr}_m^n(x)]^r))}$$

$$= -1$$

from the difference-balance property of $f(x)$.

Case 2) $i' = k'$ for nonzero i' and k' :

Obviously, $R_{i',i'}(1) = 2^n - 1$. When $\delta \notin F_{2^m}$, the correlation function is given as

$$R_{i',i'}(\delta) = 2^{n-2m} \sum_{c \in F_{2^m}^*} \sum_{x_1 \in F_{2^m}^*} \omega_4^{\{f([x_1 c]^r) + 2f([x_1 a c]^r)\} - \{f(x_1^r) + 2f([x_1 a]^r)\}}$$

$$+ \frac{2^{n-2m} - 1}{2^m - 1} \sum_{x_1 \in F_{2^m}^*} \omega_4^0$$

$$+ 2^{n-2m} \sum_{x_1 \in F_{2^m}^*} \omega_4^{\{f([x_1 c]^r) + 2f([x_1 a c]^r)\}}$$

$$+ 2^{n-2m} \sum_{x_1 \in F_{2^m}^*} \omega_4^{\{f(x_1^r c^r) + 2f([x_1 a c]^r)\}}$$

$$= 2^{n-2m} + 2^{n-2m} - 1 + 2 \times 2^{n-2m}(-1) = -1.$$

Case 3) $i' = 0$ or $k' = 0$:

In this case, it is easy to show that $R_{i',0}(\delta) = R_{0,i'}(\delta) = -1$ for $\delta \notin F_{2^m}$ and $R_{0,0}(\delta) = -1$ for $\delta \neq 1$. □

Here is an example of 64×64 quaternary Hadamard matrix constructed from the Theorem 3.

Example 2. Let α be a primitive element in F_{2^6}. Let $T = \frac{2^6 - 1}{2^3 - 1} = 9$ and $r = 5$. Using the GMW-sequence $\mathrm{tr}_1^3([\mathrm{tr}_3^6(\alpha^t)]^r)$ of period 63, we can construct quaternary sequences of period 63 as

$$s_0(t) = 2\mathrm{tr}_1^3([\mathrm{tr}_3^6(\alpha^t)]^5)$$

$$s_i(t) = \mathrm{tr}_1^3([\mathrm{tr}_3^6(\alpha^t)]^r) + 2\mathrm{tr}_1^3([\mathrm{tr}_3^6(\alpha^{t+9i})]^5), \ 1 \leq i \leq 8.$$

These sequences make a quaternary Hadamard matrix as

$$\mathcal{H}_L = (h_{ij}),$$

where h_{ij} is given as

$$h_{ij} = \begin{cases} w^0 & \text{if } i = 0 \text{ or } j = 0 \\ w^{2\mathrm{tr}_1^3([\mathrm{tr}_3^6(\alpha^{j-1+i_9})]^5)} & \text{if } 1 \leq i \leq T \text{ and } j \neq 0 \\ w^{\mathrm{tr}_1^3([\mathrm{tr}_3^6(\alpha^{j-1+i_9})]^r) + 2\mathrm{tr}_1^3([\mathrm{tr}_3^6(\alpha^{j-1+i_9+9\lfloor (i-1)/9 \rfloor})]^5)} & \text{otherwise,} \end{cases}$$

where $i_9 = (i - 1) \bmod 9$.

□

References

1. Again, S.S. : Hadamard matrices and their Applications. Lecture Notes in Mathematics, Vol. 1168, Springer-Verlag, New York (1980)
2. Boztas, S., Hammons, R., and Kumar, P.V.: 4-phase sequences with near optimum correlation properties. IEEE Trans. on Inform. theory, Vol. 38, (1992) 1101-1113
3. Craigen, R.: Hadamard matrices and designs. Chapter IV. 24. CRC Handbook of Combinatorial Designs, Edited by C. J. Colbourn and J.H. Dinitz, CRC Press, New York (1996) 370-377
4. Kim, J.-H. and Song, H.-Y.: Existence of cyclic Hadamard difference sets and its relation to binary sequences wiht ideal autocorrelation. J. Commun. Networks, Vol. 1, No. 1, (1999) 14-18
5. Kim, S.H., Jang, J.W., No, J.S., and Chung, H.: New construction of quaternary low correlation zone sequences. submitted to IEEE Trans. Inform. Theory, (2004)
6. Kim, S.-H., Chung, H., No, J.-S., and Helleseth, T.: New cyclic relative difference sets constructed from d-homogeneous functions with difference-balanced property. to appear in IEEE Trans. Inform. Theory
7. Klapper, A.: d-form sequence: Families of sequences with low correlation values and large linear spans. IEEE Trans. Inform. Theory, Vol. 41, No. 2, (1995) 423-431
8. van Lint, J.H. and Wilson, R. M.: A course in combinatorics, Cambridge Univ. Press, New York (1992)
9. Matsufuji, S. and Suehiro, N.: Complex Hadamard matrices related to bent sequences. IEEE Trans. on Inform. Theory, Vol. 42, No. 2, (1996) 637
10. No, J.-S.: New cyclic difference sets with Singer parameters constructed from d-homogeneous functions. Designs. Codes and Cryptography, Vol. 33, Issue 3, (2004) 199–213
11. No, J.-S., Yang, K., Chung, H., and Song, H.-Y.: On the construction of binary sequences with ideal autocorrelation property. Proc. IEEE Int. Symp. Inform. Theory and Its Appl. (ISITA'96), Victoria, British Columbia, Canada (1996) 837-840
12. Popovic, B.M., Suehiro, N., and Fan, P. Z.: Orthogonal sets of quadriphase sequences with good correlation properties. IEEE Trans. on Inform. Theory, Vol. 48, No. 4, (2002) 956-959
13. Seberry, J. and Yamada, M.: Hadamard matrices, sequences, and block design. Contemporary Design Theory: Collection of Surveys,(1992) 431-569
14. Simon, M. K. *et al.*: Spread Spectrum Communications, Vol. 1, Rockville, MD: Computer Science Press, 1985; revised ed., McGraw-Hill, (1994)
15. Song, H.-Y. and Golomb, S.W.: On the existence of cyclic Hadamard difference sets. IEEE Trans. Inform. Theory, Vol. IT-40, (1994) 1266-1268
16. TIA/EIA/IS-95: Mobile Station-Base Station Compatibility Standard for Dual-Mode Wideband Spread Spectrum Cellular System. Telecommunications Industry Association as a North American 1.5 MHz Cellular CDMA Air-Interface Standard, (1993)

Spectral Orbits and Peak-to-Average Power Ratio of Boolean Functions with Respect to the $\{I, H, N\}^n$ Transform

Lars Eirik Danielsen and Matthew G. Parker

The Selmer Center, Department of Informatics, University of Bergen,
PB 7800, N-5020 Bergen, Norway
{larsed, matthew}@ii.uib.no
http://www.ii.uib.no/~{larsed,matthew}

Abstract. We enumerate the inequivalent self-dual additive codes over
GF(4) of blocklength n, thereby extending the sequence A090899 in *The
On-Line Encyclopedia of Integer Sequences* from $n = 9$ to $n = 12$. These
codes have a well-known interpretation as quantum codes. They can also
be represented by graphs, where a simple graph operation generates the
orbits of equivalent codes. We highlight the regularity and structure of
some graphs that correspond to codes with high distance. The codes can
also be interpreted as quadratic Boolean functions, where inequivalence
takes on a spectral meaning. In this context we define PAR$_{IHN}$, peak-
to-average power ratio with respect to the $\{I, H, N\}^n$ transform set. We
prove that PAR$_{IHN}$ of a Boolean function is equivalent to the the size of
the maximum independent set over the associated orbit of graphs. Finally
we propose a construction technique to generate Boolean functions with
low PAR$_{IHN}$ and algebraic degree higher than 2.

1 Self-dual Additive Codes over GF(4)

A quantum error-correcting code with parameters $[[n, k, d]]$ encodes k qubits in a
highly entangled state of n qubits such that any error affecting less than d qubits
can be detected, and any error affecting at most $\frac{d-1}{2}$ qubits can be corrected. A
quantum code of the stabilizer type corresponds to a code $\mathcal{C} \subset \mathrm{GF}(4)^n$ [1]. We
denote $\mathrm{GF}(4) = \{0, 1, \omega, \omega^2\}$, where $\omega^2 = \omega + 1$. *Conjugation* in $\mathrm{GF}(4)$ is defined
by $\bar{x} = x^2$. The *trace map*, tr : $\mathrm{GF}(4) \mapsto \mathrm{GF}(2)$, is defined by $\mathrm{tr}(x) = x + \bar{x}$.
The *trace inner product* of two vectors of length n over $\mathrm{GF}(4)$, $\boldsymbol{u}$ and $\boldsymbol{v}$, is
given by $\boldsymbol{u} * \boldsymbol{v} = \sum_{i=1}^{n} tr(u_i \overline{v_i})$. Because of the structure of stabilizer codes, the
corresponding code over $\mathrm{GF}(4)$, $\mathcal{C}$, will be *additive* and satisfy $\boldsymbol{u} * \boldsymbol{v} = 0$ for
any two codewords $\boldsymbol{u}, \boldsymbol{v} \in \mathcal{C}$. This is equivalent to saying that the code must
be *self-orthogonal* with respect to the trace inner product, i.e., $\mathcal{C} \subseteq \mathcal{C}^\perp$, where
$\mathcal{C}^\perp = \{\boldsymbol{u} \in \mathrm{GF}(4)^n \mid \boldsymbol{u} * \boldsymbol{c} = 0, \forall \boldsymbol{c} \in \mathcal{C}\}$.

We will only consider codes of the special case where the dimension $k = 0$.
Zero-dimensional quantum codes can be understood as highly-entangled single
quantum states which are robust to error. These codes map to additive codes

T. Helleseth et al. (Eds.): SETA 2004, LNCS 3486, pp. 373–388, 2005.
© Springer-Verlag Berlin Heidelberg 2005

over GF(4) which are *self-dual* [2], $\mathcal{C} = \mathcal{C}^{\perp}$. The number of inequivalent self-dual additive codes over GF(4) of blocklength n has been classified by Calderbank et al. [1] for $n \leq 5$, by Höhn [3] for $n \leq 7$, by Hein et al. [4] for $n \leq 7$, and by Glynn et al. [5] for $n \leq 9$. Moreover, Glynn has recently posted these results as sequence A090899 in *The On-Line Encyclopedia of Integer Sequences* [6]. We extend this sequence from $n = 9$ to $n = 12$ both for indecomposable and decomposable codes as shown in table 1. Table 2 shows the number of inequivalent indecomposable codes by distance. The distance, d, of a self-dual additive code over GF(4), $\mathcal{C}$, is the smallest weight (i.e., number of nonzero components) of any nonzero codeword in $\mathcal{C}$. A database of orbit representatives with information about orbit size, distance, and weight distribution is also available [7].

Table 1. Number of Inequivalent Indecomposable (i_n) and (Possibly) Decomposable (t_n) Self-Dual Additive Codes Over GF(4)

n	1	2	3	4	5	6	7	8	9	10	11	12
i_n	1	1	1	2	4	11	26	101	440	3,132	40,457	1,274,068
t_n	1	2	3	6	11	26	59	182	675	3,990	45,144	1,323,363

Table 2. Number of Indecomposable Self-Dual Additive Codes Over GF(4) by Distance

$d \backslash n$	2	3	4	5	6	7	8	9	10	11	12
2	1	1	2	3	9	22	85	363	2,436	26,750	611,036
3			1	1	4	11	69	576	11,200	467,513
4				1		5	8	120	2,506	195,455
5									1	63
6										1
Total	1	1	2	4	11	26	101	440	3,132	40,457	1,274,068

2   Graphs, Boolean Functions, and LC-Equivalence

A self-dual additive code over GF(4) corresponds to a *graph state* [4] if its generator matrix, G, can be written as $G = \Gamma + \omega I$, where Γ is a symmetric matrix over GF(2) with zeros on the diagonal. The matrix Γ can be interpreted as the adjacency matrix of a simple undirected graph on n vertices. It has been shown by Schlingemann and Werner [8], Grassl et al. [9], Glynn [10], and Van den Nest et al. [11] that all stabilizer states can be transformed into an equivalent graph state. Thus all self-dual additive codes over GF(4) can be represented by graphs. These codes also have another interpretation as quadratic Boolean functions over n variables. A quadratic function, f, can be represented by an adjacency matrix, Γ, where $\Gamma_{i,j} = \Gamma_{j,i} = 1$ if $x_i x_j$ occurs in f, and $\Gamma_{i,j} = 0$ otherwise.

Example 1. A self-dual additive code over GF(4) with parameters $[[6, 0, 4]]$ is generated by the generator matrix

$$\begin{pmatrix} \omega & 0 & 0 & 1 & 1 & 1 \\ 0 & \omega & 0 & \omega^2 & 1 & \omega \\ 0 & 0 & \omega & \omega^2 & \omega & 1 \\ 0 & 1 & 0 & \omega & \omega^2 & 1 \\ 0 & 0 & 1 & \omega & 1 & \omega^2 \\ 1 & \omega^2 & 0 & \omega & 0 & 0 \end{pmatrix}.$$

We can transform the generator matrix into the following generator matrix of an equivalent code corresponding to a graph state,

$$\begin{pmatrix} \omega & 0 & 0 & 1 & 1 & 1 \\ 0 & \omega & 1 & 1 & 0 & 1 \\ 0 & 1 & \omega & 1 & 1 & 0 \\ 1 & 1 & 1 & \omega & 1 & 1 \\ 1 & 0 & 1 & 1 & \omega & 0 \\ 1 & 1 & 0 & 1 & 0 & \omega \end{pmatrix} = \Gamma + \omega I.$$

Γ is the adjacency matrix of the graph shown in fig. 1(a). It can also be represented by the quadratic Boolean function $f(x) = x_0 x_3 + x_0 x_4 + x_0 x_5 + x_1 x_2 + x_1 x_3 + x_1 x_5 + x_2 x_3 + x_2 x_4 + x_3 x_4 + x_3 x_5$.

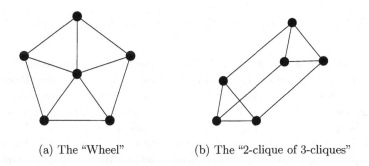

(a) The "Wheel"   (b) The "2-clique of 3-cliques"

Fig. 1. The LC Orbit of the [[6,0,4]] "Hexacode"

Recently, Glynn et al. [5, 10] has re-formulated the primitive operations that map equivalent self-dual additive codes over GF(4) to each other as a single, primitive operation on the associated graphs. This symmetry operation is referred to as *Vertex Neighbourhood Complementation* (VNC). It was also discovered independently by Hein et al. [4] and by Van den Nest et al. [11]. The identification of this problem as a question of establishing the *local unitary equivalence* between those quantum states that can be represented as graphs or Boolean functions was presented by Parker and Rijmen at SETA'01 [12]. Graphical representations have also been identified in the context of quantum codes by Schlingemann and Werner [8] and by Grassl et al. [9]. VNC is another name for *Local Complementation* (LC), referred to in the context of *isotropic systems* by Bouchet [13,14]. LC is defined as follows.

Definition 1. *Given a graph* $G = (V, E)$ *and a vertex* $v \in V$. *Let* $N_v \subset V$ *be the neighbourhood of* v, *i.e., the set of vertices adjacent to* v. *The subgraph induced by* N_v *is complemented to obtain the LC image* G^v.

It is easy to verify that $(G^v)^v = G$.

Theorem 1 (Glynn et al. [5,10]). *Two graphs* G *and* H *correspond to equivalent self-dual additive codes over* GF(4) *iff there is a finite sequence of vertices* $v_1, v_2, \ldots, v_s$, *such that* $(((G^{v_1})^{v_2})^{\cdots})^{v_s} = H$.

The symmetry rule can also be described in terms of quadratic Boolean functions.

Definition 2. *If the quadratic monomial* $x_i x_j$ *occurs in the algebraic normal form of the quadratic Boolean function* f, *then* x_i *and* x_j *are mutual neighbours in the graph represented by* f, *as described by the* $n \times n$ *symmetric adjacency matrix,* Γ, *where* $\Gamma_{i,j} = \Gamma_{j,i} = 1$ *if* $x_i x_j$ *occurs in* f, *and* $\Gamma_{i,j} = 0$ *otherwise. The quadratic Boolean functions* f *and* f' *are LC equivalent if*

$$f'(\boldsymbol{x}) = f(\boldsymbol{x}) + \sum_{\substack{j,k \in N_a \\ j<k}} x_j x_k \pmod 2,$$

where $a \in \mathbb{Z}_n$ *and* N_a *comprises the neighbours of* x_a *in the graph representation of* f.

A finite number of repeated applications of the LC operation generates the orbit classes presented in this paper and, therefore, induces an equivalence between quadratic Boolean functions. We henceforth refer to this equivalence as *LC-equivalence* and the associated orbits as *LC orbits*. If the graph representations of two self-dual additive codes over GF(4) are isomorphic, they are also considered to be equivalent. This corresponds to a permutation of the labels of the vertices in the graph or the variables in the Boolean function. We only count members of an LC orbit up to isomorphism. As an example, fig. 1 shows the graph representation of the two only non-isomorphic members in the orbit of the [[6, 0, 4]] "Hexacode".

A recursive algorithm, incorporating the package *nauty* [15] to check for graph isomorphism, was used to generate the LC orbits enumerated in table 1. Only the LC orbits of indecomposable codes (corresponding to connected graphs) were generated, since all decomposable codes (corresponding to unconnected graphs) can easily be constructed by combining indecomposable codes of shorter lengths.

Consider, (a) self-dual additive codes over GF(4) of blocklength n, (b) pure quantum states of n qubits which are joint eigenvectors of a commuting set of operators from the Pauli Group [1], (c) quadratic Boolean functions of n variables, (d) undirected graphs on n vertices. Then, under a suitable interpretation, we consider objects (a), (b), (c), and (d) to be mathematically identical.

3   Regular Graph Structures

Although a number of constructions for self-dual additive codes over GF(4) exist [5,16], it appears that the underlying symmetry of their associated graphs has not been identified or exploited to any great extent. We highlight the regularity and structure of some graphs that correspond to self-dual additive codes over GF(4) with high distance. Of particular interest are the highly regular "nested clique" graphs. Fig. 2 shows a few examples of such graphs. There is an upper

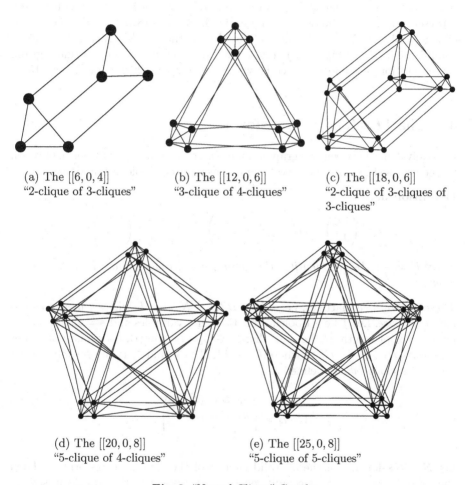

(a) The [[6, 0, 4]] "2-clique of 3-cliques"

(b) The [[12, 0, 6]] "3-clique of 4-cliques"

(c) The [[18, 0, 6]] "2-clique of 3-cliques of 3-cliques"

(d) The [[20, 0, 8]] "5-clique of 4-cliques"

(e) The [[25, 0, 8]] "5-clique of 5-cliques"

Fig. 2. "Nested Clique" Graphs

bound on the possible distance of self-dual additive codes over GF(4) [2]. Codes that meet this bound are called *extremal*. Other bounds on the distance also exist [1,17]. Of the codes corresponding to graphs shown in fig. 2, the [[6, 0, 4]], [[12, 0, 6]], and [[20, 0, 8]] codes are extremal. To find the "nested clique" graph representations, one may search through the appropriate LC orbits. Also note that all "nested clique" graphs we have identified so far have *circulant* adjacency matrices. An exhaustive search of all graphs with circulant adjacency matrices of up to 30 vertices has been performed.

If d is the distance of a self-dual additive code over $GF(4)$, then every vertex in the corresponding graph must have a vertex degree of at least $d - 1$. This follows from the fact that a vertex with degree δ corresponds to a row in the generator matrix, and therefore a codeword, of weight $\delta+1$. All the graphs shown in fig. 2 satisfy the minimum possible regular vertex degree for the given distance. Some extremal self-dual additive codes over $GF(4)$ do not have any regular graph representation, for example the unique $[[11, 0, 5]]$ and $[[18, 0, 8]]$ codes. For codes of length above 25 and distance higher than 8 the graph structures get more complicated. For example, with a non-exhaustive search, we did not find a graph representation of a $[[30, 0, 12]]$ code with regular vertex degree lower than 15.

4   The $\{I, H, N\}^n$ Transform

LC-equivalence between two graphs can be interpreted as an equivalence between the generalised Fourier spectra of the two associated Boolean functions.

Definition 3. *Let*

$$I = \begin{pmatrix} 1 & 0 \\ 0 & 1 \end{pmatrix}, \quad H = \frac{1}{\sqrt{2}} \begin{pmatrix} 1 & 1 \\ 1 & -1 \end{pmatrix}, \quad N = \frac{1}{\sqrt{2}} \begin{pmatrix} 1 & i \\ 1 & -i \end{pmatrix},$$

where $i^2 = -1$, be the Identity, Hadamard, and Negahadamard kernels, respectively.

These are *unitary* matrices, i.e., $II^\dagger = HH^\dagger = NN^\dagger = I$, where $\dagger$ means *conjugate transpose*. Let f be a Boolean function on n variables and $\boldsymbol{s} = 2^{-\frac{n}{2}}(-1)^{f(\boldsymbol{x})}$ be a vector of length 2^n. Let s_j, where $j \in \mathbb{Z}_{2^n}$, be the jth coordinate of $\boldsymbol{s}$. Let $U = U_0 \otimes U_1 \otimes \cdots \otimes U_{n-1}$ where $U_k \in \{I, H, N\}$, and $\otimes$ is the *tensor product* (or *Kronecker product*) defined as

$$A \otimes B = \begin{pmatrix} a_{00}B & a_{01}B & \cdots \\ a_{10}B & a_{11}B & \cdots \\ \vdots & \vdots & \ddots \end{pmatrix}.$$

Let $\boldsymbol{S} = U\boldsymbol{s}$ for any of the 3^n valid choices of the $2^n \times 2^n$ transform U. Then the set of 3^n vectors, $\boldsymbol{S}$, is a multispectra with respect to the transform set, U, with $3^n 2^n$ spectral points. We refer to this multispectra as the spectrum with respect to the $\{I, H, N\}^n$ transform. (Using a similar terminology, the spectrum with respect to the $\{H\}^n$ transform would simply be the well-known Walsh-Hadamard spectrum). It can be shown that the $\{I, H, N\}^n$ spectrum of an LC orbit is invariant to within coefficient permutation. Moreover if, for a specific choice of U, $\boldsymbol{S}$ is flat (i.e., $|S_i| = |S_j|, \forall i, j$), then we can write $\boldsymbol{S} = v^{4f'(\boldsymbol{x})+h(\boldsymbol{x})}$, where f' is a Boolean function, h is any function from $\mathbb{Z}_2^n$ to $\mathbb{Z}_8$, and $v^4 = -1$. If the algebraic degree of $h(\boldsymbol{x})$ is ≤ 1, we can always eliminate $h(\boldsymbol{x})$ by post-multiplication by a tensor product of matrices from $\mathcal{D}$, the set of 2×2 diagonal and anti-diagonal unitary matrices [18], an operation that will never change the spectral coefficient magnitudes. Let M be the multiset of f' existing within the $\{I, H, N\}^n$ spectrum for the subcases where $h(\boldsymbol{x})$ is of algebraic degree ≤ 1. The $\{I, H, N\}^n$-orbit of f is then the set of distinct members of M. In particular, if f is quadratic then the $\{I, H, N\}^n$-orbit is the LC orbit [18].

Example 2. We look at the function $f(\boldsymbol{x}) = x_0 x_1 + x_0 x_2$. The corresponding bipolar vector, ignoring the normalization factor, is

$$\boldsymbol{s} = (-1)^{f(\boldsymbol{x})} = (1, 1, 1, -1, 1, -1, 1, 1)^T.$$

We choose the transform $U = N \otimes I \otimes I$ and get the result

$$\boldsymbol{S} = U\boldsymbol{s} = (v, v^7, v^7, v, v^7, v, v, v^7)^T, \quad v^4 = -1.$$

We observe that $|S_i| = 1$, $\forall i$, which means that $\boldsymbol{S}$ is flat and can be expressed as

$$\boldsymbol{S} = v^{4(x_0 x_1 + x_0 x_2 + x_1 x_2) + (6x_0 + 6x_1 + 6x_2 + 1)}.$$

We observe that $h(\boldsymbol{x})$, the terms that are not divisible by 4, are all linear or constant. We can therefore eliminate $h(\boldsymbol{x})$, in this case by using the transform

$$D = \begin{pmatrix} 1 & 0 \\ 0 & i \end{pmatrix} \otimes \begin{pmatrix} 1 & 0 \\ 0 & i \end{pmatrix} \otimes \begin{pmatrix} v^7 & 0 \\ 0 & v \end{pmatrix}.$$

We get the result

$$D\boldsymbol{S} = (-1)^{x_0 x_1 + x_0 x_2 + x_1 x_2},$$

and thus $f'(\boldsymbol{x}) = x_0 x_1 + x_0 x_2 + x_1 x_2$. The functions f and f' are in the same $\{I, H, N\}^n$ orbit, and since they are quadratic functions, the same LC orbit. This can be verified by applying the LC operation to the vertex corresponding to the variable x_0 in the graph representation of either function.

5   Peak-to-Average Power Ratio w.r.t. $\{I, H, N\}^n$

Definition 4. *The peak-to-average power ratio of a vector, $\boldsymbol{s}$, with respect to the $\{I, H, N\}^n$ transform [19] is*

$$\mathrm{PAR}_{IHN}(\boldsymbol{s}) = 2^n \max_{\substack{\forall U \in \{I, H, N\}^n \\ \forall k \in \mathbb{Z}_{2^n}}} |S_k|^2, \quad \text{where } \boldsymbol{S} = U\boldsymbol{s}.$$

If a vector, $\boldsymbol{s}$, has a completely flat $\{I, H, N\}^n$ spectrum (which is impossible) then $\mathrm{PAR}_{IHN}(\boldsymbol{s}) = 1$. If $\boldsymbol{s} = 2^{-\frac{n}{2}}(1, 1, \ldots, 1, 1)$ then $\mathrm{PAR}_{IHN}(\boldsymbol{s}) = 2^n$. A typical vector, $\boldsymbol{s}$, will have a $\mathrm{PAR}_{IHN}(\boldsymbol{s})$ somewhere between these extremes.

For quadratic functions, PAR_{IHN} will always be a power of 2. The PAR of s can be alternatively expressed in terms of the *generalised nonlinearity* [19],

$$\gamma(f) = 2^{\frac{n}{2}-1}\left(2^{\frac{n}{2}} - \sqrt{PAR_{IHN}(s)}\right),$$

but in this paper we use the PAR measure. Let $s = 2^{-\frac{n}{2}}(-1)^{f(x)}$, as before. When we talk about the PAR_{IHN} of f or its associated graph G, we mean $PAR_{IHN}(s)$. It is desirable to find Boolean functions with high generalised nonlinearity and therefore low PAR_{IHN} [20]. PAR_{IHN} is an invariant of the $\{I, H, N\}^n$ orbit and, in particular, the LC orbit. We observe that Boolean functions from LC orbits associated with self-dual additive codes over GF(4) with high distance typically have low PAR_{IHN}. This is not surprising as the distance of a quantum code has been shown to be equal to the recently defined *Aperiodic Propagation Criteria distance* (APC distance) [20] of the associated quadratic Boolean function, and APC is derived from the aperiodic autocorrelation which is, in turn, the autocorrelation "dual" of the spectra with respect to $\{I, H, N\}^n$. Table 3 shows PAR_{IHN} values for every LC orbit representative for $n \leq 12$.

Table 3. PAR_{IHN} of LC Orbit Representatives

n	2	4	8	16	32	64	128	256	512	1024	2048
					Number of orbits with specified PAR_{IHN}						
1	1										
2	1										
3		1									
4		1	1								
5		1	2	1							
6		1	5	4	1						
7			6	14	5	1					
8			9	52	32	7	1				
9			2	156	212	60	9	1			
10			1	624	1,753	639	103	11	1		
11				3,184	25,018	10,500	1,578	163	13	1	
12				12,323	834,256	380,722	43,013	3,488	249	16	1

Definition 5. *Let $\alpha(G)$ be the independence number of a graph G, i.e., the size of the maximum independent set in G. Let $[G]$ be the set of all graphs in the LC orbit of G. We then define $\lambda(G) = \max_{H \in [G]} \alpha(H)$, i.e., the size of the maximum independent set over all graphs in the LC orbit of G.*

Consider as an example the Hexacode which has two non-isomorphic graphs in its orbit (see fig. 1). It is evident that the size of the largest independent set of each graph is 2, so $\lambda = 2$. The values of λ for all LC orbits for $n \leq 12$ clearly show that λ and d, the distance of the associated self-dual additive code over GF(4), are related. LC orbits associated with codes with high distance typically have small values for λ. Table 4 summarises this observation by giving the ranges of λ observed for all LC orbits associated with codes of given lengths and distances. For instance, $[[12, 0, 2]]$ codes exist with any value of λ between 4 and 11, while $[[12, 0, 5]]$ and $[[12, 0, 6]]$ codes only exist with $\lambda = 4$.

Table 4. Range of Maximum Independent Set Size

d	Range of λ for specified n										
	2	3	4	5	6	7	8	9	10	11	12
2	1	2	2,3	3,4	3–5	3–6	3–7	4–8	4–9	4–10	4–11
3				2	3	3,4	3,4	3–5	4–6	4–7	4–8
4					2		3,4	3,4	3–5	4–6	4–7
5										4	4
6											4

Definition 6. *Let Λ_n be the minimum value of λ over all LC orbits with n vertices.*

From table 4 we observe that $\Lambda_n = 2$ for n from 3 to 6, $\Lambda_n = 3$ for n from 7 to 10, and $\Lambda_n = 4$ when n is 11 or 12.

Theorem 2. $\Lambda_{n+1} \geq \Lambda_n$, *i.e., Λ_n is monotonically nondecreasing when the number of vertices is increasing.*

Proof. Consider a graph $G = (V, E)$ with $n + 1$ vertices. Select a vertex v and let G' be the induced subgraph on the n vertices $V \backslash \{v\}$. We generate the LC-orbit of G'. The LC operations may add or remove edges between G' and v, but the presence of v does not affect the LC orbit of G'. The size of the largest independent set in the LC orbit of G' is at least Λ_n. This is also an independent set in the LC orbit of G, so $\Lambda_{n+1} \geq \Lambda_n$. □

A very loose lower bound on Λ_n can also be given. Consider a graph containing a clique of size k. It is easy to see that an LC operation on any vertex in the clique will produce an independent set of size $k - 1$. Thus the maximum clique in an LC orbit, where the largest independent set has size λ, can not be larger than $\lambda + 1$. If r is the *Ramsey number* $R(k, k + 1)$ [21], then it is guaranteed that all simple undirected graphs with minimum r vertices will have either an independent set of size k or a clique of size $k + 1$. It follows that all LC orbits with at least r vertices must have $\lambda \geq k$. Thus $\Lambda_n \geq k$ for $n \geq r$. For instance, $R(3, 4) = 9$, so LC orbits with at least 9 vertices can not have λ smaller than 3.

For $n > 12$, we have computed the value of λ for some graphs corresponding to self-dual additive codes over GF(4) with high distance. This gives us upper bounds on the value of Λ_n, as shown in table 5. The bounds on Λ_{13} and Λ_{14} are tight, since $\Lambda_{12} = 4$ and $\Lambda_{n+1} \geq \Lambda_n$.

For $n = 10$, there is a unique LC orbit that satisfies, optimally, $\lambda = 3$, $PAR_{IHN} = 8$ and $d = 4$. One of the graphs in this orbit is the *graph complement* of the "double 5-cycle" graph, shown in fig. 3.

Table 5. Upper Bounds on Λ_n

n	13	14	15	16	17	18	19	20	21
$\Lambda_n \leq$	4	4	5	5	5	6	6	6	9

Fig. 3. The "Double 5-Cycle" Graph

Theorem 3 (Parker and Rijmen [12]). *Given a graph $G = (V, E)$ with a maximum independent set $A \subset V$, $|A| = \alpha(G)$. Let $\boldsymbol{s} = (-1)^{f(\boldsymbol{x})}$, where $f(\boldsymbol{x})$ is the boolean function representation of G. Let $U = \bigotimes_{i \in A} H_i \bigotimes_{i \notin A} I_i$, i.e., the transform applying H to variables corresponding to vertices $v \in A$ and I to all other variables. Then $\max_{\forall k \in \mathbb{Z}_{2^n}} |S_k|^2 = 2^{\alpha(G)}$, where $\boldsymbol{S} = U\boldsymbol{s}$.*

Arratia et al. [22] introduced the *interlace polynomial* $q(G, z)$ of a graph G. Aigner and van der Holst [23] later introduced the interlace polynomial $Q(G, z)$. Riera and Parker [24] showed that $q(G, z)$ is related to the $\{I, H\}^n$ spectra of the quadratic boolean function corresponding to G, and that $Q(G, z)$ is related to the $\{I, H, N\}^n$ spectra.

Theorem 4 (Riera and Parker [24]). *Let f be a quadratic boolean function and G its associated graph. Then PAR_{IHN} of f is equal to $2^{\deg Q(G,z)}$, where $\deg Q(G, z)$ is the degree of the interlace polynomial $Q(G, z)$.*

Theorem 5. *If the maximum independent set over all graphs in the LC orbit $[G]$ has size $\lambda(G)$, then all functions corresponding to graphs in the orbit will have $PAR_{IHN} = 2^{\lambda(G)}$.*

Proof. Let us for brevity define $P(G) = PAR_{IHN}(\boldsymbol{s})$, where $\boldsymbol{s} = 2^{-\frac{n}{2}}(-1)^{f(\boldsymbol{x})}$, and $f(\boldsymbol{x})$ is the boolean function representation of G. From theorem 3 it follows that $P(G) \geq 2^{\lambda(G)}$. Choose $H = (V, E) \in [G]$ with $\alpha(H) = \lambda(G)$. If $|V| = 1$ or 2, the theorem is true. We will prove the theorem for $n \geq 2$ by induction on $|V|$. We will show that $P(H) \leq 2^{\alpha(H)}$, which is equivalent to saying that $P(G) \leq 2^{\lambda(G)}$. It follows from theorem 4 and the definition of $Q(H, z)$ by Aigner and van der Holst [23] that $P(H) = \max\{P(H \backslash u), P(H^u \backslash u), P(((H^u)^v)^u \backslash u)\}$. (We recall that H^u denotes the LC operation on vertex u of H.) Assume, by induction hypothesis, that $P(H \backslash u) = 2^{\lambda(H \backslash u)}$. Therefore, $P(H \backslash u) = 2^{\alpha(K \backslash u)}$ for some $K \backslash u \in [H \backslash u]$. Note that $K \backslash u \in [H \backslash u]$ implies $K \in [H]$. It must then be true that $\alpha(K \backslash u) \leq \alpha(K) \leq \alpha(H)$, and it follows that $P(H \backslash u) \leq 2^{\alpha(H)}$. Similar arguments hold for $P(H^u \backslash u)$ and $P(((H^u)^v)^u \backslash u)$, so $P(H) \leq 2^{\alpha(H)}$. $\square$

As an example, the Hexacode has $\lambda = 2$ and therefore $PAR_{IHN} = 2^2 = 4$.

Corollary 1. *Any quadratic Boolean function on n or more variables must have $PAR_{IHN} \geq 2^{\Lambda_n}$.*

Definition 7. *PAR_{IH} is the peak-to-average power ratio with respect to the transform set $\{I, H\}^n$, otherwise defined in the same way as PAR_{IHN}.*

Definition 8. *PAR_l is the peak-to-average power ratio with respect to the infinite transform set $\{U\}^n$, consisting of matrices of the form*

$$U = \begin{pmatrix} \cos\theta & \sin\theta e^{i\phi} \\ \sin\theta & -\cos\theta e^{i\phi} \end{pmatrix},$$

where $i^2 = -1$, and θ and ϕ can take any real values. $\{U\}$ comprises all 2×2 unitary transforms to within a post-multiplication by a matrix from $\mathcal{D}$, the set of 2×2 diagonal and anti-diagonal unitary matrices.

Theorem 6 (Parker and Rijmen [12]). *If s corresponds to a bipartite graph, then $PAR_l(s) = PAR_{IH}(s)$.*

It is obvious that $\{I, H\}^n \subset \{I, H, N\}^n \subset \{U\}^n$, and therefore that $PAR_{IH} \leq PAR_{IHN} \leq PAR_l$. We then get the following corollary of theorems 5 and 6.

Corollary 2. *If an LC orbit, $[G]$, contains a bipartite graph, then all functions corresponding to graphs in the orbit will have $PAR_l = 2^{\lambda(G)}$.*

Thus, all LC orbits with a bipartite member have $PAR_{IHN} = PAR_l$. Note that these orbits will always have $PAR_l \geq 2^{\lceil \frac{n}{2} \rceil}$ [12] and that the fraction of LC orbits which have a bipartite member appears to decrease exponentially as the number of vertices increases. In the general case, PAR_{IHN} is only a lower bound on PAR_l. For example, the Hexacode has $PAR_{IHN} = 4$, but a tighter lower bound on PAR_l is 4.486 [12]. (This bound has later been improved to 5.103 [25].)

6   Construction for Low PAR_{IHN}

So far we have only considered *quadratic* Boolean functions which correspond to graphs and self-dual additive codes over GF(4). For cryptographic purposes, we are interested in Boolean functions of degree higher than 2. Such functions can be represented by *hypergraphs*, but they do not correspond to quantum stabilizer codes or self-dual additive codes over GF(4). A non-quadratic Boolean function, $f(x)$, can, however, be interpreted as a quantum state described by the probability distribution vector $s = 2^{-\frac{n}{2}}(-1)^{f(x)}$. A single quantum state corresponds to a quantum code of dimension zero whose distance is the APC distance [20]. The APC distance is the weight of the minimum weight quantum error operator that gives an errored state not orthogonal to the original state and therefore not guaranteed to be detectable.

We are interested in finding Boolean functions of algebraic degree greater than 2 with low PAR_{IHN}, but exhaustive searching becomes infeasible with more than a few variables. We therefore propose a construction technique for nonquadratic Boolean functions with low PAR_{IHN} using the best quadratic functions as building blocks. Before we describe our construction we must first state what we mean by "low PAR_{IHN}". For $n = 6$ to $n = 10$ we computed PAR_{IHN} for samples from the space $\mathbb{Z}_2^{2^n}$, to determine the range of PAR_{IHN} we can expect just by guessing. Table 6 summarises these results. If we can construct Boolean functions with PAR_{IHN} lower than the sampled minimum, we can consider our construction to be somewhat successful.

Table 6. Sampled Range of PAR_{IHN} for $n = 6$ to 10

n	6	7	8	9	10
# samples	50000	20000	5000	2000	1000
Range of PAR_{IHN}	6.5–25.0	9.0–28.125	12.25–28.125	14.0625–30.25	18.0–34.03

Parker and Tellambura [26, 27] proposed a generalisation of the Maiorana-McFarland construction for Boolean functions that satisfies a tight upper bound on PAR with respect to the $\{H, N\}^n$ transform (and other transform sets), this being a form of Golay Complementary Set construction and a generalisation of the construction of Rudin and Shapiro and of Davis and Jedwab [28]. Let $p(x)$ be a Boolean function on $n = \sum_{j=0}^{L-1} t_j$ variables, where $T = \{t_0, t_1, \ldots, t_{L-1}\}$ is a set of positive integers and $x \in \mathbb{Z}_2^n$. Let $y_j \in \mathbb{Z}_2^{t_j}$, $0 \le j < L$, such that $x = y_0 \times y_1 \times \cdots \times y_{L-1}$. Construct $p(x)$ as follows.

$$p(x) = \sum_{j=0}^{L-2} \theta_j(y_j)\gamma_j(y_{j+1}) + \sum_{j=0}^{L-1} g_j(y_j), \qquad (1)$$

where θ_j is a permutation: $\mathbb{Z}_2^{t_j} \to \mathbb{Z}_2^{t_{j+1}}$, γ_j is a permutation: $\mathbb{Z}_2^{t_{j+1}} \to \mathbb{Z}_2^{t_j}$, and g_j is any Boolean function on t_j variables. It has been shown [27] that the function $p(x)$ will have $PAR_{HN} \le 2^{t_{\max}}$, where $t_{\max}$ is the largest integer in T. It is helpful to visualise this construction graphically, as in fig. 4. In this example, the size of the largest partition is 3, so $PAR_{HN} \le 8$, regardless of what choices we make for θ_j, γ_j, and g_j.

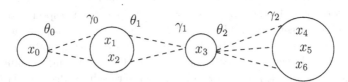

Fig. 4. Example of Construction with $PAR_{HN} \le 8$

Observe that if we set $L = 2$, $t = t_0 = t_1$, let θ_0 be the identity permutation, and $g_0 = 0$, construction (1) reduces to the Maiorana-McFarland construction over $2t$ variables. Construction (1) can also be viewed as a generalisation of the "path graph", $f(\boldsymbol{x}) = x_0 x_1 + x_1 x_2 + \cdots + x_{n-2} x_{n-1}$, which has optimal PAR with respect to $\{H, N\}^n$. Unfortunately, the "path graph" is not a particularly good construction for low PAR_{IHN}. But as we have seen, graphs corresponding to self-dual additive codes over $GF(4)$ with high distance do give us Boolean functions with low PAR_{IHN}. We therefore propose the following generalised construction.

$$p(\boldsymbol{x}) = \sum_{i=0}^{L-1} \sum_{j=i+1}^{L-1} \Gamma_{i,j}(\boldsymbol{y_i}) \Gamma_{j,i}(\boldsymbol{y_j}) + \sum_{j=0}^{L-1} g_j(\boldsymbol{y_j}), \tag{2}$$

where $\Gamma_{i,j}$ is either a permutation: $\mathbb{Z}_2^{t_i} \rightarrow \mathbb{Z}_2^{t_j}$, or $\Gamma_{i,j} = 0$, and g_j is any Boolean function on t_j variables. It is evident that Γ can be thought of as a "generalised adjacency matrix", where the entries, $\Gamma_{i,j}$, are no longer 0 or 1 but, instead, 0 or permutations from $\mathbb{Z}_2^{t_i}$ to $\mathbb{Z}_2^{t_j}$. Construction (1) then becomes a special case where $\Gamma_{i,j} = 0$ except for when $j = i + 1$ (i.e., the "generalised adjacency matrix" of the "path graph"). In order to minimise PAR_{IHN} we choose the form of the matrix Γ according to the adjacency matrix of a self-dual additive code over $GF(4)$ with high distance. We also choose the "offset" functions, g_j, to be Boolean functions corresponding to self-dual additive codes over $GF(4)$ with high distance. Finally for the non-zero $\Gamma_{i,j}$ entries, we choose selected permutations, preferably nonlinear to increase the overall degree. Here are some initial results which demonstrate that, using (2), we can construct Boolean functions of algebraic degree greater than 2 with low PAR_{IHN}. (We use an abbreviated ANF notation for some many-term Boolean functions, e.g. $012, 12, 0$ is short for $x_0 x_1 x_2 + x_1 x_2 + x_0$.)

Example 3 (n = 8). Use the Hexacode graph $f = 01, 02, 03, 04, 05, 12, 23, 34, 45, 51$ as a template. Let $t_0 = 3$, $t_1 = t_2 = t_3 = t_4 = t_5 = 1$. (See fig. 5.) We use the following matrix Γ.

$$\Gamma = \begin{pmatrix} 0 & 02,1 & 02,1 & 02,1 & 02,1 & 02,1 \\ 3 & 0 & 3 & 0 & 0 & 3 \\ 4 & 4 & 0 & 4 & 0 & 0 \\ 5 & 0 & 5 & 0 & 5 & 0 \\ 6 & 0 & 0 & 6 & 0 & 6 \\ 7 & 7 & 0 & 0 & 7 & 0 \end{pmatrix}$$

Let $g_0(\boldsymbol{y_0}) = 01, 02, 12$ and all other g_j any arbitrary affine functions. Then, using (2) to construct $p(\boldsymbol{x})$ we get $p(\boldsymbol{x}) = 023, 024, 025, 026, 027, 01, 02, 12, 13, 14, 15, 16, 17, 34, 37, 45, 56, 67$. Then $p(\boldsymbol{x})$ has $PAR_{IHN} = 9.0$.

Example 4 (n = 8). Use the Hexacode graph $f = 01, 02, 03, 04, 05, 12, 23, 34, 45, 51$ as a template. Let $t_0 = 3$, $t_1 = t_2 = t_3 = t_4 = t_5 = 1$. (See fig. 5.) We use the following matrix Γ.

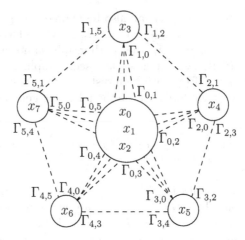

Fig. 5. Example of Construction with low PAR_{IHN}

$$\Gamma = \begin{pmatrix} 0 & 02,1 & 12,0,1,2 & 01,02,12,1,2 & 01,02,12 & 02,12,1,2 \\ 3 & 0 & 3 & 0 & 0 & 3 \\ 4 & 4 & 0 & 4 & 0 & 0 \\ 5 & 0 & 5 & 0 & 5 & 0 \\ 6 & 0 & 0 & 6 & 0 & 6 \\ 7 & 7 & 0 & 0 & 7 & 0 \end{pmatrix}$$

Let $g_0(\boldsymbol{y_0}) = 01, 12$ and all other g_j any arbitrary affine functions. Then, using (2) to construct $p(\boldsymbol{x})$ we get $p(\boldsymbol{x}) = 015, 016, 023, 025, 026, 027, 124, 125, 126, 127, 01, 04, 12, 13, 14, 15, 17, 24, 25, 27, 34, 37, 45, 56, 67$. Then $p(\boldsymbol{x})$ has $\mathrm{PAR}_{IHN} = 9.0$.

Example 5 (n = 9). Use the triangle graph $f = 01, 02, 12$ as a template. Let $t_0 = t_1 = t_2 = 3$. (See fig. 6.) Assign the permutations

$$\Gamma_{0,1} = \Gamma_{0,2} = (12, 0, 1, 2)(01, 2)(02, 1, 2),$$
$$\Gamma_{1,0} = (34, 5)(35, 4, 5)(45, 3, 4, 5),$$
$$\Gamma_{1,2} = (45, 3, 4, 5)(34, 5)(35, 4, 5),$$
$$\Gamma_{2,0} = (68, 7, 8)(78, 6, 7, 8)(67, 8),$$
$$\Gamma_{2,1} = (78, 6, 7, 8)(67, 8)(68, 7, 8).$$

Let $g_0(\boldsymbol{y_0}) = 01, 02, 12$, $g_1(\boldsymbol{y_1}) = 34, 35, 45$, and $g_2(\boldsymbol{y_2}) = 67, 68, 78$. Then, using (2) to construct $p(\boldsymbol{x})$ we get, $p(\boldsymbol{x}) = 0135, 0178, 0245, 0267, 1234, 1268, 3467, 3568, 4578, 014, 015, 016, 017, 018, 023, 024, 025, 028, 034, 068, 125, 127, 128, 134, 145, 167, 168, 234, 235, 245, 267, 268, 278, 348, 357, 358, 378, 456, 457, 458, 468, 478, 567, 568, 578, 05, 07, 08, 13, 14, 17, 23, 25, 26, 28, 36, 37, 38, 46, 56, 58, 01, 02, 12, 34, 35, 45, 67, 68, 78$. Then $p(\boldsymbol{x})$ has $\mathrm{PAR}_{IHN} = 10.25$.

The examples of our construction satisfy a low PAR_{IHN}. Further work should ascertain the proper choice of permutations. Finally, there is an even more

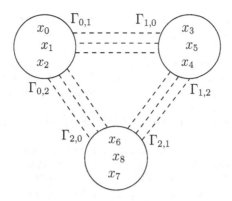

Fig. 6. Example of Construction with low PAR$_{IHN}$

obvious variation of construction (2), suggested by the graphs of fig. 2, where the functions g_j are chosen either to be quadratic cliques or to be further "nested" versions of construction (2). We will report on this variation in a future paper.

References

1. Calderbank, A.R., Rains, E.M., Shor, P.M., Sloane, N.J.A.: Quantum error correction via codes over GF(4). IEEE Trans. Inform. Theory **44** (1998) pp. 1369–1387 http://arxiv.org/quant-ph/9608006.
2. Rains, E.M., Sloane, N.J.A.: Self-dual codes. In Pless, V.S., Huffman, W.C., eds.: Handbook of Coding Theory. Elsevier (1998) 177–294 http://arxiv.org/math/0208001.
3. Höhn, G.: Self-dual codes over the Kleinian four group. Mathematische Annalen **327** (2003) pp. 227–255 http://arxiv.org/math/0005266.
4. Hein, M., Eisert, J., Briegel, H.J.: Multi-party entanglement in graph states. Phys. Rev. A **69** (2004) http://arxiv.org/quant-ph/0307130.
5. Glynn, D.G., Gulliver, T.A., Maks, J.G., Gupta, M.K.: The geometry of additive quantum codes. Submitted to Springer-Verlag (2004)
6. Sloane, N.J.A.: The On-Line Encyclopedia of Integer Sequences. Web page (2004) http://www.research.att.com/~njas/sequences/.
7. Danielsen, L.E.: Database of self-dual quantum codes. Web page (2004) http://www.ii.uib.no/~larsed/vncorbits/.
8. Schlingemann, D., Werner, R.F.: Quantum error-correcting codes associated with graphs. Phys. Rev. A **65** (2002) http://arxiv.org/quant-ph/0012111.
9. Grassl, M., Klappenecker, A., Rotteler, M.: Graphs, quadratic forms, and quantum codes. In: Proc. IEEE Int. Symp. Inform. Theory. (2002) p. 45
10. Glynn, D.G.: On self-dual quantum codes and graphs. Submitted to Elect. J. Combinatorics. http://homepage.mac.com/dglynn/.cv/dglynn/Public/SD-G3.pdf-link.pdf (2002)
11. Van den Nest, M., Dehaene, J., De Moor, B.: Graphical description of the action of local Clifford transformations on graph states. Phys. Rev. A **69** (2004) http://arxiv.org/quant-ph/0308151.

12. Parker, M.G., Rijmen, V.: The quantum entanglement of binary and bipolar sequences. In Helleseth, T., Kumar, P.V., Yang, K., eds.: Sequences and Their Applications, SETA'01. Discrete Mathematics and Theoretical Computer Science Series, Springer-Verlag (2001) Long version: http://arxiv.org/quant-ph/0107106.

13. Bouchet, A.: Isotropic systems. European J. Combin. **8** (1987) pp. 231–244

14. Bouchet, A.: Recognizing locally equivalent graphs. Discrete Math. **114** (1993) pp. 75–86

15. McKay, B.D.: nauty User's Guide. (2004) http://cs.anu.edu.au/~bdm/nauty/nug.pdf.

16. Gulliver, T.A., Kim, J.-L.: Circulant based extremal additive self-dual codes over GF(4). IEEE Trans. Inform. Theory **50** (2004) pp. 359–366

17. Grassl, M.: Bounds on d_{min} for additive $[[n, k, d]]$ QECC. Web page (2003) http://iaks-www.ira.uka.de/home/grassl/QECC/TableIII.html.

18. Riera, C., Petrides, G., Parker, M.G.: Generalized bent criteria for Boolean functions. Technical Report 285, Dept. of Informatics, University of Bergen, Norway (2004) http://www.ii.uib.no/publikasjoner/texrap/pdf/2004-285.pdf.

19. Parker, M.G.: Generalised S-box nonlinearity. NESSIE Public Document, NES/DOC/UIB/WP5/020/A. https://www.cosic.esat.kuleuven.ac.be/nessie/reports/phase2/SBoxLin.pdf (2003)

20. Danielsen, L.E., Gulliver, T.A., Parker, M.G.: Aperiodic propagation criteria for Boolean functions. Submitted to Inform. Comput. http://www.ii.uib.no/~matthew/GenDiff4.pdf (2004)

21. Radziszowski, S.P.: Small Ramsey numbers. Elect. J. Combinatorics (2002) pp. 1–42 Dynamical Survey DS1, http://www.combinatorics.org/Surveys/ds1.pdf.

22. Arratia, R., Bollobás, B., Sorkin, G.B.: The interlace polynomial of a graph. J. Combin. Theory Ser. B **92** (2004) pp. 199–233 http://arxiv.org/math/0209045.

23. Aigner, M., van der Holst, H.: Interlace polynomials. Linear Algebra and its Applications **377** (2004) pp. 11–30

24. Riera, C., Parker, M.G.: Spectral interpretations of the interlace polynomial. Submitted to WCC2005. http://www.ii.uib.no/~matthew/WCC4.pdf (2004)

25. Parker, M.G., Gulliver, T.A.: On graph symmetries and equivalence of the six variable double-clique and wheel. Unpublished (2003)

26. Parker, M.G., Tellambura, C.: A construction for binary sequence sets with low peak-to-average power ratio. In: Proc. IEEE Int. Symp. Inform. Theory. (2002) p. 239 http://www.ii.uib.no/~matthew/634isit02.pdf.

27. Parker, M.G., Tellambura, C.: A construction for binary sequence sets with low peak-to-average power ratio. Technical Report 242, Dept. of Informatics, University of Bergen, Norway (2003) http://www.ii.uib.no/publikasjoner/texrap/pdf/2003-242.pdf.

28. Davis, J.A., Jedwab, J.: Peak-to-mean power control in OFDM, Golay complementary sequences and Reed-Muller codes. IEEE Trans. Inform. Theory **45** (1999) pp. 2397–2417

New Constructions and Bounds for 2-D Optical Orthogonal Codes

Reza Omrani, Petros Elia, and P. Vijay Kumar[*]

Communication Sciences Institute, Department of Electrical Engineering - Systems,
University of Southern California,3740 McClintock Ave.,
Los Angeles, CA 90089-2565, USA
{omrani, elia, vijayk}@usc.edu

Abstract. Some efficient constructions and bounds for 2-D optical orthogonal codes in which the spreading is carried out over both wavelength and time are provided. Such codes are of current practical interest as they enable optical communication at lower chip rate.

The bounds provided include 2-D versions of the Johnson bound as well as a novel bound that makes use of the properties of a maximum-distance separable codes.

1 Introduction

An (n, ω, κ) OOC $\mathcal{C}$, $1 \leq \kappa \leq \omega \leq n$, is a family of $\{0,1\}$-sequences of length n and Hamming weight ω satisfying:

$$\sum_{k=0}^{n-1} x(k)y(k \oplus_n \tau) \leq \kappa \tag{1}$$

for $\{x, y\}$ in $\mathcal{C}$, either $x \neq y$ or $\tau \neq 0$, where $\oplus_n$ denotes addition modulo n and κ is the maximum collision parameter(MCP).

Let $\Phi(n, \omega, \kappa)$ denote the largest possible cardinality of an (n, ω, κ) OOC code. Then by the Johnson bound [1][2]

$$\Phi(n, \omega, \kappa) \leq \left\lfloor \frac{1}{\omega} \left\lfloor \frac{n-1}{\omega-1} \left\lfloor \frac{n-2}{\omega-2} \cdots \left\lfloor \frac{n-\kappa}{\omega-\kappa} \right\rfloor \right\rfloor \right\rfloor \right\rfloor := J(n, w, \kappa). \tag{2}$$

An OOC $\mathcal{C}$ of size $|\mathcal{C}|$ is said to be optimal when $|\mathcal{C}| = \Phi(n, \omega, \kappa)$ and asymptotically optimal if

$$\lim_{n \to \infty} \frac{|\mathcal{C}|}{\Phi(n, \omega, \kappa)} = 1.$$

Some optimal or asymptotically optimal constructions can be found in [2][3][4][5].

---

[*] Work partially supported by DARPA OCDMA Program Grant No. N66001-02-1-8939.

T. Helleseth et al. (Eds.): SETA 2004, LNCS 3486, pp. 389–395, 2005.

With 1-D OOCs, the code size grows relatively slowly with increase in sequence length n. As a result, for the kinds of code size desired in practice, the code length is necessarily large, causing an excessively high chip rate. The advent of WDM technology has made it possible to spread in both wavelength and time [6]. The corresponding codes, termed 2-D Optical Orthogonal Codes(2-D OOC), tend to require smaller code lengths and hence lower chip rates.

2   2-D OOCs

A 2-D $(\Lambda \times T, \omega, \kappa)$ OOC $\mathcal{C}$ is a family of $\{0, 1\}$ $\Lambda \times T$ arrays of constant weight ω. Each pair $\{A, B\}$ in $\mathcal{C}$ is required to satisfy:

$$\sum_{\lambda=1}^{\Lambda} \sum_{t=0}^{T-1} A(\lambda, t) B(\lambda, (t \oplus_T \tau)) \leq \kappa \tag{3}$$

whenever either $A \neq B$ or $\tau \neq 0$.

To simplify implementation, additional restrictions on the codewords may be placed such as:

– one-pulse per wavelength(OPPW): each row of every $(\Lambda \times T)$ code array in $\mathcal{C}$ has Hamming weight $= 1$.
– at most one-pulse per wavelength(AM-OPPW): each row of each $(\Lambda \times T)$ code array in $\mathcal{C}$ has Hamming weight ≤ 1.

Figure 1 shows an example of a 2-D OOC with at most one-pulse per wavelength (AM-OPPW) restriction.

The corresponding Johnson bound on code size in the 2-D case is given by [7]:

$$\Phi(\Lambda \times T, \omega, \kappa) \leq \left\lfloor \frac{1}{T} \left\lfloor \frac{\Lambda T}{\omega} \left\lfloor \frac{\Lambda T - 1}{\omega - 1} \cdots \left\lfloor \frac{\Lambda T - \kappa}{\omega - \kappa} \right\rfloor \right\rfloor \right\rfloor \right\rfloor$$

$$= \left\lfloor \frac{\Lambda}{\omega} \left\lfloor \frac{\Lambda T - 1}{\omega - 1} \cdots \left\lfloor \frac{\Lambda T - \kappa}{\omega - \kappa} \right\rfloor \right\rfloor \right\rfloor := J(\Lambda \times T, \omega, \kappa) \tag{4}$$

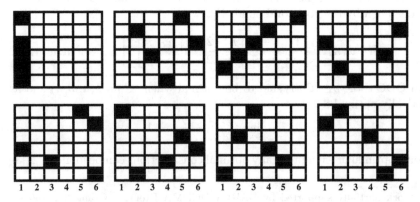

Fig. 1. A $(6 \times 6, 5, 1)$ AM-OPPW 2-D OOC, with 8 codewords

Theorem 1. *It can be shown[7]:*

$$\Lambda J(\Lambda T, \omega, \kappa) \leq J(\Lambda \times T, \omega, \kappa) \leq \Lambda J(\Lambda T, \omega, \kappa) + (\Lambda - 1) \ .$$

3 Generalization of the Johnson Bound to Nonbinary Constant Weight Codes

To our knowledge, the following result has not previously appeared:

Theorem 2. *If $A(\Lambda, \omega, \kappa)$ is the maximum possible size of a constant-weight code of length Λ, Hamming weight ω, maximum number of nonzero agreements between any two codewords less than or equal to κ over an alphabet of size $(T+1)$ which contains 0 then $A(\Lambda, \omega, \kappa)$ is bounded by the following upper bound:*

$$A(\Lambda, \omega, \kappa) \leq \left\lfloor \frac{T\Lambda}{\omega} \left\lfloor \frac{T(\Lambda - 1)}{\omega - 1} \cdots \left\lfloor \frac{T(\Lambda - \kappa)}{\omega - \kappa} \right\rfloor \right\rfloor \right\rfloor$$

Since an optical code with at-most one-pulse-per-wavelength restriction can be regarded as a constant-weight code over an alphabet of size $(T + 1)$, The above bound can be adapted to the following bound on the size of AM-OPPW OOCs.

Corollary 1.

$$\Phi_{AM-OPPW}(\Lambda \times T, \omega, \kappa) \leq \left\lfloor \frac{\Lambda}{\omega} \left\lfloor \frac{T(\Lambda - 1)}{\omega - 1} \cdots \left\lfloor \frac{T(\lambda - \kappa)}{\omega - \kappa} \right\rfloor \right\rfloor \right\rfloor$$

Applying the new bound to a $(6 \times 6, 5, 1)$ AM-OPPW 2-D OOC shows that the AM-OPPW 2-D OOC shown in Figure 1 is an optimal AM-OPPW 2-D OOC.

Remark 1. For the exactly one pulse per wavelength, in which case, $\Lambda = \omega$, the bound in the Corollary above reduces to the Singleton bound:

$$\Phi(\Lambda \times T, \omega, \kappa) \ \leq \ T^{\kappa}.$$

Remark 2. For the binary case the bound reduces to the well-known Johnson bound[1].

4 Constructions of 2-D OOCs

4.1 Chinese-Remainder-Theorem (CRT) Construction

Let $\{s_k\}$ be a 1-D periodic sequence of period $n = n_1 n_2$ with $(n_1, \ n_2) = 1$. Let the 2-D doubly periodic binary array $S(k_1, k_2)$ of period $(n_1 \times n_2)$ be defined by

$$S(k \pmod{n_1}, \ k \pmod{n_2}) \ = \ s(k).$$

Then it can be shown that there is a 1-1 correspondence between the 2-D doubly-periodic correlation values of $\{S\}$ and the 1-D correlation values of $\{s\}$.

From this it follows that if we have an (n, ω, κ)-OOC $\mathcal{S} = \{\{s_k^{(p)}\}|1 \leq p \leq P\}$ with $n = \Lambda T$ where $(\Lambda, T) = 1$, we can generate a 2-D OOC of size ΛP by associating with each 1-D OOC $\{s_k^{(p)}\}$, the Λ 2-D OOCs corresponding to the Λ distinct vertical cyclic shifts $\{S^{(p)}(k_1 \oplus_\Lambda \tau_1, k_2)\}$, $0 \leq \tau_1 < \Lambda$ of $\{S^{(p)}(k_1, k_2)\}$. Thus, the 2-D code family has Λ times the size of the 1-D family[7].

Theorem 3. *Let $n = \Lambda T$, $(\Lambda, T) = 1$. A 1-D OOC of Johnson-bound-achieving size $J(n, w, \kappa)$ gives rise to a 2-D OOC whose size differs at most by $(\Lambda - 1)$ from the corresponding 2-D Johnson bound $J(\Lambda \times T, w, \kappa)$. If in addition, the 1-D OOC is of optimal size $\frac{1}{\Lambda T}\left\lfloor \frac{\Lambda T}{w} \left\lfloor \frac{\Lambda T - 1}{w - 1} \cdots \left\lfloor \frac{\Lambda T - \kappa}{w - \kappa} \right\rfloor \right\rfloor \right\rfloor$ then the 2-D OOC resulting from the above method is also optimal with respect to the Johnson bound[7].*

4.2   Function-Plot Construction

A 2-D OOC can be regarded as the graph of a function $\lambda = f(t)$, $0 \leq t \leq T - 1$, $0 \leq \lambda \leq \Lambda - 1$ mapping time into wavelength or vice-versa, $t = f(\lambda)$. All of the constructions below employ polynomial functions whose degree is bounded above by the desired value of MCP κ.

Polynomial Constructions When Either Λ or T Is Prime

[P1] *Mapping Wavelength to Time, $T = p$ is prime:*
Let $1 \leq \Lambda \leq p$, and $\lambda \in \{1, 2, \cdots, \Lambda\}$. Here we consider polynomials $f(\lambda)$ mapping wavelength into time. For any $0 \leq \delta \leq p - 1$, we declare two polynomials $f(\lambda), f(\lambda) + \delta$ to be equivalent. Then the different code matrices correspond to choosing precisely one polynomial from each equivalence class. For each polynomial $f(\cdot)$ the $(\Lambda \times T)$ code array C is given by $C(\lambda, t) = 1$ iff $f(\lambda) = t$, where $0 \leq \lambda \leq \Lambda - 1$. This results in a $(\Lambda \times p, \Lambda, \kappa)$ 2-D OOC with size p^κ, and $\kappa \leq \Lambda \leq p$ [7]. When $\kappa = 1$, this construction is equivalent to the carrier-hopping prime codes of length p by Kwong and Yang[8].

[P2] *Mapping Time to Wavelength, $\Lambda = p$ is prime:*
Let $T \mid (p-1)$. Here we consider polynomials mapping time into wavelength. Let $\alpha \in Z_p$ have multiplicative order T. Let us associate to time slot t, the element α^t. We define two polynomials $f(x), g(x)$ in $\mathcal{F}_\kappa$ to be equivalent if $f(\alpha^i x) = g(x)$ for some $i \in \mathbb{Z}_T$. We construct a 2-D OOC by discarding all polynomials $f(x)$, which satisfy $f(\alpha^i x) = f(x)$ for $i \neq 0$, and choosing one function $f(\cdot)$ from each of the remaining equivalence classes and associating to it, the $(\Lambda \times T)$ code array C by letting $C(\lambda, t) = 1$ iff $f(\alpha^t) = \lambda$ where $t \in \mathbb{Z}_T$ and $\Lambda \in Z_p$. This results in a $(p \times T, T, \kappa)$ 2-D OOC of size $\frac{1}{T}\sum_{d|(p-1)} \left(p^{\lceil \frac{\kappa+1}{d} \rceil} - 1\right)\mu(d)$ [4][7].

Polynomial Constructions When T or $\Lambda = p^m - 1$, p Prime
Let α be a primitive element of $GF(p^m)$.

[P3] *Mapping Wavelength to Time, $T = p^m - 1$:*
Let $1 \leq \Lambda \leq p^m$ and associate to each time slot t, the element α^t. We define two polynomials $f(x), g(x)$ in $\mathcal{F}_\kappa$ to be equivalent if $\alpha^i f(x) = g(x)$

for some $i \in \mathbb{Z}$. After choosing one function from each equivalence class, we proceed to construct a $(\Lambda \times (p^m - 1), w, \kappa)$ 2-D OOC with $w \geq \Lambda - \kappa$ and size $\frac{q^{\kappa+1}-1}{q-1}$ by assigning one code array to each of the chosen functions as described above. We have $w \geq \Lambda - \kappa$ because for some values within the domain $f(x) = 0$ which is not within the range. To keep the weight constant for any codeword we delete some arbitrary ones to keep the weight equal to $\Lambda - \kappa$ for all codewords. This leads to a $(\Lambda \times (p^m - 1), \Lambda - \kappa, \kappa)$ 2-D OOC with size $\frac{q^{\kappa+1}-1}{q-1}$ [7].

[P4] *Mapping Time to Wavelength,* $\Lambda = p^m - 1$:
Let us associate to each wavelength λ, the element α^λ. Let $T \mid (p^m - 1)$, and $\beta \in GF(p^m)$ have multiplicative order T. We define two polynomials $f(x), g(x)$ in $\mathcal{F}_\kappa$ to be equivalent if $f(\beta^i x) = g(x)$ for some $i \in \mathbb{Z}_T$. By discarding all polynomials $f(x)$, which satisfy $f(\beta^i x) = f(x)$ for $i \neq 0$, and choosing one function from each of the remaining equivalence classes, and assigning one code array to each of these functions as described above, we can construct a $(p^m - 1 \times T, T - \kappa, \kappa)$ 2-D OOC of size
$$\frac{1}{T} \sum_{d|(p^m-1)} \left(p^{m\lceil \frac{\kappa+1}{d} \rceil} - 1 \right) \mu(d) \ [4][7].$$

Remark 3. In the function plot constructions if one maps wavelength into time, the resulting 2-D OOC will be of maximally OPPW-type. The same is the case if $\kappa = 1$ and the mapping proceeds from time to wavelength.

Theorem 4. *All of the above constructions are asymptotically optimal with respect to the Johnson bound.*

Construction P1 above can be shown to be optimal by Corollary 1.

4.3   Construction P1 – A Reed-Solomon Code Construction

Consider a $(\Lambda \times T, \omega, \kappa)$ 2-D OOC $\mathcal{C}$ with exactly one pulse per wavelength. Using the new bound of Corollary 1, it follows that when the 2-D OOC code is required to place precisely one pulse per wavelength, the code size $\mathcal{C}$ is upper bounded by:
$$|\mathcal{C}| \leq T^\kappa.$$

With each code array $A(\lambda, t)$ in $\mathcal{C}$, we can identify a column vector $\underline{a}$ of length Λ whose symbols are drawn from the set $\mathbb{Z}_T$ of integers modulo T as follows:
$$a_\lambda = t \quad \text{iff} \quad A(\lambda, t) = 1, 1 \leq \lambda \leq \Lambda.$$

A $(\Lambda \times T, \omega, \kappa)$ 2-D OOC can now be constructed using this identification. The $(\Lambda \times T, \omega, \kappa)$ 2-D OOC is equivalent to a $\mathcal{C}_{\mathbb{Z}_T}$ code of length Λ over the alphabet $\mathbb{Z}_T$ such that (a) $d_{\min}(\mathcal{C}_{\mathbb{Z}_T}) \geq \Lambda - \kappa$ and (b) the added condition $\underline{a} \in \mathcal{C}_{\mathbb{Z}_T}$, implies $\underline{a} + b\underline{1} \in \mathcal{C}_{\mathbb{Z}_T}$ for any $b \in \mathbb{Z}_T$, where $\underline{1}$ is the all-1 vector.

Theorem 5. *Let T be prime and $\Lambda \leq T$. Let $\mathcal{C}_T$ be a Reed-Solomon (RS) code over the field Z_T having minimum distance*

$$d_{\min} = \Lambda - \kappa .$$

Then the corresponding $(\Lambda \times T, \omega, \kappa)$ is an optimal code under the one-pulse-per wavelength restriction.

This construction can be shown to be identical to construction P1 above. The new viewpoint however, will be of use in the construction below.

4.4   At Most One Pulse per Wavelength – Mixing RS and Constant Weight Codes

Consider a $(\Lambda \times T, \omega, \kappa)$ 2-D OOC $\mathcal{C}$ under the requirement that there is at most one pulse per wavelength. We propose a construction that makes use of a combination of a constant weight binary code and a maximal near-MDS code. Let $\mathcal{C}_{cw}$ be a constant weight binary code of maximum size with following parameters: length$= \Lambda$, weight$= \omega$, and maximum real inner product of any two codes $\leq \kappa$. The size of $\mathcal{C}_{cw}$ is upper bounded by $| \mathcal{C}_{cw} | \leq \prod_{l=0}^{\kappa} \lfloor \frac{(\Lambda-l)}{(\omega-l)} \rfloor$.

The idea is to construct a 2-D OOC whose code arrays are partitioned into $| \mathcal{C}_{cw} |$ subsets with each subset associated to a distinct codeword in $\mathcal{C}_{cw}$. Consider a codeword in $\mathcal{C}_{cw}$ where the 1's in this binary codeword, appear in the ω symbol locations $i_1, i_2, \cdots, i_w$. We associate with this codeword, a maximal collection of 2-D code arrays with MCP κ which are such that only the wavelengths associated to rows $i_1, \cdots, i_w$ contain a pulse. No pulse is sent along any of the other wavelengths. Now for any of $| \mathcal{C}_{cw} |$ choices of ω wavelengths we can use the construction of the previous part to generate $| \mathcal{C}_{oppw} |$ 2-D OOCs with exactly one pulse per wavelength.

It is easy to see that the union of these 2-D code array subsets forms an overall 2-D OOC code with parameters $(\Lambda \times T, \omega, \kappa)$, and size $| \mathcal{C}_{cw} \| \mathcal{C}_{oppw} |$.

For the case when T is prime and $\omega \leq T$, each subset can be constructed using Reed-Solomon codes and is therefore of maximal possible size T^κ. The overall size of the 2-D OOC for this case is given by $| \mathcal{C}_{cw} | T^\kappa \leq T^\kappa \prod_{l=0}^{\kappa} \lfloor \frac{(\Lambda-l)}{(\omega-l)} \rfloor$. This should be compared against the bound from Corollary 2. From this we see that if the underlying binary constant weight code is asymptotically optimal, the same holds for our overall constructions.

This construction generalizes a previous construction by Yang, Kwong, and Chang [9].

References

1. S. M. Johnson:"A New Upper Bound for Error-Correcting Codes". IEEE Trans. Information Theory, Vol. 8, 203-207, April 1962
2. F. Chung, J. A. Salehi, and V. K. Wei:"Optical Orthogonal Codes: Design. Analysis, and Applications". IEEE Trans. Information Theory, Vol. 35, 596-604, May 1989

3. H. Chung and P. V. Kumar:"Optical Orthogonal Codes - New Bounds and an Optimal Construction". IEEE Trans. Information Theory, Vol. 36, 866-873, July 1990

4. O. Moreno, Z. Zhang, P. V. Kumar, and V. Zinoviev:"New Constructions of Optimal Cyclically Permutable Constant Weight Codes". IEEE Trans. Information Theory, Vol. 41, 448-455, March 1995

5. O. Moreno, P. V. Kumar, H. Lu, and R. Omrani:"New Construction of Optical orthogonal Codes, Distinct Difference Sets and Synchronous Optical orthogonal Codes". Proc. Int. Symposium on Information Theory, 2003

6. G. C. Yang, and W. C. Kwong:"Performance Comparison of Multiwavelength CDMA and WDMA+CDMA for Fiber-Optic Networks". IEEE Trans. Communications, Vol. 45, 1426-1434, Nov. 1997

7. R. Omrani, and P. V. Kumar:"2-D Optical orthogonal Codes". Proc. 41st Allerton Conf. Comm., Control, Comp., 2003

8. W. C. Kwong, and G. C. Yang:"Extended Carrier-Hopping Prime Codes for Wavelength-Time Optical Code-Division Multiple Access". IEEE Trans. Communications, Vol. 52, 1084-1091, July 2004

9. G. C. Yang, W. C. Kwong, and C. Y. Chang:"Multiple-Wavelength Optical Orthogonal Codes Under Prime-Sequence Permutations". Proc. Int. Symposium on Information Theory, 2004

Topics on Optical Orthogonal Codes

Reza Omrani[1], Oscar Moreno[2,*], and P. Vijay Kumar[1,**]

[1] Communication Sciences Institute, Department of Electrical Engineering - Systems,
University of Southern California,3740 McClintock Ave.,
Los Angeles, CA 90089-2565, USA
{omrani, vijayk}@usc.edu
[2] Gauss Research Lab, University of Puerto Rico,
PO Box 23334, San Juan, PR 00931-3334, Puerto Rico
moreno@uprr.pr

Abstract. Recently, there has been an upsurge of interest in using Code-Division Multiple-Access communication over optical fiber channels (OCDMA). In this paper we provide a new Johnson-bound-optimal construction of OOC with parameter $\lambda = 1$. We use the idea of the same construction to generate OOCs with $\lambda > 1$. A new bound for optical orthogonal codes based on a known bound for constant weight codes is introduced. This bound is used to prove the optimality of our constructions. We also present a recursive technique for generating OOCs with $\lambda = 1$ that makes use of a recursive construction for cyclic block designs by Colbourn and Colbourn [1]. This technique has yielded several new optimal constructions for Optical Orthogonal Codes.

1 Introduction

Recently there has been an upsurge of interest in applying Code Division Multiple Access (CDMA) techniques to optical networks (OCDMA) [2]. The spreading codes used in an OCDMA system are called optical orthogonal codes(OOC).

An (n, ω, λ) Optical Orthogonal Code (OOC) $\mathcal{C}$ where $1 \leq \lambda \leq \omega \leq n$, is a family of $\{0,1\}$-sequences of length n and Hamming weight ω satisfying

$$\sum_{k=0}^{n-1} x(k)y(k \oplus_n \tau) \leq \lambda \tag{1}$$

for $\{x, y\} \in \mathcal{C}$, either $x \neq y$ or $\tau \neq 0$, where $\oplus_n$ denotes addition modulo n. We will refer to λ as the maximum correlation parameter(MCP).

---

* Work partially supported by NSF CISE Grant No. EIA-0080926.
** Work partially supported by DARPA OCDMA Program Grant No. N66001-02-1-8939.

T. Helleseth et al. (Eds.): SETA 2004, LNCS 3486, pp. 396–405, 2005.

Let $\Phi(n, \omega, \lambda)$, denote the largest possible cardinality of an (n, ω, λ) OOC code. Then by the Johnson upper bound [3]:

$$\Phi(n, \omega, \lambda) \leq \left\lfloor \frac{1}{\omega} \left\lfloor \frac{n-1}{\omega-1} \left\lfloor \frac{n-2}{\omega-2} \cdots \left\lfloor \frac{n-\lambda}{\omega-\lambda} \right\rfloor \right\rfloor \right\rfloor \right\rfloor . \tag{2}$$

An OOC of size P is optimal when $P = \Phi(n, \omega, \lambda)$ and asymptotically optimal if:

$$\lim_{n \to \infty} \frac{P}{\Phi(n, w, \lambda)} = 1.$$

There are many optimal and asymptotically optimal constructions in the literature satisfying the Johnson bound with $\lambda = 1$ [3][4][5]. Even there are some constructions with $\lambda = 2$ which satisfy the Johnson bound [6]. We have not found any construction with $\lambda > 2$ satisfying the Johnson bound in the literature, in addition we couldn't find any optimal construction for $\lambda > 2$ either.

It is known that the Johnson Bound is not always achievable, for example [7] gives some bounds for constant weight codes which are tighter than Johnson bound in some regions, hence their equivalent bound for OOC are tighter than the above bound. It seems that, Johnson bound is not very tight for special cases when λ is larger than 1.

A (v, k, t)-DDS is a family $(B_i | i \in I, t = |I|)$ of subsets of $\mathbb{Z}_v$ each of cardinality k, such that among the $tk(k-1)$ differences $\{a - b (mod\ v) | a, b \in B_i; a \neq b; i \in I\}$, each nonzero element $g \in \mathbb{Z}_v$ occurs at most once. A (v, k, t)-DDS is called a perfect DDS if the $tk(k-1)$ differences give all the nonzero elements of $\mathbb{Z}_v$. Therefore $tk(k-1) = v - 1$.

Lemma 1. *The concept of an $(n, \omega, 1)$-OOC is equivalent to a (v, k, t)-DDS with $n = v$, $k = \omega$, and $P = t$, where P is the size of $(n, \omega, 1)$ OOC.* □

Based on the definition of a perfect DDS, a perfect OOC can be defined as one that attains the Johnson bound without using the brackets.

In Section 2 a new optimal construction for $\lambda = 1$ is generated, and then generalized to $\lambda > 1$. In Section 3 a new upper bound on the size of OOC is derived, and then used to prove the optimality of some OOC constructions which were not know to be optimal previously. In Sections 4 and 5 a recursive construction for OOCs is introduced, and used to construct families of optimal and perfect OOCs.

While most of the results of this paper can be found in [5][12][13][14], by the same authors of this paper, it should be mentioned that these results have been presented here from a new perspective, and most of the proofs are new proofs, and they are more related here.

2   New Optimal Constructions for OOCs

In this section we need to use some properties of affine geometries. In following we are going to introduce the necessary concepts:

The points of $EG(a, q)$, the affine geometry of dimension a over $GF(q)$, consist of all elements of $GF(q^a)$. Let $\xi_0, \xi_1, \cdots, \xi_d$ be $d + 1$ linearly independent elements of $GF(q^a)$. The q^d points of the form:

$$\xi_0 + v_1\xi_1 + v_2\xi_2 + \cdots + v_d\xi_d$$

with $v_i \in GF(q)$ for $1 \le i \le d$, constitute a d-flat in $EG(a, q)$ passing through the point ξ_0[8][9][10].Moreover each d-flat can be shown as a subset of size q^d of the set $\{-\infty, 0, 1, \cdots, q^a - 2\}$ using the mapping $\log_\beta$ when β is a primitive element of $GF(q^a)$ over $GF(q)$. A d-flat, for $d = 1$ is called a line, and for $d = a - 1$ is called a hyperplane. Any two d- flat either are not intersecting, or their intersection is a $(d - 1)$-flat.

Lemma 2. *Let $\{c_1, c_2, \cdots, c_{q^d}\}$ represent the points on a d-flat , then the set $\{c_1 + i, c_2 + i, \cdots, c_{q^d} + i\}$ mod $(q^a - 1)$, for any given integer i is also defines some d-flat, which we call it a cyclic shifted version of the original d-flat.*

Proof. If $\xi_0, \xi_1, \cdots, \xi_d$ are linearly independent points generating the d-flat:

$$\forall c_j, \exists\ v_{1j}, \cdots v_{dj} \in GF(q) : \beta^{c_j} = \xi_0 + \sum_{k=1}^{d} v_{kj}\xi_k$$

$$\Rightarrow \beta^{c_j+i} = \beta^{c_j}\beta^i = \xi_0\beta^i + \sum_{k=1}^{d} v_{kj}\xi_k\beta^i$$

All we need is to show that $\xi_0\beta^i, \cdots \xi_d\beta^i$ are linearly independent, which is obvious.

$\square$

If we start with a d-flat and start cyclically shift it by one unit, the d-flat is called cyclic if at some point we return to the original d-flat. The smallest number of shifts to return to the original d-flat is called the cycle of d-flat.

Theorem 1. *In $EG(a, q)$, every d-flat is cyclic, and the cycle of any d-flat not passing through origin, is equal to $q^a - 1$.*

Proof. Let the integers $C = \{c_1, c_2, \cdots, c_{q^d}\}$ represent the points on a d-flat. Since $c_i + (q^a - 1) = c_i$ mod $(q^a - 1)$, obviously every d-flat is cyclic. If we assume that the cycle of C is equal to $g, C = C + g$ mod $(q^a - 1)$, then obviously $g|(q^a - 1)$. Let's assume $g < (q^a - 1)$. Let's add all the elements of C and $C + g$ mod $(q^a - 1)$:

$$\sum_{i=1}^{q^d} c_i = \sum_{i=1}^{q^d}(c_i + g) = \sum_{i=1}^{q^d} c_i + q^d g \quad \text{mod } (q^a - 1)$$

$$\Rightarrow q^d g = 0 \quad \text{mod } (q^a - 1) \Rightarrow (q^a - 1)|q^d g \Rightarrow \frac{q^a - 1}{g}|q^d$$

Since we assumed $g < (q^a - 1)$, then $\frac{q^a - 1}{g}$ is an integer greater than 1, which means that there is a common factor greater than 1 between $q^a - 1$ and q^d which is impossible. So $g = (q^a - 1)$, and we are done. Note that the above argument is not true when the d-flat is passing through origin. That is because, when the d-flat is passing through origin one of the c_is is equal to $-\infty$, and $-\infty + g = -\infty$. □

The above theorem which was originally stated in [9], is the key property of affine geometries we use to construct OOCs.

2.1 $\lambda = 1$

A new construction for OOC is presented in this section, which is obtained by generalizing the earlier construction of DDS due to Bose and Chowla [11]:

Theorem 2. *Take α a primitive element of $GF(q^a)$ over $GF(q)$. For any vector $\underline{\ell_i} = (\ell_1, \cdots, \ell_{i-1}, \ell_i = 1)$ with all $\ell_j \in GF(q)$ and $i \leq a - 1$, define $P_{\underline{\ell_i}}(x) = \ell_i x^{i+1} + \ell_{i-1} x^i + \cdots \ell_1 x = x^{i+1} + \ell_{i-1} x^i + \cdots \ell_1 x$. Each of these polynomials is generating one codeword of the code. The codeword corresponding to the $P_{\underline{\ell_i}}(x)$ has a 1 precisely in the coordinates corresponding to $\log_\beta(P_{\underline{\ell_i}}(\alpha) + v)$ for any $v \in \mathbb{F}_q$, where β is any primitive element of $GF(q^a)$ over $GF(q)$.*

The above construction gives us an $(n = q^a - 1, \omega = q, \lambda = 1)$ OOC with $q^{a-2} + q^{a-3} + \ldots + 1$ codewords, for any q which is a power of a prime[12].

Proof. Take two elements $P_{\underline{\ell_i}}(\alpha)$ and 1 in $GF(q^a)$. These two elements are linearly independent so all the elements of the form $P_{\underline{\ell_i}}(\alpha) + v$ for $v \in FG(q)$ are making a line not passing through the origin, in an $EG(a, q)$ affine geometry. Using the Theorem 1 we know that each of these lines has a cycle of length $q^a - 1$, so any cyclic shift of these lines is another line in the geometry.

Now we need to prove that none of these lines is a cyclic shift of some other line in this set:

Assume that $P_{\underline{\ell_i}}(\alpha) + v$ is a cyclic shift of $P_{\underline{\ell'_j}}(\alpha) + v'$, so:

$$\alpha^\tau(P_{\underline{\ell_i}}(\alpha) + v) = P_{\underline{\ell'_j}}(\alpha) + v'$$

If $\alpha^\tau \in GF(q)$ this is impossible, so we assume $\alpha^\tau \in GF(q^a) \setminus GF(q)$. So:

$$\alpha^\tau P_{\underline{\ell_i}}(\alpha) - P_{\underline{\ell'_j}}(\alpha) = v' - \alpha^\tau v \tag{3}$$

Depending on τ, α^τ can be written as a linear combination of $\alpha^{a-1}, \cdots, \alpha, 1$, with at least one nonzero power of α. Assume $\alpha^\tau = \sum_{i=0}^{a-1} c_i \alpha^i$ then, the left hand side of the Equation 3 can be written as as a fixed expression in power of α say $\sum_{i=0}^{a-1} c'_i \alpha^i$:

$$\sum_{i=0}^{a-1} c'_i \alpha^i = \sum_{i=1}^{a-1} (-v) c_i \alpha^i + (v' - v)$$

Since we know at least one of c_is for $i > 0$ is nonzero then Equation 3 can have maximally one solution . So the two lines have maximally one point in common, and are not identical.

We can conclude that each of the lines $P_{\underline{\ell_i}}(\alpha) + v$ is a representative of a distinct cyclic orbit of length $q^a - 1$. Since each two distinct lines can intersect maximally at one point, so this set of representatives is satisfying all the conditions of a $(q^a - 1, q, 1)$ OOC.

$\square$

Remark 1. The case for $a = 2$ is the Bose(-Chowla) construction. It is easy to check that our new construction of Theorem 2 is optimal with respect to the Johnson bound.

Remark 2. The above Theorem is proved in [12] directly without using the properties of affine geometries.

2.2 $\lambda > 1$

The generalized Bose-Chowla OOC construction of Theorem 2 can be used to construct a new class of OOC with $\lambda \geq 2$ [13][14]:

Theorem 3. *Let $\mathbb{F}_{q^a}$, be a finite field with q^a elements, with α and β primitive elements of $\mathbb{F}_{q^a}$ over $\mathbb{F}_q$. Now construct a single codeword of length $q^a - 1$ which has 1 precisely in the coordinates corresponding to $\log_\beta(\alpha^{a-1} + \ell_{a-2}\alpha^{a-2} + \cdots + \ell_1\alpha + \ell_0)$ for all $\ell_i \in \mathbb{F}_q$. This gives rise to an OOC with parameters $(q^a - 1, q^{a-1}, q^{a-2})$ of size $\Phi = 1$.*

Proof. The elements $\alpha^{a-1}, \alpha^{a-2}, \cdots, \alpha, 1$ are linearly independent elements of $GF(q^a)$. By the definition of an $(a-1)$-flat(hyperplane), all the points of the form $\alpha^{a-1} + \ell_{a-2}\alpha^{a-2} + \cdots + \ell_1\alpha + \ell_0$ are making a hyperplane for all $\ell_i \in GF(q)$. Since this hyperplane is not passing through the origin, by Theorem 1, it is cyclic of length $q^a - 1$. Using the fact that any two $(a - 1)$-flats can intersect maximally in q^{a-2} points, it can be concluded that this hyperplane makes a $(q^a - 1, q^{a-1}, q^{a-2})$ OOC of size 1.

$\square$

Theorem 4. *If the set S consists of the support of all ones of the new code, then amongst all the differences $(s_i - s_j | s_i, s_j \in S; i \neq j)$, every $\log_\beta \theta$ where $\theta \in \mathbb{F}_{q^a} \setminus \mathbb{F}_q$ occurs exactly q^{a-2} times, and none of $\log_\beta \theta$ with $\theta \in \mathbb{F}_q$ occurs[13][14].*

Proof. The number of occurrence of each difference r is actually equal to the number of intersecting points between the original hyperplane, and its cyclic shifted version by r. If $\beta^r \neq 1 \in GF(q)$ the two hyper planes are not intersecting. That is because, if we have any intersection, we should have a nonzero linear combination of $\alpha^{a-1}, \cdots, \alpha, 1$ equal to 0, which is impossible. From the remaining differences each can occur either q^{a-2} times(the number of points in intersection of two hyperplanes), or not occurring at all. Since there exist totally $q^{a-1}(q^{a-1}-1)$ differences between the elements of S, and each can happen maximally q^{a-2} times, there should be $q^a - q$ different families of differences, so any difference r, with $r \in GF(q^a) \setminus GF(q)$ should happen exactly q^{a-2} times.

□

The above theorem says that the construction in Theorem 3 corresponds to a relative difference set [15] as well. While the existence of a relative difference set with these parameters was previously known, mentioning it here is important, since we have given a direct construction of it.

Remark 3. References [13][14] give an alternate proof of the Theorems 3 and 4 without using the properties of affine geometries, and only using the algebraic properties of the structure. To our knowledge the proof in [13][14] is a new proof.

3   New Bound on OOC, and Some Optimal Constructions

Theorem 5. *Based on a bound on the size of constant weight codes derived by Johnson[16] the following upper bound on the size of OOCs can be found[13][14]:*

$$\Phi(n,\omega,\lambda) \leq \left\lfloor \frac{\omega - \lambda}{\omega^2 - \lambda n} \right\rfloor, \quad when \lambda < \frac{\omega^2}{n} \tag{4}$$

Proof. Johnson has the following bound on the size of constant weight codes [16]:

$$A(n,\omega,\lambda) \leq \left\lfloor \frac{n(\omega - \lambda)}{\omega^2 - \lambda n} \right\rfloor, \quad when \lambda < \frac{\omega^2}{n}$$

Where $A(n,\omega,\lambda)$ is the maximum achievable size of a constant weight code of length n, weight ω, and maximum pairwise real inner product λ. Since by adding all n cyclic shifts of an OOC to it, we will end up with an (n,ω,λ) constant weight code, the bound in Equation 4 is straight forward.

□

Theorem 6. *Any (v,k,λ) difference set is an optimal (v,k,λ) OOC of size $\Phi = 1$[13][14].*   □

Corollary 1. *The following optimal OOCs of size 1 can be constructed using difference sets[14]:*

1. *Singer Difference Set:*
$$(\tfrac{q^{n+1}-1}{q-1}, \tfrac{q^n-1}{q-1}, \tfrac{q^{n-1}-1}{q-1}), q \ a \ prime \ power.$$
2. *Quadratic Residues in $GF(p^r)$:*
$$(4t - 1, 2t - 1, t - 1), \ 4t - 1 = p^r = 3(mod \ 4), \ p \ a \ prime.$$
3. *Biquadratic Residues of Primes:*
$$(4x^2 + 1, x^2, \tfrac{x^2-1}{4}), \ p = 4x^2 + 1, \ p \ a \ prime, \ x \ odd.$$

4. *Biquadratic Residues and Zero Modulo Primes:*
$$(4x^2 + 9, x^2, \tfrac{x^2+3}{4}), \ p = 4x^2 + 9, \ p \ a \ prime, \ x \ odd.$$
5. *Octic Residues of Primes:*
$$(p, a^2, b^2), \ p = 8a^2 + 1 = 64b^2 + 9, \ p \ a \ prime, a \ and \ b \ odd \ integers.$$
6. *Octic Residues and Zero for Primes:*
$$(p, a^2 + 6, b^2 + 7), \ p = 8a^2 + 49 = 64b^2 + 441, \ p \ a \ prime, \ a \ odd, \ b \ even.$$

□

Remark 4. While the fact that difference sets are OOCs of size 1 is known for a long time, their optimality wasn't proved previously, and even they have been assumed not to be optimal.

The bound in Theorem 5 can be mixed with a bound from [6] to give the following bound:

Theorem 7. *If $\lambda n < \omega^2$ then [14]:*

$$\Phi(n, \omega, \lambda) \le min(1, \left\lfloor \frac{\omega - \lambda}{\omega^2 - \lambda n} \right\rfloor), \quad when \lambda < \frac{\omega^2}{n}$$

□

Remark 5. Substituting the parameters of the new construction from Theorem 3 in the above bound: $\Phi(q^a - 1, q^{a-1}, q^{a-2}) \le 1$, so the new code from Theorem 3 is always optimal.

4   Recursive Construction for OOC and DDS's

M.J.Colbourn and C.J.Colbourn proposed two recursive constructions for cyclic BIBD's [1]. Their Construction A was generalized [17] to construct DDS recursively as follows:

Construction A': Given a (v, k, t)-DDS, $v \ne 0 \pmod{k}$ if $gcd(r, (k-1)!) = 1$, a (vr, k, rt)-DDS may be constructed as follows. For each $D = \{0, d_1, ..., d_{k-1}\}$, take the r difference sets $\{0, d_1 + iv, d_2 + 2iv, ..., d_{k-1} + (k-1)iv\}, 0 \le i < r$, with addition performed modulo vr. If furthermore, there exists an (r, k, t')-DDS D', then a $(vr, k, rt + t')$-DDS can be constructed by adding the t' difference sets $\{0, vs_1, ..., vs_{k-1}\}$ for each $D'_i = \{0, s_1, ..., s_{k-1}\}$ of $D' = \{D'_i | 1 \le i \le t'\}$.

This result translates to OOC using Lemma 1 as follows:

Theorem 8. *Given an $(n, \omega, 1)$, OOC of size $\Phi(n, \omega, 1)$, $n \ne 0 \pmod{\omega}$, and r a natural number with all its prime factors greater than or equal to ω, we can construct an $(nr, \omega, 1)$, OOC of size $\Phi(nr, \omega, 1) = r\phi(n, \omega, 1)$. In addition if there is an $(r, \omega, 1)$, OOC of size $\Phi'(r, \omega, 1)$, then we can construct an $(nr, \omega, 1)$, OOC of size $\Phi(nr, \omega, 1) = r\Phi(n, \omega, 1) + \Phi'(r, \omega, 1)$ [12].*

□

Theorem 9. *If the base OOC used in the recursive construction of Theorem 8 is optimal with respect to the Johnson bound in Equation 2, the OOC resulting from applying the recursive construction in Theorem 8 is at least asymptotically optimal in the same bound[12].* ☐

As it is shown in following, in some cases the recursive construction of theorem 8 can generate optimal OOCs:

Theorem 10. *Given an $(n, \omega, 1)$ perfect OOC, let r be a natural number such that its prime factors are $\geq \omega$ then using the first part of Theorem 8 we obtain an (nr, ω, r)-OOC. Whenever $r < \omega^2 - \omega$ then this OOC is optimal, using the Johnson Bound in Equation 2 with brackets[12].* ☐

Theorem 11. *Applying the recursive construction of theorem 8 with $r = p$, a prime, to the OOC construction of theorem 3 with $q = p$, we obtain a $(p(p^a - 1), p, 1)$ optimal OOC of size $\frac{p^a - 1}{p - 1} - 1$[12].* ☐

5 Perfect Optical Orthogonal Codes

Using theorem 8, for any two OOCs, A and B satisfying the conditions of this theorem, we obtain a new OOC that we call $A * B$. If A, B are two OOCs where A has parameters $(n, \omega, 1)$ and size $\Phi(n, \omega, 1)$ and B has parameters $(r, \omega, 1)$ and size $\Phi(r, \omega, 1)$, then we say that $A, B \in P(\omega)$. Whenever r and ω satisfy the conditions of Theorem 8(the prime factors of r are bigger than or equal to ω) we will call B admissible. The new OOC $A * B$ has parameters $(nr, \omega, 1)$ and size $\Phi(nr, \omega, 1) = r\Phi(n, \omega, 1) + \Phi(r, \omega, 1)$. Applying this method recursively we obtain the following construction:

Theorem 12. *If A, B are perfect OOCs in $P(\omega)$, then if B is admissible $A * B^j$ is also a perfect OOC, for any perfect B in $P(\omega)$, and $j \geq 0$[12].*

Proof. Since both A and B are perfect OOCs then for $A * B$:

$$\Phi(nr, \omega, 1) = r\Phi(n, \omega, 1) + \Phi(r, \omega, 1) =$$
$$r\frac{n - 1}{\omega(\omega - 1)} + \frac{r - 1}{\omega(\omega - 1)} = \frac{nr - 1}{\omega(\omega - 1)}$$

So $A * B$ is satisfying Johnson bound without brackets and is perfect too. We can extend this property inductively to $A * B^j$.

☐

Theorem 13. *The following two families of perfect OOCs can be generated using Theorem 12 [12]:*

1. *For $A = (\left(\frac{q^{d'+1} - 1}{q - 1}\right), q + 1, 1)$ and $B = (\left(\frac{q^{d+1} - 1}{q - 1}\right), q + 1, 1)$ two OOCs from projective geometry construction [3], with both d' and d even integers, and B admissible then we can construct the family of $(\left(\frac{q^{d+1} - 1}{q - 1}\right)^i \cdot \left(\frac{q^{d'+1} - 1}{q - 1}\right), q + 1, 1)$ perfect OOCs for any positive integer i.*

2. For $A = (n_1, \omega, 1)$ and $B = (n_2, \omega, 1)$ two OOCs based on Wilson's BIBD's [6] then we can construct the family of $(n_1^i.n_2^j, \omega, 1)$ perfect OOCs for $i, j \geq 0$.

Proof. By considering that $\left(\binom{\frac{q^{d'+1}-1}{q-1}}{}, q+1, 1 \right)$ OOCs from lines of a projective geometry [3] for d even ,and OOCs based on Wilson's BIBD's from [6] are perfect OOCs, and using Theorem 12 the results are obvious. It should be noted that in the OOCs based on Wilson's BIBD, the length of the OOC is always a prime, so these OOCs are always admissible.

$\square$

Theorem 14. *Any perfect OOC is equivalent to an optimal cyclic constant weight code satisfying the Johnson bound for constant weight codes too[12].*

Proof. Adding all the cyclic shifts of an OOC to it generated a cyclic constant weight code. If the OOC is perfect then $\Phi(n, \omega, \lambda) = \frac{(n-1)\cdots(n-\lambda)}{\omega(\omega-1)\cdots(\omega-\lambda)}$, and size of the corresponding constant weight code is $n\frac{(n-1)\cdots(n-\lambda)}{\omega(\omega-1)\cdots(\omega-\lambda)}$, which is an upper bound for $A(n, \omega, \lambda)$ by Johnson bound [16].

$\square$

Acknowledgment

The authors are grateful to the referees for several useful comments and corrections including some hints about the connection with affine geometries.

References

1. M. J. Colbourn and C. J. Colbourn:"Recursive Constructions for Cyclic Block Designs". Journal of Statistical Planning and Inference, Vol. 10, 97-103, 1984
2. J. A. Salehi:"Code Division Multiple-Access Techniques in Optical Fiber Networks-Part I: Fundamental Principles". IEEE Trans. Communications, Vol. 37, 824-833, Aug. 1989
3. F. Chung, J. A. Salehi, and V. K. Wei:"Optical Orthogonal Codes: Design. Analysis, and Applications". IEEE Trans. Information Theory, Vol. 35, 596-604, May 1989
4. O. Moreno, Z. Zhang, P. V. Kumar, and V. Zinoviev:"New Constructions of Optimal Cyclically Permutable Constant Weight Codes". IEEE Trans. Information Theory, Vol. 41, 448-455, March 1995
5. O. Moreno, P. V. Kumar, H. Lu, and R. Omrani:"New Construction of Optical Orthogonal Codes, Distinct Difference Sets and Synchronous Optical orthogonal Codes". Proc. Int. Symposium on Information Theory, 2003
6. H. Chung and P. V. Kumar:"Optical Orthogonal Codes - New Bounds and an Optimal Construction". IEEE Trans. Information Theory, Vol. 36, 866-873, July 1990
7. E. Agrell, A. Vardy, and K. Zeger:"Upper Bounds for Constant-Weight Codes". IEEE Trans. Information Theory, Vol. 46, 2373-2395, Nov. 2000

8. S. Lin, and D. J. Costello:"Error Control Coding: Fundamentals and Applications". Prentice-Hall, Inc, 1983

9. C. R. Rao:"Cyclical Generation of Linear Subspaces in Finite Geometries". Proc. Combinatorial Mathematics and its Applications,515-535, 1969

10. F. J. MacWilliams, and N. J. A. Sloane:"The Theory of Error-Correcting Codes". New York: North-Holland, 1977

11. R. C. Bose and S. Chowla:"On the Construction of Affine Difference Sets". Bull. Calcutta Math. Soc., Vol. 37, 107-112, 1945

12. O. Moreno, R. Omrani, P. V. Kumar, and H. Lu:"New Construction for Optical Orthogonal Codes, and Distinct Difference Sets". Submitted to IEEE Trans. Information Theory

13. R. Omrani, O. Moreno, and P. V. Kumar:"Optimal Optical Orthogonal Codes with $\lambda > 1$". Proc. Int. Symposium on Information Theory, 2004

14. O. Moreno, R. Omrani, and P. V. Kumar:"Optimal Optical Orthogonal Codes with $\lambda > 1$". CSI Publication, CSI-04-09-01, to be submitted to IEEE Trans. Information Theory

15. J. E. H. Elliott, and A. T. Butson:"Relative Difference Sets". Illinois Journal of Mathematics, Vol. 10, 517-531, 1966

16. S. M. Johnson:"A New Upper Bound for Error-Correcting Codes". IEEE Trans. Information Theory, Vol. 8, 203-207, Apr. 1962

17. C. Zhi, F. Pingzhi, and J. Fan:"Disjoint Difference Sets, Difference Triangle Sets and Related Codes". IEEE Trans. Information Theory, Vol. 38, 518-522, March 1992

Weighted Degree Trace Codes for
PAPR Reduction

Patrick Solé[1] and Dmitrii Zinoviev[2]

[1] CNRS-I3S, ESSI, Route des Colles, 06 903 Sophia Antipolis, France
ps@essi.fr
[2] Institute for Problems of Information Transmission, Russian Academy of Sciences,
Bol'shoi Karetnyi, 19, GSP-4, Moscow, 101447, Russia
dzinov@iitp.ru

Abstract. Trace codes over the rings $\mathbb{Z}_{2^l}$, are used to construct spherical codes with controlled peak to average power ratios (PAPR). The main proof technique is the local Weil bound on hybrid character sums over Galois rings.

Keywords: CDMA, Correlation, Galois Rings, hybrid character sums, PAPR, PSK.

1   Introduction

In a recent paper [8] weighted degree trace codes of length $2^m - 1$ have been proposed to reduce the peak to average power ratios (PAPR) of signals used in an orthogonal frequency division multiplexing (OFDM) environnment. In that setting the PAPR of an individual codeword c of period T

$$PAPR(c) \; = \; \max_{0 \leq t \leq 1} \frac{\left| \Re\left(\sum_{k=0}^{n-1} c_i e(-kt) \right) \right|}{||c||},$$

where $e(x) := exp(2\pi i x)$ for x real ($i = \sqrt{-1}$), (viewed as a complex vector by Phase Shift Keying (PSK) modulation) is controlled by the maximum in module of its z-transform $\hat{c}(z)$ on the unit circle. By use of Lagrange polynomial interpolation ([8–§VI.A]) the problem reduces to producing an upper bound for the quantity

$$M_d(c) \; = \; \max_{j=0}^{T-1} \left| \hat{c}\left(e(\frac{j}{T}) \right) \right|.$$

The RHS of this bound turns out, in the case of trace codes, to be an hybrid character sum (combined additive and multiplicative characters) over a Galois ring of characteristic 2^l. It can be handled using the local Weil bounds of [10]. Furthermore, weighted degree driven trace codes can be defined following [9].

T. Helleseth et al. (Eds.): SETA 2004, LNCS 3486, pp. 406–413, 2005.

The aim of the present note[1] is to make more explicit, and in places correct this approach and to extend it to trace codes of length $2^{l-1}(2^m - 1)$. These generalize to a more complex polynomial trace argument the maximum length sequences over rings due to Dai [1] and further investigated in [2].

The material is organized as follows. Section II contains definitions and notation on Galois rings. Section III collects the bounds we need on characters sums. Section IV studies an enumerative problem on polynomials, and allows us to define the family of codes $S_{l,m,D}$. Section V and VI contain bounds on, respectively, the PAPR of codewords and the minimum Euclidean distance of the codes we construct.

2   Preliminaries

Let $R = GR(2^l, m)$ denote the Galois ring of characteristic 2^l with 2^{lm} elements. Let ξ be an element in $GR(2^l, m)$ that generates the Teichmüller set $\mathcal{T}$ of $GR(2^l, m)$. Specifically, let $\mathcal{T} = \{0, 1, \xi, \xi^2, \ldots, \xi^{2^m-2}\}$ and $\mathcal{T}^* = \{1, \xi, \xi^2, \ldots, \xi^{2^m-2}\}$. The 2-adic expansion of $x \in GR(2^l, m)$ is given by

$$x = x_0 + 2x_1 + \cdots + 2^{l-1}x_{l-1},$$

where $x_0, x_1, \ldots, x_{l-1} \in \mathcal{T}$. The Frobenius operator F is defined for such an x as

$$F(x_0 + 2x_1 + \cdots + 2^{l-1}x_{l-1}) = x_0^2 + 2x_1^2 + \cdots + 2^{l-1}x_{l-1}^2,$$

and the trace Tr, from $GR(2^l, m)$ downto $\mathbb{Z}_{2^l}$, as

$$\text{Tr} := \sum_{j=0}^{m-1} F^j.$$

3   Local Weil Bound

Let l be a positive integer ≥ 4, and $\omega = e^{2\pi i/2^l}$ be a primitive 2^l-th root of 1 in $\mathbb{C}$.

Let $f(x)$ denote a polynomial in $R[x]$ and let

$$f(x) = F_0(x) + 2F_1(x) + \ldots + 2^{l-1}F_{l-1}(x)$$

denote its 2-adic expansion. Let d_i be the degree in x of F_i. Let $\Psi(x)$ be the standard additive character of R, applied to a typical $x \in R$:

$$\Psi(x) = \omega^{Tr(x)}.$$

---

[1] The paper has been written under the partial financial support of the Russian fund for fundamental research (under project No. 03 - 01 - 00098).

Let χ be a multiplicative character whose order divides $2^m - 1$. Set D_f to be the *weighted degree* of f, defined as

$$D_f = \max(d_0 2^{l-1}, d_1 2^{l-2}, \ldots, d_{l-1}).$$

With the above notation, we have (under mild technical conditions) the bound

$$\left| \sum_{x \in T} \Psi(f(x)) \chi(x) \right| \leq D_f 2^{m/2}. \tag{1}$$

See [10] for details.

We will need the following property of the weighted degree:

Lemma 1. *Let $f(x) \in R[x]$ and $\alpha \in R^* = R \backslash 2R$ be a unit of R and let $g(x) = f(\alpha x) \in R[x]$. Then*

$$D_g = D_f,$$

where D_f, D_g are respectively the weighted degrees of the polynomials $f(x)$ and $g(x)$.

Proof. Due to the linearity, we can assume that $f(x)$ is of the form $2^i F(x)$, where $F(x) \in T$ is of degree d. Thus

$$F(x) = c_0 + c_1 x + \ldots + c_d x^d,$$

where $c_j \in T$, $j = 0, 1, \ldots, d$ and the weighted degree D_f of f is equal to $2^{l-1-i} d$. Suppose that

$$\alpha^k = \sum_{j=0}^{l-1} \alpha_{jk} 2^j.$$

Substituting αx into $F(x)$, and using the above expansion we obtain that $F(\alpha x)$ equals

$$\sum_{k=0}^{d} c_k \alpha^k x^k = \sum_{k=0}^{d} c_k \left(\sum_{j=0}^{l-1} \alpha_{jk} 2^j \right) x^k.$$

Changing the order of summation, this is

$$\sum_{j=0}^{l-1} 2^j \sum_{k=0}^{d} \alpha_{jk} c_k x^k = \sum_{j=0}^{l-1} 2^j F_j(x),$$

where $F_j(x)$ are polynomials in $T[x]$ of degree at most d. Since α is a unit and $\alpha_{0k} \neq 0$ ($k = 0, \ldots, d$), the polynomial

$$F_0(x) = \sum_{k=0}^{d} \alpha_{0k} c_k x^k,$$

is of degree d. Thus the weighted degree of $f(\alpha x)$ equals $2^{l-1-i} d$. $\square$

4   Polynomials over the Galois Ring $GR(2^l, m)$

Recall that $R = GR(2^l, m)$. A polynomial

$$f(x) = \sum_{j=0}^{d} c_j x^j \in R[x]$$

is called **canonical** if $c_j = 0$ for all even j.

Given an integer $D \geq 4$, define

$$S_D = \{f(x) \in R[x] \mid D_f \leq D, f \text{ is canonical}\},$$

where D_f is the weighted degree of f. Observe that S_D is an $GR(2^l, m)$-module. We have the following (stated in [5–p.459]; see [11–Lemma 4.1] for a detailed proof):

Lemma 2. *For any integer $D \geq 4$, we have:*

$$|S_D| = 2^{(D - \lfloor D/2^l \rfloor)m},$$

where $\lfloor x \rfloor$ is the largest integer $\leq x$.

Throughout this note, we let $n = 2^m$ and $R^* = R\backslash 2R$. Let $\beta = \xi(1+2\lambda) \in R$, where $\xi \in T$ and $\lambda \in R^*$. Assume $1 + 2\lambda$ is of order 2^{l-1}. Since ξ is of order $2^m - 1$ then β is an element of order $N = 2^{l-1}(2^m - 1)$. Following [2–Lemma 2], we define the code of length N:

$$S_{l,m,D} = \{(\mathrm{Tr}(f(\beta^t)))_{t=0}^{N-1} \mid f \in S_D\}. \tag{2}$$

5   Peak-to-Average Power Ratios

We prepare the upper bound on the PAPR of codewords by a result on a character sum.

Theorem 3. *Let $\beta = \xi(1 + 2\lambda) \in R$, where $\xi \in T^*$, $\lambda \in R^* = R\backslash 2R$, and $1 + 2\lambda \in R$ is an element of order 2^{l-1}. Set $N = 2^{l-1}(2^m - 1)$. Then for any $j \in [0, 2^{l-1} - 1]$, we have*

$$\left| \sum_{k=0}^{N-1} \Psi(f(\beta^k)) e^{2\pi i k j / N} \right| \leq 2^{l-1}[D_f \sqrt{2^m} + 1]. \tag{3}$$

Proof. Since $(2^{l-1}, 2^m - 1) = 1$, as j ranges over $\{0, 1, \ldots, N - 1\}$, the set of pairs

$$\{j \,(\mathrm{mod}\, 2^m - 1), \ j \,(\mathrm{mod}\, 2^{l-1})\}$$

runs over all pairs (k_1, k_2), where $k_1 \in \{0, 1, \ldots, 2^m - 2\}$ and $k_2 \in \{0, 1, \ldots, 2^{l-1} - 1\}$. Thus the set

$$\{\beta^k; \ k = 0, 1, \ldots, N - 1\} = \{\xi^k(1 + 2\lambda)^k; \ k = 0, 1, \ldots, N - 1\}$$

is equal to the direct product of sets:

$$\{\xi^{k_1};\ k_1 = 0,1,\dots,2^m - 2\} \times \{(1+2\lambda)^{k_2};\ k_2 = 0,1,\dots,2^{l-1} - 1\}, \quad (4)$$

where

$$k \equiv k_1(\mathrm{mod}\,2^m - 1),\ k \equiv k_2(\mathrm{mod}\,2^{l-1}).$$

By the Chinese Remainder Theorem there exist integers c_1, c_2 such that for all $k = 0,1,\dots,N-1$ we can write

$$k = 2^{l-1}c_1 k_1 + (2^m - 1)c_2 k_2.$$

Consequently since $\beta = \xi(1 + 2\lambda)$, where $\xi^{2^m-1} = 1$ and $(1+2\lambda)^{2^{l-1}} = 1$, the sum of the left hand side of (3) is equal to:

$$\sum_{k_2=0}^{2^{l-1}-1} \sum_{k_1=0}^{2^m-2} \Psi(f(\xi^{k_1}(1+2\lambda)^{k_2}))e^{2\pi i k j/N}. \quad (5)$$

Set $f_{k_2}(x) = f(x(1+2\lambda)^{k_2})$. Then, expressing k as a function of k_1 and k_2, the above sum is equal to:

$$\sum_{k_2=0}^{2^{l-1}-1} \sum_{k_1=0}^{2^m-2} \Psi(f_{k_2}(\xi^{k_1}))e^{2\pi i c_1 k_1 j/(2^m-1)}e^{2\pi i c_2 k_2 j/2^{l-1}}.$$

Using Lemma 1 the weighted degree of f_{k_2} is bounded by D_f. Applying (1) to each of the 2^{l-1} inner sums yields:

$$\left| \sum_{t=0}^{2^m-2} \Psi(f_{k_2}(\xi^t)e^{2\pi i c_1 k_1 j/(2^m-1)} \right| \leq D_f\sqrt{2^m} + 1.$$

Thus, the absolute value of (5) can be estimated from above by

$$2^{l-1}[D_f\sqrt{2^m} + 1].$$

The result follows.   □

This character sum estimate translates immediately in terms of PAPR.

Corollary 4. *For every $c \in S_{l,m,D}$ the PAPR is at most*

$$\frac{2^{l-1}}{2^m - 1}(1 + D\sqrt{2^m})^2\left(\frac{2}{\pi}\log(2N) + 2\right)^2,$$

where log *stands for the natural logarithm.*

Proof. Follows by combining the preceding result and the definition of N into the Lagrange bound [8–Lemma 9].   □

6   Euclidean Distance

We prepare the bound on the minimum Euclidean distance of the code $S_{l,m,D}$ by a correlation approach.

Theorem 5. *With notation as above, and for all phase shifts τ, in the range $0 < \tau < N$, let*

$$\Theta(\tau) = \sum_{t=0}^{N-1} \omega^{c_t - c'_{t+\tau}},$$

where $c_t = Tr(f_1(\beta^t))$ and $c'_t = Tr(f_2(\beta^t))$. We then have the bound ($l \geq 4$):

$$|\Theta(\tau)| \leq 2^{l-1}[(D_f - 1)\sqrt{2^m} + 1].$$

Proof. As we have $c_t = Tr(f_1(\beta^t))$ and $c'_t = Tr(f_2(\beta^t))$, where t ranges between 0 and $N - 1$ and $\beta = \xi(1 + 2\lambda)$ is of order $N = 2^{l-1}(2^m - 1)$.

We obtain that:

$$\Theta(\tau) = \sum_{t=0}^{N-1} \Psi(f_1(\beta^t) - f_2(\beta^{t+\tau})). \tag{6}$$

By definition of Ψ, we have:

$$\Psi(f_1(\beta^t) - f_2(\beta^{t+\tau})) = \Psi(f_3(\beta^t)),$$

where $f_3(x) = f_1(x) - f_2(x\beta^\tau)$. Note that if $f(x) \in S_D$ then by Lemma 1 $f(x\beta^\tau) \in S_D$ since the change of variables $x \to x\beta^\tau$ does not increase the weighted degree. Moreover S_D is an R-linear space. Thus the polynomial $f_3(x)$ belongs to S_D along with f_1 and f_2. Further, as t ranges between 0 and $N - 1$, the set $\{\beta^t \; ; \; t = 0, 1, \ldots, N - 1\}$ is equal to the product

$$\{\xi^{k_1}; \; k_1 = 0, 1, \ldots, 2^m - 2\} \times \{(1 + 2\lambda)^{k_2}; \; k_2 = 0, 1, \ldots, 2^{l-1} - 1\}.$$

Thus, in (6), the sum over t is equal to

$$\sum_{k_2=0}^{2^{l-1}-1} \sum_{k_1=0}^{2^m-2} \Psi(f_3(\xi^{k_1}(1 + 2\lambda)^{k_2})) = \sum_{k_2=0}^{2^{l-1}-1} \sum_{k_1=0}^{2^m-2} \Psi(g_{k_2}(\xi^{k_1})), \tag{7}$$

where $g_{k_2}(x) = f_3(x(1 + 2\lambda)^{k_2})$ and for any k_2, $g_{k_2} \in S_D$. Applying (1) to each of the 2^{l-1} sums:

$$\left| \sum_{t=0}^{2^m-2} \Psi(f_{k_2}(\xi^t)) \right| \leq (D_f - 1)\sqrt{2^m} + 1.$$

Thus, the absolute value of (7) can be estimated above by

$$2^{l-1}[(D_f - 1)\sqrt{2^m} + 1]. \tag{8}$$

The result follows.   □

We are now in a position to estimate the minimum Euclidean distance d_E of our code $S_{l,m,D}$.

Corollary 6. *For all integers $l \geq 4$ and $m \geq 3$ we have*

$$d_E \geq 2^l(2^m - 2 - (D-1)\sqrt{2^m}).$$

Proof. Viewed as a vector of $\mathbb{C}^N$ by exponentiating by ω, every codeword of $S_{l,m,D}$ is of (squared norm) N. The result follows now from the preceding theorem by the identity on the standard hermitian inner product $< ., . >$ in $\mathbb{C}^N$:

$$< x - y, x - y >^2 = < x, x >^2 + < y, y >^2 - 2\Re(< x, y >).$$

$\square$

7   Conclusions and Perspective

In this note we constructed a linear code $S_{l,m,D}$ over the ring $\mathbb{Z}_{2^l}$ with the following parameters

- length $N = 2^{l-1}(2^m - 1)$
- size $|S_{l,m,D}| = 2^{(D - \lfloor D/2^l \rfloor)m}$
- PAPR at most $\leq \frac{2^{l-1}}{2^m-1}(1 + D\sqrt{2^m})^2 \left(\frac{2}{\pi}\log(2N) + 2\right)^2$
- minimum Euclidean distance at least $\geq 2^l(2^m - 2 - (D-1)\sqrt{2^m})$

By using similar arguments it can be shown that the parameters of the primitive length trace codes C_t^{-1} in [8] satisfy the following (with $D = 2t-1$) estimates

- length $n = 2^m - 1$
- size $|C_t^{-1}| = 2^{(D - \lfloor D/2^e \rfloor)m}$
- PAPR at most $\leq \frac{(1+D\sqrt{2^m})^2}{2^m-1} \left(\frac{2}{\pi}\log(2n) + 2\right)^2$
- minimum Euclidean distance at least $\geq 2(2^m - 2 - (D-1)\sqrt{2^m})$

In particular this simplifies and corrects the estimate of [8–VII.C] on the number of codewords. The minimum distance estimate is new.

References

1. Zong-Duo Dai, Binary sequences derived from ML-sequences over rings I: period and minimal polynomial, J. Cryptology (1992) 5:193–507.
2. S. Fan, W. Han, Random properties of the highest level sequences of primitive Sequences over $\mathbb{Z}_{2^e}$, IEEE Trans. Inform. Theory vol. IT-**49**,(2003) 1553–1557.
3. T. Helleseth, P.V. Kumar, Sequences with low Correlation, in *Handbook of Coding theory*, vol. II, V.S. Pless, W.C. Huffman, eds, North Holland (1998), 1765–1853.
4. P.V. Kumar, T. Helleseth, An expansion of the coordinates of the trace function over Galois rings, AAECC **8**, 353–361, (1997).

5. P.V. Kumar, T. Helleseth and A.R. Calderbank, An upper bound for Weil exponential sums over Galois rings and applications, IEEE Trans. Inform. Theory, vol. IT-**41**, pp. 456–468, May 1995.
6. J. Lahtonen, S. Ling, P. Solé, D. Zinoviev, $\mathbb{Z}_8$-Kerdock codes and pseudo-random binary sequences, J. of Complexity, in press.
7. F.J. MacWilliams, N.J.A. Sloane, *The Theory of Error-Correcting Codes*, North-Holland (1977).
8. K.G. Paterson, V. Tarokh, On the existence and construction of good codes with low peak to average power ratios, IEEE Trans. Inform. Theory, vol. IT-**46**, pp. 1974–1987, September 2000.
9. A. Shanbhag, P. V. Kumar and T. Helleseth, Improved Binary Codes and Sequence Families from $\mathbb{Z}_4$-Linear Codes, IEEE Trans. on Inform. Theory, vol. IT-**42**, pp. 1582-1586, Sep. 1996.
10. A. Shanbhag, P. V. Kumar and T. Helleseth, "Upper bound for a hybrid sum over Galois rings with applications to aperiodic correlation of some q-ary sequences," IEEE Trans. Inform. Theory, vol. IT-42, pp. 250-254, Jan. 1996.
11. P. Solé, D. Zinoviev, Low Correlation, high non linearity sequences for multi code CDMA, Submitted.

Which Irreducible Polynomials Divide Trinomials over GF(2)?

Solomon W. Golomb and Pey-Feng Lee

University of Southern California,
Department of Electrical Engineering-Systems,
Los Angeles, California 90089-2565 U.S.A
{c/o milly, peyfeng1}@usc.edu

Abstract. The output sequence of a binary linear feedback shift register with k taps corresponds to a polynomial $f(x) = 1 + x^{a_1} + x^{a_2} + \ldots + x^{a_k}$, where the exponents $a_1, a_2, \ldots, a_k = n$ are the positions of the taps, and n, the degree of $f(x)$, is the length of the shift register. Different initial states of the shift register may give rise to different output sequences. The simplest shift registers to implement involve only two taps ($k = 2$). It is therefore of interest to know which irreducible polynomials $f(x)$ divide trinomials, over GF(2), since the output sequences corresponding to $f(x)$ can be obtained from a two-tap linear shift register (with a suitable initial state) if and only if $f(x)$ divides some trinomial $t(x) = x^m + x^a + 1$ over GF(2). In this paper we develop the theory of which irreducible polynomials over GF(2) do, or do not, divide trinomials.

1   Introduction

Linear feedback shift register sequences have been widely used in many important applications, such as wireless communications, bit error rate measurements, error correcting codes and stream ciphers. There are two main advantages for using linear feedback shift register sequences: they are extremely fast and easy to implement both in hardware and software, and they can be readily analyzed by using algebraic techniques. A thorough introduction to the theory of shift register sequences is in the book by Golomb [1].

The output sequence of a binary linear feedback shift register with k taps corresponds to a polynomial $f(x) = 1 + x^{a_1} + x^{a_2} + \ldots + x^{a_k}$, where the exponents $a_1, a_2, \ldots, a_k = n$ are the positions of the taps, and n, the degree of $f(x)$, is the length of the shift register. The period of $f(x)$ is defined as the smallest integer t such that $f(x)|(x^t - 1)$. As shown in [1], the output sequence of a binary linear feedback shift register is periodic, and if $f(x)$ is irreducible, the period t of $f(x)$ will always be a divisor of $2^n - 1$. When $t = 2^n - 1$, $f(x)$ is called a primitive polynomial over GF(2), which corresponds to a "maximal-length sequence" (or "m-sequence" for short). It is a fact that every primitive polynomial is also irreducible over GF(2), but the converse is not true. Generally, different "initial" states of the shift register may give rise to different output sequences. When the corresponding

T. Helleseth et al. (Eds.): SETA 2004, LNCS 3486, pp. 414–424, 2005.

polynomial $f(x)$ is irreducible, the period of the shift register sequence does not depend on the initial state, excepting only the initial condition "all 0's". Specifically, if $f(x)$ is a primitive polynomial, all the non-zero initial states of the shift register will generate the very same m-sequence except for phase shifts.

For certain values of n, there are primitive trinomials $x^n + x^a + 1$ of degree n. These correspond to shift registers in which only two "taps" are involved in the feedback mod 2 adder. This is the simplest way to generate an m-sequence, which is why primitive (and irreducible) trinomials have been of special interest among the set of all primitive (and irreducible) polynomials. Tables of primitive (and irreducible) trinomials can be found in [1] [2] [3] [4]. It is easy to show that there are irreducible (though not primitive) trinomials over GF(2) for infinitely many different degrees n. It is also conjectured (and highly likely) but still unproved that there are primitive trinomials for infinitely many degrees n. However, by a theorem of R. Swan [5], there are infinitely many degrees n (including all multiples of 8) for which there are no irreducible trinomials, and a fortiori no primitive trinomials, over GF(2). When a primitive trinomial of degree n does not exist, an "almost primitive" trinomial may be used as an alternative. (A polynomial $p(x)$ of degree n is **almost primitive** if $p(x) \neq 0$ and $p(x)$ has a primitive factor of degree $> n/2$.) Algorithms for finding almost primitive trinomials can be found in [6]. Moreover, as shown in [7], the moments of the partial-period correlation of an m-sequence are related to the number of trinomials of bounded degree (determined by the particular partial period under consideration) that the characteristic polynomial of the m-sequence divides.

It is therefore of interest to know which irreducible polynomials $f(x)$ divide trinomials over GF(2), since the output sequence corresponding to $f(x)$ can be obtained from a two-tap linear feedback shift register (with a suitable initial state) if and only if $f(x)$ divides some trinomial $t(x) = x^m + x^a + 1$ over GF(2). In this paper we develop the theory of which irreducible polynomials do, or do not, divide trinomials over GF(2). In Section 2, some basic results and theorems are presented relating to the "primitivity" t, and a clever criterion is introduced to determine whether a given irreducible polynomial $f(x)$ divides trinomials, for every odd primitivity $t > 1$. In Section 3, further results are presented in terms of the index r, which is the number of irreducible factors of the "t^{th} cyclotomic polynomial" $\Phi_t(x)$ over GF(2). We will see how the index r contributes to determining whether a given irreducible polynomial $f(x)$ of prime primitivity t divides trinomials over GF(2). In Section 4, the "multiplicative module" M is introduced, which is the set of positive (odd) integers t such that the irreducible polynomials of odd primitivity $t > 1$ divide trinomials over GF(2). The set G of generators of M is discussed. Then the computational results are presented.

2   Basic Theorems

In this section, $f(x)$ denotes an irreducible polynomial of degree $n > 1$ over GF(2) having a root α and "primitivity" t, which means that $\alpha^t = 1$ (i.e. that $f(x)$ divides $x^t - 1$), where t is the smallest positive integer with this property.

The following facts, many of which depend on the primitivity t of $f(x)$, have been established.

Theorem 1. $f(x)$ *divides some trinomial iff there exist distinct positive integers i and j with $\alpha^i + \alpha^j = 1$.*

Proof. The trinomial $h(x) = x^i + x^j + 1$ is divisible by $f(x)$ if and only if $h(\alpha) = \alpha^i + \alpha^j + 1 = 0$, i.e. $\alpha^i + \alpha^j = 1$.   □

Theorem 2. *If $f(x)$ divides any trinomial, then $f(x)$ divides infinitely many trinomials.*

Proof. Suppose $f(x)$ divides $x^m + x^a + 1$. Then $f(x)$ also divides $x^{m+st} + x^{a+rt} + 1$ for all positive integers r and s.   □

Theorem 3. *If $f(x)$ divides any trinomials, then $f(x)$ divides some trinomial of degree $< t$.*

Proof. If $f(x)$ divides $x^m + x^a + 1$, we have $\alpha^m + \alpha^a + 1 = 0$. Since $\alpha^t = 1$, this gives $\alpha^{m'} + \alpha^{a'} + 1 = 0$ where $m' \equiv m(mod\ t)$ and $a' \equiv a(mod\ t)$, from which we can pick m' and a' on the range from 0 to $t - 1$. Then $f(x)$ must divides some trinomial $x^{m'} + x^{a'} + 1$ of degree $< t$.   □

Hence, if $f(x)$ divides no trinomial of degree $< t$, then $f(x)$ will never divide any trinomials. This provides a finite decision procedure for whether a given irreducible polynomial ever divides trinomials. Also, an irreducible polynomial $f(x)$ either divides infinitely many trinomials, or divides no trinomials. In the following, we show that every primitive polynomial divides infinitely many trinomials, and present a whole family of irreducible polynomials $x^{t-1} + x^{t-2} + \ldots + x + 1$ which divide no trinomials.

Theorem 4. *If $f(x)$ is a primitive polynomial of degree n (i.e. if $t = 2^n - 1$), then $f(x)$ divides trinomials.*

Proof. Since α is a root of $f(x)$, the powers $1, \alpha^1, \alpha^2, \alpha^3, \ldots, \alpha^{t-1}$ are all distinct, and constitute all the non-zero elements of the field $\mathrm{GF}(2^n)$. Hence, for all $i, 0 < i < t$, $1 + \alpha^i = \alpha^j$ for some $j \neq i$, $0 < j < t$. Thus, $f(x)$ divides $x^i + x^j + 1$ for each such pair (i, j).   □

In fact, when $f(x)$ is a primitive polynomial, it divides exactly $(t - 1)/2$ trinomials of degree $< t$. Since the primitive polynomials precisely correspond to m-sequences, this theorem says that every m-sequence can be obtained from a two-tap linear shift register. This is a very simple and efficient way to generate an m-sequence.

Theorem 5. *For odd $t > 3$, if $f(x) = \frac{(x^t - 1)}{(x - 1)} = x^{t-1} + x^{t-2} + \ldots + x + 1$ is irreducible, then $f(x)$ divides no trinomials.*

Proof. The lowest degree polynomial having the root α of $f(x)$ as a root is the irreducible polynomial $f(x) = \frac{(x^t-1)}{(x-1)} = x^{t-1} + x^{t-2} + \ldots + x + 1$, which has $t > 3$ terms. Suppose $f(x)$ divides some trinomial $x^m + x^a + 1$, so that $\alpha^m + \alpha^a + 1 = 0$. Then $x^{m'} + x^{a'} + 1 = 0$, where m' and a' are m and a reduced modulo t, respectively, and are less than t. Thus α is a root of $x^{m'} + x^{a'} + 1$, a trinomial of degree $\leq t - 1$. But the only polynomial of degree $\leq t - 1$ with α as a root is its minimal polynomial, the t-term irreducible polynomial $f(x)$ of degree $t - 1$. □

This occurs iff 2 is a "primitive root" modulo t with t prime. Examples include $t = 5, 11, 13, 19, 29, 37, 53, 59, 61, 67, 83, 101, \ldots$ By Artin's Conjecture [8], there are infinitely many primes p for which 2 is a primitive root (i.e. where 2 is a generator of the multiplicative group of $GF(p)$). Hence, we have infinitely many irreducible polynomials of this kind, which never divide trinomials.

Definition 1. *The "t^{th} cyclotomic polynomial" is given by*

$$\Phi_t(x) = \prod_{d|t}(x^{t/d} - 1)^{\mu(d)} \tag{1}$$

where $\mu(d)$ is the Möbius mu-function.

For any odd integer $t > 3$, let $\Phi_t(x) = f_1(x)f_2(x) \ldots f_r(x)$ be the factorization over GF(2) of the "t^{th} cyclotomic polynomial" into irreducible factors. It is known that all the $f_i(x)$'s have the same degree (say, n) and the same primitivity t. These factors are **all** the irreducible polynomials having primitivity t.

Theorem 6. *If any one of the $f_i(x)$'s divides a trinomial, then all r of the $f_i(x)$'s divide trinomials.*

Proof. Collectively, the roots of the polynomials $f_1(x), f_2(x), \ldots, f_r(x)$ are all the powers $\alpha, \alpha^2, \alpha^3, \ldots, \alpha^{t-1}$, of a single root α of $\Phi_t(x)$, which can be taken to be a root of any one of the polynomials $f_i(x)$. Also, the roots of $\alpha^t = 1$ always form a cyclic group under multiplication. If α is a primitive root of $\alpha^t = 1$, then every other primitive root will be a power of α. Suppose $f_i(x)$ divides the trinomial $x^m + x^a + 1$. Then $\alpha^m + \alpha^a + 1 = 0$, where we selected α to be a root of $f_i(x)$. For any other polynomial $f_j(x)$ from the set of divisors of $\Phi_t(x)$, let one of its roots be $\beta = \alpha^u$, with $GCD(t, u) = 1$ (i.e. $rt + su = 1$ for some r, s). Then for some s, $1 \leq s \leq t - 1$, we have $\alpha = \beta^s$, from which $(\beta^s)^m + (\beta^s)^a + 1 = \beta^{sm} + \beta^{sa} + 1 = 0$, whereby $f_j(x)$ divides the trinomial $x^{sm} + x^{sa} + 1 = 0$. □

Specifically, if any one of the $f_i(x)$'s is already a trinomial, then all the $f_i(x)$'s divide trinomials. The theorem provides that for any odd t, either all the $f_i(x)$'s divide trinomials or none divides trinomials. According to this elegant result, a clever criterion for testing whether an irreducible polynomial divides trinomials is the following.

Theorem 7 (Welch's Criterion). *For any odd integer t, the irreducible polynomials of primitivity t divide trinomials iff $GCD(1+x^t, 1+(1+x)^t)$ has degree greater than 1.*

Proof. Let $c_t(x) = \frac{(x^t-1)}{(x-1)} = f_1(x)f_2(x)\ldots f_r(x)$ (not necessary the tth cyclotomic polynomial). Then $(1 + x^t) = (1 + x)c_t(x)$, and $(1 + (1 + x)^t) = xc_t(1+x)$. Thus, except for possible linear factors, $GCD(1+x^t, 1+(1+x)^t) = GCD(c_t(x), c_t(1+x))$. Let $c_t(x) = \frac{(x^t-1)}{(x-1)} = f_1(x)f_2(x)\ldots f_r(x)$ be the factorization of $c_t(x)$ into irreducible factors. Then the roots of $f_1(x), f_2(x), \ldots, f_r(x)$ collectively are $\alpha, \alpha^2, \alpha^3, \ldots, \alpha^{t-1}$ where $\alpha \neq 1$ and $\alpha^t = 1$. Thus, the roots of the irreducible factors of $c_t(1+x)$ are $1+\alpha, 1+\alpha^2, 1+\alpha^3, \ldots, 1+\alpha^{t-1}$. Hence, the GCD in question has degree > 1 if and only if one of the roots (say $1 + \alpha^j$) from $c_t(1 + x)$ equals to one of the roots (say α^i) from $c_t(x)$ (i.e. $1 + \alpha^j = \alpha^i$). This is the precise condition that a factor of $c_t(x)$ with α as a root divides the trinomial $x^i + x^j + 1$. $\qquad\square$

This computationally useful criterion, due to L.R. Welch, determines whether the irreducible polynomials of primitivity t divide trinomials, for any odd integer $t > 3$, without directly identifying which irreducible polynomial divides which trinomial.

3   Further Results

When the primitivity t is a prime p, let $\Phi_p(x) = \frac{(x^p-1)}{(x-1)} = f_1(x)f_2(x)\ldots f_r(x)$ be the factorization of the "p^{th} cyclotomic polynomial" into irreducible factors over GF(2). Here the index $r = \phi(p)/n = (p-1)/n$ is the number of irreducible factors of $\Phi_p(x)$ over GF(2), and $(p-1)/r$ is the order of 2 in the multiplicative group modulo p. Again, all the $f_i(x)$'s have the same degree (say, n) and the same primitivity p, and they are **all** the irreducible polynomials having primitivity p over GF(2). Let α be a root of $f_i(x)$. The following results relate to the index r.

Definition 2. *Let $f(x)$ be an irreducible polynomial of degree n. The reciprocal of $f(x)$ is defined as $f^*(x) = x^n f(\frac{1}{x})$. When $f^*(x) = f(x)$, then we say $f(x)$ is "self-reciprocal" (i.e. if α is a root of $f(x)$, then α^{-1} is also a root of $f(x)$).*

Lemma 1. *For prime values $p > 3$ with $\Phi_p(x) = \frac{(x^p-1)}{(x-1)} = f_1(x)f_2(x)\ldots f_r(x)$ as a product of $r > 1$ irreducible polynomials, if any of the $f_i(x)$'s is self-reciprocal, then $f_i(x)$ cannot be a trinomial.*

Proof. Since $f_i(x)$ is self-reciprocal, we have $f_i(x) = x^{(p-1)/r} f_i(\frac{1}{x})$. If $f_i(x)$ is a trinomial, it must be $x^{(p-1)/r} + x^{(p-1)/2r} + 1$, which divides $x^{3(p-1)/2r} + 1$, whereby $\alpha^{3(p-1)/2r} = 1$, but $3(p-1)/2r < p$ for all $r > 1$, which contradicts p being the smallest positive exponent with $\alpha^p = 1$. $\qquad\square$

Theorem 8. *For prime $p > 3$, if any of the $f_i(x)$'s is self-reciprocal, then none of the $f_i(x)$'s divide trinomials.*

Proof. Since $f_i(x)$ is self-reciprocal, it cannot be a trinomial by the Lemma. Suppose $f_i(x)$ divides some trinomial. WLOG, $f_i(x)$ divides a trinomial $t_i(x) = x^m + x^a + 1$ with $1 \le a < m < p$. Write $t_i(x) = x^m + x^a + 1 = f_i(x)g_i(x)$. Then $t_i^*(x) = x^m + x^{m-a} + 1 = f_i(x)g_i^*(x)$ where $g_i^*(x) = x^d g_i(\frac{1}{x})$ and $d = $ degree$(g_i(x))$. Thus $t_i(x) + t_i^*(x) = x^{m-a} + x^a = f_i(x)(g_i(x) + g_i^*(x))$ and $f_i(x)$ divides $x^{m-a} + x^a = x^a(1 + x^{|m-2a|})$, where $0 \le |m - 2a| < p$, which contradicts α, the root of $f_i(x)$, having primitivity p, unless $m - 2a = 0$. In this case, $g_i^*(x) = g_i(x)$ and $t_i^*(x) = t_i(x)$, so that $t_i(x) = x^m + x^{m/2} + 1$, which divides $x^{3m/2} + 1$, so that $\alpha^{3m/2} = 1$, whereas also $\alpha^p = 1$. This requires $\alpha^{|p - 3m/2|} = 1$, but since $(p - 1)/r < m < p$ and $r > 1$, we have $3(p - 1)/2r < 3m/2 < 3p/2$, from which $0 < |p - 3m/2| < p$, contradicting α having primitivity p. □

Corollary 1. *Let $p > 3$ be a prime and $\Phi_p(x) = \frac{(x^p - 1)}{(x-1)} = f_1(x)f_2(x) \ldots f_r(x)$ be a product of r irreducible polynomials. If $r > 1$ is an odd number, then the $f_i(x)$'s divide no trinomials.*

Proof. When $r > 1$ is odd, then at least one of the $f_i(x)$'s is self-reciprocal. The result follows from the theorem. □

Hence, if the index r of the pth cyclotomic polynomial $\Phi_p(x)$ is any odd number, then all the irreducible factors of $\Phi_p(x)$ divide no trinomials, except for the trinomial $x^2 + x + 1$ with $r = 1$ and $p = 3$, which is already a trinomial. When r is an even number, it no longer guarantees that at least one of the $f(x)$'s is self-reciprocal. But there are only two cases: either at least one of the $f(x)$'s is self-reciprocal, or none of them is self-reciprocal. In the former case, all the $f(x)$'s divide no trinomials by the above theorem. In the latter case, the polynomials appear in pairs (i.e. if α is a root of $f_i(x)$, then α^{-1} is a root of $f_i^*(x)$). Then the $f_i(x)$'s may or may not divide trinomials.

Theorem 9. *Let $p > 7$ be a prime and $\Phi_p(x) = \frac{(x^p - 1)}{(x-1)} = f_1(x)f_2(x)$ be a product of two irreducible polynomials (i.e. $r = 2$). Then the $f_i(x)$'s divide no trinomials.*

Proof. If either one of the $f_i(x)$'s is self-reciprocal, then the $f_i(x)$'s divide no trinomials by the previous theorem. Otherwise, the $f_i(x)$'s form a reciprocal pair. If α is a root of $f_1(x)$, then α^{-1} is a root of $f_2(x)$. Suppose $f_1(x)$ divides some trinomial $t_1(x)$ (including the case that $f_1(x)$ itself is a trinomial). Then we can write $t_1(x) = x^m + x^a + 1 = f_1(x)g_1(x)$, with $1 \le a < m < p$, and replacing α by α^{-1} for all roots of $t_1(x)$, we get $f_2(x)g_1^*(x) = t_1^*(x) = x^m + x^{m-a} + 1$. Then the product $t_1(x)t_1^*(x) = f_1(x)f_2(x)g_1(x)g_1^*(x) = \Phi_p(x)g_1(x)g_1^*(x) = x^{2m} + x^{2m-a} + x^{m+a} + x^m + x^a + x^{m-a} + 1$, and this 7-nomial must have both α and α^{-1} as roots, which must satisfy both $\alpha^p = 1$ and $\alpha^{-p} = 1$. We may therefore reduce all exponents in $t_1(x)t_1^*(x)$ modulo p, to get an at most 7-term polynomial of degree $< p$, but which has all $p - 1$ roots of $\Phi_p(x)$ as roots, contradicting the fact that the only (non-zero) polynomial of degree $< p$ with all roots of $\Phi_p(x)$ as roots is $\Phi_p(x)$ itself, which has $p > 7$ terms. □

Note that when $r = 2$ and $p = 7$, neither of the $f_i(x)$'s is self-reciprocal, where $\Phi_7(x) = x^6 + x^5 + x^4 + x^3 + x^2 + x + 1 = (x^3 + x^2 + 1)(x^3 + x + 1)$ has only 7 terms, and the two factors of $\Phi_7(x)$ are already trinomials.

Theorem 10. *Let p be a prime and $\Phi_p(x) = \frac{(x^p - 1)}{(x - 1)} = f_1(x)f_2(x)f_3(x)f_4(x)$ be a product of four irreducible polynomials (i.e. $r = 4$). Then the $f_i(x)$'s divide no trinomials.*

Proof. If any one of the $f_i(x)$'s is self-reciprocal, then the $f_i(x)$'s divide no trinomials by Theorem 8. Otherwise, the $f_i(x)$'s come in pairs. Let α be a root of $f_1(x)$ and α^{-1} be a root of $f_2(x)$. Similarly, let $\beta = \alpha^u$ be a root of $f_3(x)$ and β^{-1} be a root of $f_4(x)$. Suppose $f_1(x)$ divides some trinomial $t_1(x)$ (including the case that $f_1(x)$ itself is a trinomial). Then we can write $t_1(x) = x^m + x^a + 1 = f_1(x)g_1(x)$, with $1 \le a < m < p$. Replacing α by α^{-1} for all roots of $t_1(x)$, we get $f_2(x)g_1^*(x) = t_1^*(x) = x^m + x^{m-a} + 1$. Then the product $t_1(x)t_1^*(x) = f_1(x)f_2(x)g_1(x)g_1^*(x) = x^{2m} + x^{2m-a} + x^{m+a} + x^m + x^a + x^{m-a} + 1$ is a 7-term polynomial which has both α and α^{-1} as roots. Similarly, suppose $f_3(x)$ divides some trinomial $t_3(x)$ (including the case that $f_3(x)$ itself is a trinomial). Then we can write $t_3(x) = x^k + x^b + 1 = f_3(x)g_3(x)$, with $1 \le b < k < p$. Replacing β by β^{-1} for all roots of $t_3(x)$, we get $f_4(x)g_3^*(x) = t_3^*(x) = x^k + x^{k-b} + 1$. Then the product $t_3(x)t_3^*(x) = f_3(x)f_4(x)g_3(x)g_3^*(x) = x^{2k} + x^{2k-b} + x^{k+b} + x^k + x^b + x^{k-b} + 1$ is also a 7-term polynomial which has both β and β^{-1} as roots. Then $t_1(x)t_1^*(x)t_3(x)t_3^*(x) = f_1(x)f_2(x)f_3(x)f_4(x)g_1(x)g_1^*(x)g_3(x)g_3^*(x) = \Phi_p(x)g_1(x)g_1^*(x)g_3(x)g_3^*(x)$ has at most 49 terms, and has α, α^{-1}, β and β^{-1} as roots, satisfying $\alpha^p = 1$, $\alpha^{-p} = 1$, $\beta^p = 1$ and $\beta^{-p} = 1$. We may therefore reduce all exponents in $t_1(x)t_1^*(x)t_3(x)t_3^*(x)$ modulo p, to get an at most 49-term polynomial of degree $< p$, but which has all $p - 1$ roots of $\Phi_p(x)$ as roots, contradicting the fact that the only (non-zero) polynomial of degree $< p$ with all roots of $\Phi_p(x)$ as roots is $\Phi_p(x)$ itself, which has $p > 49$ terms. Since the first $\Phi_p(x) = \frac{(x^p - 1)}{(x - 1)} = f_1(x)f_2(x)f_3(x)f_4(x)$, with four irreducible factors, happens at $p = 113 > 49$, the result follows. $\square$

Theorem 11. *Let p be a prime and $\Phi_p(x) = \frac{(x^p - 1)}{(x - 1)} = f_1(x)f_2(x)\ldots f_r(x)$ be a product of r irreducible polynomials. If the index r is any even number, then the $f_i(x)$'s divide no trinomials if $p > 7^{r/2}$.*

Proof. If any one of the $f_i(x)$'s is self-reciprocal, then the $f_i(x)$'s divide no trinomials by the above theorem. Otherwise, the $f_i(x)$'s appear in pairs. By using similar arguments, let α_i be a root of $f_i(x)$ and α_i^{-1} be a root of $f_{i+1}(x)$ with $1 \le i < r$, where i is odd. Suppose $f_i(x)$ divides some trinomial $t_i(x)$ (including the case that $f_i(x)$ itself is a trinomial). Then we can write $t_i(x) = x^m + x^a + 1 = f_i(x)g_i(x)$, with $1 \le a < m < p$. Replacing α_i by α_i^{-1} for all roots of $t_i(x)$, we get $f_{i+1}(x)g_i^*(x) = t_i^*(x) = x^m + x^{m-a} + 1$. Then the product $t_i(x)t_i^*(x) = f_i(x)f_{i+1}(x)g_i(x)g_i^*(x) = x^{2m} + x^{2m-a} + x^{m+a} + x^m + x^a + x^{m-a} + 1$ is a 7-term polynomial which has both α_i and α_i^{-1} as roots. Therefore, $t_1(x)t_1^*(x)t_3(x)t_3^*(x)\ldots t_{r-1}(x)t_{r-1}^*(x)$

$$= f_1(x)f_2(x)f_3(x)f_4(x)\ldots f_r(x)g_1(x)g_1^*(x)g_3(x)g_3^*(x)\ldots g_{r-1}(x)g_{r-1}^*(x)$$
$$= \Phi_p(x)\cdot g_1(x)g_1^*(x)g_3(x)g_3^*(x)\ldots g_{r-1}(x)g_{r-1}^*(x)$$

has at most $7^{r/2}$ terms, and has α_i's and α_i^{-1}'s as roots, satisfying $\alpha_i^p = 1$ and $\alpha_i^{-p} = 1$ for every odd integer $i < r$. We may therefore reduce all exponents in $t_1(x)t_1^*(x)t_3(x)t_3^*(x)\ldots t_{r-1}(x)t_{r-1}^*(x)$ modulo p, to get an at most $7^{r/2}$-term polynomial of degree $< p$, but which has all $p-1$ roots of $\Phi_p(x)$ as roots, contradicting the fact that the only (non-zero) polynomial of degree $< p$ with all roots of $\Phi_p(x)$ as roots is $\Phi_p(x)$ itself, which has $p > 7^{r/2}$ terms. □

The above theorem also provides a finite decision procedure for whether a given irreducible polynomial of prime primitivity p divides trinomials, in terms of the index r. Among the first 1,000,000 odd primes (i.e. 3 to 15,485,867), about 98.9% have their index $r \le 100$; 95.7% have $r \le 24$; and 92.6% have $r \le 15$. As predicted by Artin's Conjecture, about 37.4% have $r = 1$. Also, about 28.1% have $r = 2$; 6.6% have $r = 3$; 4.7% have $r = 4$; 1.9% have $r = 5$; 5.0% have $r = 6$; 0.9% have $r = 7$; 3.5% have $r = 8$; 0.7% have $r = 9$; and 1.4% have $r = 10$. (The probabilistic argument used in Artin's Conjecture for $r = 1$ can be modified for these larger values of r.)

According to the index r ($1 \le r \le 15$), we summarize the test results for whether the irreducible polynomials of prime primitivity p divide trinomials as follows:

1. When $r = 1$, $f_1(x)$ divides trinomials only at $p = 3$.
2. When $r = 2$, the $f_i(x)$'s divide trinomials only at $p = 7$.
3. When $r = 4$, the $f_i(x)$'s divide no trinomials.
4. When $r = 6$, the $f_i(x)$'s divide trinomials only at $p = 31$.
5. When $r = 8$, the $f_i(x)$'s divide trinomials only at $p = 73$.
6. When $r = 10$, the $f_i(x)$'s divide no trinomials.
7. When $r = 12$, the $f_i(x)$'s divide no trinomials.
8. When $r = 14$, the $f_i(x)$'s divide no trinomials.
9. When $r > 1$ is an odd number, the $f_i(x)$'s divide no trinomials.

4 The Multiplicative Module M

In this section, a "multiplicative module" M is introduced. It is a fact that if the irreducible polynomials of primitivity t divide trinomials, then the irreducible polynomials of primitivity mt also divide trinomials for every odd integer $m \ge 1$. Therefore, let M be the set of positive (odd) integers t such that the irreducible polynomials of odd primitivity $t > 1$ divide trinomials. Then, in view of the closure property, we call M a "multiplicative module". That is, for every $t \in M$, we also have $mt \in M$ for every odd integer $m \ge 1$. An element g of M is a **generator** of M iff $g \in M$ but no proper factor h of g is in M. Let G be the subset of M consisting of the generators of M. From Theorem 4, the polynomials of primitivity $t = 2^n - 1$ divide trinomials for every integer $n > 1$, each of these numbers $(3, 7, 15, 31, 63, 127, 255, 511, \ldots)$ is in M, and each has (at least) one factor in G.

Hence, all the "Mersenne primes" ($2^n - 1$ being prime) are automatically in G. These include $\{3; 7; 31; 127; 8,191; 131,071; 524,287; 2,147,483,647; \ldots\}$. Aside from the Mersenne primes, there are other primes in G. The first non-Mersenne-prime generator is 73, corresponding to eight irreducible polynomials of degree 9 and primitivity $t = 73$, which do divide trinomials. (In fact, two of these eight irreducible polynomials are already trinomials; therefore all eight of them must divide trinomials). By complete computer search for all odd primes $t < 3,000,000$, only five other prime elements of G (not Mersenne primes) exist: $\{73; 121,369; 178,481; 262,657;$ and $599,479\}$. It was mentioned that among the eight irreducible factors of $\Phi_{73}(x)$, two of them are already trinomials. However, none of the irreducible factors of $\Phi_t(x)$ are trinomials for $t = 121,369$ or $178,481$ or $262,657$ or $599,479$. It is not necessary that if the irreducible factors of $\Phi_t(x)$ divide trinomials, then at least one of the factors has to be a trinomial.

Let $\Phi_n(2)$ denote the nth cyclotomic polynomial evaluated at 2. All the elements of G currently known can be expressed fairly simply in terms of the numbers $\Phi_n(2)$. The Mersenne primes are precisely the numbers $\Phi_n(2)$ when n is prime and $2^n - 1$ is prime. Of the other five known prime numbers in G, three are values of $\Phi_n(2)$: $73 = \Phi_9(2)$, $262,657 = \Phi_{27}(2)$, and $599,479 = \Phi_{33}(2)$. This suggests the possibility that whenever $\Phi_n(2)$ is prime, where n is an odd prime, then $\Phi_n(2) \in G$. From [9], which lists the factorizations of $2^n - 1$ for all odd integers $n < 1200$, the first counterexample occurs at $151 = \Phi_{15}(2)$, which is not in G, and the next non-Mersenne-prime case $4,432,676,798,593 = \Phi_{49}(2)$ is too large to test. Besides $73 = \Phi_9(2)$, $262,657 = \Phi_{27}(2)$, and $599,479 = \Phi_{33}(2)$, the other two cases result from dividing $\Phi_n(2)$ by a "small" prime factor: $121,369 = \Phi_{39}(2)/79$, and $178,481 = \Phi_{23}(2)/47$. These two are both instances where $\Phi_n(2)$ has two prime factors, and t is the much larger of these two factors. While this may be a good way to look for likely values of t, it is not a reliable indicator. For example, $\Phi_{35}(2) = 71 \cdot 122,921$, but $t = 122,921$ is not in G, and the next non-Mersenne-prime cases, $\Phi_{37}(2) = 223 \cdot 1,616,318,177$ and $\Phi_{41}(2) = 13,367 \cdot 164,511,353$, are also too large to test by current methods.

If $g \in G$ is composite, then (by the definition of G) no prime factor of g is in G. The smallest composite $g \in G$ is $85 = 5 \cdot 17$, where $85 \in G$ but $5 \notin G$ and $17 \notin G$. All the eight irreducible factors of $\Phi_{85}(x)$ divide trinomials, even though none of them is a trinomial. By complete computer search, there are ten composite elements of G up to $t < 800,000$. These are: $\{85; 2,047; 3,133; 4,369; 11,275; 49,981; 60,787; 76,627, 140,911;$ and $486,737\}$. Seven other larger composite elements of G are currently known: $\{1,826,203; 2,304,167; 2,528,921; 8,727,391; 14,709,241; 15,732,721;$ and $23,828,017\}$. Among these seventeen composite elements of G known so far, most of them are either divisors of $\Phi_n(2)$ or of $\Phi_n(2)\Phi_{2n}(2)$ for various values of n. For example, $2,047 = 23 \cdot 89 = \Phi_{11}(2)$, $8,727,391 = 71 \cdot 122,921 = \Phi_{35}(2)$, and $14,709,241 = 631 \cdot 23,311 = \Phi_{45}(2)$ are of the form $\Phi_n(2)$; $2,304,167 = 1,103 \cdot 2,089 = \Phi_{29}(2)/233$, $23,828,017 = 11,119 \cdot 2,143 = \Phi_{51}(2)/103$, and $486,737 = 233 \cdot 2,089 = \Phi_{29}(2)/1103$ are of the form $\Phi_n(2)/c$, where c is a prime factor of $\Phi_n(2)$; $85 = 5 \cdot 17 = \Phi_4(2)\Phi_8(2)$, $3,133 = 13 \cdot 241 = \Phi_{12}(2)\Phi_{24}(2)$, $4,369 = 17 \cdot 257 =$

$\Phi_8(2)\Phi_{16}(2)$, $49,981 = 151 \cdot 331 = \Phi_{15}(2)\Phi_{30}(2)$, $140,911 = 43 \cdot (29 \cdot 113) = \Phi_{14}(2)\Phi_{28}(2)$, and $15,732,721 = 241 \cdot (97 \cdot 673) = \Phi_{24}(2)\Phi_{48}(2)$ are of the form $\Phi_n(2)\Phi_{2n}(2)$; $60,787 = 89 \cdot 683 = \Phi_{11}(2)\Phi_{22}(2)/23$, $76,627 = 19 \cdot (37 \cdot 109) = \Phi_{18}(2)\Phi_{36}(2)/3$, $1,826,203 = 337 \cdot 5,419 = \Phi_{21}(2)\Phi_{42}(2)/7$, and $2,528,921 = 41 \cdot 61,681 = \Phi_{20}(2)\Phi_{40}(2)/5$ are of the form $\Phi_n(2)\Phi_{2n}(2)/c$, where c is a prime factor of $\Phi_n(2)$. The only exception so far, which is neither a divisor of $\Phi_n(2)$ nor of $\Phi_n(2)\Phi_{2n}(2)$, is $11,275 = 11 \cdot (5 \cdot 41) \cdot 5 = \Phi_{10}(2)\Phi_{20}(2)\Phi_5(2)$. However, all these composite elements in G suggest testing the following kinds of numbers for membership in G.

1. If $\Phi_n(2)$ is prime and both of $\Phi_n(2)$ and $\Phi_{2n}(2) \notin M$, then $\Phi_n(2)\Phi_{2n}(2) \in G$, though this is not true for $n = 10$.
2. If $\Phi_n(2)$ is composite and $\Phi_{2n}(2) \notin M$, then $\Phi_n(2)\Phi_{2n}(2)/c \in G$, where c is a prime factor of $\Phi_n(2)$.

Also, it seems possible that $\Phi_{4n}(2) \notin M$ for all integers $n > 0$ (all n up to 15 have been verified); and $\Phi_n(2) \notin M$ implies $\Phi_{2n}(2) \notin M$ (all n up to 20 have been verified).

5   Conclusions

The "theory" of the irreducible polynomials which divide trinomials over GF(2) has an interesting structure. All the polynomials of (odd) primitivity t are the r irreducible factors of $\Phi_t(x)$ over GF(2), and either all or none of them divide trinomials. Whether it is "all" or "none" depends to a considerable extent on r. (For all odd $t > 3$ and all odd r, the answer is "none". For $r = 4, 10, 12$, and 14, the answer is "none" for all odd primes t. There is only one prime value of t, for each of $r = 2, 6$, and 8, for which the answer is "all" rather than "none". For each even $r > 14$, there is only a finite range, $t \leq 7^{r/2}$, for prime values of t where the answer **might** be "all".) The odd values of $t > 1$ such that polynomials of primitivity t divide trinomials form a "multiplicative module" M, which is closed with respect to multiplication by odd numbers. The set G of "generators" of M is quite sparse, and its members seem related to numbers of the form $\Phi_n(2)$.

References

1. S.W. Golomb, Shift Register Sequences, Holden-Day, Inc. 1967; Second Edition, Aegean Park Press 1982.
2. N. Zierler and J. Brillhart, On Primitive Trinomials (mod 2), Information and Control 13, 1968, 541-554.
3. N. Zierler and J. Brillhart, On Primitive Trinomials (mod 2) II, Information and Control 14, 1969, 566-569.
4. G. Seroussi, Table of Low-Weight Binary Irreducible Polynomials, Computer Systems Laboratory, HPL-98-135, 1998.
5. R.G. Swan, Factorization of Polynomials over Finite Fields, Pacific J. Mathematics 12 (1962)pp. 1099-1106.

6. R. P. Brent and P. Zimmermann, Algorithms for Finding Almost Irreducible and Almost Primitive Trinomials, Proceedings of a Conference in Honor of Professor H.C. Williams, The Fields Institute, May 2003.

7. J. H. Lindholm, An Analysis of the Pseudo-Randomness Properties of Subsequences of Long m-Sequences, IEEE Transactions on Information Theory, Vol. IT-14, No. 4, July, 1968, 569-576.

8. C. Hooley, Artin's conjecture for primitive roots, J. Reine Angew. Math. 225 (1967), 209-220.

9. J. Brillhart, D.H. Lehmer, J.L. Selfridge, B. Tuckerman, and S.S. Wagstaff, Jr., Factorizations of $b^n \pm 1 (b = 2, 3, 5, 6, 7, 10, 11, 12)$ up to high powers, Contemporary Mathematics, Third Edition, Vol. 22, American Math. Soc. 2003.

Autocorrelation Properties of Resilient Functions and Three-Valued Almost-Optimal Functions Satisfying PC(p)[*]

Seunghoon Choi[1],[**] and Kyeongcheol Yang[2]

[1] Samsung Electronics Co. Ltd.,
Suwon, Gyeonggi 442-600, Korea
seunghoon.choi@samsung.com
[2] Dept. of Electronic and Electrical Engineering,
Pohang University of Science and Technology (POSTECH),
Pohang, Gyungbuk 790-784, Korea
kcyang@postech.ac.kr

Abstract. The absolute indicator and the sum-of-squares indicator are used as a measure of global avalanche criterion (GAC) to evaluate the propagation characteristics of Boolean functions in a global manner. In this paper, we derive a new lower bound on the absolute indicator of resilient functions and three-valued almost-optimal functions satisfying the propagation criterion of degree p (or PC(p)).

1   Introduction

The strict avalanche criterion (SAC) or the propagation criterion of degree p (or PC(p), for short) are employed as a measure of the propagation characteristics of Boolean functions [8], [15]. Even though a Boolean function satisfies SAC or PC(p), it may have a cryptographically unwelcome property called a linear structure. Zhang et al. [17] introduced the global avalanche criterion (GAC) to compensate for the weak points of SAC and PC(p). The absolute indicator and the sum-of-squares indicator are used as a measure of the GAC. Son et al. [11] derived a lower bound on the sum-of-squares indicator of the balanced functions and Sung et al. [12] improved their results. The GAC of correlation-immune functions is analyzed in [14], [20]. Recently, Maitra presented better results on the GAC for correlation-immune and resilient functions in [5].

In this paper, we consider correlation-immune and resilient functions satisfying PC(p) and derive a new lower bound on their absolute indicators. We give a counter-example against Theorem of Charpin and Pasalic (Theorem 5, [4]).

---

[*] This work was supported by Grant No. R01-2003-000-10330-0 from the Basic Research Program of the Korea Science and Engineering Foundation (KOSEF).

[**] S. Choi was with the Dept. of Electronic and Electrical Engineering, Pohang University of Science and Technology (POSTECH), while he was doing this research.

Their corrected bound in [7] is numerically compared with our bound in order to show that these bounds are complementary to each other. We also consider the autocorrelation properties of three-valued almost-optimal functions satisfying PC(p) and derive a new lower bound on their absolute indicators.

The paper is organized as follows. In Section 2, we introduce some definitions and notation for our presentation. In Section 3, we give a counter-example against Theorem of Charpin and Pasalic [4] and then derive a new lower bound on the absolute indicator of correlation-immune and resilient functions satisfying PC(p). In Section 4, we derive a lower bound on the absolute indicator of three-valued almost-optimal functions satisfying PC(p). Finally, we give concluding remarks in Section 5.

2   Preliminaries

Let $\mathbb{F}_2^n$ be the set of binary n-tuple vectors, where $\mathbb{F}_2 = \{0, 1\}$. A Boolean function f with n variables is a function from $\mathbb{F}_2^n$ to $\mathbb{F}_2$ and is uniquely represented by a polynomial in $x_1, \ldots, x_n$:

$$f(x_1, \ldots, x_n) = a_0 + a_1 x_1 + \cdots + a_n x_n + a_{12} x_1 x_2 + \cdots + a_{1 \ldots n} x_1 \ldots x_n \quad (1)$$

where $a_0, a_1, \ldots, a_{1 \ldots n} \in \mathbb{F}_2$ and $+$ denotes the modulo-2 addition. The form in (1) is called the *algebraic normal form* (ANF) of f. The *degree* of f, denoted by $\deg(f)$, is the number of variables in the highest order product term with a nonzero coefficient. A Boolean function of degree at most 1 is said to be *affine*. An affine function whose constant equals to zero is said to be *linear*. Every linear function can be expressed by $\varphi_\alpha(x) \triangleq \alpha \cdot x$ for some $\alpha = (\alpha_1, \ldots, \alpha_n) \in \mathbb{F}_2^n$, where $\alpha \cdot x = \alpha_1 x_1 + \cdots + \alpha_n x_n$ for $x = (x_1, \ldots, x_n) \in \mathbb{F}_2^n$.

For simple notation, let $\mathcal{B}_n$ be the set of all Boolean functions with n variables. For a Boolean function $f \in \mathcal{B}_n$, let

$$\mathcal{F}(f) \triangleq \sum_{x \in \mathbb{F}_2^n} (-1)^{f(x)}. \quad (2)$$

Note that $\mathcal{F}(f)$ is the difference between the occurrences of one and the occurrences of zero in $f(x)$ when x runs through all the elements of $\mathbb{F}_2^n$. In particular, f is said to be *balanced* if $\mathcal{F}(f) = 0$. The *Walsh transform* of a function $f \in \mathcal{B}_n$ with respect to α is defined by $\mathcal{F}(f + \varphi_\alpha)$, where

$$\mathcal{F}(f + \varphi_\alpha) = \sum_{x \in \mathbb{F}_2^n} (-1)^{f(x) + \varphi_\alpha(x)}. \quad (3)$$

The set $\{\mathcal{F}(f + \varphi_\alpha) \mid \alpha \in \mathbb{F}_2^n\}$ is called the *Walsh spectrum* of f.

Let $\mathcal{L}(f)$ be the maximum magnitude of the Walsh transform of $f \in \mathcal{B}_n$, defined by

$$\mathcal{L}(f) \triangleq \max_{\alpha \in \mathbb{F}_2^n} |\mathcal{F}(f + \varphi_\alpha)|. \quad (4)$$

It is easily checked that $\mathcal{L}(f) \geq 2^{n/2}$ for any function $f \in \mathcal{B}_n$ with equality if and only if f is bent when n is even. The *nonlinearity* $\mathcal{N}(f)$ of $f \in \mathcal{B}_n$ is defined by $\mathcal{N}(f) = 2^{n-1} - \mathcal{L}(f)/2$. Note that a bent function has the largest nonlinearity but is not balanced. f is said to be *almost-optimal* if $\mathcal{L}(f) \leq 2^{(n+1)/2}$ when n is odd, and $\mathcal{L}(f) \leq 2^{(n+2)/2}$ when n is even. f is said to be *three-valued* if its Walsh spectrum takes at most three values $0, \pm\mathcal{L}(f)$.

The correlation-immunity of $f \in \mathcal{B}_n$ is characterized by the set of zero values in its Walsh spectrum [10], [16]. f is m-th order *correlation-immune* if $\mathcal{F}(f + \varphi_\alpha) = 0$ for any $\alpha \in \mathbb{F}_2^n$ such that $1 \leq \mathrm{wt}(\alpha) \leq m$, where $\mathrm{wt}(\alpha)$ is the Hamming weight of $\alpha \in \mathbb{F}_2^n$ which is the number of ones in α. An m-th order correlation-immune function is said to be m-*resilient* if it is balanced. Note that 0-resilient functions are balanced functions [13]

The propagation characteristics of f are concerned with its derivatives $D_\alpha f$ given by $D_\alpha f(x) = f(x) + f(x + \alpha)$ with respect to any direction $\alpha \in \mathbb{F}_2^n$. When $D_\alpha f$ is a constant, such $\alpha \neq 0$ is called a *linear structure* of f. Note that the set of all linear structures of f forms a subspace of $\mathbb{F}_2^n$, called the *linear space* of f. f is said to *satisfy propagation criterion of degree* p, i.e., PC(p) if $\mathcal{F}(D_\alpha f) = 0$ for any $\alpha \in \mathbb{F}_2^n$ such that $1 \leq \mathrm{wt}(\alpha) \leq p$.

The absolute indicator and the sum-of-squares indicator are measures to estimate the propagation characteristics in a global manner (i.e., GAC) [17]. For a function $f \in \mathcal{B}_n$, the absolute indicator of f is defined by

$$\Delta_f = \max_{\alpha \in \mathbb{F}_2^n, \ \alpha \neq 0} |\mathcal{F}(D_\alpha f)|$$

and the sum-of-squares indicator of f is defined by

$$\sigma_f = \sum_{\alpha \in \mathbb{F}_2^n} \mathcal{F}^2(D_\alpha f).$$

Note that the smaller Δ_f and σ_f, the better the GAC of a function. In [17], Zhang and Zheng showed that $0 \leq \Delta_f \leq 2^n$ and $2^{2n} \leq \sigma_f \leq 2^{3n}$.

3   The Absolute Indicator of Correlation-Immune Functions Satisfying PC(p)

Charpin and Pasalic [4] first presented a lower bound on the absolute indicator of m-resilient functions of degree d satisfying PC(p) in the following.

Theorem 1 ([4]). *Let $f \in \mathcal{B}_n$ be an m-resilient function of degree d satisfying PC(p). Set $\epsilon = \lfloor \frac{n-m-2}{d} \rfloor$. Then for any $\alpha \in \mathbb{F}_2^n$, we have*

$$\mathcal{F}(D_\alpha f) \equiv 0 \pmod{2^{2m+p+2\epsilon+5-n}}.$$

This property is significant only for $2m + p + 2\epsilon + 2 > n$.

Based on the result, Charpin and Pasalic [4] claimed that $\Delta_f \geq 2^{2m+p+2\epsilon+5-n}$. However, we show that their claim is incorrect by giving a counter-example in the following.

Example: Consider the function $f \in \mathcal{B}_9$ given by

$$f(x_1, \ldots, x_9) = x_1(1 + x_2 + x_3 + x_4 + x_5 + x_6 + x_7 + x_8 + x_9)$$
$$+ x_2 x_3 + x_4 x_5 + x_6 x_7 + x_8 x_9.$$

Note that f has degree $d = 2$ and is a 0-resilient function. Furthermore, f also satisfies PC(8). Thus, the claim by Charpin and Pasalic implies that $\Delta_f \geq 1024$. On the other hand, f has the all-one vector as its linear structure, since $f(x_1, \ldots, x_9) + f(x_1 + 1, \ldots, x_9 + 1) = 1$. Therefore, $\Delta_f = 512$, which is a contradiction. Note that $\Delta_f = 2^n$ for any quadratic non-bent function $f \in \mathcal{B}_n$ [17]. $\qquad \square$

Remark: Charpin and Pasalic missed the assumption that the function admits no linear structure in their theorem [4]. Our observation in the above example was informed to Charpin by Prof. Helleseth after his visit to the authors at POSTECH in the fall of 2002. Their theorem has been corrected later by Pasalic in his Ph. D. thesis by adding the condition that $2m + p + 2\epsilon + 5 - n \leq n$ (see Theorem 7.11, [7]).

Zheng and Zhang derived a lower bound on the sum-of-squares indicator of a Boolean function [18] as follows:

Theorem 2 ([18]). *Let $f \in \mathcal{B}_n$ and $N_f \triangleq \#\{\alpha \in \mathbb{F}_2^n \mid \mathcal{F}(f + \varphi_\alpha) \neq 0\}$. Then $\sigma_f \geq \frac{2^{3n}}{N_f}$. Moreover, if f has a three-valued Walsh spectra 0, $\pm 2^i$, then $\sigma_f = \frac{2^{3n}}{N_f}$.*

Maitra [5] presented an improved lower bound on the sum-of-squares indicator of m-th order correlation-immune and m-resilient functions directly from the definition of correlation-immunity and resiliency and the bound in Theorem 2 in the following:

Theorem 3 ([5]). *Let $f \in \mathcal{B}_n$ be an m-th order correlation-immune function. Then $\sigma_f \geq \frac{2^{3n}}{2^n - \sum_{i=1}^m \binom{n}{i}}$. Moreover, if f is m-resilient, then $\sigma_f \geq \frac{2^{3n}}{2^n - \sum_{i=0}^m \binom{n}{i}}$.*

Following the results by Maitra, we derive a new lower bound on the absolute indicator of m-th order correlation-immune and m-resilient functions of degree d satisfying PC(p).

Theorem 4. *Let $f \in \mathcal{B}_n$ be an m-th order correlation-immune function of degree d satisfying PC(p). Then the absolute indicator of f is bounded by*

$$\Delta_f \geq 2^{\lfloor \frac{n-2}{d-1} \rfloor + 2} \cdot k_1,$$

where k_1 is the least integer among all positive integers i_1 satisfying

$$i_1^2 \geq \frac{1}{2^n - 1 - \sum_{j=1}^{p} \binom{n}{j}} \left(\frac{2^{3n-4-2\lfloor \frac{n-2}{d-1} \rfloor}}{2^n - \sum_{l=1}^{m} \binom{n}{l}} - 2^{2n-4-2\lfloor \frac{n-2}{d-1} \rfloor} \right).$$

Similarly, when $f \in \mathcal{B}_n$ is an m-resilient function of degree d satisfying $PC(p)$, the absolute indicator of f is bounded by

$$\Delta_f \geq 2^{\lfloor \frac{n-2}{d-1} \rfloor + 2} \cdot k_2,$$

where k_2 is the least integer among all positive integers i_2 satisfying

$$i_2^2 \geq \frac{1}{2^n - 1 - \sum_{j=1}^{p} \binom{n}{j}} \left(\frac{2^{3n-4-2\lfloor \frac{n-2}{d-1} \rfloor}}{2^n - \sum_{l=0}^{m} \binom{n}{l}} - 2^{2n-4-2\lfloor \frac{n-2}{d-1} \rfloor} \right).$$

Proof. Since f is an m-th order correlation immune function, we have from Theorem 3

$$\sigma_f = \sum_{\alpha \in \mathbb{F}_2^n \setminus \{0\}} \mathcal{F}^2(D_\alpha f) + 2^{2n} \tag{5}$$

$$\geq \frac{2^{3n}}{2^n - \sum_{l=1}^{m} \binom{n}{l}}. \tag{6}$$

Since any derivative $D_\alpha f$ of f is a function of degree $d-1$ with a linear structure,

$$\mathcal{F}(D_\alpha f) \equiv 0 \ (\bmod\ 2^{\varepsilon+2})$$

where $\varepsilon = \lfloor \frac{n-2}{d-1} \rfloor$ (see [1], [6]). For a nonnegative integer i, set $\lambda_i \triangleq \#\{\alpha \in \mathbb{F}_2^n \setminus \{0\} : |\mathcal{F}(D_\alpha f)| = i 2^{\varepsilon+2}\}$. Then

$$\sum_{\alpha \in \mathbb{F}_2^n \setminus \{0\}} \mathcal{F}^2(D_\alpha f) = \sum_i \lambda_i i^2 2^{2\varepsilon+4} = 2^{2\varepsilon+4} \sum_i \lambda_i i^2.$$

On the other hand, let λ be the number of $\alpha \in \mathbb{F}_2^n \setminus \{0\}$ such that $\mathcal{F}(D_\alpha f) \neq 0$. Since f satisfies $PC(p)$,

$$\lambda \leq 2^n - 1 - \sum_{j=1}^{p} \binom{n}{j}.$$

So there exists an integer i_1 such that

$$2^{2\varepsilon+4} \sum_i \lambda_i i^2 \leq 2^{2\varepsilon+4} \lambda i_1^2 \leq 2^{2\varepsilon+4} i_1^2 \left(2^n - 1 - \sum_{j=1}^{p} \binom{n}{j} \right) \tag{7}$$

and by (5), (6) and (7), we can define the smallest positive integer k_1 among integers i_1 satisfying

$$\frac{2^{3n}}{2^n - \sum_{l=1}^{m} \binom{n}{l}} - 2^{2n} \leq 2^{2\varepsilon+4} i_1^2 \left(2^n - 1 - \sum_{j=1}^{p} \binom{n}{j} \right).$$

Table 1. Comparison of Lower bounds on the absolute indicator of m-resilient functions of degree d satisfying $PC(p)$ when $n = 8, 9, 10$ and 11. Here, CP denotes the correct version of Charpin and Pasalic's theorem

n	d	m	p	Thm 4	CP [7]	n	d	m	p	Thm 4	CP [7]
8	3	0	3	**32**	16	10	3	3	2	**64**	32
8	3	0	4	32	32	10	3	3	3	64	64
8	3	0	5	32	**64**	10	3	3	4	64	**128**
8	6	1	5	16	16	10	4	2	2	**16**	8
8	6	1	6	24	**32**	10	4	2	3	16	16
8	7	0	7	**24**	16	10	4	2	4	16	**32**
9	4	1	3	**16**	8	11	5	1	5	**16**	8
9	4	1	4	16	16	11	5	1	6	16	16
9	4	1	5	16	**32**	11	5	1	7	16	**32**
9	7	1	6	16	16	11	8	2	5	**16**	8
9	7	1	7	24	**32**	11	8	2	6	16	16
9	8	0	8	**24**	16	11	8	2	7	32	32

From the definition of k_1, there exists an element $\alpha \in \mathbb{F}_2^n \backslash \{0\}$ such that

$$|\mathcal{F}(D_\alpha f)| \geq k_1 \cdot 2^{\varepsilon+2}$$

and we get a lower bound on the absolute indicator given by

$$\Delta_f \geq 2^{\varepsilon+2} \cdot k_1.$$

A similar approach may apply to the case that f is an m-resilient function of degree d satisfying $PC(p)$. □

Our lower bound on the absolute indicator of m-resilient functions of degree d satisfying $PC(p)$ is numerically compared with the correct version of Charpin and Pasalic's theorem in Tables 1 and 2 for $n = 8, \ldots, 12$. Note that the meaningful ranges of n, m, p and d may be determined by the Siegenthaler inequality $m+d \leq n - 1$ [10] and the Zheng-Zhang inequality $m + p \leq n - 1$ [19]. Tables 1 and 2 show that our lower bound becomes tighter than the correct version of Charpin and Pasalic's theorem when both m and p are smaller and n gets larger.

In order to compare the actual value of the absolute indicator with its bound given by Theorem 4 for m-resilient functions of degree d satisfying $PC(p)$, we give an example in the following.

Example: Consider the function $f \in \mathcal{B}_8$ given by

$$f(x_1, \ldots, x_8) = x_1(x_2 + x_3 + x_6 + x_5 x_6 + x_5 x_8 + x_6 x_8)$$
$$+ x_3(x_2 + x_5 + x_5 x_6 + x_5 x_8 + x_6 x_8)$$
$$+ x_4(x_5 + x_6) + x_5(1 + x_6 + x_7 + x_8) + x_7 x_8.$$

Note that f has $n = 8$, $d = 3$, $m = 0$ and $p = 3$. It is easily checked by computer that $\Delta_f = 256$. Theorem 4 gives $\Delta_f \geq 32$, while the correct version of Charpin and Pasalic's theorem gives $\Delta_f \geq 16$. □

Table 2. Comparison of Lower bounds on the absolute indicator of m-resilient functions of degree d satisfying $PC(p)$ when $n = 12$. Here, CP denotes the correct version of Charpin and Pasalic's theorem

d	m	p	Thm 4	CP [7]	d	m	p	Thm 4	CP [7]
3	0	7	**128**	64	5	3	8	80	**512**
3	0	8	**128**	128	5	4	1	**32**	16
3	0	9	128	**256**	5	4	2	32	32
3	1	5	**128**	64	6	1	5	**16**	4
3	1	6	**128**	128	7	0	10	24	**32**
3	1	7	128	**256**	7	0	11	**72**	64
4	1	9	32	**256**	7	4	4	**40**	32
4	1	10	96	**512**	7	4	5	48	**64**
4	2	1	**32**	4	8	0	10	24	**32**
4	2	2	**32**	8	8	0	11	**72**	64

4 The Absolute Indicator of Three-Valued Almost-Optimal Functions Satisfying PC(p)

The sum-of-squares indicator of a Boolean function is related to its nonlinearity or its Walsh spectrum through the following theorem.

Theorem 5 ([18]). Let $f \in \mathcal{B}_n$ and let $\mathcal{L}(f) = \max_{\alpha \in \mathbb{F}_2^n} |\mathcal{F}(f + \varphi_\alpha)|$. Then we have

$$\sigma_f \leq 2^n \mathcal{L}(f)^2$$

with equality if and only if the Walsh Spectrum of f takes at most three values, 0, $\mathcal{L}(f)$, $-\mathcal{L}(f)$.

Applying Theorem 5 to the almost-optimal functions leads directly to the following corollary.

Corollary 6 ([1]). Let $f \in \mathcal{B}_n$ be an almost-optimal function. Then the sum-of-squares indicator of f satisfies $\sigma_f \leq 2^{2n+1}$ when n is odd and $\sigma_f \leq 2^{2n+2}$ when n is even. All cases have equality if and only if the Walsh Spectrum of f takes at most three values, 0, $\mathcal{L}(f)$, $-\mathcal{L}(f)$.

The next theorem gives the distribution of the Walsh spectra for a three-valued Boolean function [1].

Theorem 7 ([1]). Let $f \in \mathcal{B}_n$. Assume that the Walsh spectrum of f takes at most three values, 0, $\pm \mathcal{L}(f)$. Then $\mathcal{L}(f) = 2^i$ with $i \geq n/2$ and

$$N_f = \frac{2^{2n}}{\mathcal{L}(f)^2} = 2^{2n-2i},$$

$$Z_f = 2^n - 2^{2n-2i}$$

where $N_f = \#\{\alpha \in \mathbb{F}_2^n \mid \mathcal{F}(f + \varphi_\alpha) \neq 0\}$, $Z_f = \#\{\alpha \in \mathbb{F}_2^n \mid \mathcal{F}(f + \varphi_\alpha) = 0\}$.

An *almost-optimal* function with three-valued Walsh spectra is called a *three-valued almost-optimal* function [2]. Applying Theorem 7 to this case, we have

$$N_f = 2^{n-1}, \quad Z_f = 2^{n-1} \qquad \text{if } n \text{ is odd;} \tag{8}$$
$$N_f = 2^{n-2}, \quad Z_f = 3 \cdot 2^{n-2} \qquad \text{if } n \text{ is even.} \tag{9}$$

When n is odd, it is easily checked from the definition of resiliency that if a *three-valued almost-optimal* function has m-resiliency, we have $\sum_{i=0}^{m} \binom{n}{i} \leq 2^{n-1} = Z_f$ and therefore $m \leq (n-1)/2$.

A simple lower bound on the GAC of three-valued almost-optimal functions can be easily obtained from the above results and Lemma 2 [5] in the following.

Corollary 8. *Let $f \in \mathcal{B}_n$ be a three-valued almost-optimal function. The absolute indicator of f is bounded by*

$$\Delta_f > \begin{cases} 2^{n/2}, & \text{if } n \text{ is odd;} \\ \sqrt{3} \cdot 2^{n/2}, & \text{if } n \text{ is even.} \end{cases}$$

Proof. From Lemma 2 [5] and Equation (8), we have

$$\Delta_f \geq \sqrt{\frac{1}{2^n - 1} \frac{2^{2n} \cdot 2^{n-1}}{2^n - 2^{n-1}}} > \sqrt{\frac{1}{2^n} \cdot 2^{2n}} = 2^{n/2}.$$

when n is odd. Similarly, when n is even,

$$\Delta_f \geq \sqrt{\frac{1}{2^n - 1} \frac{2^{2n} \cdot 3 \cdot 2^{n-2}}{2^n - 3 \cdot 2^{n-2}}} > \sqrt{\frac{1}{2^n} \cdot 3 \cdot 2^{2n}} = \sqrt{3} \cdot 2^{n/2}.$$

$\square$

In the case of three-valued almost-optimal functions of degree d satisfying $PC(p)$, a stronger bound on their absolute indicator can be derived as follows.

Theorem 9. *Let $f \in \mathcal{B}_n$ be a three-valued almost-optimal function of degree d satisfying $PC(p)$. When n is odd,*

$$\Delta_f \geq 2^{\lfloor \frac{n-2}{d-1} \rfloor + 2} \cdot l_1,$$

where l_1 is the least integer among all positive integers i satisfying

$$i^2 \geq \frac{2^{2n-4-2\lfloor \frac{n-2}{d-1} \rfloor}}{2^n - 1 - \sum_{j=1}^{p} \binom{n}{j}}.$$

Similarly, when n is even,

$$\Delta_f \geq 2^{\lfloor \frac{n-2}{d-1} \rfloor + 2} \cdot l_2,$$

where l_2 is the least integer among all positive integers i satisfying

$$i^2 \geq \frac{3 \cdot 2^{2n-4-2\lfloor \frac{n-2}{d-1} \rfloor}}{2^n - 1 - \sum_{j=1}^{p} \binom{n}{j}}.$$

Proof. For simplicity, consider only the case where n is odd. Because f is a three-valued almost-optimal function, we have from Corollary 6

$$\sigma_f = \sum_{\alpha \in \mathbb{F}_2^n \setminus \{0\}} \mathcal{F}^2(D_\alpha f) + 2^{2n} = 2^{2n+1}$$

Since any derivative $D_\alpha f$ of f is a function of degree $d-1$ with a linear structure,

$$\mathcal{F}(D_\alpha f) \equiv 0 \pmod{2^{\varepsilon+2}}$$

where $\varepsilon = \lfloor \frac{n-2}{d-1} \rfloor$ (see [1], [6]). For any $\alpha \in \mathbb{F}_2^n \setminus \{0\}$, there exists a positive integer i such that $0 \leq i \leq 2^{n-2-\varepsilon}$ and $\mathcal{F}^2(D_\alpha f) = i^2 2^{2(\varepsilon+2)}$. Set $\lambda_i = \#\{\alpha \in \mathbb{F}_2^n \setminus \{0\} : |\mathcal{F}(D_\alpha f)| = i\,2^{\varepsilon+2}\}$. Then,

$$\sum_{\alpha \in \mathbb{F}_2^n \setminus \{0\}} \mathcal{F}^2(D_\alpha f) = \sum_{i=1}^{c} \lambda_i i^2 2^{2\varepsilon+4} = 2^{2n}, \tag{10}$$

where $c = 2^{n-2-\varepsilon}$. On the other hand, let λ be the number of $\alpha \in \mathbb{F}_2^n \setminus \{0\}$ such that $\mathcal{F}(D_\alpha f) \neq 0$. Since f satisfies PC(p),

$$\lambda \leq 2^n - 1 - \sum_{j=1}^{p} \binom{n}{j}.$$

The same procedure as in the Proof of Theorem 4 leads to the result.   □

It is known in [3] that if the Walsh spectrum of $f \in \mathcal{B}_n$ takes at most three values $0, \pm \mathcal{L}(f)$ and $\mathcal{L}(f) = 2^i$, then the degree d of f satisfies $d \leq n - i + 1$. Therefore, for any three-valued almost-optimal function

$$d \leq \begin{cases} (n+1)/2, & \text{if } n \text{ is odd}; \\ n/2, & \text{if } n \text{ is even}. \end{cases}$$

Based on Theorem 9, lower bounds on the absolute indicator of three-valued almost-optimal functions of degree d satisfying PC(p) are listed in Tables 3, 4 and 5, when $n = 6$ and 7, respectively. Note that $\Delta_f > 2^n$ implies that such function does not exist, as in the case of $n = 6$, $d = 2$ and $p = 5$. Also, note that if a Boolean function $f \in \mathcal{B}_n$ has $\Delta_f = 2^n$, then it has a linear structure as in the cases of $n = 7, d = 3, p = 6$ and $n = 7, d = 4, p = 6$ in Table 4.

In the followings we give some examples to compare the actual value of the absolute indicator with its bound given by Theorem 9 for three-valued almost-optimal functions of degree d satisfying PC(p).

Example: Consider the function $f \in \mathcal{B}_7$ given by

$$f(x_1, \ldots, x_7) = (x_1 + x_4 + x_6)(x_4 + x_5 + x_6)(x_2 + x_3 + x_5 + x_7)(x_5 + x_6 + x_7)$$
$$+ (x_1 + x_4 + x_6)(x_2 + x_3 + x_5 + x_7)(x_6 + x_7)(x_2 + x_4 + x_5 + x_6)$$

Table 3. Lower bounds on the absolute indicator of three-valued almost-optimal functions of degree d satisfying PC(p) when $n = 6$

d	p	Δ_f	d	p	Δ_f
2	1	64	3	1	16
2	2	64	3	2	32
2	3	64	3	3	32
2	4	64	3	4	48
2	5	128	3	5	112

Table 4. Lower bounds on the absolute indicator of three-valued almost-optimal functions of degree d satisfying PC(p) when $n = 7$

d	p	Δ_f	d	p	Δ_f
2	1	128	3	4	32
2	2	128	3	5	48
2	3	128	3	6	128
2	4	128	4	1	16
2	5	128	4	2	16
2	6	128	4	3	16
3	1	16	4	4	24
3	2	16	4	5	48
3	3	16	4	6	128

$$+(x_1 + x_4 + x_6)(x_4 + x_5 + x_6)(x_2 + x_3 + x_5 + x_7)$$
$$+(x_1 + x_4 + x_6)(x_2 + x_3 + x_5 + x_7)(x_2 + x_7)$$
$$+(x_1 + x_4 + x_6)(x_4 + x_5 + x_6)$$
$$+(x_2 + x_3 + x_5 + x_7)(x_5 + x_6 + x_7)$$
$$+(x_6 + x_7)(x_2 + x_4 + x_5 + x_6).$$

f is a three-valued almost-optimal function of degree $d = 4$ satisfying PC(1). It is easily checked by computer that $\Delta_f = 64$. Theorem 9 gives $\Delta_f \geq 16$. □

Example: Consider the function $f \in \mathcal{B}_8$ given by

$$f(x_1,\ldots,x_8) = x_1(x_2 + x_3 + x_3x_4x_7 + x_3x_4x_8 + x_3x_6x_7 + x_3x_6x_8)$$
$$+x_3(1 + x_2 + x_4 + x_6 + x_4x_7 + x_4x_8 + x_6x_7 + x_6x_8)$$
$$+x_4x_6(x_7 + x_8) + x_5(x_7 + x_8) + x_8.$$

f is a three-valued almost-optimal function of degree $d = 4$ satisfying PC(1). It is easily checked by computer that $\Delta_f = 256$. Theorem 9 gives $\Delta_f \geq 32$. □

Table 5. Lower bounds on the absolute indicator of three-valued almost-optimal functions of degree d satisfying PC(p) when $n = 8$

d	p	Δ_f	d	p	Δ_f
2	1	256	3	3	64
2	2	256	3	4	64
2	3	256	4	1	32
2	4	256	4	2	32
2	5	256	4	3	48
3	1	32	4	4	48
3	2	32	4	5	80

5   Conclusion

We derived a new lower bound on the absolute indicator of m-th order correlation-immune or m-resilient functions of degree d satisfying PC(p). In a similar way, a lower bound on the absolute indicator of three-valued almost-optimal functions was also presented.

Acknowledgments

The authors wish to thank the anonymous referees for their valuable comments to improve the presentation of the paper.

References

1. A. Canteaut, C. Carlet, P. Charpin and C. Fontaine, "Propagation Characteristics and Correlation-Immunity of Highly Nonlinear Boolean Functions," In *Advances in Cryptology - EUROCRYPT 2000*, LNCS 1807, pp. 507-522, Springer-Verlag, 2000.
2. A. Canteaut, C. Carlet, P. Charpin and C. Fontaine, "On Cryptographic Properties of the Cosets of $\mathcal{R}(1, m)$," In *IEEE Transactions on Information Theory*, IT-47, pages 1494-1513, May 2001.
3. C. Carlet, "Two new classes of bent functions," In *Advances in Cryptology - EUROCRYPT'93*, LNCS 765, pp. 77-101, Springer-Verlag, 1994.
4. P. Charpin and E. Pasalic, "On propagation characteristics of resilient functions," In *Selected Areas in Cryptography, SAC 2002*, LNCS, Springer-Verlag, 2002.
5. S. Maitra, "Autocorrelation Properties of Correlation Immune Boolean Functions," In *Advances in Cryptology - INDOCRYPT 2001*, LNCS 2247, pp. 242-253, Springer-Verlag, 2001.
6. R. J. McEliece, "Weight congruence for p-ary cyclic codes," *Discrete Mathematics*, 3:177-192, 1972.
7. E. Pasalic, "On boolean functions on symmetric-key ciphers," Ph. D. thisis, Lund University, Sweden, pp. 112-113, Feb. 2003.
8. B. Preneel, W. Van Leekwijck, L. Van Linden, R. Govaerts and J. Vandewalle, "Propagation Characteristics of Boolean Functions," In *Advances in Cryptology - EUROCRYPT'90*, LNCS 437, pp. 155-165, Springer-Verlag, 1990.

9. O. S. Rothaus, "On bent functions," *J. Combin. Theory Ser. A*, 20:300-305, 1976.

10. T. Siegenthaler, "Correlation-Immunity of Nonlinear Combining Functions for Cryptographic Applications," In *IEEE Transactions on Information Theory*, IT-30, pages 776-780, Sept. 1984.

11. J. J. Son, J. I. Lim, S. Chee and S. H. Sung, "Global avalanche characteristics and nonlinearity of balanced Boolean functions," *Information Processing Letters*, 65:139-144, 1998.

12. S. H. Sung, S. Chee and C. Park, "Global avalanche characteristics and propagation criterion of balanced Boolean functions," *Information Processing Letters*, 69:21-24, 1999.

13. Y. Tarannikov, "On Resilient Boolean Functions with Maximum Possible Nonlinearity," In *Progress in Cryptology - INDOCRYPT 2000*, LNCS 1977, pp. 19-30, Springer-Verlag, 2000.

14. Y. Tarannikov, P. Korolev and A. Botev, "Autocorrelation Coefficients and Correlation Immunity of Boolean Functions," In *Advances in Cryptology - ASIACRYPT 2001*, LNCS 2248, pp. 460-479, Springer-Verlag, 2001.

15. A. F. Webster and S. E. Tavares, "On the design of S-boxes," In *Advances in Cryptology - CRYPTO'85*, LNCS 219, pp. 523-534, Springer-Verlag, 1985.

16. G. Xiao and J. L. Massey, "A Spectral Characterization of Correlation-Immune Combining Functions," In *IEEE Transactions on Information Theory*, IT-34(3), pages 569-571, 1988.

17. X. M. Zhang and Y. Zheng, "GAC - the Criterion for Global Avalanche Characteristics of Cryptographic Functions," *Journal of Universal Computer Science*, 1(5):316-333, 1995.

18. Y. Zheng and X. M. Zhang, "On Plateaud Functions," In *IEEE Transactions on Information Theory*, IT-47(3), pages 1215-1223, March 2001.

19. Y. Zheng and X. M. Zhang, "On Relationships among Avalanche, Nonlinearity, and Correlation Immunity," In *Advances in Cryptology - ASIACRYPT'00*, LNCS 1976, pp. 470-482, Springer-Verlag, 2000.

20. Y. Zheng and X. M. Zhang, "New Results on Correlation Immunity," In *ICISC 2000*, LNCS 2015, pp. 49-63, Springer-Verlag, 2001.

Group Algebras and Correlation Immune Functions

Alexander Pott

Institute for Algebra and Geometry,
Otto-von-Guericke University Magdeburg,
39016 Magdeburg, Germany
alexander.pott@mathematik.uni-magdeburg.de

1   Introduction

In this paper we consider functions $F : \mathbb{F}_2^m \to \{\pm 1\}$ which satisfy certain linear and differential properties. The investigation of these properties is motivated by applications in cryptography.

The linear property that we are interested in is "correlation immunity", the differential properties are known under the name of "avalanche criteria". It is not our purpose to construct new correlation immune functions or new functions with good differential properties, but we will describe known constructions (Maiorana-McFarland construction) and its variations in terms of group rings. This is (notationally) a quite useful description since it yields immediate further generalizations and it gives easy ways to obtain bounds on the maximum nonlinearity of the functions.

Moreover, the group ring approach is suitable if one is interested in the connection between differential and linear properties of functions. We will demonstrate this by giving very much simplified proofs of main results in [11]. Moreover, we can correct a (minor) mistake in that paper and show an interesting connection with divisible difference sets.

We note that our viewpoint is not completely new, see [2], for instance.

This paper is organized as follows: In the next section we give the main definitions (avalanche criteria, correlation immunity, nonlinearity). Then we describe the group ring approach and introduce (divisible) difference sets. In Section 4, we analyze the Maiorana-McFarland construction and its variations using the notion of group rings. The final section deals with the connection between "avalanche critera" and "correlation immunity".

2   Basic Definitions

Let $F : \mathbb{F}_2^m \to \mathbb{C}$. The **Walsh transform** $\mathcal{W}(F)$ of F is a function $\mathbb{F}_2^m \to \mathbb{C}$ defined as follows:

$$\mathcal{W}(F)(v) = \sum_{x \in \mathbb{F}_2^m} F(x) \cdot (-1)^{\langle v, x \rangle}.$$

T. Helleseth et al. (Eds.): SETA 2004, LNCS 3486, pp. 437–450, 2005.
© Springer-Verlag Berlin Heidelberg 2005

In many cases, F is a Boolean function, i.e. $F(x) \in \mathbb{F}_2$ for all $x \in \mathbb{F}_2^m$. In this case we replace the function $F(x)$ by $(-1)^{F(x)}$ (in order to get a complex-valued function) and we obtain the Walsh transform

$$\mathcal{W}(F)(v) = \sum_{x \in \mathbb{F}_2^m} (-1)^{F(x) + \langle v, x \rangle}. \tag{1}$$

Throughout this paper, when we speak about Boolean functions $F : \mathbb{F}_2^m \to \mathbb{F}_2$, we always interprete F as indicated above, and the Walsh transform is always computed according to (1). The number

$$\mathcal{L}(F) = \max_{v \in \mathbb{F}_2^m} |\mathcal{W}(F)(v)|$$

is called the **linearity** of F. In the literature, the notion of the **nonlinearity** $\mathcal{N}(F)$ is more common. The connection between $\mathcal{N}(F)$ and $\mathcal{L}(F)$ is simply

$$\mathcal{N}(F) = 2^{m-1} - \frac{1}{2}\mathcal{L}(F).$$

We are interested in functions F where $\mathcal{L}(F)$ is small, hence $\mathcal{N}(F)$ is large. It is well known that the maximum Walsh coefficient is $\geq 2^{m/2}$ with "=" if and only if F is a bent function, which is a Boolean function F such that

$$\mathcal{W}(F)(v) = \pm 2^{m/2} \quad \text{for all } v \in \mathbb{F}_2^m,$$

see [10], for instance.

We call the collection of values $\mathcal{W}(F)(v)$, $v \in \mathbb{F}_2^m$, the **Walsh spectrum** of F. One sometimes refers to properties of the Walsh spectrum as the **linear properties** of F.

An important property of the Walsh transformation is the **inversion formula** which roughly means that we know F if we know its Walsh transform:

$$F(v) = \frac{1}{2^m} \sum_{x \in \mathbb{F}_2^m} \mathcal{W}(F)(x) \cdot (-1)^{\langle v, x \rangle}.$$

The **differential properties** of F are also important. We define

$$\mathcal{D}(F)(v) = \sum_{x \in \mathbb{F}_2^m} F(x) \cdot F(v - x).$$

Again, in the Boolean case this may change to

$$\mathcal{D}(F)(v) = \sum_{x \in \mathbb{F}_2^m} (-1)^{F(x) + F(v - x)}.$$

The set of values $\mathcal{D}(F)(v)$ is called the **differential spectrum** of F.

The following connection between the linear and differential properties of a function F is well known, see also Section 3:

$$\mathcal{D}(F)(v) = \frac{1}{2^m} \sum_{x \in \mathbb{F}_2^m} |\mathcal{W}(F)(x)|^2 \cdot (-1)^{\langle v, x \rangle}. \tag{2}$$

Using the inversion formula, this shows

$$[\mathcal{W}(F)(v)]^2 = \sum_{x \in \mathbb{F}_2^m} \mathcal{D}(F)(x) \cdot (-1)^{\langle v, x \rangle}.$$

Therefore, the linear properties of F are completely determined by the differential properties, and vice versa.

We note that the bound $\mathcal{L}(F) \geq 2^{m/2}$ is an easy consequence of (2): just put $v = 0$ and note $\mathcal{D}(F)(0) = 2^m$.

Sometimes we are interested only in "partial" properties of the Walsh and the differential spectrum (correlation immunity, avalanche criteria). The **weight** $\text{wt}(v)$ of a vector $v = (v_1, \ldots, v_m)$ is the number of entries $v_i \neq 0$. We say that a function f is k-**correlation immune** if $\mathcal{W}(F)(v) = 0$ whenever $1 \leq \text{wt}(v) \leq k$. The function is called k-**resilient** if it is k-immune and $\mathcal{W}(F)(0) = 0$, i.e. the function is **balanced** (the value 1 occurs the same number of times as 0 in the image of f).

A function F on $\mathbb{F}_2^m$ is p-**avalanche** if $\mathcal{D}(F)(v) = 0$ for all v with $1 \leq \text{wt}(v) \leq p$. This is often called "propagation criterion of degree p".

3   Discrete Fourier Transform, Group Rings and Difference Sets

The Walsh transform is a special case of the discrete Fourier transform (DFT) for abelian groups that we describe next.

Let G be a multiplicatively written <u>abelian</u> group of order n, and let $\mathbb{K}$ be a field that contains a primitive n-th root of unity. In many cases, we have $\mathbb{K} = \mathbb{C}$. Then there are n different homomorphisms $G \to \mathbb{K}^*$ (where $\mathbb{K}^*$ is the multiplicative group of $\mathbb{K}$). We call these homomorphisms **characters**. They form a group $\hat{G}$ which is isomorphic to G. The identity element is called the **principal character**. In the special case that $G = \mathbb{F}_2^m$, characters χ_v are the mappings $\chi_v(x) = (-1)^{\langle v, x \rangle}$.

Now we consider the group algebra $\mathbb{K}[G]$. This may be viewed as the set of mappings $G \to \mathbb{K}$ or, equivalently, the set of formal sums

$$\sum_{g \in G} a(g)\, g$$

where $a(g) \in \mathbb{K}$. We may add group ring elements componentwise

$$\left(\sum_{g \in G} a(g)\, g \right) + \left(\sum_{g \in G} b(g)\, g \right) = \sum_{g \in G} (a(g) + b(g))\, g.$$

The multiplication (convolution) is defined as follows:

$$\left(\sum_{g \in G} a(g)\, g \right) \cdot \left(\sum_{g \in G} b(g)\, g \right) = \sum_{g \in G} \left(\sum_{h \in G} (a(h)b(gh^{-1})) \right) g.$$

Throughout this paper we will not distinguish between the mapping $F : G \to \mathbb{K}$ and the corresponding group ring element $\sum_{g \in G} F(g)\, g$ in $\mathbb{K}[G]$.

If $F = \sum_{g \in G} F(g)\, g$, we define $F^{(-1)} = \sum_{g \in G} F(g)\, g^{-1}$. We obtain

$$F \cdot F^{(-1)} = \sum_{g \in G} \mathcal{D}(F)(g)\, g$$

where $\mathcal{D}(F)(g)$ is defined as in Section 2 where we replace the additively written group $\mathbb{F}_2^m$ by the multiplicatively written group G.

Characters can be extended by linearity to homomorphisms

$$\chi : \quad \mathbb{K}[G] \quad \to \mathbb{K}$$
$$\sum a(g)\, g \mapsto \sum a(g)\, \chi(g).$$

For $A \in \mathbb{K}[G]$, the element

$$\mathcal{W}(A) = \sum_{\chi \in \hat{G}} \chi(A)\, \chi \quad \text{in } \mathbb{K}[\hat{G}]$$

is called the **Fourier transform** of A, and the mapping

$$\mathcal{W} : \mathbb{K}[G] \to \mathbb{K}[\hat{G}]$$
$$A \ \mapsto \mathcal{W}(A)$$

is called the discrete Fourier transform (DFT). The most important properties of the DFT are the inversion formula

$$F(g) = \frac{1}{|G|} \sum_{\chi \in \hat{G}} \mathcal{W}(F)(\chi) \cdot \chi(g^{-1})$$

and the convolution property

$$\mathcal{W}(F \cdot H)(\chi) = \mathcal{W}(F)(\chi) \cdot \mathcal{W}(H)(\chi).$$

Moreover, we have

$$\sum_{g \in G} \chi(g) = \begin{cases} |G| \text{ if } \chi = \chi_0 \\ 0 \ \text{ if } \chi \neq \chi_0 \end{cases}$$

which are called the **orthogonality relations** for characters.

If G is the additive group of $\mathbb{F}_2^m$, the DFT reduces to the Walsh transform. Group rings in general have been studied extensively in the mathematical literature. This is the reason why it is worth to point out the close connection between group rings and (Boolean) functions. A lot of theory has been developed about group rings and the DFT that may be applicable to the very special case of Boolean functions. As an example, we describe the well known concept of the "partial DFT".

We assume that $G = G_1 \times G_2$ is the direct product of two abelian groups. Let

$$A = \sum_{g \in G} a(g)\, g = \sum_{h \in G_1} \left(\sum_{k \in G_2} a(h\,k)\, k \right) h.$$

We define

$$A_h := \sum_{k \in G_2} a(h\,k)\, k \text{ in } \mathbb{K}[G_2],$$

hence we have $A = \sum_{h \in G_1} A_h\, h$. If $\chi \in \hat{G}_2$ is a character of G_2, then we define

$$A_\chi := \sum_{h \in G_1} \chi(A_h) h \tag{3}$$

which is an element in $\mathbb{K}[G_1]$. We may apply characters μ of G_2 and get, using the inversion formula,

$$\chi(A_h) = \frac{1}{|G_1|} \sum_{\mu \in \hat{G}_1} (\mu\chi)(A)\mu(h^{-1}).$$

These concepts will be used to prove the results in the final part of this paper.

Another reason for introducing group rings is that usually the literature on difference sets uses this language, and difference sets occur quite naturally if one investigates differential properties of functions.

We assume that $F : G \to \{0,1\}$ is the characteristic function of a set $D = \{g \in G : F(g) = 1\}$. We say that D (or the correpsonding characteristic function) is a (v, k, λ)-**difference set** in G if

$$D \cdot D^{(-1)} = (k - \lambda) + \lambda G \quad \text{in } \mathbb{C}[G].$$

Here we identify subsets T of a group with the corresponding group ring element $T = \sum_{g \in T} g$. Moreover, $k - \lambda$ is an abbreviation for $k - \lambda$ times the identity element of G.

Using characters and the inversion formula, it is easy to see that $D \subseteq G$ is a (v, k, λ)-difference set if and only if

$$|\chi(D)|^2 = \begin{cases} k^2 & \text{if } \chi = \chi_0 \\ k - \lambda & \text{if } \chi \neq \chi_0. \end{cases}$$

This is true only in the abelian case since the concept of the DFT as described above works only in that case. We say that $D \subseteq G$ is an $(m, n, k, \lambda_1, \lambda_2)$-**divisible difference set** in G relative to N if G contains a normal subgroup N of order n such that

$$D \cdot D^{(-1)} = (k - \lambda_1) + \lambda_1 N + \lambda_2(G - N) \tag{4}$$

for some integers k, λ_1, λ_2 (here $k = |D|$ and $|G| = mn$). In character terms, this means (again, only in the abelian case)

$$|\chi(D)|^2 = \begin{cases} k^2 & \text{if } \chi = \chi_0 \\ k - \lambda_1 & \text{if } \chi_{|N} \neq \chi_0 \\ k - \lambda_1 + n(\lambda_1 - \lambda_2) & \text{if } \chi_{|N} = \chi_0. \end{cases}$$

Here $\chi_{|N}$ denotes the restriction of χ to the subgroup N. Note that $\chi_{|N}$ is a character of N.

The case of divisible difference sets with $k = \lambda_1$ is trivial: In this case, D must be a subset of G which is a union of cosets of N. Therefore, we may apply the canonical epimorphism $\Psi : G \to G/N$ to D to obtain an element $\Psi(D) \in \mathbb{C}[G/N]$. The element $\Psi(D)$ has coefficients 0 and n, therefore $D' := \Psi(D)/n$ has coefficients 0 and 1. The following equation follows from (4):

$$D'(D')^{(-1)} = \frac{k - \lambda_2}{n} + \frac{\lambda_2}{n} G/N.$$

Therefore, $\Psi(D)/n$ is a difference set. Conversely, any difference set in G/N can be lifted to a divisible difference set in G, see [8], for instance:

Theorem 1. (1) *If D is an (m, n, k, k, λ)-divisible difference set in G relative to a normal subgroup N then $\Psi(D)/n$ is an $(m, k/n, \lambda/n)$-difference set in G/N (where Ψ is the canonical epimorphism $G \to G/N$).*
(2) *If D is an (m, k, λ) difference set in a group H, and if G is a group containing N as a normal subgroup such that $G/N = H$, then the pre-image of D under the epimorphism Ψ is an $(m, n, nk, nk, \lambda n)$-divisible difference set in G relative to N.*

Another important class of divisible difference sets that we encounter in Section 4 is the following, see [9], for instance:

Theorem 2. *Let N be any subgroup of order 2 in $\mathbb{F}_2^m$. Then there exists a divisible difference sets in $\mathbb{F}_2^m$ relative to N with parameters $(2^{m-1}, 2, 2^{m-1}, 0, 2^{m-2})$ whenever m is odd.*

We note that such difference sets cannot exist if m is even: Just apply a character χ which is nontrivial on N to the divisible difference set D. We have $|\chi(D)|^2 = 2^{m-1}$ which is impossible if m is even since $\chi(D) \in \mathbb{Z}$.

4   Correlation Immune Functions

We are now going to show how our group theoretic approach can be used to describe known constructions of correlation immune functions. First of all, we want to describe the most important classical construction of correlation immune functions (Theorem 3), often called the **Maiorana-McFarland** construction.

Let $\mathbb{F}_2^m = \mathbb{F}_2^s \times \mathbb{F}_2^t$. We are going to construct a Boolean function F on $\mathbb{F}_2^m$ such that $\chi(F) = 0$ for all characters of weight between 0 and k.

Let $H(v)$ be the linear subspace $H(v) = \{x : (-1)^{\langle v,x \rangle} = 1\}$ in $\mathbb{F}_2^t$, $v \in \mathbb{F}_2^t$. We call $\mathrm{wt}(v)$ the **weight of the hyperplane** $H(v)$. The affine subspaces are $H^a(v) = \{x : (-1)^{\langle v,x \rangle} = -1\}$.

Theorem 3. *Let* H_g, $g \in \mathbb{F}_2^s$, *be linear hyperplanes in* $\mathbb{F}_2^t$, *where* $\mathbb{F}_2^m = \mathbb{F}_2^s \times \mathbb{F}_2^t$ *(as above). The Boolean function*

$$D = \sum_{g \in \mathbb{F}_2^s} \pm g \left(H_g - H_g^a \right) \quad in \; \mathbb{C}[\mathbb{F}_2^s \times \mathbb{F}_2^t] \tag{5}$$

is k-resilient provided that the weight of each H_g is at least $k+1$.

Proof. Let $\chi_w : x \mapsto (-1)^{\langle x,w \rangle}$ be a character of $\mathbb{F}_2^t$. Then $\chi_w(H_g) = 0$ if $w \neq v$ where $H_g = H(v)$. We have $\chi_v(H_g - H_g^a) = 2^t$. Since the weights of all hyperplanes in (5) are $\geq k+1$, the case $\chi_v(H_g - H_g^a) = 2^t$ does not occur. $\qquad \square$

An immediate generalization is in [6]:

Theorem 4. *Let* $\tau : \mathbb{F}_2^s \to \pm 1$. *Then*

$$D = \sum_{g \in \mathbb{F}_2^s} \tau(g) g (H_g - H_g^a) \quad in \; \mathbb{C}[\mathbb{F}_2^s \times \mathbb{F}_2^t]$$

is k-resilient if and only if the weight of each hyperplane H_g is at least k, and for all subspaces H of weight k, we have

$$\sum_{g \in \mathbb{F}_2^s : H_g = H} \tau(g) = 0.$$

It is also possible to permit subspaces of weight $< k$:

Theorem 5. *Let* $\tau : \mathbb{F}_2^s \to \pm 1$. *The set*

$$D = \sum_{g \in \mathbb{F}_2^s} \tau(g) g (H_g - H_g^a) \quad in \; \mathbb{C}[\mathbb{F}_2^s \times \mathbb{F}_2^t]$$

is k-resilient if and only if the following holds: If $j \leq k$ is the weight of a subspace H, then the element

$$F := \sum_{g \in \mathbb{F}_2^s : H_g = H} \tau(g) g$$

satisfies $\chi(F) = 0$ for all characters of $\mathbb{F}_2^s$ of weight $\leq k - j$.

Apparently, Theorem 5 has been also known to other people working in this area, see, for instance, [3]. The main goal of this section is not to present a new deep theorem but to show that group ring notation is appropriate for studying correlation immune functions: Theorem 4 and Theorem 5 hardly need a proof!

It has been noted that the elements $H_g - H_g^a$ can be replaced by arbitrary Boolean functions H with the property that the character values $\chi(H)$ are 0 for all characters of weight $\leq k$ or, as indicated in Theorems 4 and 5, $\chi(H) \neq 0$ for only few characters of weight $\leq k$. The functions corresponding to hyperplanes are not the only functions that can be used, see [4].

A quite general version of the Maiorana-McFarland-construction may be formulated as follows:

Theorem 6. *Let* $\tau : \mathbb{F}_2^s \to \pm 1$. *Let* $\mathcal{T}$ *be a set of Boolean functions on* $\mathbb{F}_2^t$ *(i.e. group ring elements with coefficients* ± 1*), such that for each character* $\gamma \in \hat{\mathbb{F}}_2^t$ *of weight* $\leq k$ *there is at most one* $T \in \mathcal{T}$ *such that* $\gamma(T) \neq 0$. *Let* T_g *(*$g \in \mathbb{F}_2^s$*) be a family of Boolean functions* $T_g \in \mathcal{T}$. *Then the sets* $S_\gamma := \{g \in \mathbb{F}_2^s : \gamma(T_g) \neq 0\}$ *are pairwise disjoint where* γ *runs through the characters of weight* $\leq k$, *and* $T_g = T_h$ *if* $g, h \in S_\gamma$. *Then the function*

$$D = \sum_{g \in \mathbb{F}_2^s} \tau(g) g T_g \quad in \; \mathbb{C}[\mathbb{F}_2^s \times \mathbb{F}_2^t]$$

is k*-resilient if and only if the following holds for all characters* γ *of weight* $\leq k$:

$$F(\gamma) := \sum_{g \in S_\gamma} \tau(g) g \tag{6}$$

satisfies $\chi(F(\gamma)) = 0$ *for all characters* χ *of* $\mathbb{F}_2^s$ *of weight* $\leq k - wt(\gamma)$.

Proof. Obvious.   □

Note that you may take

$$\mathcal{T} = \{H(v) - H^a(v) \; : \; v \in \mathbb{F}_2^t\}.$$

If we omit the condition in Theorem 6 that there is at most one T_g such that $\gamma(T_g) \neq 0$, then the Theorem remains true if we replace $F(\gamma)$ in (6) by $\sum_{g \in S_\gamma} \tau(g) g \gamma(T_g)$. However, such a statement seems to be uninteresting: It only reformulates the definition of correlation immune functions in $\mathbb{F}_2^m$ when $\mathbb{F}_2^m$ is replaced by $\mathbb{F}_2^s \times \mathbb{F}_2^t$.

Let us briefly look at the linearity of functions constructed according to Theorem 6. It is possible to give some bounds if the $T_g's$ correspond to hyperplanes, see also [3]. The proof given here is a slightly reformulated version of the proof in [3].

Theorem 7. *Let* D *be constructed as in Theorem 6, where* $T_g = H_g - H_g^a$ *for some hyperplane* H_g *in* $\mathbb{F}_2^t$. *Then*

$$\mathcal{L}(D) = 2^t \cdot \max_{\chi \in \hat{\mathbb{F}}_2^s, \gamma \in \hat{\mathbb{F}}_2^t} \left| \chi(F(\gamma)) \right|.$$

If e_γ *is the number of characters* χ *such that* $\chi(F(\gamma)) \neq 0$, *then*

$$\max_{\chi \in \hat{\mathbb{F}}_2^s} \left| \chi(F(\gamma)) \right| \geq \sqrt{\frac{2^s \cdot |S_\gamma|}{e_\gamma}} \geq \sqrt{|S_\gamma|}. \tag{7}$$

Proof. We have $[F(\gamma)]^2(0) = |S_\gamma|$. The inversion formula gives

$$[F(\gamma)]^2(0) = \frac{1}{2^s} \sum_{\chi \in \hat{\mathbb{F}}_2^s} \mathcal{W}([F(\gamma)]^2)(\chi)$$

which shows (7), which is the only nontrivial part in the theorem.   □

In Theorem 7, the nonempty sets S_γ form a partition of $\mathbb{F}_2^s$ if γ runs through all characters of $\mathbb{F}_2^t$. Hence there is at least one character γ with $|S_\gamma| \geq 2^{s-t}$. In the classical case of the Maiorana-McFarland construction, we have $S_\gamma \neq \emptyset$ at most for characters γ of weight $> k$. The number of such characters is $\sum_{i=k+1}^t \binom{t}{i}$. Therefore, in the classical case there is at least one character γ such that

$$|S_\gamma| \geq \frac{2^s}{\sum_{i=k+1}^t \binom{t}{i}}.$$

This is precisely the bound in [3].

5   Avalanche and Correlation Criteria

We will now investigate the connection between the p-avalanche property and q-correlation immunity. Most of this is contained in [11], except the bound $p + q \leq m - 1$ in Theorem 8 for the balanced case. Erroneously, the authors in [11] stated $p + q \leq m - 2$. The proofs given here are, again, different and more transparent than in the original paper.

Before we continue, let us define **linear structures**. Again, this is a concept in cryptography, see [7]. We say that $v \in G$ is a linear structure of $D \in \mathbb{K}[G]$ if $Dv = \pm D$. The linear structures form a subgroup of G. They help to prove the connection between p-avalanche and q-immunity.

The following Proposition is obvious; it is also contained in [7] for the special case of elementary abelian groups.

Proposition 1. *The element $v \in G$ is a linear structure of $D = \sum d(g) g \in \mathbb{C}[G]$ if and only if $\chi(v) = 1$ for all characters $\chi \in \hat{G}$ with $\chi(D) \neq 0$, or $\chi(v) = -1$ for all these characters.*

Proof.   Note $\chi(Dv) = \chi(D)\chi(v)$, hence $Dv = D$ (resp. $Dv = -D$) if and only if $\chi(v) = 1$ (resp. $\chi(v) = -1$) for all characters with $\chi(D) \neq 0$.   □

Note that v is also a linear structure of $DD^{(-1)}$. If D has coefficients ± 1, we can say more, see also [7].

Proposition 2. *An element D with coefficients ± 1 has a linear structure v if and only if the coefficient of v in $DD^{(-1)}$ is $\pm|G|$.*

Proof.   Observe that the coefficient of the identity in $(Dv)D^{(-1)}$ is the same as the coefficient of v^{-1} in $DD^{(-1)}$ which is also the coefficient of v.   □

Elements D in $\mathbb{F}_2^m$ satisfy $D = D^{(-1)}$. Since we only investigate functions on $\mathbb{F}_2^m$, we may also replace $D^{(-1)}$ by D in the proposition above.

Bent functions D on $\mathbb{F}_2^m$ are m-avalanche (which is well known) and they have constant absolute character value $|\chi(D)| = 2^{m/2}$ (as noted earlier). They have no linear structure $v \neq 0$, otherwise $D^2 v = \pm D^2$, but we have $D^2 = 2^m$. Therefore, it is natural to ask about the connection between the avalanche criteria, character values and linear structures. The following proposition gives a partial answer and is implicitely contained in [11].

Proposition 3. *Assume that F is p-avalanche with $p > 0$ and $|\tau(F)| = 2^{m-\frac{p}{2}}$ for some character τ. If F is not bent (i.e. $p < m$), then every vector of weight $p + 1$ is a linear structure. Moreover, we must have $p = m - 1$ and therefore $Fw = \pm F$ for the all-one-vector w in $\mathbb{F}_2^m$.*

Proof.   We may assume $p \neq m$ since F is not bent. Put $S = F^2$. We decompose

$$\mathbb{F}_2^m = G_1 \times G_2 \tag{8}$$

where G_1 is generated by p vectors of the standard basis, and G_2 is generated by the remaining vectors in the standard basis. We write

$$S = \sum_{h \in G_2} S_h h, \quad S_h \in \mathbb{C}[G_1]. \tag{9}$$

The avalanche property implies $S_0 = 2^m$ and $S_u = x \cdot j$ $(x \in \mathbb{C})$, where u is any vector of weight 1 in G_2 and j is the all-one-vector in G_1. Note that x (which may depend on u) is the coefficient of $j + u$ in S. The partial DFT shows

$$2^m = \chi(S_0) = \frac{1}{2^{m-p}} \sum_{\mu \in \hat{G}_2} \chi\mu(S).$$

Let $|\chi\mu(F)| = 2^{m-\frac{p}{2}}$, hence $\chi\mu(S) = 2^{2m-p}$ and therefore $\chi'\mu(S) = 0$ for $\chi' \neq \chi$ (note that $\chi\mu(S) \geq 0$ for all characters χ, μ). Similarly,

$$\chi(S_u) = \frac{1}{2^{m-p}} \sum_{\mu \in \hat{G}_2} \chi\mu(S)\mu(u)$$

and therefore $|x| = |\chi(S_u)| = 2^m$. Since the decomposition in (8) is arbitrary, all vectors of weight $p + 1$ have coefficients $\pm 2^m$ in S, hence all vectors of weight $p + 1$ are linear structures. Since the linear structures form a group, all vectors would be linear structures provided that $p < m - 1$. But if every vector is a linear structure, the function cannot be p-avalanche for some $p \geq 1$ since F^2 has no coefficients 0, see Proposition 2.   □

Theorem 8. *Let F be a q-correlation immune function on $\mathbb{F}_2^m$. If F is p-avalanche and not bent (which excludes the case $p = m$), then the following holds:*

1. If F is balanced, then $p + q \leq m - 1$ with equality if and only if $p = m - 1$ or $p = 0$.
2. If F is unbalanced, then $p + q \leq m$ with equality if and only if $p = m - 1$ or $p = 0$.

The case of equality $p = m - 1$ may occur only if m is odd. The case $p = 0$ in the balanced case occurs precisely for the functions

$$F(v) = (-1)^{\langle v, j \rangle} \quad \text{or} \quad F(v) = -(-1)^{\langle v, j \rangle},$$

where j is the all-one-vector. In the unbalanced case, we have $p = 0$ if and only if F is constant, i.e.

$$F(v) = 1 \quad \text{or} \quad F(v) = -1.$$

This theorem covers basically all of Section 5 in [11]. The case $p = m - 1$ in the balanced case is missing in [11].

Proof. Without loss of generality, let us assume that $p + q \leq m$ with $p \neq m$. We decompose $\mathbb{F}_2^m$ as in (8). We apply the epimorphism $\Psi : G \to G/G_1$ to F^2 and obtain

$$[\Psi(F)]^2 = 2^m + \sum_{\substack{g \in G/G_1 \\ g \neq 1}} a(g)\, g.$$

The characters of G/G_1 are precisely the characters χ_w of G which are principal on G_1, thus $\langle v, w \rangle = 0$ for all $v \in G_1$. If G_1 is the subgroup of vectors $(x_1, \ldots, x_m)$ such that $x_{p+1} = \ldots = x_m = 0$, then we must have $w = (0, \ldots, 0, y_{p+1}, \ldots, y_m)$, hence $\mathrm{wt}(w) \leq q$.

We investigate the <u>balanced</u> case first. We have $\chi([\Psi(F)]^2) = 0$ for <u>all</u> characters χ of G/G_1, a contradiction to $[\Psi(F)]^2 \neq 0$. If $p + q = m - 1$, then there is precisely one character χ of G/G_1 with $\chi([\Psi(F)]^2) \neq 0$. The inversion formula shows

$$[\Psi(F)]^2 = \gamma \cdot \sum_{g \in G/G_1} \chi(g)\, g$$

and

$$\gamma = \frac{\chi([\Psi(F)]^2)}{2^{m-p}}.$$

Since the coefficient of the identity in $[\Psi(F)]^2$ is 2^m, we have $\chi([\Psi(F)]^2) = 2^{2m-p}$, hence $|\chi(\Psi(F))| = 2^{m-\frac{p}{2}}$. The mapping $\gamma = \chi \circ \Psi$ is a character of G such that $|\gamma(F)| = 2^{m-\frac{p}{2}}$. Proposition 3 shows $p = m - 1$.

In the <u>unbalanced</u> case, $[\Psi(F)]^2 = 2^m \cdot G/H$ and therefore $\chi_0([\Psi(F)]^2) = 2^{2m-p}$ for the principal character χ_0. This shows $p = m - 1$, using Proposition 3. $\qquad\square$

We can characterize the case $p = m - 1$ in terms of divisible difference sets:

Theorem 9. A function $F : \mathbb{F}_2^m \to \{\pm 1\}$ is $(m-1)$-avalanche if and only if $D_F := \{g \in \mathbb{F}_2^m : F(g) = 1\}$ is a $(2^{m-1}, 2, k, \lambda_1, \lambda_2)$-divisible difference set in $\mathbb{F}_2^m$ relative to $N = \{0, j\}$, where $2^m = 4(k - \lambda_2)$ and $j = (1, \ldots, 1)$ is the all-one-vector.

Proof. Assume that D_F is a divisible difference set with the parameters above. We have $F = 2D_F - G$ (this transforms the element D_F with coefficients 0 and 1 into the element F with coefficients ± 1), hence

$$F^2 = 4[D_F]^2 + 2^m G - 4kG.$$

This shows that the coefficients of all the elements $g \neq 0, j$ in F^2 are $2^m - 4(k - \lambda_2) = 0$, hence the function F is $(m-1)$-avalanche.

Conversely, let F be an $(m-1)$-avalanche function, hence

$$F^2 = 2^m + \lambda \cdot j$$

for some $\lambda \in \mathbb{Z}$. We put $D_F = \frac{F+G}{2}$ which is a group ring element with coefficients 0 and 1, hence it describes a subset of G which is the desired difference set: We compute

$$[D_F]^2 = \frac{1}{4}F^2 + 2^{m-2}G + \frac{1}{2}\, FG = 2^{m-2} + \frac{\lambda}{4}j + 2^{m-2}G + \frac{1}{2}\, FG.$$

We have $FG = [\chi_0(F)]G$ (where χ_0 is the trivial character) since $Gh = G$ for all $h \in G$. This shows that D_F is a divisible $(2^m, 2, k, \lambda_1, \lambda_2)$-difference set relative to N, where

$$k = 2^{m-1} + \frac{\chi_0(F)}{2}$$

$$\lambda_1 = \frac{\lambda}{4} + 2^{m-2} + \frac{\chi_0(F)}{2}$$

$$\lambda_2 = 2^{m-2} + \frac{\chi_0(F)}{2}.$$

We have $2^m = 4(k - \lambda_2)$, which proves the Theorem.   □

There are only two possible types of divisible difference sets that may occur in this theorem:

Corollary 1. (1)   *If F is balanced then D_F is a*

$$(2^{m-1}, 2, 2^{m-1}, 0, 2^{m-2}) - \textit{divisible difference set.} \tag{10}$$

(2)   *If F is unbalanced, then $\Psi(D_F)/2$ is a*

$$(2^{m-1}, 2^{m-2} \pm 2^{(m-3)/2}, 2^{m-3} \pm 2^{(m-3)/2}) - \textit{difference set,} \tag{11}$$

where Ψ is the canonical epimorphism $\mathbb{F}_2^m \to \mathbb{F}_2^m/N$.

Proof. (1) If F is balanced, then $k = 2^{m-1}$, hence $\lambda_2 = 2^{m-2}$ which implies $\lambda_1 = 0$ (by counting differences).

(2) In this case, $Fj = F$ using Proposition 3, hence $\Psi(D_F)$ is constant on cosets of N. This is equivalent to saying $k = \lambda_1$. Therefore, $\Psi(D_F)/2$ is a difference set in an elementary-abelian 2-group (Theorem 1). It is well known that such difference sets have the parameters (11), see [1], for instance.   □

Two remarks are in order: The Boolean functions corresponding to the difference sets with parameters (11) are bent functions, see [10], for instance. They exist for all odd m. Using Theorem 2, we can construct the functions F in (2) of Corollary 1 for all odd m. Examples of type (1) in Corollary 1 exist for all odd m (Theorem 2), too.

We note that an example of a correlation immune function corresponding to a difference set with parameters (10) is also contained in [5].

6   Conclusion

In this paper, we have described known variations of the Maiorana-McFarland construction of correlation immune functions in terms of the unified notion of group rings. This notion has the advantage of being quite clear, it works for arbitrary abelian groups, and it immediately yields further generalizations.

In particular, we looked at the connection between correlation immunity and avalanche properties. We pointed out that certain extremal cases correspond to difference sets.

Acknowledgements

The author thanks an anonymous referee for his valuable comments, in particular for pointing out a mistake in the formulation of Theorem 6.

References

1. T. BETH, D. JUNGNICKEL, AND H. LENZ, *Design Theory*, Cambridge University Press, Cambridge, 2 ed., 1999.
2. A. CANTEAUT, C. CARLET, P. CHARPIN, AND C. FONTAINE, *On cryptographic properties of the cosets of R(1, m)*, IEEE Trans. Inform. Theory, 47 (2001), pp. 1494–1513.
3. C. CARLET, *A larger class of cryptographic boolean functions via a study of the maiorana-mcfarland construction*, in CRYPTO 2002, Advances in Cryptology, vol. 2442 of Lecture Notes in Computer Science, 2002, pp. 549–564.
4. C. CARLET, *On the confusion and diffusion properties of Maiorana-McFarland's and extended Maiorana-McFarland's functions*, J. Complexity, 20 (2004), pp. 182–204.
5. S. CHOI AND K. YANG, *Autocorrelation properties of resilient functions and three-valued almost-optimal functions satisfying PC(p)*, in Proceedings (Extended Abstracts) of SETA04, T. Helleseth, D. Sarwate (Eds.), 2004, pp. 167–171.
6. T. W. CUSICK, *On constructing balanced correlation immune functions*, in Sequences and their applications (Singapore, 1998), Springer Ser. Discrete Math. Theor. Comput. Sci., Springer, London, 1999, pp. 184–190.

7. S. DUBUC, *Characterization of linear structures*, Des. Codes Cryptogr., 22 (2001), pp. 33–45.

8. D. JUNGNICKEL, *On automorphism groups of divisible designs*, Canad. J. Math., 34 (1982), pp. 257–297.

9. A. POTT, *Finite Geometry and Character Theory*, vol. 1601 of Lecture Notes in Mathematics, Springer-Verlag, Berlin, Heidelberg, 1995.

10. J. WOLFMANN, *Bent functions and coding theory*, in Difference sets, sequences and their correlation properties (Bad Windsheim, 1998), vol. 542 of NATO Adv. Sci. Inst. Ser. C Math. Phys. Sci., Kluwer Acad. Publ., Dordrecht, 1999, pp. 393–418.

11. Y. ZHENG AND X.-M. ZHANG, *On relationships among avalanche, nonlinearity, and correlation immunity*, in Advances in cryptology—ASIACRYPT 2000 (Kyoto), vol. 1976 of Lecture Notes in Comput. Sci., Springer, Berlin, 2000, pp. 470–482.

Author Index

Lecture Notes in Computer Science

For information about Vols. 1–3400

please contact your bookseller or Springer